ACID DEPOSITION Long-Term Trends

Committee on Monitoring and Assessment of Trends in Acid Deposition
Environmental Studies Board
Commission on Physical Sciences, Mathematics, and Resources
National Research Council

NATIONAL ACADEMY PRESS
Washington, D.C. 1986

NATIONAL ACADEMY PRESS 2101 Constitution Avenue, NW Washington, DC 20418

NOTICE: The project that is the subject of this report was approved by the Governing Board of the National Research Council, whose members are drawn from the Councils of the National Academy of Sciences, the National Academy of Engineering, and the Institute of Medicine. The members of the committee responsible for the report were chosen for their special competences and with regard for appropriate balance.

This report has been reviewed by a group other than the authors according to procedures approved by a Report Review Committee consisting of members of the National Academy of Sciences, the National Academy of Engineering, and the Institute of Medicine.

The National Research Council was established by the National Academy of Sciences in 1916 to associate the broad community of science and technology with the Academy's purposes of furthering knowledge and of advising the federal government. The Council operates in accordance with general policies determined by the Academy under the authority of its congressional charter of 1863, which establishes the Academy as a private, nonprofit, self-governing membership corporation. The Council has become the principal operating agency of both the National Academy of Sciences and the National Academy of Engineering in the conduct of their services to the government, the public, and the scientific and engineering communities. It is administered jointly by both Academies and the Institute of Medicine. The National Academy of Engineering and the Institute of Medicine were established in 1964 and 1970, respectively, under the charter of the National Academy of Sciences.

Although the information in this publication has been funded in part by the U.S. Environmental Protection Agency under Assistance Agreement No. CR-811626-01-0 and the U.S. Geological Survey under Grant No. 14-08-0001-G-954 to the National Academy of Sciences, it may not necessarily reflect the views of these agencies, and no official endorsement should be inferred.

Library of Congress Catalog Card Number 86-70311

International Standard Book Number 0-309-03647-X

Printed in the United States of America

9/87

Committee on Monitoring and Assessment of Trends in Acid Deposition

JAMES H. GIBSON, Colorado State University, *Chairman*
ANDERS W. ANDREN, University of Wisconsin
RAYMOND S. BRADLEY, University of Massachusetts
DONALD F. CHARLES, Indiana University
TERRY A. HAINES, University of Maine
RUDOLF B. HUSAR, Washington University
ARTHUR H. JOHNSON, University of Pennsylvania
JAMES R. KRAMER, McMaster University
SAMUEL B. MCLAUGHLIN, Oak Ridge National Laboratory
STEPHEN A. NORTON, University of Maine
GARY OEHLERT, University of Minnesota
GARY J. STENSLAND, Illinois State Water Survey
JOHN TRIJONIS, Santa Fe Research Corporation
DOUGLAS M. WHELPDALE, Atmospheric Environment Service, Canada

Staff

WILLIAM M. STIGLIANI, *Staff Officer*
JOYCE E. FOWLER, *Administrative Secretary*

Commission on Physical Sciences, Mathematics, and Resources

Preface

Recognizing that deposition of chemical pollutants from the atmosphere constitutes one of the more significant environmental issues of our time, both nationally and internationally, the National Research Council has undertaken a series of studies designed to summarize what is known about this complex phenomenon and to correct many of the misunderstandings surrounding it. A central objective of this activity has been to delineate more clearly the role of our industrial society in contributing to the chemical composition of the atmosphere as well as the consequences for the environment of the ultimate return of the pollutants to the Earth's surface.

The first report in the series—*Atmosphere-Biosphere Interactions: Toward a Better Understanding of the Consequences of Fossil Fuel Combustion* (1981)—provides a comprehensive synthesis of our knowledge of the aquatic and terrestrial effects of the deposition of substances derived from fossil fuel combustion. The second report—*Acid Deposition: Atmospheric Processes in Eastern North America* (1983)—addressed the relationships between average rates of emission of sulfur oxides and deposition of acid sulfate. The third report—*Acid Deposition: Processes of Lake Acidification* (1984)—reviewed alternative hypotheses regarding geochemical processes in watersheds responding to acid inputs from the atmosphere. Another report—*Acid Deposition: Effects on Geochemical Cycling and Biological Availability of Trace Elements* (1985)—was prepared jointly with the Royal Society of Canada (RSC) and the Mexican Academy of Scientific Research (AIC). It examined the influences of atmospheric deposition on biogeochemical processes involving trace elements.

The current report addresses another dimension of the relationships among emissions, deposition, and environmental effects. If industrial activity is implicated in atmospheric deposition, then the phenomenon is a relatively recent one and should be related to changes in the levels and kinds of such activity. It is logical, then, to seek historical data or other reliable information that may be available to assess whether emissions, deposition, and effects are temporally linked over the period since the beginning of industrialization (or over a portion of the period), thus suggesting possible cause-effect relationships. The original suggestion for the study came from the Tri-Academy Committee on Acid Deposition, a joint activity of the NAS, RSC, and AIC.

The Committee on Monitoring and Assessment of Trends in Acid Deposition was organized by the National Research Council to carry out the study. The committee, comprising experts in the disciplines of atmospheric chemistry, visibility, environmental monitoring, analytical chemistry, aquatic chemistry, geochemistry, geology, forestry, fisheries resources, climatology, and statistics, conducted its work under the auspices of the Council's Environmental Studies Board. The committee reviewed and summarized the available literature and data on historical trends in phenomena related to the emission and deposition of acidic materials. Other chemical substances are recognized to play an important role in certain effects, but insufficient information existed to incorporate their assessment in this report.

In the process of reviewing the literature and available data, the committee found that a number of the potentially informative sets of data that exist had previously not been analyzed for this purpose. We concluded that in order to conduct a comprehensive study, it would be necessary for us to obtain and analyze these data. The report therefore contains not only our review of the literature but also original analyses of a number of data sets. The analyses were prepared by members of the committee, often with the assistance of consultants and colleagues, and are contained in the respective chapters of the report. Members of the committee and their colleagues who prepared the respective chapters are identified. The analyses, which commanded considerable commitments of time and energy from the members, permitted us to assess the implications of findings for a number of diverse phenomena together, which greatly enhances the strength of our conclusions.

The committee is indebted to its colleagues and consultants for their contributions and to the reviewers for their helpful comments and suggestions. We are particularly indebted to Richard A. Smith and Richard B. Alexander of the U.S. Geological Survey for their significant contributions to the analysis of surface water chemistry data and to the preparation of Chapter 7. We wish to single out for special recognition William M. Stigliani, professional staff officer for the project; without his tireless work, this complex and difficult project could not have been accomplished. Other members of the NRC staff also made important contributions to the effort, including Myron F. Uman, staff director of the Environmental Studies Board, who provided administrative leadership; Joyce Fowler and Carolyn Stewart, project secretaries; Roseanne Price and Joanne B. Sprehe, editors; and the manuscript processing unit headed by Estelle Miller.

The report required an unusual level of effort and commitment by members of the committee. It is not possible to commend them adequately for their enthusiasm, dedication, and cooperation. They were presented with the difficult challenge and, in performing original analyses, undertook more than most NRC committees are asked to do. Although assisted by staff, colleagues, and consultants, and having the benefit of helpful comments from more than 30 anonymous reviewers, the committee bears full responsibility for the report.

Finally, the project would not have been possible without the generous support of its sponsors, the U.S. Environmental Protection Agency, the U.S. Geological Survey, and a consortium of private foundations: the Carnegie Corporation of New York, the Charles E. Culpepper Foundation, Inc., the William and Flora Hewlett Foundation, the John D. and Kathryn T. MacArthur Foundation, the Andrew W. Mellon Foundation, the Rockefeller Foundation, and the Alfred P. Sloan Foundation.

J. H. GIBSON, *Chairman*

Contents

APPENDIXES

1

Summary and Synthesis

Over the past decade or so, the phenomenon of acid rain, or more properly, acid deposition, has evolved in the United States from a scientific curiosity to an issue of considerable public concern and controversy. The issues raised by its possible adverse effects are not confined to specific localized areas, but are regional, national, and even international in scope. A wide variety of effects have been attributed to acid deposition, its gaseous precursors, and certain products of their chemical reactions including ozone. Possible environmental consequences include adverse effects on human health, acidification of surface waters with subsequent decreases in fish populations, the acidification of soils, reduced forest productivity, erosion and corrosion of engineering materials, degradation of cultural resources, and impaired visibility over much of the United States and Canada.

To evaluate these possibilities, scientific hypotheses have been formulated linking postulated or observed effects to acid deposition and/or its precursors. For example, acidification of lakes is thought to be the result of the deposition of acidifying substances, either directly as deposition to the water surface or indirectly by interaction with soils in the watershed to enhance transport of hydrogen and aluminum ions to surface waters. Fish are adversely affected by acidification and the increased concentrations of aluminum that frequently accompany it. Alternatively, other hypotheses involving both natural acidification processes and/or other factors related to human activity have been proposed to explain the same effects. For example, land use practices such as timber harvesting, agriculture, and residential development are known to affect surface water chemistry and in specific circumstances might be more important for water quality than acid deposition. Similarly, changes in fish populations may be influenced by changes in stocking policies, the introduction of competing fish species, commercial or sport fishing, and pollution from pesticides or other commercial chemicals.

It appears that an alternative explanation based on other human activities or natural phenomena can be or has been proposed for every proposed link between the deposition of airborne chemicals and an adverse environmental effect. In many instances several alternatives are plausible.

This study was organized to investigate spatial patterns and temporal trends in acid deposition and its gaseous precursors in eastern North America and patterns and trends in environmental parameters that might result from acid deposition. The Committee on Monitoring and Assessment of Trends in Acid Deposition was asked not only to review previous efforts in this regard, but also to extend the analyses if there were approriate data. To meet these requirements we had to perform new analyses and make extensive checks on the quality of original data.

We assumed that if a mechanism existed linking acid deposition to an environmental effect, it should be possible to demonstrate that acid deposition is associated spatially and/or temporally with the effect. If such an association cannot be established, then either a cause-and-effect relationship does not exist or our understanding of the mechanisms and rate-governing factors is not adequate. Despite imperfect knowledge of the relationships between emissions and deposition and rates of responses of ecosystems, careful evaluation of data on phenomena that are linked to acid deposition by plausible mechanisms should provide additional insight into the relationships among emissions, deposition, and effects.

Some of the questions asked about the spatial relations of acid deposition and related phenomena were the following: Is the deposition of acidic sulfates and nitrates highest in areas where densities of emissions of sulfur and nitrogen oxides are highest? Are patterns of ecosystem changes attributed to acid deposition also found in regions of low deposition? With regard to temporal associations, we asked: How has the chemical composition of precipitation changed with time? How acidic was precipitation or dry deposition 30 years ago? 50 years ago? 100 years ago? Does the timing of postulated changes in aquatic and terrestrial ecosystems coincide with changes in patterns of sulfur and nitrogen oxide emissions?

Simultaneous examination of multiple patterns of spatial distributions and temporal trends provides a more robust test for the existence of linkages between emissions, deposition, and environmental effects than inferences based on data for pairs of phenomena. In conducting the study, it quickly became apparent that answers to questions about acid deposition have proved elusive because historical data from which to judge trends and hypothesize environmental responses are scarce. The earliest records of the direct measurement of levels of acidity, sulfate, and nitrate in deposition in North America date from as early as 1910 (McIntyre and Young 1923, Harper 1942, Hidy et al. 1984), one year after the invention of the pH scale for measurements of the acidity of aqueous solutions. Although they are informative, data obtained before 1950 are sketchy and of limited value because they pertain only to a few locations over short periods of time and in many cases are not reliable. Only since the late 1970s have extensive deposition monitoring networks been established to gather quality-assured data in a systematic way.

In some cases however, longer and more extensive records do exist for systems thought to be affected by acid deposition or its airborne precursors. These records include long-term data on visibility, chemical composition of waters in lakes and streams, chemical and biological composition of lake sediments, fish populations, growth patterns in trees as evidenced in ring widths, distributions of lichens, erosion of tombstones, and chemical composition of glacial ice cores, groundwater, and soils.

Unfortunately, investigations in a number of these areas conducted before the early 1970s were not designed to study acid deposition per se, and hence the original records usually do not include all the information required for a definitive intepretation of temporal changes. For example, the interpretation of earlier lake and stream chemical data is difficult because the records frequently fail to include adequate documentation of the sampling procedures and chemical analytical methods employed. Also often missing are quantitative descriptions of the variability in the data that may have been introduced by analogous changes in climate or weather or by human activities such as changes in land use patterns.

On the other hand, the establishment of spatial relationships among current values for emissions, deposition, and environmental responses appears to be more straightforward because information about factors that might bias the analyses is generally more readily available.

To determine which types of phenomena to study and which data were appropriate and feasible to include in the study, we developed certain criteria, the most important of which were the following:

1. A documented or postulated relationship, either direct or indirect, to acid deposition.
2. Availability of published data or original data with sufficient documentation to permit peer review.
3. Availability of data representative of broad geographical regions and/or temporal data with unambiguous dating.

Consequently, we selected the following for inclusion in the study:

1. Emissions of sulfur and nitrogen oxides (Chapter 2). The major contributors to atmospheric deposition of sulfur and nitrogen compounds in North America are anthropogenic emissions of sulfur and nitrogen oxides. Estimates of emissions have been compiled in this report based on data on the production and use of fossil fuels, estimates of their sulfur content, and emission factors for nitrogen oxides released during combustion.
2. Precipitation chemistry (Chapter 5). Quality-assured data on precipitation chemistry for a broad region in eastern North America have been available since about 1978. These data permit spatial analyses of the chemistry of wet deposition in the region, but the time period of this record is not sufficiently long to establish statistically significant temporal trends. Time series of longer duration, dating from the early to the middle 1960s, are available at a few sites, however, and we have performed trend analyses on some of these data.
3. Atmospheric sulfates and visibility (Chapter 4). A direct effect of sulfur dioxide emissions is the production of atmospheric sulfate aerosols that reduce visibility. Historical and spatial data on atmospheric sulfate and visibility are available from a number of stations.
4. Surface water chemistry (Chapter 7). One of the most studied effects of the deposition of acidic chemical species is the acidification of surface waters. Many data are available for analysis. We selected what we judged to be key data sets and included in our analysis

those that were amenable to rigorous assessment of their reliability. In all cases checks for internal consistency of the original historical data had to be performed before data were incorporated into the analyses.

5. Sediment chemistry and abundance of diatom taxa (Chapter 9). Changes in watershed and lake chemistry are recorded in lake sediments, which provide historical data of the longest time series. Dating and chemical analysis of successive intervals of sediment cores provide chronological information on changes in water chemistry. Assemblages of diatoms and chrysophytes in sediments can be analyzed to reconstruct historical lake water acidity.

6. Fish populations (Chapter 8). Fish populations are hypothesized to decline in acidified lakes. Some records are available relating fluctuations in populations of selected fish species to historical records of lake and stream chemistry.

7. Tree rings (Chapter 6). The decline of forest trees is perhaps the most controversial phenomenon that some researchers have attributed to acid deposition. Some tree ring data are available for red spruce populations in the higher elevation forests of the northern Appalachian Mountains.

It became apparent during the evaluation processes that we would not be able to rely exclusively on the published literature if we were to meet our goal of examining spatial patterns or temporal trends of multiple phenomena. In some cases extensive evaluations of spatial distributions or temporal trends had not been performed for even a single phenomenon and thus our evaluation depended, at least in part, on our own analysis of unpublished data. Owing to limitations in available data, the effects of oxidants and other air pollutants are not considered in this report.

The committee's findings and conclusions are listed below. In subsequent sections of this chapter we describe the rationale employed in drawing these conclusions. First we present our methodology for assessing the likelihood of a cause-and-effect relationship based on the criteria of mechanism and consistency of data. We then briefly discuss the specific mechanisms that may link acid deposition to related phenomena and some factors that complicate our analysis. We follow this section with a detailed analysis of the degree of consistency in spatial patterns and temporal trends among acid deposition, emissions, and the environmental changes often attributed to acid deposition.

FINDINGS AND CONCLUSIONS

When trends and patterns are found that establish temporal and spatial consistency in cases for which plausible mechanisms link acid deposition to other phenomena, cause-and-effect relationships can be postulated with some degree of confidence. Previous attempts to evaluate temporal trends have been limited because of large uncertainties inherent in historical data bases. Our premise was that through careful selection of a number of types of data and a number of quality-assured data bases, a more robust analysis for consistency and associations among the data might be possible. We believe that the results of our analyses, presented in later sections of this chapter and in the following chapters of this report, demonstrate the validity of this approach, and that we can formulate the following major findings and conclusions:

1. Through statistical analysis of regional spatial patterns, we find a strong association among the following five parameters: (a) emission densities of sulfur dioxide (SO_2), (b) concentrations of sulfate aerosol, (c) ranges of visibility, (d) sulfate concentrations in wet precipitation, and (e) sulfate fluxes in U.S. Geological Survey Bench-Mark streams. From this result and because of the existence of plausible mechanisms linking the phenomena, we conclude that in eastern North America a causal relationship exists between anthropogenic sources of emissions of SO_2 and the presence of sulfate aerosol, reduced visibility, and wet deposition of sulfate. Our analysis also indicates that for Bench-Mark streams in watersheds showing no evidence of dominating internal sources of sulfate there is a cause-and-effect relationship between SO_2 emissions and stream sulfate fluxes. Magnitudes of sulfur emissions and deposition of sulfur oxides are highest in a region spanning the midwestern and northeastern United States.

2. Based on data on fossil fuel production and consumption, we conclude that acid precursors, particularly SO_2, have been emitted in substantial quantities in the atmosphere over eastern North America

since the early 1900s. In particular, SO_2 emissions in the northeastern quadrant of the United States have fluctuated near current amounts since the 1920s. These conclusions are supported by limited data on long-term trends in visibility and the presence in lake sediments of chemicals emitted during combustion of coal and other fossil fuels.

3. Substantial differences in temporal trends in SO_2 emissions among regions of the United States have emerged since about 1970. Before 1970, temporal trends in SO_2 emissions in the various regions were congruent, although the amounts of emissions were of different magnitudes. From data on SO_2 emissions, reduction in visibility, and sulfate in Bench-Mark streams since about 1970, we conclude that the southeastern United States has experienced the greatest rates of increase in parameters related to acid deposition. The midwestern United States has experienced rates of increase somewhat lower than the Southeast. In the northeastern United States the trend has been one of modest decreases.

4. The record of the chemistry of Bench-Mark streams suggests that changes in stream sulfate flux determine changes in stream water alkalinity and base cation concentrations in drainage basins that have acid soils and low-alkalinity waters. Increases in stream sulfate flux are associated with decreases in alkalinity and/or increases in amounts of base cation in surface waters. The change in alkalinity per unit change in sulfate depends on site-specific characteristics. Changes in sulfate observed in Bench-Mark streams are consistent with changes in SO_2 emissions on a regional basis. Analysis of a sulfur mass balance for 626 lakes in the northeastern United States and southeastern Canada demonstrates that the sulfate output from lakes in general is proportional to sulfate inputs in wet deposition. The ratio of output to input decreases with distance from major source regions, suggesting that dry deposition is an important contributor to sulfate flux inputs, especially near major source regions.

5. Data on alkalinity of some lakes in New York, New Hampshire, and Wisconsin suggest that changes in alkalinity greater in magnitude than about 100 μeq/L can occur over time periods of about 50 years. Changes of this magnitude are too large to be caused by acid deposition alone and may result from other human activities or natural causes. We have not attempted to identify the exact nature of the causes of these large changes.

6. Analysis of diatom and chrysophyte stratigraphy for sediments in 10 low-alkalinity Adirondack Mountain lakes studied indicates that 6 of them became increasingly acidic between 1930 and 1970. Because the trend in acidification is consistent with both the presence of other substances in sediments that indicate fossil fuel combustion and current lake acidification models, and because the observed acidification cannot be explained by known disturbances of the watersheds or by other natural processes, acid deposition is the most probable causal agent. These findings are supported by fish population data for 9 of the 10 lakes in the Adirondacks for which concurrent data exist. Diatom data from lakes in New England indicate slight or no decrease in pH. Data for southeastern Canada are insufficient to examine trends in acidification.

7. Based on comparisons of historical data on alkalinity and pH recorded in the 1920s, 1930s, and 1940s with recent data for several hundred lakes in Wisconsin, New Hampshire, and New York, we find that many lakes have decreased in pH and alkalinity and many have increased in pH and alkalinity. On average, lakes sampled in Wisconsin have increased in alkalinity and pH. The New Hampshire lakes on average show no overall change in alkalinity and a small increase in pH. Interpretation of changes in the New York lakes is sensitive to assumptions about the application of colorimetric techniques in the historical survey and the selection of recent data bases. Depending on the assumptions, New York lakes on average either experienced no changes in alkalinity and pH or have decreased in alkalinity and pH. In the judgment of the committee, the weight of the evidence indicates that the atmospheric deposition of sulfate has caused some lakes in the Adirondack Mountains to decrease in alkalinity. We base this conclusion on three types of evidence: (a) Sulfate concentrations in wet-only deposition in the region of the Adirondack Mountains and sulfate concentrations in Adirondack lakes are relatively high in comparison with those in other areas in the northeastern United States and southeastern Canada. We have demonstrated that increasing sulfate in surface waters is associated with decreasing alkalinity in low-alkalinity surface waters. (See Conclusion 4.) (b) Diatom-inferred pH and other supporting evidence provide a strong indication of acidification from acid deposition in low-alkalinity lakes. (See Conclusion 6.) (c) We calculated alkalinity changes in New York lakes four

different ways to account for different assumptions. Three of the results indicate, on average, a decrease in alkalinity (median values of -28, -44, and -69 μeq/L), and one result shows no overall change (median value of +1 μeq/L).

Because of ambiguities regarding the assumptions employed in the historical New York survey, we cannot currently determine which of these results is most accurate, and hence we cannot quantify the number of New York lakes that have been affected by acid deposition.

8. Emissions of oxides of nitrogen (NO_x) are estimated to have increased steadily since the early 1900s, with an accelerated rate of increase in the Southeast since about 1950. Reliable data do not exist to determine historical trends of nitrate concentrations in the atmosphere, precipitation, or surface waters.

9. Although high-quality data to assess trends in fish populations as a function of surface water acidity are sparse, the data that are available indicate that fish populations decline concurrently with acidification. The strongest evidence in support of this finding comes from some Adirondack Mountain lakes. These data demonstrate declines in acid-sensitive fish species populations over the past 20 to 40 years in lakes thought to have been acidified over the same time period. (See Conclusion 6.) The number of cases studied is too small to permit any projections of the total number of fish populations that may have been affected by acidification.

10. Geographically widespread reductions in tree ring width and increased mortality of red spruce in high-elevation forests of the eastern United States began in the early 1960s and have continued to the present. The changes occurred about the same time as important climatic anomalies and in areas subject to comparatively high rates of acid deposition. The roles of competition, climatic and biotic stresses, and acid deposition and other pollutants cannot be adequately evaluated with currently available data.

METHODS

It is important to establish the requirements for inferring that a relationship between data sets implies causality. Two variables can be considered to be associated if their values are paired in some related way across a population, and they are unassociated if a special pairing does not exist. To establish that an association exists, it is necessary only to show that the variables do not appear to be paired at random. Thus, as we will see in Chapter 5, sulfate and nitrate in wet deposition are associated across eastern North America; regions with high wet sulfate deposition also tend to have high wet nitrate deposition, and vice versa.

Association between variables is necessary but not sufficient to infer the existence of a causal relationship. Mosteller and Tukey (1977) list three criteria--consistency, responsiveness, and mechanism--at least two of which are usually needed to support causation. Consistency implies that (all other things being equal) the relationship between the variables is consistent across populations in direction, or perhaps even in amount. If a relationship between the variables holds in each data set, then the relationship is consistent. Responsiveness involves experimentation. If we can manipulate a system by changing one variable, does the other variable also change appropriately? Mechanism means a step-by-step path from the "cause" to the "effect," with the ability to establish linkage at each step.

Observation of a correlation between two variables can establish consistency, but it cannot establish either responsiveness or the mechanism of possible causation. Thus, correlation is not adequate to prove a cause-and-effect relationship. Continuing the earlier example, sulfate and nitrate in wet deposition have a clear, consistent relationship, but neither experiment nor mechanism implicates changes in one as causing changes in the other. In fact, we know that they both arise from a common source, the high-temperature combustion of fossil fuels (National Research Council 1983).

In this report, we use the criteria of mechanism and consistency for suggesting cause-and-effect relationships. Some controlled studies of responsiveness are discussed, but field experiments on most of the variables are generally not considered practicable. In the next section of this chapter, we describe a number of conceptual linkages among the various types of data. These linkages are in fact mechanisms, stepping from one variable, e.g., emissions of SO_2, to another variable, e.g., visibility, with causation natural at each step. Ideas about mechanisms then motivated analyses to determine whether consistent relationships exist among the variables over space and time.

MECHANISMS

In this section, we summarize the conceptual mechanisms that could link emissions, atmospheric deposition, and environmental responses; the analyses of trends and spatial patterns are discussed in the following section. The general conceptual relationships are shown schematically in Figure 1.1 and are described further in the following chapters. The figure does not depict all the possible environmental interactions of sulfur and nitrogen oxides or the many possibilities for their ultimate fates. It does, however, indicate relationships among the phenomena examined in this report: emissions, visibility, chemistry of precipitation, chemistry of lake and stream waters, fish populations, forests, and chemical and biological stratigraphy of lake sediments. Dry deposition is shown in this general diagram and is discussed in various chapters of this report, but lack of data precluded any detailed quantitative analysis of its temporal trends.

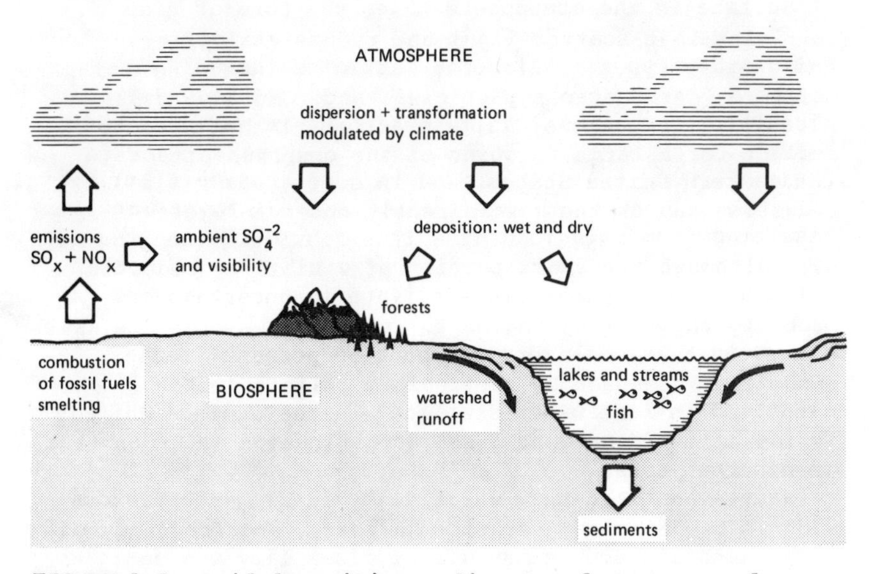

FIGURE 1.1 Acid deposition: diagram of sources and affected ecosystems.

Mechanisms of Linkage Among Atmospheric Processes

Emissions

In eastern North America more than 90 percent of the emissions of sulfur and nitrogen oxides into the atmosphere are the result of combustion of fossil fuels (Robinson 1984). These oxides (SO_2 and NO_x) are transformed in the atmosphere into sulfate (SO_4^{2-}) and nitrate (NO_3^-), respectively, by complex series of catalytic and photochemically initiated reactions (National Research Council 1983). The transformation processes and rates depend on factors such as sunlight, temperature, humidity, clouds, and the presence of the various catalysts. The sulfates, nitrates, and other reaction products are subsequently removed from the atmosphere by both wet- and dry- deposition processes.

Visibility

Sulfate in the atmosphere takes the form of fine particles that scatter light and reduce visibility. While other types of airborne particles including moisture, carbonaceous particles, and road dust cause visibility reductions, sulfate aerosol is believed to account for a large fraction of the observed effect in the eastern United States, and in some areas visibility reduction may be the most directly observable effect resulting from emissions of sulfur dioxide. (See Chapter 4.) Although the exact portion of visibility reduction attributable to particulate sulfate is uncertain and probably varies from region to region, investigators have estimated that, on average, about 50 percent of the reduction is caused by particulate sulfate (often dissolved in fine water droplets) in the eastern United States, with a somewhat lesser contribution in urban than in nonurban areas.

Short- and long-term variability in climatic factors such as temperature, relative humidity, and frequency of air stagnation events may affect visibility and can complicate the association between trends in visibility and trends in SO_2 emissions.

Deposition

Historically, concern about acid deposition has focused on the acidity (pH) of precipitation as the important factor in assessing potential environmental effects. The terms acid rain and acid deposition reflect this focus. More recently, as the understanding of the phenomena has improved, attention has shifted to the deposition of sulfur and nitrogen compounds (as well as other chemicals) from the atmosphere.

Concentrations of sulfate and nitrate ions (SO_4^{2-} and NO_3^-, respectively) in wet and dry deposition are related to sulfur and nitrogen oxide emissions, while the "acidity" of wet deposition is the result of concentrations of both acidic compounds (i.e., strong acids (nitric and sulfuric acids) as well as weak organic acids) and alkaline materials incorporated in deposition. Thus, in establishing the relationship between emissions and environmental effects, knowledge of the fate of sulfur and nitrogen compounds is critical, and in this report we have focused attention on the emissions, deposition, and fates of sulfur and nitrogen oxides.

Sulfur and nitrogen oxides and their transformation products are deposited from the atmosphere onto the Earth's surface as acids or acidifying substances. The processes that remove the constituents from the atmosphere are complex (National Research Council 1983). How quickly a given substance is removed from the atmosphere depends on its physical form, meteorological events, and the characteristics of the surface. Average residence times in the atmosphere of SO_2, NO_x, and the important reaction products (SO_4^{-2}, NO_3^-, etc.) range from hours to weeks. During this time they may travel either short distances (0 to 100 km) or longer distances (100 to 1000 km) from their sources before they are removed from the atmosphere. Average distances of transport are generally a few tens to a few hundred kilometers.

Wet deposition is only one of two major pathways of atmospheric deposition. The other pathway, dry deposition, is an important process for the removal of gases and airborne fine and coarse particles. Wet and dry deposition combined provide the bases for interactions of the compounds involved in acid deposition with the terrestrial and aquatic ecosystems. Data on dry deposition, however, are sparse, and our analyses of temporal trends and spatial distributions of acid deposition are based primarily on wet-only deposition. (See Chapter 5.)

Mechanisms of Linkage Between Acid Deposition and Surface Water Chemistry

Most of the water in lakes and streams has flowed through the forest canopy, soils, and geologic materials of the watershed before entering the lake or stream. Because terrestrial watersheds are chemically reactive media, any model linking acid deposition to surface water acidification must account for the nature of the interactions among the soils and soil solutions in watersheds. Similarly, the analysis of trends must be sensitive to the uncertainties inherent in quantifying the processes. (See Chapter 7.)

The "mobile anion" hypothesis implicates sulfate as the major determinant in lake acidification (Seip 1980). Sulfate in soil solution moves through soils as mobile (negatively charged) anions. When this occurs, an equivalent amount of (positively charged) cations must also be transported. This equivalence is necessary to satisfy the requirement for a charge balance in all aqueous solutions. The common cations in acidic soils are aluminum (Al^{3+}), hydrogen (H^+), calcium (Ca^{2+}), magnesium (Mg^{2+}), potassium (K^+), and sodium (Na^+). As the concentration of sulfate in the soil solution and soil leachate increases, the concentrations of dissolved cations must also increase. In general, lakes and streams in watersheds having hydrologic flow paths through highly acidic soils are expected to match an increase in sulfate concentration with a higher proportion of acid cations (H^+ and Al^{+3}), thereby reducing surface water alkalinity. Less acid soils will tend to counter an increased sulfate flux with a higher proportion of base cations (Ca^{2+}, Mg^{2+}, or Na^+).

Sulfates are not mobile in all soils. Watersheds may retain or release sulfate by adsorption and reduction-oxidation (redox) reactions that are often associated with biological fixation. These reactants may cause increases in acidity (oxidation) or decreases in acidity (reduction) (Wagner et al. 1982). Some watersheds may contain weatherable sulfur-bearing minerals in the soil, subsoil, and bedrock that can act as sulfur sources.

The fate of nitrate is also complicated because it is biologically active. During the growing season, nitrate introduced into the terrestrial environment is likely to be converted to organic nitrogen in plant material or microbial biomass by a number of processes. As with sulfur, most of these processes either produce acid (oxidation) or consume acid (reduction). Nitrate entering surface waters through either runoff or direct deposition may be rapidly consumed by aquatic biota with the result that excess concentrations of nitrate are eliminated. However, for a very brief period in the early spring, nitric acid released from the watershed can be the major cause of a sharp, but temporary decrease in pH in lakes and streams. This nitrate may be released from the snowpack, the soil, or both.

Organic soils of geologic terrain that are resistant to weathering are naturally acidic. Waters moving through these soils, if not already acidic, may acidify, and receiving lakes naturally tend to become more acidic with time. This process is usually slow (i.e., on the order of centuries to millennia for changes of the order of 1 pH unit). If the hydrologic pathways to surface waters through humic soils are direct, acidic surface waters dominated by organic acids may result. Furthermore, changes in land use practices can cause changes in inputs and outputs of acids in a watershed with subsequent changes in surface water chemistry. The effects of land use practices on the pH of surface waters are not well known, but lakes may become more or less acidic, depending on site-specific conditions.

Differences in the geologies of watersheds and in the histories of land use practices within watersheds can cause surface waters in a region to respond differently even while they are receiving equivalent amounts of acidity from atmospheric inputs. Other factors, such as changes in the volume of water, can cause seasonal and interannual changes in alkalinity. Thus there are likely to be a number of contributing causes for trends in surface water alkalinity.

Mechanisms of Linkage Between Acid Deposition and Lake Sediments

The chemical composition of lake sediments reflects the net results of all chemical inputs and outputs over time. Changes in atmospheric chemical inputs to soils can modify leaching rates of certain chemical species from the terrestrial environment that serve as inputs to surface waters. Additionally, many constituents deposited from the atmosphere, both solids and liquids, may be incorporated directly into accumulating sediment. Thus, lake sediments provide, directly or indirectly, a natural stratigraphic record of the changes in the chemical composition of surface waters and may reflect changes in atmospheric deposition. In addition, the sedimentary record may preserve information on the populations of biological organisms that can be related to water quality at the time when sediment was deposited. (See Chapter 9.) However, in the case of both the chemical and the biological record, older sediment from shallower water may, with their respective characteristics, be incorporated into younger accumulating sedimentary sequences. Thus changes in atmospheric deposition of metals to the lake or watershed may not be directly recorded by the sediments.

Sediment Chemistry

Sulfur deposited from the atmosphere may be added to sediment in a variety of ways, including incorporation into sedimenting organic matter, adsorption onto the surface of particles, chemical migration into sediment as sulfate, reduction of sulfate to sulfide, and precipitation from solution as sulfide. Thus, it is not possible to draw conclusions about the precise chronology of the deposition of sulfur compounds from analysis of concentration as a function of depth in a sediment core. No data exist to suggest that the nitrogen content of sediments reflects the deposition rates of nitrogen compounds.

However, the industrial processes from which SO_2 and NO_x are emitted also produce emissions of a number of heavy metals (for example, lead, zinc, copper, and vanadium). Concentrations of these metals typically are low in lake sediments in unpolluted regions and increase in more densely populated regions. Net accumulation in sediment of certain metals (for example, zinc, manganese, and calcium) depends in part on pH. Because these metals are more soluble at lower pH values, they may be preferentially lost (redissolved into the lake water) from sediment as the pH of the lake water decreases. This loss caused by decreasing pH may be reflected in the sediment record and may provide information on the period of time over which acidification has occurred. However, leaching of sediments by acidified lake water may affect the chemistry of sediments deposited up to a few tens of years prior to acidification. Acidic components (H^+, SO_4^{2-}, etc.) can diffuse centimeters into the sediment and desorb or dissolve metal that diffuses out of the sediment.

Other materials deposited in sediment that are indicative of fossil fuel combustion include polycyclic aromatic hydrocarbons (PAHs) and components of fly ash. Although these trace substances are not so direct a measure of acid deposition as are the measurement of sulfur and nitrogen oxides, their presence in elevated concentrations in lake sediments is indicative of emissions from the same sources.

Diatoms and Chrysophytes in Sediments and pH of Lake Water

Diatoms and chrysophytes are single-celled algae that are abundant in most lakes. The relative abundance of specific taxa can be related to the lake water pH because most species are sensitive to the pH of the water in which they live. For a given region (such as the Adirondack Mountains of the northeastern United States), a calibration data set can be established by correlating the current chemistry (especially the pH) of lake waters to the populations of different diatoms species in the surface (recent) sediment. Because diatoms are well preserved in sediments, an analysis of their distribution and abundance in deeper (older) sediments may be used to reconstruct a quantitative estimate of historical trends in pH.

Mechanisms of Linkage Between Acid Deposition and Biologic Effects

Forests

From a wide variety of theoretical and experimental studies, a considerable amount of information exists on the effects that acid deposition and other airborne chemicals might have on trees in the field. Researchers have investigated the possibility of adverse effects on foliar integrity, foliar leaching, root growth, soil properties, microbial activity in the soil, resistance to pests and pathogens, germination of seeds, and establishment of seedlings and have shown a variety of positive and negative effects. Effects of acid deposition combined with natural abiotic stresses and with gaseous pollutants such as ozone are also possible. There are many ways in which acid deposition might affect forests, but to date it is not clear from the existing field evidence in North America that any of the mechanisms have caused changes in the forest or in the growth of trees. Thus, although there are many possible pathways that might link acid deposition to changes in tree growth, no proposed linkages between deposition and effects on trees in North America are adequately supported by the mechanistic data currently available. (See Chapter 6.)

Recent field studies have shown that red spruce have died in unexpected numbers in high-elevation forests of the Appalachians. Because of the interception of highly acidic cloud waters at these elevations, the deposition of acidic substances and other airborne chemicals is high in comparison with that in the rest of the forests of eastern North America. The lack of any obvious natural cause and the high rates of acid deposition make it logical to examine the timing of this phenomenon to determine the extent to which it corresponds with trends in deposition of acidic substances.

It is well known that climatic factors can induce decline diseases of forest trees (complex diseases related to a multiplicity of factors). Examining trends in climate in relation to trends in tree ring widths and mortality of forest trees is critical in the context of this report.

Fish Populations

One of the most widely publicized of the biotic responses attributed to acid deposition has been the decline of fish populations in certain freshwater lakes and streams in portions of the eastern United States. Declining populations have also been noted in the Scandinavian countries and Canada. Many species of fish cannot tolerate surface waters in which the pH has dropped below about 5, and surface waters attaining this level of acidity, from whatever the cause, could experience fish population losses. Simultaneously, trivalent aluminum ion (Al^{3+}) can reach toxic concentrations in sufficiently acidified surface waters.

Factors unrelated to acidity such as changes in stocking policy, beaver activity, or intensified sport fishing may affect fish populations. Even when fish declines have been directly linked to increasing acidity of the aqueous environment, linking fish loss to acid deposition requires the further step of linking the increasing acidity to acid deposition.

Relationships Among Mechanisms

The several mechanisms described above provide the basis for exploring further whether various biological and geochemical processes are linked to emissions of SO_2 and NO_x. Thus, for example, a sustained increase in SO_2 and NO_x emissions on a wide geographic basis may result in a number of responses, including reduced visibility, increased concentrations of sulfate and nitrate in precipitation and surface waters, and increased deposition in lake sediments of other pollutants associated with these same emissions sources. A premise of this study is that simultaneous examination of trends and spatial patterns in all the relevant phenomena may help to establish further evidence of linkages despite the complicating factors that may frustrate interpretation of data about a single phenomenon.

Complicating Factors

The mechanisms described above suggest that certain relationships exist among the data variables. In a perfect world, perhaps we could establish the relationships with rigor. In the practical world, however, various factors complicate the analysis, raising the possibility that the relationships may not be discernible in existing data. We discuss some of those complicating factors here.

1. Atmospheric processes that transport, chemically convert, and remove acidic or acidifying substances from the atmosphere are subject to meteorological and climatological variability. The distance and direction of the flow of gaseous emissions and their transformation products from their sources depend on meteorological conditions at the time of emission and in the subsequent few days. The conditions can be expected to vary seasonally and interannually. On longer time scales, mild winters with small amounts of snow may lessen the usually sharp rise in acidity observed in lakes and streams during spring snowmelt in areas receiving acid deposition. In addition, long-term climatological

changes in patterns of atmospheric circulation may result in changes in patterns of acid deposition. Climatic variations also complicate attempts to assess the effects of acid deposition since ecosystems may be responding to changes in both climate and deposition. Separating an "acid deposition" signal from a "climatic" signal is a major problem in evaluating trends in effects of acid deposition in natural systems.

2. The relative importance to atmospheric deposition of distant and nearby sources is uncertain. Sulfur compounds emitted from a stack may be deposited near the stack, a few tens or hundreds of kilometers away, or even at distances of thousands of kilometers (National Research Council 1983). On the other hand, the amount deposited generally decreases with distance from the source. Thus, variables related to acid deposition at any given location reflect both nearby and distant sources, but in proportions that are not precisely known and that may change over time and space. This complication makes it more difficult to establish associations between emissions and variables thought to be related to atmospheric deposition since we do not have a precise way to determine the relative contributions of local and distant sources.

3. The effect observed may not be synchronous with the onset of deposition. For example, the response times of lakes receiving acid deposition may vary widely, depending on characteristics of the lake and watershed and on the rate of deposition (National Research Council 1984). This consideration is important in the evaluation of trends because it suggests, for example, that lakes with different watersheds may respond to changes in deposition on different time scales.

4. Frequently, data are not available for all the relevant variables in a hypothetical relationship. Studies of mass balance of sulfur and nitrogen are vulnerable to this complication, since there are no extensive data on dry deposition (Chapter 7). The flux of sulfate into a lake can be approximated by data on wet-only deposition, but the contribution of dry deposition to the watershed can vary significantly from year to year and place to place. Measurements have shown that dry deposition is an important fraction of total deposition in areas close to point sources. It is suspected that in some ecosystems, particularly in forested areas where the forest canopy may serve as an efficient collector of dry sulfate, net dry sulfate deposition may be equal to or greater than wet-deposited

sulfate. Furthermore, depending on the type of soil, sulfate may be adsorbed by the soil or additional sulfate may be added to the soil solution by weathering of internal mineral sources. Biological processes also play a role in sulfur cycling in watersheds. The lack of data for some of the important processes affecting sulfur mass balances complicates the interpretation of results and introduces uncertainties.

5. Analytical methods have changed over time. To obtain historical data that can be compared with more recent data, it often is necessary to adjust the data for systematic bias resulting in changes in the analytical techniques. In the case of the chemical analysis of lake waters, scientists have applied different assumptions to derive different adjustment factors. Such disparity may cause confusion in interpreting historical trends.

SPATIAL PATTERNS AND TEMPORAL TRENDS

Regions

It was helpful in our analyses to define six specific geographical regions in eastern North America (Figure 1.2). The regions were established as contiguous groups of states and provinces that in our judgment have experienced similar temporal trends in SO_2 emissions over the past 50 years. Regions A (the Canadian provinces) and F (the western United States) were not included in all the analyses.

We have analyzed emissions and environmental responses in each region independently. Thus, our analyses do not specifically account for interregional dependences as these might be affected, for example, by long-range transport from one region to another. The approach is valid for our purposes since we are interested in describing trends averaged over broad geographical areas rather than focusing on specific source-receptor relationships. In general, comparisons of trends cannot be used to deduce relative contributions to deposition at one location by a number of sources at diverse, distant locations. The results of the analyses--that trends in environmental indices appear to respond to trends in sulfur emissions on a regional basis--do not imply that long-range transport between the regions is necessarily unimportant. Clearly, interregional transport will be important, especially near the boundaries of the regions.

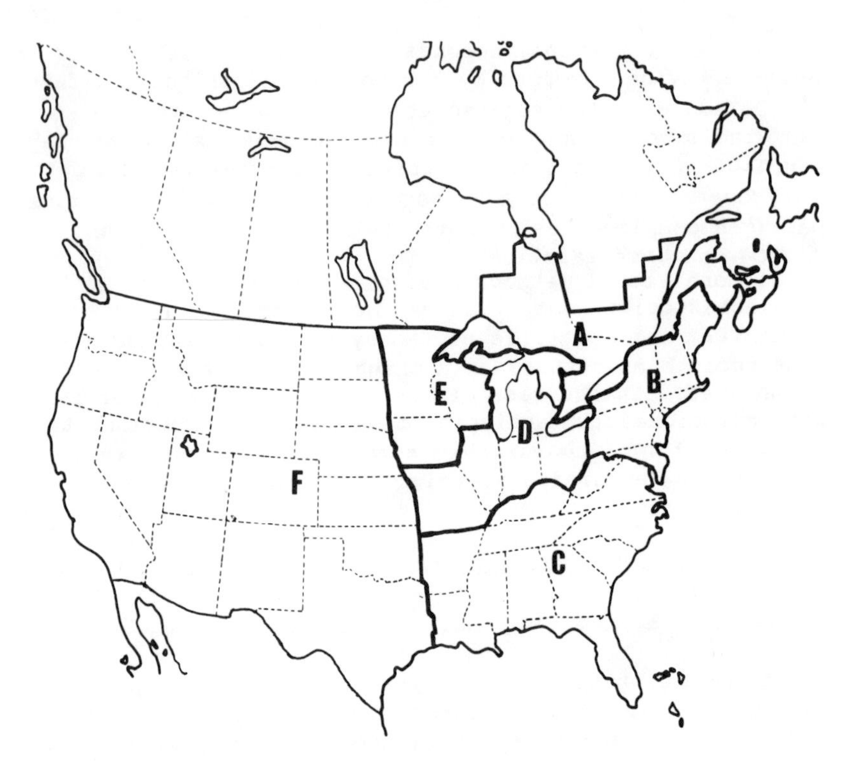

Region

A (Eastern Canada): Eastern Ontario, New Brunswick, Southern Quebec

B (Northeastern U.S.): Connecticut, Delaware, Maine, Maryland, Massachusetts, New Hampshire, New Jersey, New York, Pennsylvania, Rhode Island, Vermont

C (Southeastern U.S.): Alabama, Arkansas, Florida, Georgia, Kentucky, Louisiana, Mississippi, North Carolina, South Carolina, Tennessee, Virginia, West Virginia

D (Midwestern U.S.): Illinois, Indiana, Michigan, Missouri, Ohio

E (North central U.S.): Iowa, Minnesota, Wisconsin

F (Western U.S.): All states in the contiguous U.S. not included in Regions B to E

FIGURE 1.2 Designated regions for analysis of temporal trends and regional patterns.

Spatial Patterns

Spatial associations among different types of data can be established by testing for similarities in the geographical distribution of density, concentration, flux, or the magnitude of environmental indices. Because historical data are too sparse for developing spatial patterns, we relied on data from the recent past, for which more extensive quality-assured data sets are available.

To test for possible spatial associations, we first compared the magnitudes of various parameters on a regional scale. For example, Figure 1.3 shows that the areas of highest SO_2 emission density are, in general, areas where the concentrations of sulfate and hydrogen ions in wet deposition are highest. In a similar way, regional contours of range of visibility and concentration of atmospheric sulfate can be compared with emission density patterns.

Although such visual presentations may generally suggest spatial associations among phenomena, we also applied Friedman's test (Lehmann 1975) to data from all regions except Region A--eastern Canada. (See Figure 1.2.) This test allowed a more quantitative and objective appraisal of spatial associations among the following variables: emission density of SO_2 (1980), concentration of sulfate aerosol (1972), concentration of sulfate in wet-only deposition (1980), flux of sulfate in U.S. Geological Survey Bench-Mark streams (1980), and visibility (1974-1976). Each variable was ranked from 1 to 5 across the five regions. Rankings were based on the relative magnitude of each variable, expressed as a density, concentration, and flux parameter or, in the case of visibility, as visual range; except for visibility which is an inverse relationship, a ranking of 5 was assigned to the region where the magnitude was greatest, a ranking of 4 assigned to the region where the magnitude was next greatest, etc. The results of the regional rankings are shown in Table 1.1. In cases when differences in variables between regions are too small to be distinguished, equal rankings are assigned to the similar regions (for example, 4.5 for sulfate in aerosol in Regions B and D).

The null hypothesis tested in the analysis is that there are no relationships among the variables. The p value obtained with this array of rankings is 0.009, which implies that if there were in fact no true association among the variables, an apparent association to this degree or greater would occur by chance only 0.9 percent of the time. The result indicates of a strong spatial association among SO_2 emissions, atmospheric sulfate aerosol, sulfate in precipitation, sulfate in Bench-Mark streams, and visibility, corroborating conclusions reached by visual comparisons of the relationships displayed in contoured maps.

On the basis of the rankings of the five variables given in Table 1.1, it appears that, overall, Region D experiences the highest amounts of SO_2 emissions and related effects, Region F experiences the lowest amounts, and Region E experiences the next-to-lowest amounts. Regions B and C are generally more affected than Region E but less affected than Region D.

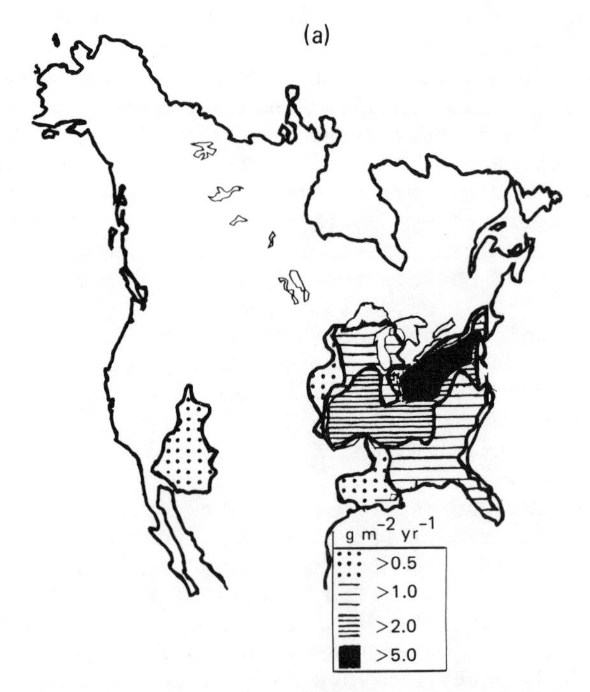

FIGURE 1.3 Comparison of spatial patterns between (a) SO_2 emission density (g $m^{-2}yr^{-1}$); (b) concentration of H^+ (μmoles/L) in wet deposition; (c) concentration of SO_4^{2-} (μmoles/L) in wet deposition; (d) concentration of sulfate aerosol (μg/m^3); and (e) visible range (in miles).

(b)

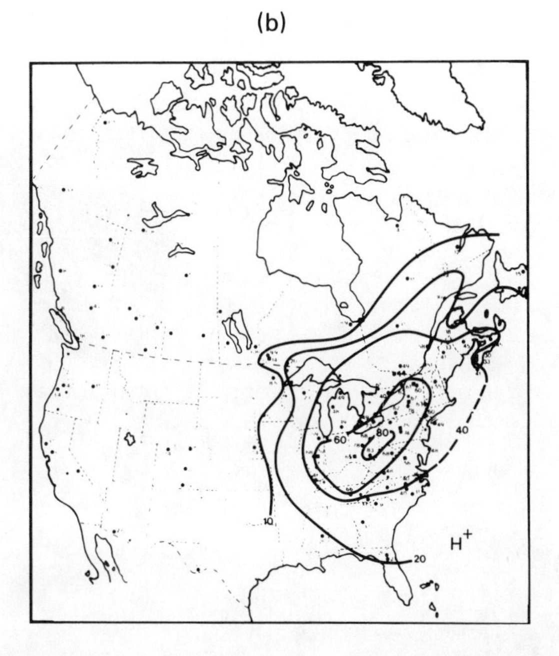

(c)

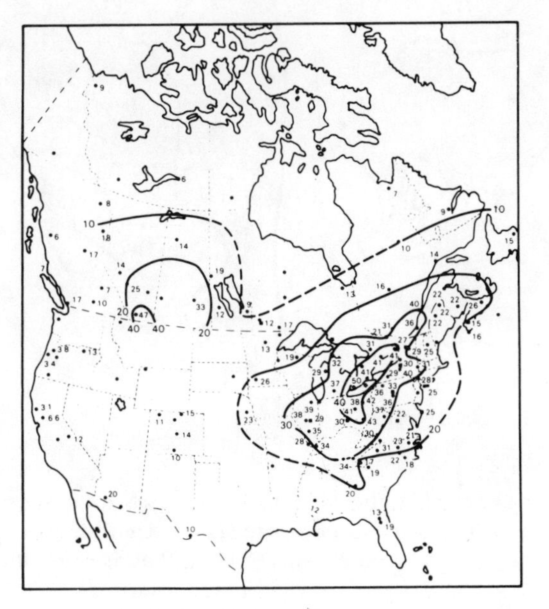

FIGURE 1.3 (continued).

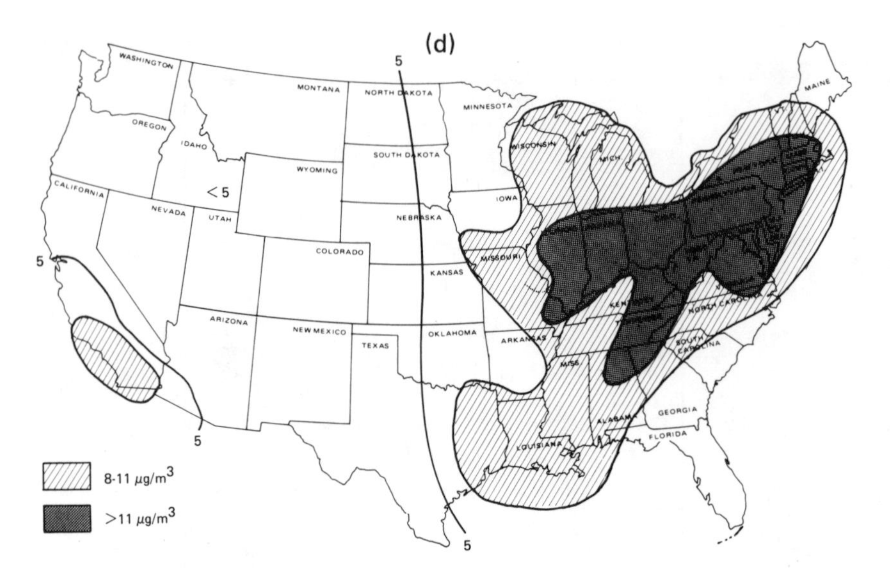

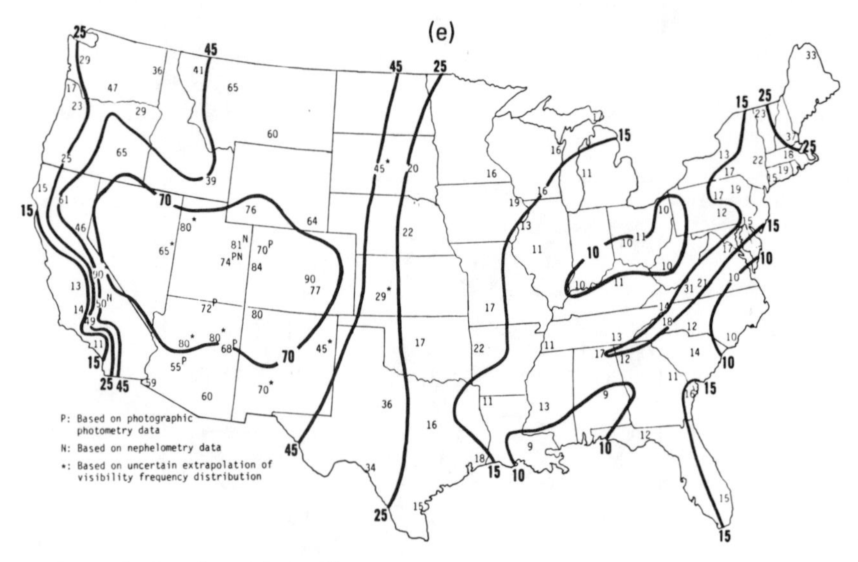

FIGURE 1.3 (continued).

In a different type of analysis, we examined the spatial association between sulfate in wet precipitation and sulfate in 626 lakes in the northeastern United States and southeastern Canada (Chapter 7). We tested the assumption that sulfate in wet deposition was the

TABLE 1.1 Regional Rankings by Magnitude of SO_2 Emissions Densities and Related Variables

Region	SO_2 Emissions	Sulfate Aerosol[a]	Visibility[b]	Sulfate Precipitation	Sulfate Streams
B	3	4.5	2	4	4
C	4	3	4	3	1
D	5	4.5	5	5	5
E	2	2	3	2	2
F	1	1	1	1	3

NOTE: p value of Friedman's test, 0.009.

[a]The difference in concentration of sulfate aerosol in Regions B and D is too small to be distinguished. In order to avoid possible bias toward maximum association, we assigned a 5 to Region B and a 4 to Region D, because this ranking (i.e., 5, 3, 4, 2, 1) shows a lesser association than would the alternative ranking (i.e., 4, 3, 5, 2, 1).

[b]Rankings for visibility are inversely related; i.e., the greater the value, the worse the visibility.

only source of sulfate to the lake. Sulfate output from the lake was calculated from the product of the sulfate concentration in the lake and the net precipitation (i.e., precipitation minus evaporation). These simplifying assumptions were necessary because additional data required for a more complete analysis of sulfate mass balance, such as inputs through dry deposition or chemical weathering of sulfate minerals in the watersheds, were not available.

The lakes were grouped into six subregions: Connecticut-Massachusetts-Rhode Island (CT-MA-RI), Maine-Vermont-New Hampshire (ME-VT-NH), New York (NY), Quebec (QU), Labrador (LB), and Newfoundland (NF). Data for each of the subregions were plotted on a log-log scale (Figure 1.4). The figure shows each of the regression lines having equal slopes of about one, but with different intercepts, m_i'. On the natural scale, these data correspond to a linear model, $y = m_i x$, where y (output flux of sulfate from lakes) is proportional to x (flux of sulfate into lakes from wet deposition) by the subregional multiplying factor m_i (the antilogarithm of m_i'). The values of m_i for the six subregions are given in Table 1.2.

Regions where lakes demonstrate the greatest excess of sulfate output ($m_i > 1$), suggesting that wet deposition accounts for only part of the sulfate input, are those nearest to large emission sources (CT-MA-RI, NY, ME-NH-VT); those with values of m_i near or less than

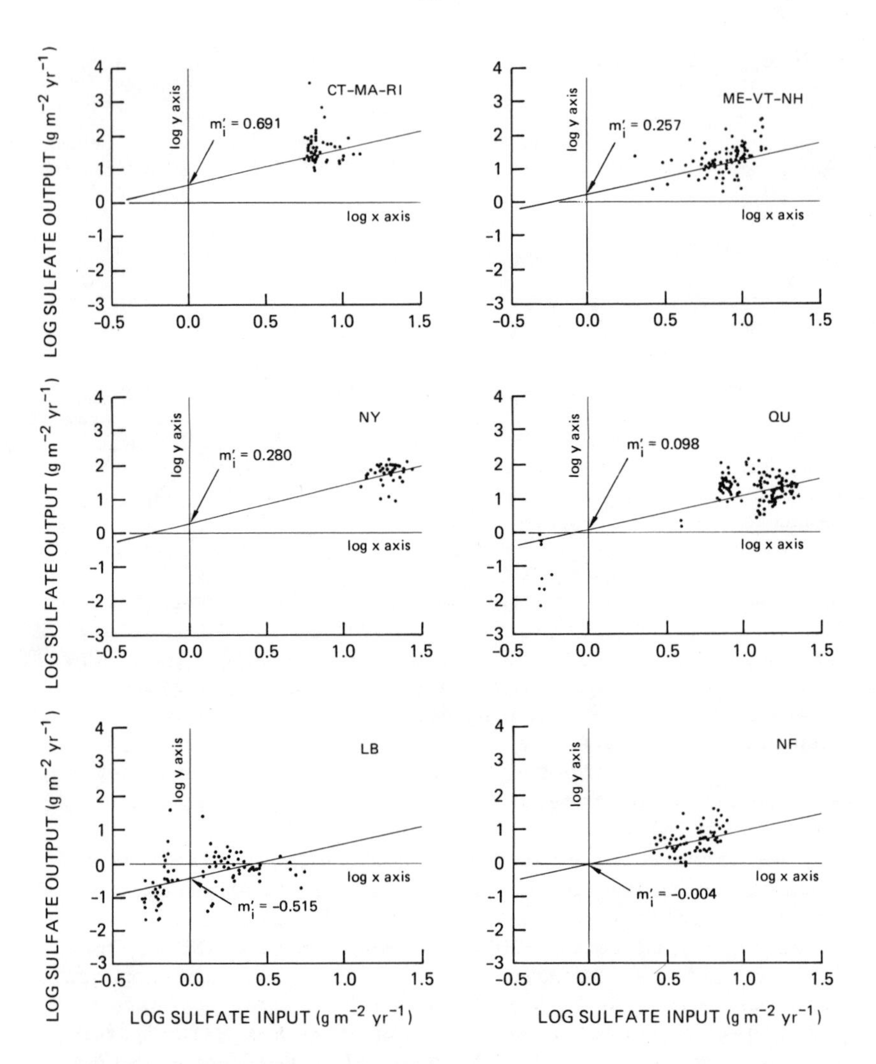

FIGURE 1.4 Regional relationships between the log of sulfate flux into lakes from wet deposition and the log of sulfate flux out of lakes from outflow. m_i' denotes the log y axis intercept.

1.0 are more distant (QU, NF, LB). This pattern of association is consistent with the hypothesis that dry deposition is a major contributor to the sulfate mass balance near emission sources. The result is also consistent with hypotheses linking emissions, sulfate in deposition, and sulfate in surface waters.

TABLE 1.2 Regional Multiplying Factors, m_i

Region	m_i	Standard Error
CT-MA-RI	2.00	0.16
ME-VT-NH	1.29	0.14
NY	1.32	0.09
QU	1.10	0.06
LB	0.60	0.02
NF	1.00	0.09

NOTE: m_i is the antilogarithm of m_i'. See Figure 1.4.

Data on fish population records and tree ring widths were not available for conducting detailed spatial analyses. However, declines in fish populations in selected lakes and streams have been documented in watersheds extending from Pennsylvania to Nova Scotia. Surface waters in these areas tend to have low alkalinity and receive relatively high inputs of atmospheric deposition of sulfate and nitrate. Although data are sparse, well-documented fish declines have occurred in a few locations in lakes of the Adirondack Mountains and in rivers of Nova Scotia, waters that have experienced increases in acidity thought to be caused by acid deposition.

Although the mechanisms of cause and effect between acid deposition and forest decline have not been established, the areas experiencing the most noticeable declines of red spruce--the high-elevation areas of New York, Vermont, and New Hampshire--receive large amounts of acid deposition as well as other pollutants.

Only limited data exist to evaluate spatial patterns relating nitrogen oxide emissions and nitrate in the environment. Both stationary and mobile sources emit NO_x, which is formed mainly by the oxidation of nitrogen in the air during combustion. The formation of NO_x depends strongly on conditions in the combustion zone, such as temperature. These uncertainties limited our ability to estimate NO_x emissions with a high degree of confidence. As a result, we can make only the general statement that areas of the country with the highest NO_x emissions are generally those in which nitrate concentrations in precipitation are greatest.

In summary, statistical tests of spatially differentiated data demonstrate regional associations among sulfur dioxide emissions, sulfates in aerosol, visibility, sulfates in precipitation, and sulfate fluxes in Bench-Mark stream waters. Spatial analysis of sulfate mass balance in 626 lakes further strengthens the postulated relationship between sulfate deposition and surface water concentrations of sulfate. The analysis is also consistent with the ideas that dry deposition of sulfate decreases with distance from regions of highest sulfur emissions and that in some cases sources of sulfur within the watershed also contribute to the sulfate input of lakes.

These findings do not resolve questions about specific source-receptor relationships. However, they demonstrate strong regional association in spatial patterns among density of SO_2 emissions, atmospheric concentrations of sulfate, range of visibility, concentration of sulfate in wet precipitation, and fluxes of sulfate in Bench-Mark streams.

Temporal Trends

The Record Since Industrialization (about 1880)

Only a few pieces of evidence exist for estimating the acidity of atmospheric deposition of 100 years ago. Hidy et al. (1984) have reviewed much of the early literature on the chemical composition of precipitation in the United States. Records of sulfate concentrations are available from a few locations over periods ranging from about 1910 to 1950. These data, however, are not of sufficient quality to permit a definitive determination of long-term trends or comparisons with more recent data. (See Chapter 5.) Hence we examined a larger pool of historical data that may be associated with acid deposition to determine whether a consistent and clearer picture of long-term trends emerges.

Examples of the data examined are shown in Figure 1.5 and include (a) estimates of coal consumption in the eastern United States (Regions B, C, D, and E) and in Region B alone (Chapter 2); (b) annual emissions of SO_2 for the eastern United States and for Region B alone (Chapter 2); (c) light extinction (an inverse measure of visibility) at the Blue Hill Observatory in Massachusetts (Chapter 4); (d) lead, vanadium, and PAHs in sediments in Big Moose Lake in New York (Chapter 9); (e) diatom-inferred pH (Chapter 9) and fish populations for the same lake (Chapter 8); (f) actual and estimated annual tree ring widths (expressed as indices) for red spruce in the Adirondack Mountains (Chapter 6); and (g)

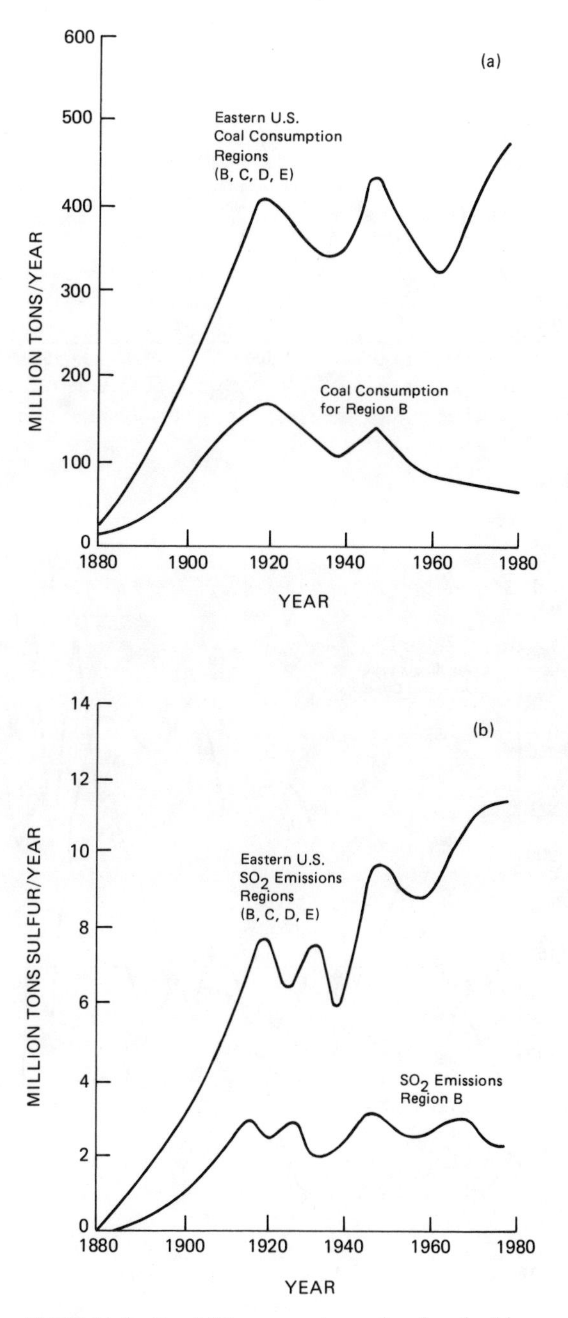

FIGURE 1.5 100-year trends in indices that may be related to acid deposition.

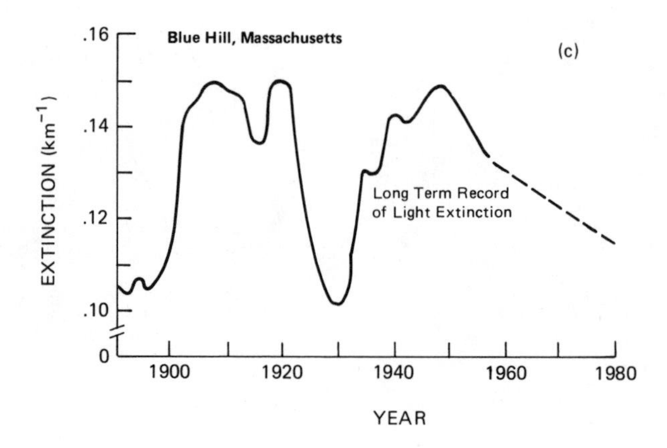

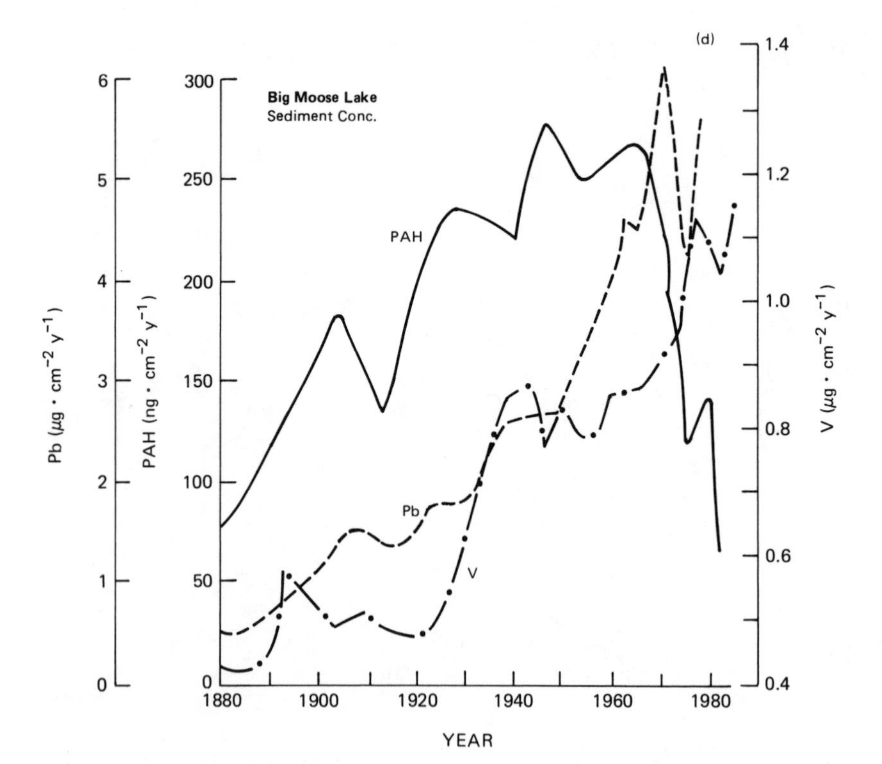

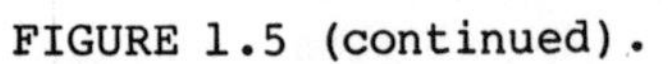
FIGURE 1.5 (continued).

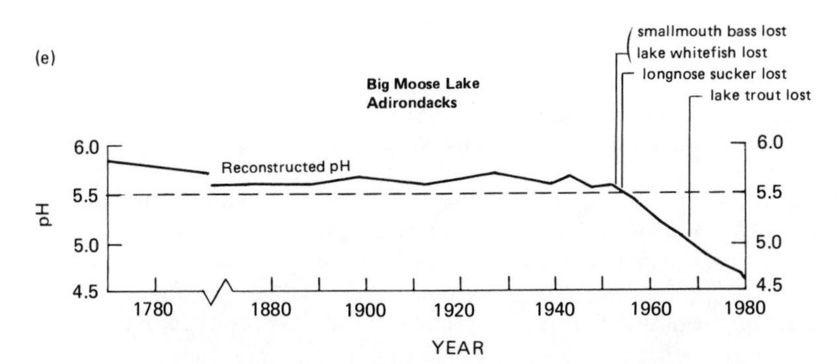

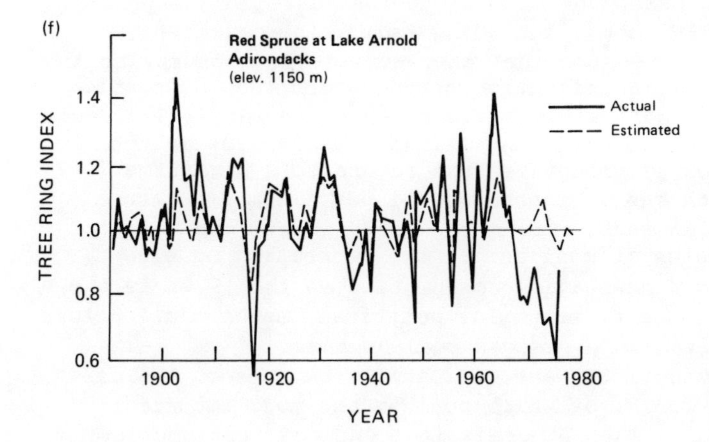

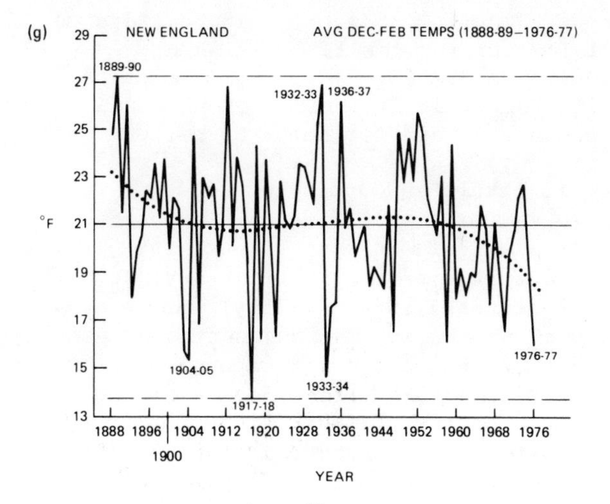

FIGURE 1.5 (continued).

the 100-year record of winter temperature in New England (Chapter 6).

Figure 1.5(a) demonstrates that the most rapid rate of increase in coal consumption in the eastern United States (Regions B through E) and the northeastern United States (Region B) occurred between about 1880 and 1920. Since 1920, coal consumption in the eastern United States has fluctuated between 300 and 500 million tons, depending on social, political, and economic factors, but in general has shown no substantial increase since the 1920s. However, coal consumption in Region B has decreased since 1920. The dramatic rate of increase in consumption around the turn of the twentieth century can be attributed to the increased use of coal to fuel a rapidly expanding industrial economy, including rail transportation. It is reasonable to assume that the decrease in consumption in the 1930s was a reflection of the widespread depression in economic activity, the increase in the early 1940s was associated with World War II, and the decline in the postwar years accompanied the return to a peacetime economy and the increased use of petroleum and natural gas. The increase during the 1960s is attributable to the expansion of coal use in the generation of electricity, and moderating consumption in the 1970s is the result of several economic, political, and social factors that resulted in a reduced coal demand.

Emissions of SO_2 are estimated from data on the use of fossil fuels, of which coal is the most important contributor. Thus SO_2 emissions and coal consumption show similar trends (Figures 1.5(a,b)). The difference in trends, especially after about 1950, is mainly attributable to emissions of SO_2 from oil consumption, a source accounting for about 10 percent of national energy consumption in 1920 and for about 50 percent in 1980.

Although there is little supporting evidence to verify the estimates of SO_2 emissions of a century ago, the evidence that is available is in concert with the general pattern. A rapid rate of increase in light extinction (corresponding to a decrease in visibility) was recorded at the observatory at Blue Hill, Massachusetts, at about the same time period as the most rapid expansion of coal use (Figure 1.5(c)). Since the second decade of the twentieth century, the trend in light extinction at this site is somewhat similar to the trends of coal consumption and SO_2 emissions for Region B.

Multiple lines of chemical evidence from lake sediments of Big Moose Lake in New York indicate a history of changing chemistry paralleling the pattern of changing fuel use over the past 100 years (Figure 1.5(d)). Sediment constituents related to emissions from coal burning (including fly ash, PAHs, certain trace metals, and excess sulfur) dramatically increased in concentration through the first half of the twentieth century. Trace metals associated with combustion of fuel oil (vanadium) and gasoline (lead) increased relatively slowly over the first half of the century and rapidly thereafter, reflecting large-scale changes in fuel use patterns around mid-century.

The only data available for reconstructing the historical record of lake-water pH on a 100-year time scale come from measurements of the distributions of diatom and chrysophte taxa present in lake sediments (Figure 1.5(e), for example). The record suggests that after remaining nearly constant over the previous two centuries, the pH of Big Moose Lake decreased rapidly beginning about 30 years ago. Corresponding data on populations of acid-sensitive fish support this conclusion.

Figure 1.5(f) shows the trend for values of the tree ring indices for red spruce at Lake Arnold in the Adirondack Mountains (elevation, 1150 m). The dashed line indicates values calculated using a model based on responses to average monthly temperatures. (See Chapter 6.) The model fits the data reasonably well until about 1967, after which index values are considerably less than predicted. This result suggests that although monthly temperature was a major factor governing tree ring width before about 1967, other factors were important after that time. It is possible that a temperature-related factor initiated the decline. Ring width in red spruce is affected by winter temperatures, and the onset of the decline at Lake Arnold and elsewhere coincided with the occurrence of cold winters in the northeastern United States, as demonstrated in the temperature record in New England in Figure 1.5(g). The unusual response in the case of red spruce has been a failure to recover as winter temperatures became more moderate. The cause of the prolonged nature of the tree-ring-width decline and extensive mortality is not clearly understood.

The Half-Century Record

<u>Lake-Water Chemistry</u> Beginning in the 1920s and 1930s, extensive surveys of the chemistry of lake waters were undertaken in several states. Most of the data were used to monitor water quality of lakes in rural areas used for fishing or other recreational activities. Unfortunately, the surveys were conducted for only a few years, so that there are no reliable long-term continuous records of this type. Coupled with recent data, however, the historical surveys allow us to compare water quality in these lakes at two points in time separated by about 50 years.

The parameters of lake-water quality that are most important for acidification are pH and alkalinity. The value of pH gives the concentration of hydrogen ion but does not provide information about the inputs of hydrogen ion that may have been neutralized by basic substances. Thus, it is possible for a highly buffered lake to receive a substantial input of acidity but show little, if any, change in pH. The response of a lake to acid deposition is likely to be more apparent as a change in alkalinity.

We reviewed and analyzed available historical data on the alkalinity and pH of lakes in New Hampshire (1930s and 1940-1950), New York (1929 to 1936), and Wisconsin (1925 to 1932) and compared them with recent data (late 1970s and early 1980s). The results of the comparison, detailed in Chapter 7, suggest that some lakes have experienced decreases in alkalinity and pH while others currently have amounts of alkalinity and values of pH higher than they were 50 to 60 years ago. In Wisconsin, more lakes experienced increases in pH and alkalinity than have experienced decreased values. In New Hampshire, the median change in alkalinity is not statistically different from zero, but the change in median pH suggests that more lakes have increased in pH than have decreased.

Any interpretation of changes in alkalinity and pH in New York lakes based only on water chemistry data is somewhat equivocal because the calculations are sensitive to the value assumed for the pH of the endpoint of the methyl orange (MO) indicator, a standard chemical used in the determination of historical lake alkalinities. Evidence is substantial in the well-documented Wisconsin survey that the endpoint using standard methods was pH 4.19. Similar documentation does not exist for the New York survey, but a review of the historical literature suggests that the endpoint was bounded in the pH range of 4.19 and 4.04.

We compared the 1930s data with two recent New York lake surveys (Pfeiffer and Festa 1980, Colquhoun et al. 1984) and determined that the results depended somewhat on which of the two surveys were included in the calculations. Thus, there are four possible pairings of assumptions as follows: (1) an MO endpoint of 4.04 pH units and comparison of historical data with the 1980 data; (2) an MO endpoint of 4.04 and comparison with the 1984 data; (3) an MO endpoint of 4.19 units and comparison with the 1980 data; and (4) an MO endpoint of 4.19 units and comparison with the 1984 data. Calculations applying assumptions 3 and 4 yield median changes in alkalinity of -69 and -44 μeq/L, and median changes in pH of -0.74 and -0.63, respectively. Applying assumptions 1 and 2, the calculations yield median changes in alkalinity of -28 and +1 μeq/L, and median changes in pH of -0.14 and -0.12 units, respectively.

In each state there are numerous examples of lakes with changes in alkalinity greater in magnitude than 100 μeq/L. Since the magnitude of change from acid deposition is estimated at about 100 μeq/L or less (Galloway 1984), it is likely that these lakes were affected by factors other than, or in addition to, acid deposition.

Relationship of Trends in Diatom-Inferred pH and Fish Populations Analysis of diatoms in sediments of selected lakes offers another indication of long-term acidification. Unlike the lake surveys in Wisconsin, New York, and New Hampshire, the lakes from which diatom data were obtained were selected on the basis of acid sensitivity (i.e., alkalinity less than 200 μeq/L) and either little or no disturbance of the watershed or good documentation of disturbance. (See Chapter 9.)

Based on paleoecological analysis of the entire history of currently acidic lakes, rates of long-term natural acidification are slow--with decreases of 1 pH unit (from 6.0 to 5.0) occurring over periods of hundreds to thousands of years. In contrast, some lakes in the Adirondack Mountains have apparently experienced decreases in pH on the order of 0.5 to 1.0 pH unit over a 20- to 40-year period in the middle of this century.

Diatom data for 31 lakes were evaluated to assess regional trends in lake acidification; data of sufficient quality are available for 11 lakes in the Adirondacks, 10 in New England, 6 in eastern Canada, and a reference set of 4 lakes in the Rocky Mountains. The evaluation suggests that certain poorly buffered eastern lakes have become substantially more acidic during the past 20 to 40 years.

The lakes for which evidence is strongest are in the Adirondack Mountains. Of the 11 lakes for which diatom data are available, 6 of the 7 lakes with a current pH at or below about 5.2 show evidence of recent acidification; the seventh is a bog lake. Analyses of chrysophyte scales (mallomonadaceae) agree well with interpretation of the diatom data. None of the 4 lakes with current values of pH above about 5.2 showed strong evidence of a pH decline. Where dating of sediments is available, the most rapid diatom-inferred pH changes (decreases of 0.4 to 1.0 pH units) occurred between 1930 and 1970 beginning in the 1930s to 1950s.

Diatom data for the 10 lakes in New England indicate either a slight decrease or no change in pH over the past century. Diatom data for the 6 lakes in eastern Canada indicate no change in pH (4 lakes with current pH greater than 6.0) or a significant decrease in pH that is probably caused by local smelting operations (2 lakes). Rocky Mountain data do not show decreases in pH.

Before 1800, several lakes in the Adirondacks and New England had diatom-inferred pH values less than 5.5. These lakes now have a pH of only 0.1 to 0.3 pH units lower, and total aluminum concentrations greater than 100 μg/L. Because of the potential importance of buffering by organic acids, a small decline in pH could be associated with significant decreases in acid neutralizing capacity.

Analysis of the Adirondack data indicates that no other acidifying process except acid deposition has been identified to explain the rapid declines in lake water pH during the past 20 to 40 years. However, watershed disturbances may also play a role, but probably a minor one for the lakes evaluated in this study.

Further evidence of acidification trends in lakes of the Adirondacks is provided by comparing measured pH, diatom-inferred pH, sediment chemistry, and fish population data on nine Adirondack Lakes (Table 1.3). They show consistent trends with some exceptions. The lakes with current values of pH of about 5.2 or less have become more acidic in recent times and have lost fish populations, whereas lakes with higher current values of pH show no obvious trend toward acidification or fish declines.

In a number of cases (e.g., Woods Lake, Upper Wallface Pond) the observed change in diatom-inferred pH is small, from about pH 5.2 to about pH 4.8, yet major changes in fish populations have occurred. There are two plausible explanations for this phenomenon. First, fish are sensitive to pH in this range, with survival decreasing abruptly over the pH range 5.2 to 4.7. Many lakes in New York and New England with pH 5.0 to 5.2 currently support fish, whereas lakes in this region with pH below 5.0 rarely support fish (Chapter 8). Second, acidification of surface waters from pH 5.2 to 4.7 may be accompanied by a decrease in dissolved organic carbon (Davis et al. 1985), and an increase in dissolved uncomplexed aluminum, which is highly toxic to fish.

In summary, we analyzed historical and recent data on pH and alkalinity from three large lake surveys in Wisconsin, New York, and New Hampshire. In all cases, some lakes decreased in alkalinity and pH since the 1930s and some increased. In some lakes the magnitude of the change in alkalinity appears to be too large to be explained solely by acid deposition. The evidence further indicates that on average Wisconsin lakes have increased in pH and alkalinity since about 1930, New Hampshire lakes show no obvious change in alkalinity but may have increased in pH, and New York lakes have shown decreases in alkalinity and pH in three of four possible combinations of assumptions, and show no change if one accepts the fourth assumption.

Data on diatom-inferred pH for 11 lakes in the Adirondack Mountains (one of them was a bog lake and was disregarded) indicate that 6 of these lakes have become more acidic over the period from about 1930 to the 1970s. All these lakes have current pH values of about 5.2 or lower. The 4 lakes with current pH values above 5.2 show no trend in pH. For 9 of the lakes, concurrent data exist on measured pH, sediment chemistry, and fish populations. The data are generally consistent and support the findings based on diatom analysis. Data for New England indicate slight or no increase in diatom-inferred pH. There is insufficient information to evaluate trends in eastern Canada. The lakes selected in the diatom studies have low alkalinities (less than 200 μeq/L) and little disturbance of their watersheds, and hence may be the most likely to show responses to acid deposition.

TABLE 1.3 Comparison of Data on Measured pH, Diatom-Inferred pH, Sediment Chemistry, and Fish Populations for Nine Lakes in the Adirondack Mountains, New York

Lake	Measured pH				Diatom-Inferred pH (Present/Pre-1800)	Sediment Chemistry	Fish Survey Data	Comments/Conclusions
	Year	pH	Method[a]	Reference[b]				
Panther	1975	6.88	F	1	6.1/6.4 Steady decrease toward surface from 5-10 cm in core	No acidification indicated	Fish population eradicated in 1951 and restocked; no changes in any species other than that caused by direct action	Possible slight reduction in pH but lake not now acidic and fish not affected
	1978-80	6.2	L	1				
Sagamore	1933	7.0	U	1	6.3/6.1 No overall significant trend	No acidification indicated	9 species in 1933, 8 species in 1976; no changes in any major species	All data except measured pH are consistent and indicate no acidification
	1961	5.6	U	1				
	1976	6.0	U	1				
	1978-80	5.6	L	2				
Woodhull	1954	5.0	F	1	6.0/6.1 No obvious trend	Dating poor. Possible slight acidification indicated by decline in Zn in surface sediment	11 species in 1954, 7-8 species in 1979-81; smallmouth bass last collected 1974, but lake trout and lake whitefish still present	Possible slight acidification indicated by two of four data sources
	1969	6.5	F	1				
	1974	6.0	F	1				
	1975	5.8	F	1				
	1976	5.1-5.3	L	3				
	1978	5.7	L	1				
	1981	5.4	L	1				
	1982	5.6	L	1				
Big Moose	1948	6.0	F	1	4.9/5.8 Marked diatom-inferred pH decrease and floristic change beginning about 1950	Acidification indicated by declining relative concentration of Zn, Ca, and Mn as well as declining deposition rates relative to TiO_2	10 species in 1948, 5-6 species since 1962; smallmouth bass and lake whitefish last collected in 1948; lake trout last collected in 1966	All data are consistent and indicate acidification
	1960	5.2	F	1				
	1965	5.2	F	1				
	1971	5.5	F	1				
	1974	4.9	L	4				
	1978-79	4.6-5.0	L	5				
	1981	5.3	L	1				
	1982	4.9	L	1				
Upper Wallface	1963	5.6	C	1	4.7/5.1 Most rapid pH and floristic change	Slight acidification indicated by relative decline of deposi-	Brook trout moderately abundant in 1963-68, gone in 1975	All data are consistent and indicate acidification
	1968	4.9	M	1				
	1975	4.9	L	1				

	1978	5.0	L	1	starting in 1945-55	tion rates for Ca and Zn		
	1979	5.0	L	1				
Honnedaga	1959-60	3.9-6.0	M	6	5.2/6.1 Insufficient data to determine pattern of change; major floristic change from bottom to top of core		Lake trout became rare and emaciated in early 1950s, last collected 1954; brook trout declined in mid 1960s and were very rare by the mid 1970s	Fish and diatom data indicate acidification but changes appear to have occurred gradually
	1975	4.91	L	1				
	1976	4.7-4.8	L	3				
	1981	4.92	L	1				
	1982	4.84	L	1				
Woods	1968	5.0	U	1	4.8/5.2 Dating information not available. More low pH indicators at top of core	Dating poor. Only trace metal data available. Acidification not indicated by Zn. Ca and Mn data not available	Lake trout reported present previously but never collected. Formerly known for large brook trout, abundant in 1961 scarce since 1966, and all of hatchery origin; golden shiner last collected in 1966	Fish and diatom data indicate acidification but data quality poor
	1973	6.0	U	1				
	1975	5.2	U	1				
	1976	5.8	U	1				
	1978-80	4.7	L	2				
	1982	5.12	L	1				
Deep	1954	4.70	U	1	4.8/5.0 Decrease in diatom-inferred pH and major floristic changes beginning 1940-50	Slight acidification indicated by relative decline in Ca, Mn, and Zn concentrations and deposition rates	Brook trout present, moderately abundant in 1954, absent in 1964; no survival of stocked fish since 1964	Three of four data sources indicate slight acidification
	1968	4.70	U	1				
	1978	4.62-4.76	L	1				
	1979	4.62-4.72	L	1				
Little Echo	1954	6.0	U	1	4.1-4.7/4.7 Recent diatom-inferred pH cannot be clearly determined. Major increase in a single taxon at top of core	Acidification indicated by relative declines in concentrations and deposition rates of Ca and Mn. Zn is extremely high	Stocked with brook trout annually from 1942-1975; surveyed five times between 1954 and 1982 but no fish ever caught	Colored bog-pond. Data inconsistent
	1976	4.4	M	1				
	1984	4.3	L	7				

[a]Key: L, laboratory meter, air-equilibrated; F, field measurement, method unknown; U, unknown; C, Hellige colorimetric; M, field meter, not air-equilibrated.

[b]Key: 1, Fish Information Network (FIN) database (J. Baker, Kilkelly Environmental Associates, personal communication, 1985); 2, Galloway et al. 1983; 3, Del Prete and Schofield 1981; 4, D. F. Charles, Indiana University, unpublished data; 5, Driscoll 1980; 6, Schofield 1965; 7, Driscoll and Newton 1985.

The Record Since 1950

General Trends The decades after World War II are characterized by rapidly changing patterns of fuel consumption and fuel use in eastern North America. Before 1945 coal was the dominant source of fuel and consumption was divided among railroads, residential and commercial heating, oven coke and other industrial processes, and electric utilities. The demand for rail transport was particularly high during the war years of the early 1940s. By the end of the 1950s, however, coal consumption by railroads and by the residential-commercial sector essentially vanished. Overall, coal use declined by about 30 percent from 1945 to 1960. However, coal for electric power more than doubled over this same period, and doubled again for the period from 1960 to 1975. Concurrent with the expansion of coal use for electricity was the construction of new power plants with increasingly higher smokestacks, resulting in more than a doubling in average stack height from the mid-1950s to the mid-1970s (Sloane 1983). The seasonal pattern of consumption also changed during this period. In the early 1950s coal consumption in winter, the peak season, was about 27 percent higher than in summer, the season of lowest consumption. By the mid-1970s both winter and summer were peak periods of comparable consumption. Total coal consumption in the eastern United States is currently comparable to consumption during the peak years of the early 1940s, owing to increased consumption after the decline in the 1950s. However, there has been a considerable shift in the regional patterns of consumption over the past two decades; some areas currently consume far greater amounts of coal than they did in the early 1940s and some areas consume far less. (See discussion of regional trends below.)

Coal provided about 50 percent of the energy needs of the United States in 1945. Currently, it provides about 20 percent as the total energy consumption has more than doubled over this period. The increasing demand for energy has been met largely by natural gas and oil. Natural gas contains little or no sulfur, and oil, after refinement, is of relatively low sulfur content. Thus, SO_2 emissions have not increased in proportion to energy consumption over the past four decades. Nitrogen oxides, however, are formed as a by-product of any high-temperature combustion process in the atmosphere, regardless of the cleanness of the fuel. Consequently, nitrogen oxide emissions in eastern North America have more than doubled from the period 1945 to 1980, and have become an increasingly important component of acid deposition.

The changes in fuel consumption, fuel use, and fuel type that have occurred over the past four decades have undoubtedly affected the geographic distribution and the composition of acid precursors in the atmosphere in subtle or more obvious ways. Environmental effects (e.g., lake acidification) have occurred in sensitive ecosystems over this same time period. However, if acid deposition is a cause, it may be difficult to determine whether the effect is a consequence of relatively recent changes in fuel use or consumption since 1945, or whether it is a result of cumulative exposure to acid deposition over many decades.

Regional Trends Estimates of SO_2 emissions in eastern North America suggest that the decade of the 1950s was a period of constant or declining emissions in all of the designated Regions A through E. In contrast, during the 1960s SO_2 emissions rose sharply in all regions. The 1970s are characterized by strong regional differences in trends of SO_2 emissions. There were differences not only in the magnitude of trends but also (for the first time) in their direction. In the northeastern states (Region B) the trend was distinctly downward. In the southeastern states (Region C) the trend was rapidly upward, continuing the trend that began in the 1960s. In the midwestern states (Region D) the trend was upward, but not so rapidly as in Region C. Emissions in the north central states (Region E) remain consistently low.

The divergence of trends in SO_2 emissions along regional lines in the 1970s provides the opportunity of testing whether other types of data also reveal consistent regional differences. We analyzed two different types of data that have continuous records for a number of years and were collected at numerous sites in different regions. One is the record of light extinction (an inverse measure of visibility) at 35 airports from 1949 to 1983 (Chapter 4). The other is the record of sulfate in Bench-Mark streams from 1965 to 1983 (Chapter 7). We have also analyzed the record of pH, sulfate, nitrate, hydrogen ion, and other ions in bulk precipitation at the Hubbard Brook Experimental Forest, New Hampshire, for the period from 1963 to 1979 (Chapter 5), but we do not include these data in our regional analysis because they represent only one site in one region. However, we note that the observed trend of decreasing sulfate at Hubbard Brook is consistent with our observation of decreasing emissions of SO_2 in Region B over this time period.

The mechanisms outlined earlier suggest that SO_2 emissions, atmospheric sulfate, light extinction, and stream sulfate are related and may exhibit similar temporal trends. We examined this suggestion on a regional basis by testing for associations among the regional trends of these variables by using Friedman's test.

For the period 1950 to 1980, only data on SO_2 emissions and light extinction are available; the regional rankings for these trends are given in Table 1.4. A ranking of 1 in Region B for the trends of both SO_2 emissions and light extinction signifies that this region experienced the lowest rate of increase in each of these parameters over the period from 1950 to 1980. Higher rankings signify greater rates of increase. The p value for Friedman's test in this case is 0.042, signifying that if there were in fact no true association between the parameters, then an apparent association to this degree or greater would occur by chance only 4.2 percent of the time. The result is indicative of a strong temporal association between SO_2 emissions and light extinction on a regional basis.

We applied the same method to test for possible associations among trends in light extinction, sulfate fluxes from Bench-Mark streams, and emissions of SO_2 for the period from 1965 to 1980. The regional rankings are shown in Table 1.5. The p value of 0.054 again gives evidence of temporal associations for these data on a regional level.

TABLE 1.4 Regional Rankings by Rate of Change in SO_2 Emission Densities and Light Extinction for the Period 1950-1980

Region	SO_2 Emissions	Light Extinction
B	1	1
C	4	4
D	3	3
E	2	2

NOTE: p value of Friedman's test, 0.042.

TABLE 1.5 Regional Rankings by Rate of Change of SO_2 Emission Densities, Light Extinction, and Stream Sulfate for the Period 1965-1980

Region	SO_2 Emissions	Light Extinction	Stream Sulfate
B	1	1	1
C	4	2	3
D	3	4	4
E	2	3	2

NOTE: p value of Friedman's test, 0.054.

As demonstrated in Chapter 7, changes in the flux of sulfate in soft-water Bench-Mark streams were balanced by changes in alkalinity and base cations. The regional pattern of trends in stream alkalinity for the period 1965 to 1983 was the approximate inverse of that of stream sulfate (see Chapter 7, Figures 7.6 and 7.7): decreases (or no increases) have occurred at several stations in Region C and Region F while increases (and no decreases) have occurred at stations in Region B. Station-by-station comparisons of alkalinity and sulfate trends, however, do not always show a consistent inverse relationship.

Beginning in the 1960s, ring widths of red spruce at high elevations throughout its range in the eastern United States decreased significantly, a change that has persisted to the present. Important regional climatic anomalies occurred when the red spruce decline began and may have been a factor in triggering the response. There is currently no direct evidence linking acid deposition to mortality and decreases in ring width, although this effect has occurred in areas that are receiving relatively large amounts of acidic substances and other types of air pollutants.

In summary, based on statistical tests of regional temporal trends, a strong association exists between SO_2 emissions and light extinction (30-year records, 1950 to 1980). A similar result is obtained for SO_2 emissions, visibility, and sulfate concentrations in Bench-Mark streams (15-year records, 1965 to 1980).

Since 1950, the northeastern United States (Region B) has experienced the smallest rates of change in these parameters. The southeastern United States (Region C) has, in general, experienced the greatest rates of change, and the Midwest (Region D) has experienced rates of change greater than Region B but less than Region C.

REFERENCES

Colquhoun, J., W. Kretzer, and M. Pfeiffer. 1984. Acidity status update of lakes and streams in New York State. New York Department of Environmental Conservation, report no. WM P-83(6/84). Albany, New York.

Davis, R., D. Anderson, and F. Berge. 1985. Loss of organic matter, a fundamental process in lake acidification: paleolimnological evidence. Nature 316:436-438.

Del Prete, A., and C. Schofield. 1981. The utility of diatom analysis of lake sediments for evaluating precipitation effects on dilute lakes. Arch. Hydrobiol. 91:332-340.

Driscoll, C. 1980. Chemical characteristics of some dilute acidified lakes and streams in the Adirondack region of New York state. Doctoral dissertation, Cornell University, Ithaca, N.Y.

Driscoll, C., and R. Newton. 1985. Chemical characteristics of Adirondack lakes. Environ. Sci. Technol. 19:1018-1023.

Galloway, J. N. 1984. Long-term acidification. Pp. 4-48--4-53 in The Acidic Deposition Phenomenon and Its Effects. Critical Assessment Review Papers, Volume II, Effects Sciences. A. P. Altshuller and R. A. Linthurst, eds. Environmental Protection Agency report no. EPA-600/8-83-016BF.

Galloway, J. N., C. Schofield, N. Peters, G. Hendrey, and E. Altwicker. 1983. Effects of atmospheric sulfur on the composition of three Adirondack lakes. Can. J. Fish. Aquat. Sci. 40:799-806.

Harper, H. J. 1942. Sulfur content of Oklahoma rainfall. Okla. Acad. Sci. Proc. 23:73-82.

Hidy, G. M., D. A. Hansen, R. C. Henry, K. Ganeson, and J. Collins. 1984. Trends in historical and precursor emissions and their airborne and precipitation products. J. Air Pollut. Control Assoc. 31:333-354.

Lehmann, E. L. 1975. Nonparametrics. San Francisco, Calif.: Holden-Day.

MacIntire, W. H., and J. B. Young. 1923. Sulfur, calcium, magnesium, and potassium content and reaction of rainfall at different points in Tennessee. Soil Sci. 15:205-227.

Mosteller, F., and J. W. Tukey. 1977. Data Analysis and Regression: A Second Course in Statistics. Reading, Mass.: Addison-Wesley.

National Research Council. 1983. Acid Deposition: Atmospheric Processes in Eastern North America. Washington, D.C.: National Academy Press.

National Research Council. 1984. Acid Deposition: Processes of Lake Acidification. Washington, D.C.: National Academy Press.

Pfeiffer, M. H., and P. J. Festa. 1980. Acidity status of lakes in the Adirondack region of New York in relationship to fish sources. New York Department of Environmental Conservation, Albany, N.Y. FW-P168(10/80). 36 pp.

Robinson, E. 1984. Natural emission sources. Pp. 2-1--2-52 in The Acidic Deposition Phenomenon and Its Effects. Critical Assessment Review Papers, Volume I, Atmospheric Sciences. A. P. Altshuller and R. A. Linthurst, eds. Environmental Protection Agency report no. EPA-600/8-83-016BF.

Schofield, C. L. 1965. Water quality in relation to survival of brook trout, *Salveilnus fontinalis* (Mitchell). Trans. Am. Fish. Soc. 94:227-235.

Seip, H. M. 1980. Acidification of freshwater--sources and mechanisms. Pp. 358-366 of Ecological Impacts of Acid Precipitation, D. Drablos and A. Tollan, eds. Proceedings of an international congress in Sandefjord, Norway. Sur Nedbors Virking Pa Skog Og Fisk (SNSF) Project, Oslo.

Sloan, C. S. 1983. Seasonal acid precipitation and emission trends in the northeastern United States. General Motors Research Laboratories, Warren, Michigan. Research publication no. GMR-4456, ENV#16.

Wagner, D. P., D. S. Fanning, J. E. Foss, M. S. Patterson, and P. A. Snow. 1982. Morphological and mineralogical features related to sulfide oxidation under natural and disturbed land surfaces in Maryland. Chapter 7 of Acid Sulfate Weathering. J. A. Kittrick, D. S. Fanning, and L. R. Hossner, eds. SSSA Special publication 10. Soil Society of America, Madison, Wis.

2

Emissions of Sulfur Dioxide and Nitrogen Oxides and Trends for Eastern North America

Rudolf B. Husar

INTRODUCTION

Air pollution is caused when an element or a compound is removed from its long-term geochemical reservoir and, with the help of the atmosphere, dispersed and transferred to another long-term geochemical reservoir. The elements redistributed in this manner are carbon, sulfur, nitrogen, and the crustal elements, including metals. Another type of air pollution occurs when human activities produce and disperse compounds that are completely foreign to nature, such as DDT, polychlorinated biphenyls, dioxin, and some nuclear fission products. The production and fate of these compounds will not be discussed further in this chapter.

We continue removing or mining carbon, sulfur, nitrogen, and crustal material from the Earth's long-term reservoir for two purposes: to "produce" energy from the fossil fuels and to manufacture from extracted minerals disposable or permanent objects for societal use. Most pollutants are emitted from the land into the air, with the atmosphere serving as a medium for dispersion and chemical reactions during the transport. The atmosphere redistributes the emitted materials onto the biota, land, lakes, and oceans. The role of the rivers is primarily in gathering these materials from the land and transferring them to their long-term storage in the oceans. Certain volatile pollutants may also be re-emitted from land or from lakes.

Examples of the Flow Rates of Sulfur and Nitrogen from Natural Processes and from Human Activities

It is worth considering a few estimates of flow rates of sulfur and nitrogen from simple back-of-the-envelope calculations. If all the coal consumed in the United States over the past century (500 million tons/yr) were piled into a single mountain, it would occupy 25 cubic kilometers, i.e., a pyramid with a base area of 6 km x 6 km, 2 km high. Spread over the entire eastern half of the United States, it would be a layer of about 1-cm thickness. This means that from a geophysical weathering perspective, about 150 g m^{-2} yr^{-1} of crustal mineral material currently goes up in smoke, considering only coal consumption in the eastern United States. By comparison, the natural weathering of the Earth's surface minerals as they are carried by rivers to the oceans is about 20 g m^{-2} yr^{-1}. Thus human activities significantly enhance the natural redistribution of the Earth's crustal material over eastern North America.

Over the past 100 years the average emission density over eastern North America was about 1-2 g sulfur m^{-2} yr^{-1}. Most of that sulfur returns to the ground somewhere over eastern North America, resulting in an average sulfur flow of 100 g sulfur m^{-2} 100 yr^{-1}. By comparison, the total sulfur content of soils is 3-30 g sulfur/m^2. Hence, the yearly sulfur flow through soils from human activities is comparable to the natural sulfur content of the soil itself.

It is also instructive to consider the sulfur and nitrogen flow from a biological perspective: in the United States, the average daily per capita emission rate to the atmosphere is roughly 200 g sulfur (400 g sulfur dioxide) and 100 g nitrogen (300 g nitrogen dioxide). This is comparable in weight to the daily per capita food consumption. The sulfur emission rate from anthropogenic sources in the United States is about 15×10^{12} g sulfur yr^{-1}. Distributing these emissions uniformly over the contiguous United States with an area of 8×10^{12} m^2, we arrive at an average emission density of 2 g sulfur m^{-2} yr^{-1}. This is comparable to the density of sulfur contained and removed yearly from the soil by harvesting agricultural products, such as corn and wheat (Beaton et al. 1974).

The biosphere is a thin shell of living matter on the Earth's surface. It is responsible for a grand-scale cycling of energy and chemical elements. Functionally, this biological cycling is maintained by three groups: producers, consumers, and decomposers. The producers are plants and some bacteria capable of producing their own food photosynthetically or by chemical synthesis. The consumers are animals that obtain their energy and protein directly by grazing, feeding on other animals, or both. The decomposers are fungi and bacteria that decompose the organic matter of producers and consumers into inorganic substances that can be reused as food by the producers; they are the recyclers of the biosphere. Nature is capable of sustaining the producer-consumer-decomposer cycle indefinitely with the Sun as the energy source. The smallest such entity that is self-sufficient is an ecosystem.

Functionally, human activities that perturb the natural environment can be divided into similar components (Figure 2.1). Producing activities include energy production (fossil fuels), manufacturing (nonfuel minerals), and growing food. The consumers are humans and their domestic animals. Decomposing or recycling activities include treatment of waste water, recycling of metals, and the burning of refuse. However, whereas an ecosystem relies on its decomposers for a complete recycling of its elements, the system created by human activity lacks such efficient decomposers. As such, manufactured materials that are no longer needed and waste by-products of industrial activity are disposed into the physical environment. The process of adding unwanted material to the environment is called pollution. The material that is not recycled is distributed by the atmosphere and the hydrosphere and delivered to the biological and geochemical receptors.

The above scheme permits a convenient accounting for the flow of materials in society from the producers to consumers to the receptors.

Pollutant Flow Diagram

The flow of matter from the producers to consumers and subsequently to the receptors is depicted schematically in Figure 2.2. Most of the production of potential pollutants begins with mining, that is, removing a substance from its long-term geochemical reservoir. The amount of pollutant mass, M_i, mobilized by mining (tons/yr) is the production rate P_i (tons/yr) of the raw material (coal, oil, smelting ore, etc.) multiplied by the concentration c_i (grams/ton) of the impurity (sulfur, mercury, lead, etc.): $M_i = c_i P_i$.

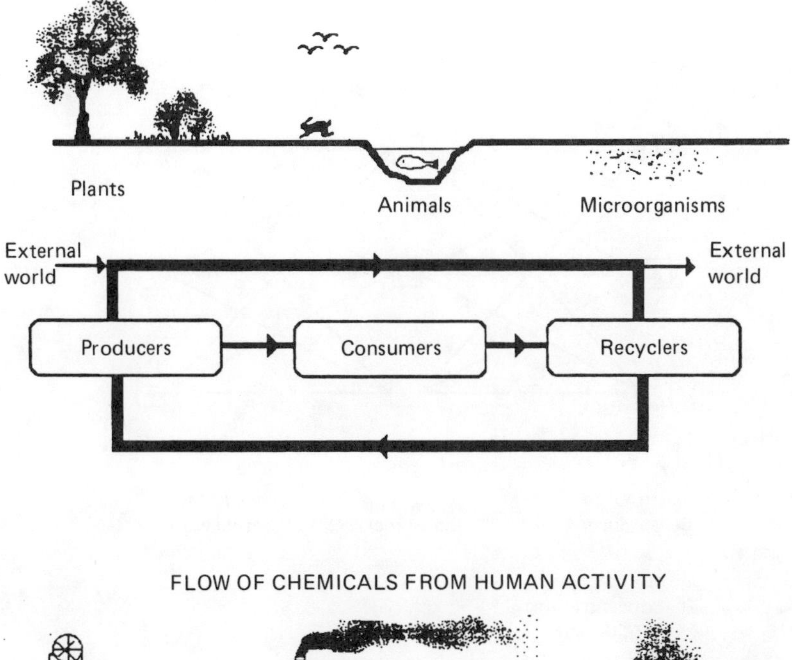

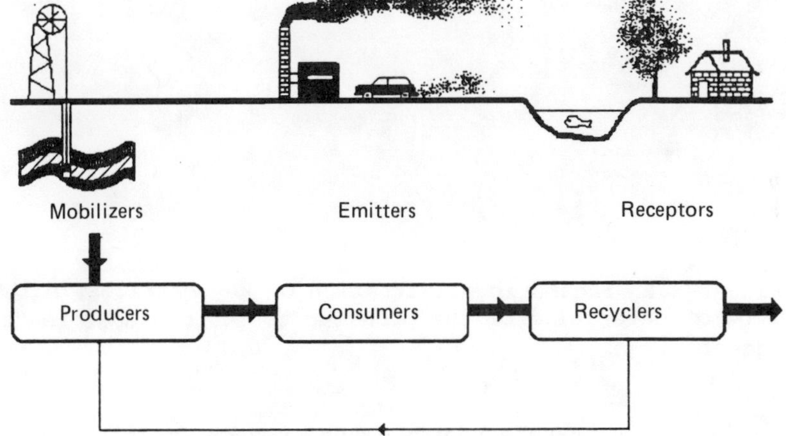

FIGURE 2.1 Diagrams of the movement of chemicals and materials through (top) the natural ecosystem and (bottom) a system resulting from human activity.

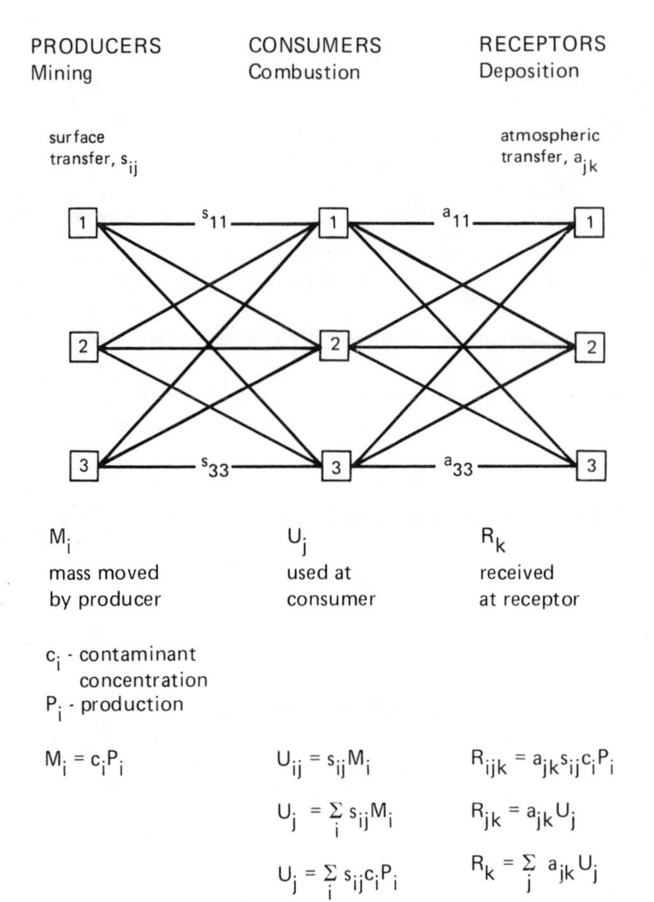

FIGURE 2.2 Schematic illustration of key matrices in the flow of material from the producer to the consumer to the receptor.

Matter is transferred from the producer to the consumer by ground transportation, including railroads, trucks, and barges. Functionally, surface transport redistributes the mined substances over a large geographical area. In principle, every producer, i, may deliver its product to any consumer, j. Mathematically, this producer-consumer transfer is characterized by a surface transfer matrix, s_{ij}.

The amount of matter, U_{ij}, originating from producer i and used at consumer j is $s_{ij}M_i$. The total amount of matter reaching the consumer j is the sum of the matter produced by all producers multiplied by their respective surface transfer matrix elements.

The next transfer occurs between the consumer, or emitter, and the environmental receptors. The consumer is located where the combustion or smelting occurs, and the receptor is located where the pollutant falls following its atmospheric transit. Again, in principle all emitters j can transfer matter through the atmosphere to all receptors k. Hence, the matter received at receptor k that originated at consumer (emitter) j, R_{jk}, is the product of the use rate U_j times the atmospheric transfer matrix, a_{jk}, from emitter j to receptor k. The total amount of matter deposited at receptor k is the sum of the use rates U_j at each emitter weighted by its atmospheric transfer matrix element. In this chapter, the numeric values of the M_i, c_i, and s_{ij} are discussed in detail; discussion of the atmospheric transfer matrix a_{jk} is beyond the scope of this report.

The emission estimate at any given emission site U_j (tons/yr) is calculated as follows: $U_j = s_{ij}c_iP_i$, where P_i (tons/yr) is the fuel and metals production rate at a given mining region, c_i (weight fraction) is the concentration of sulfur in fuel and ore, and s_{ij} is a dimensionless transfer matrix element between producers and consumers of fuel and ores.

PRODUCTION AND CONSUMPTION OF FUELS AND METALS

Combustion of coal and oil products, along with the smelting of metals, produces the bulk of the anthropogenic sulfur and nitrogen emissions to the atmosphere over North America. The driving force for fuel combustion is the demand for energy by the different economic sectors.

Energy Demand of the United States

From the turn of the century to the 1970s, U.S. energy consumption has been characterized by a steady increase in total consumption and shifts from one fuel to another (Figure 2.3). From 1850 to about 1880 wood was the primary energy source. By 1900, and during the first quarter of this century, rising energy demand was matched

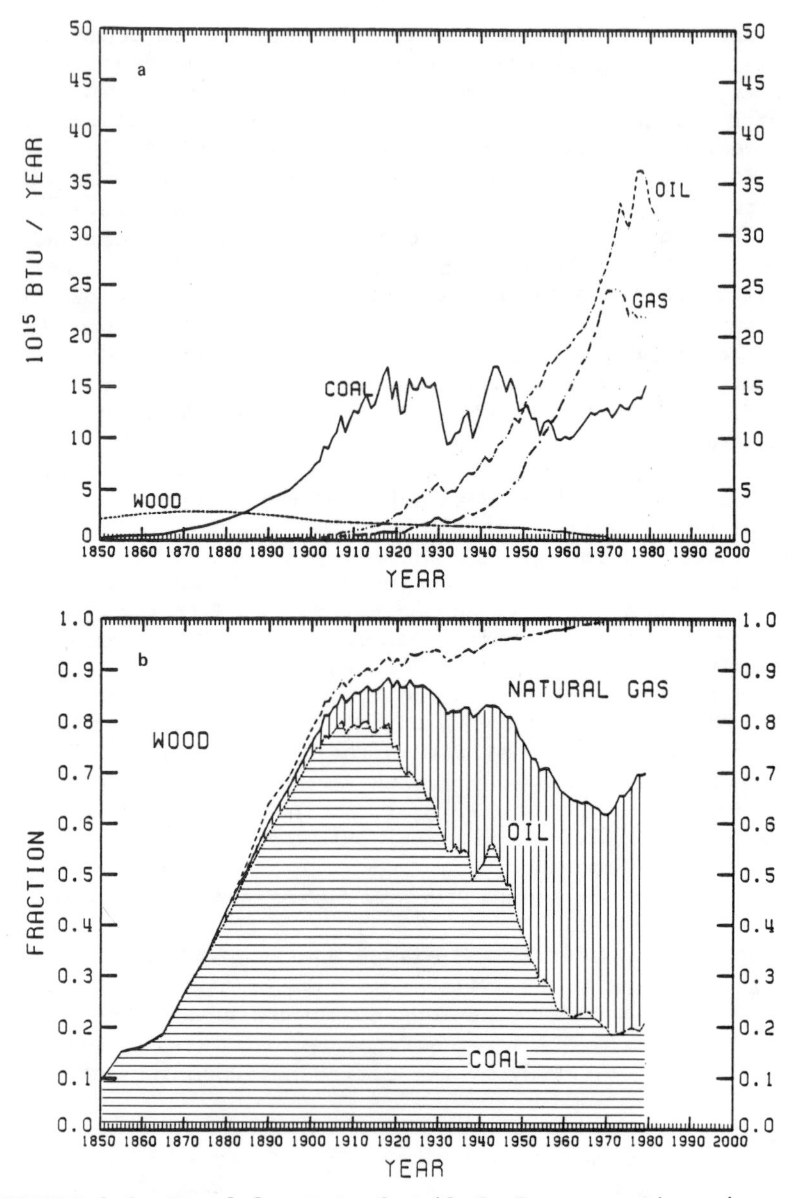

FIGURE 2.3 Trend for U.S. fossil fuel consumption since 1850. (a) Consumption by fuel type; (b) fraction of total energy by fuel type. Data for 1850 to 1880 from U.S. Bureau of the Census (1975); data from 1880 to 1932 from U.S. Geological Survey, Yearbooks (1880-1932); data from 1933 to 1980 from U.S. Bureau of Mines, Mineral Yearbooks (1933-1980).

by the increasing use of coal. The depression years of the early 1930s are reflected in the sharp drop of coal consumption, which increased again during the war years in the early to mid-1940s. Coal consumption declined to another minimum in 1960, because the increasing energy demands were supplied by cleaner fuels, natural gas and petroleum. Accelerated oil and gas consumption began in the late 1930s and 1940s, such that by 1950 the energy supplied by oil exceeded that of coal and maintained its rise up to the early 1970s. The 1973 oil embargo is reflected as a small ripple. The second dip, in the late 1970s and early 1980s, reflects another oil embargo and reduced industrial activity.

The 1950s and 1960s were the years of strong increase in natural gas consumption, which by 1960 also surpassed coal as an energy source. Nuclear energy began to supply a detectable fraction of the total energy consumption in the United States only after 1970.

Coal

Reserves and Production

In the United States, coal is mined in three regions: Appalachia, the Midwest (Interior), and the West. The coals in the regions differ in quality and concentration of impurities such as sulfur. The coal production data used in this chapter were obtained from Mineral Yearbooks (U.S. Bureau of Mines 1933-1980) and from Mineral Resources of the United States Yearbooks (U.S. Geological Survey 1880-1932).

Figure 2.4(a) shows the time-dependent contributions of the three regions to the national production of coal. The output of the Appalachian districts spanning Pennsylvania to Alabama is shown on the lowest curve. The curve shows that Appalachian production has remained at about 300 million tons/yr since about 1920. The second shaded area in Figure 2.4(a) is the contribution of the Interior region, which includes western Kentucky, Indiana, Illinois, Missouri, and Texas. The third area, negligible until around 1970, represents the contribution from the West. These curves reveal that a major shift in the coal production occurred at around 1970, when the production of Western coal became significant. Remarkably, within the span of a decade, low-sulfur Western coal captured a quarter of the United States coal market.

The coal production within the Appalachian region has also shifted substantially since 1900, as depicted in Figure 2.4(b). From about 1910 to the early 1920s the Pennsylvania coal production was about 160 million tons/yr, which has dropped to about 80 million tons in the post-World War II period. West Virginia production peaked during World War II and declined in the late 1960s. East Kentucky, on the other hand, had low production rates up to about 1960, and now it exceeds that of Pennsylvania. Similar trends are shown in the production data for three districts in the Interior region: Illinois, Indiana, and western Kentucky (Figure 2.4(c)). The significance of these shifts to sulfur emissions is that each coal-producing district has its own range of sulfur content: a shift in the relative production rate results in a change of the average sulfur content and sulfur production. The coal-production data described above define the raw material production rate P_i shown in Figure 2.2.

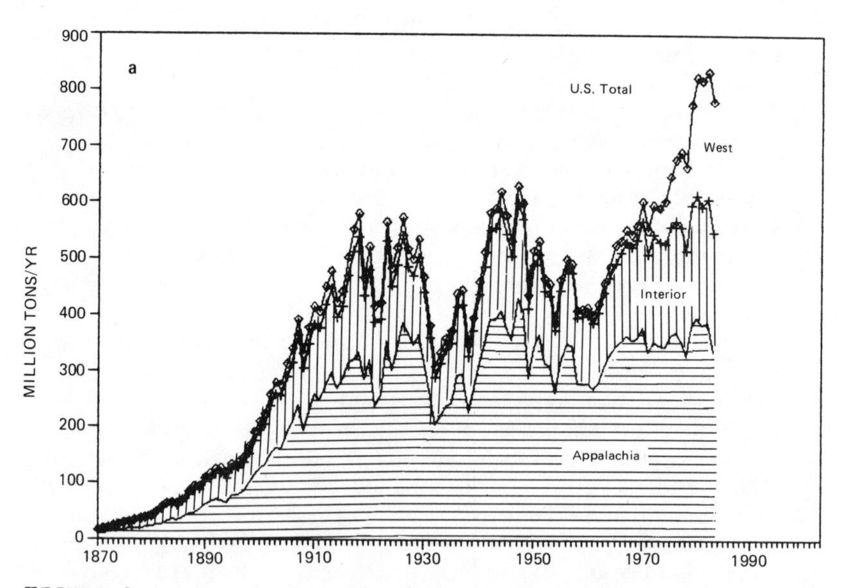

FIGURE 2.4 (a) Coal production in the three U.S. coal-producing regions: Appalachia, Interior, and West; (b) trends in Appalachian coal production; (c) trends in Interior coal production. Data from U.S. Geological Survey Yearbooks (1880-1932); U.S. Bureau of Mines, Mineral Yearbooks (1933-1980); Energy Information Administration, 1983.

Sulfur Content

The next parameter that will be examined is c_i, the concentration of the contaminant sulfur, for each coal-producing region. Knowing the production rate P_i and concentration c_i permits the calculation of the mass of contaminant, $M_i = c_i P_i$, that is mobilized by each producer.

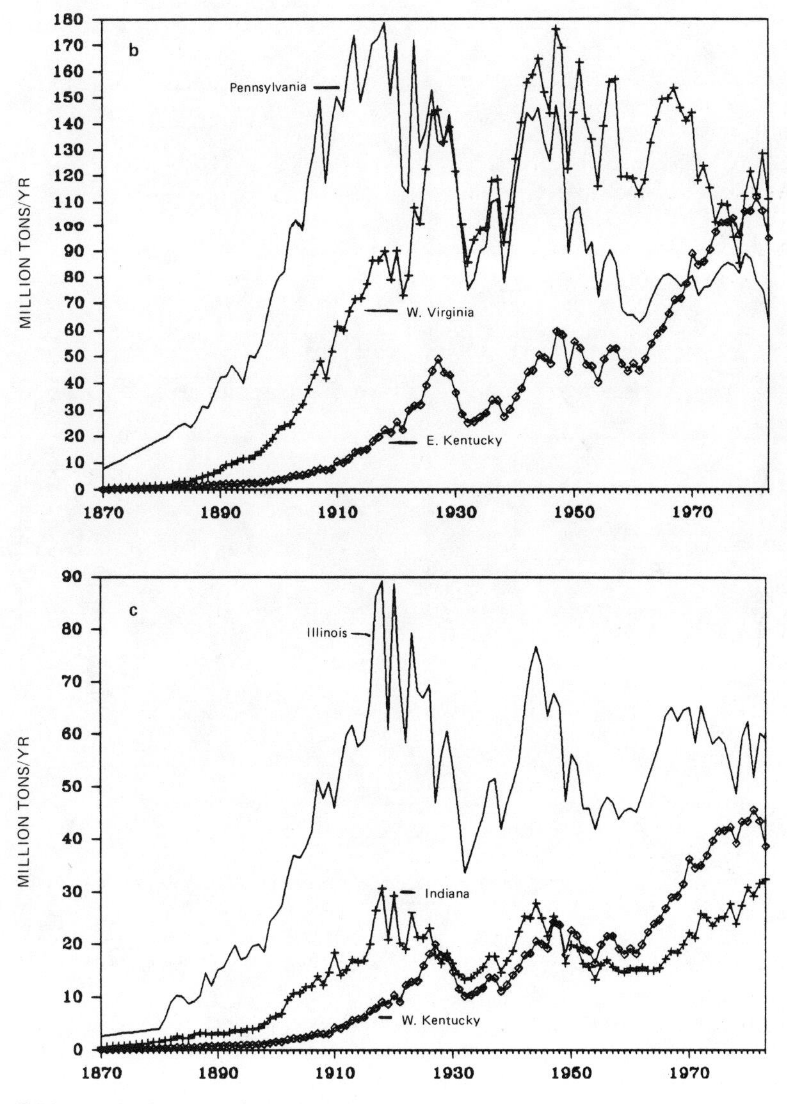

FIGURE 2.4 (continued).

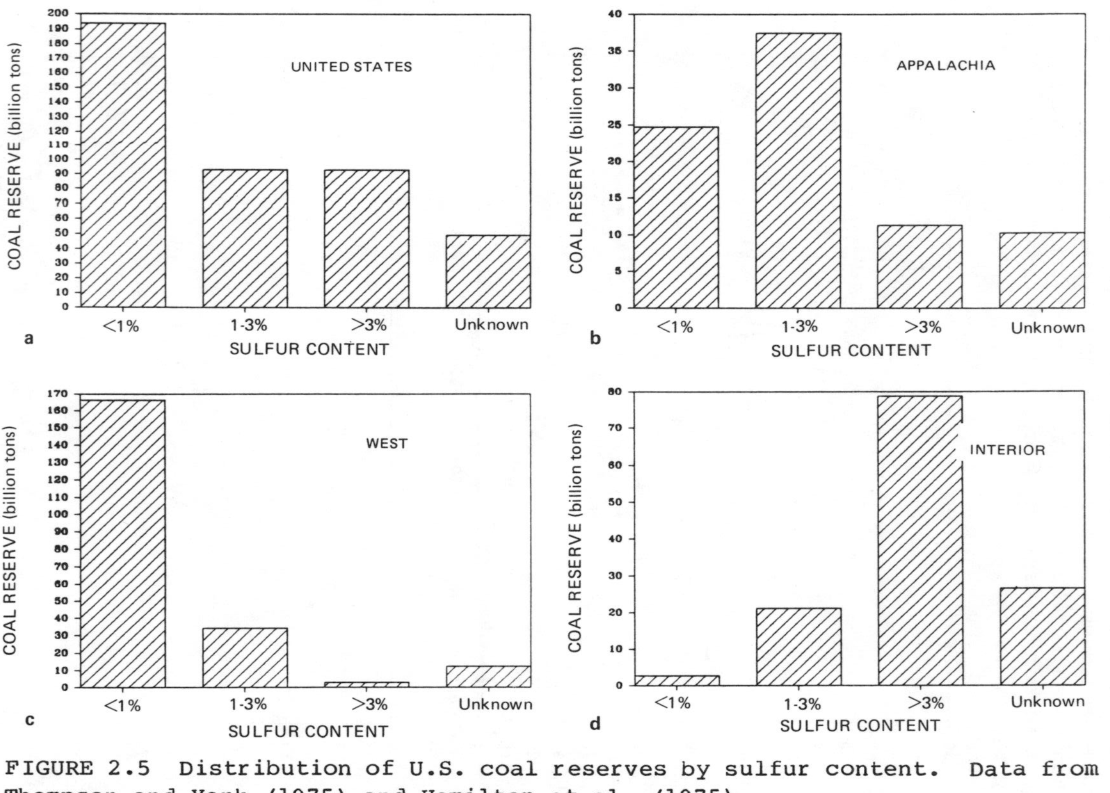

FIGURE 2.5 Distribution of U.S. coal reserves by sulfur content. Data from Thompson and York (1975) and Hamilton et al. (1975).

Each coal-producing district has a geologically defined range for the sulfur content of its coal. Western coal, for instance, is called low-sulfur since it contains less than 1 percent sulfur. On the other hand, the districts in the Midwest produce coal ranging from 2 to 4 percent sulfur, with little production outside this range. Thompson and York (1975) and Hamilton et al. (1975) have compiled an extensive data base for the sulfur content of coal reserves in the United States. The reserves are divided into four classes by sulfur content: less than 1 percent, 1 to 3 percent, greater than 3 percent, and unknown. About half of the coal reserves are of less than 1 percent sulfur content, as shown in Figure 2.5(a). The Appalachian districts have most of their coal in the 1- to 3-percent range, the Midwestern coals are mostly more than 3 percent sulfur and most of the Western coals contain less than 1 percent sulfur (Figure 2.5(b-d)).

Another estimate of the amount of sulfur mobilized by coal mining is obtained from data on the sulfur content of coal shipments. For 1970 to 1978, the U.S. Department of Energy (Energy Information Administration 1981) maintained a data base on the sulfur content of coal shipments from each coal-producing district. This was further broken down by consuming sector, such as coke ovens and electric utilities. Examples of sulfur content of coals shipped from various states are given in Figures 2.6(a) and 2.6(b). Evidently, there is a sizable distribution in sulfur content from each state, which arises from the heterogeneity of the coal beds themselves. Furthermore, there is also about a 10 percent year-to-year variation in the sulfur content from each state for the coal shipments between 1970 and 1978.

The spatial distribution of average coal sulfur content in the various states is shown in Figure 2.7, which shows that the content is highest in the Interior region, reaching 4.5 percent for Missouri. The content in Western coals ranges from 0.6 to 0.9 percent, and the coal in Appalachian districts is between 1 and 2 percent sulfur.

The distribution of sulfur that has been mobilized in coals from the three regions is shown in Figure 2.8. The area under each curve represents the tonnage of sulfur mobilized in the respective regions. The data show that most of the sulfur is from Interior coals, while the sulfur contribution of the Western coals is minimal.

Most of the coal undergoes various degrees of processing. Raw coals of different qualities can be blended at the mines or at loading and storage facilities

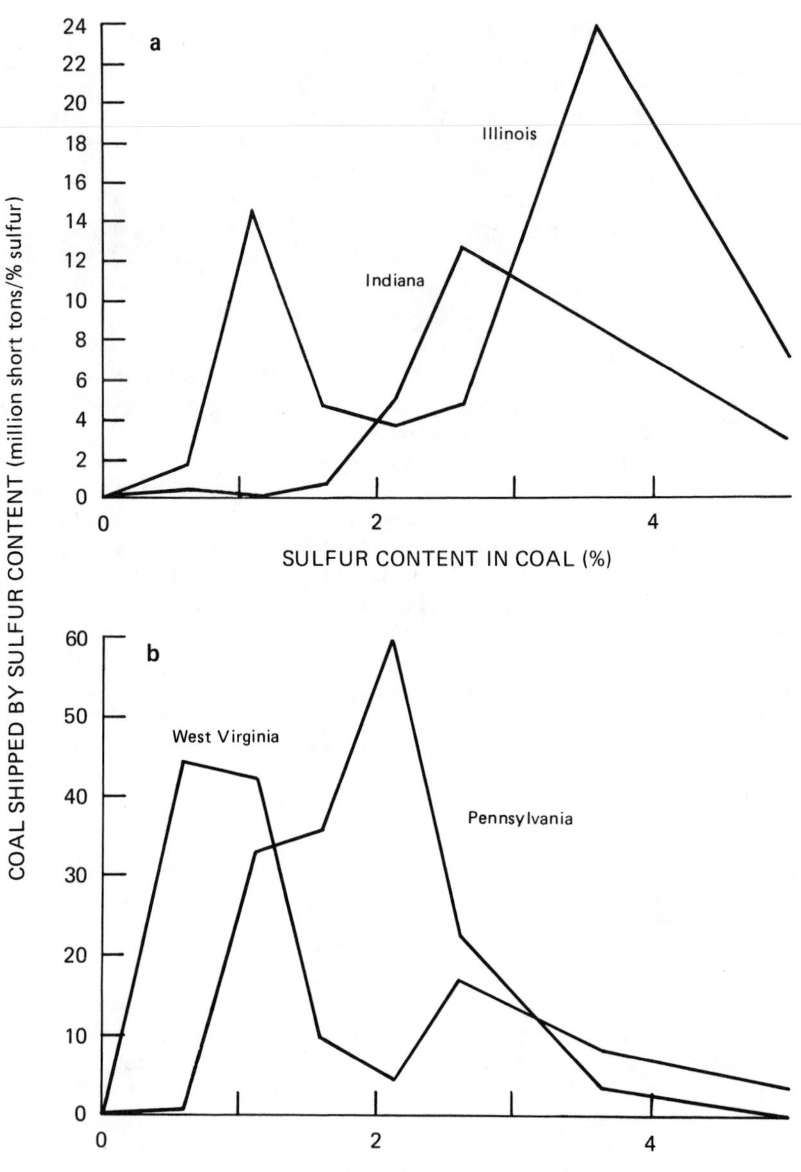

FIGURE 2.6 Distribution of coal sulfur content shipped from several coal-producing states. The data are averaged for the 9-year period from 1970 to 1978. SOURCE: Energy Information Administration 1981.

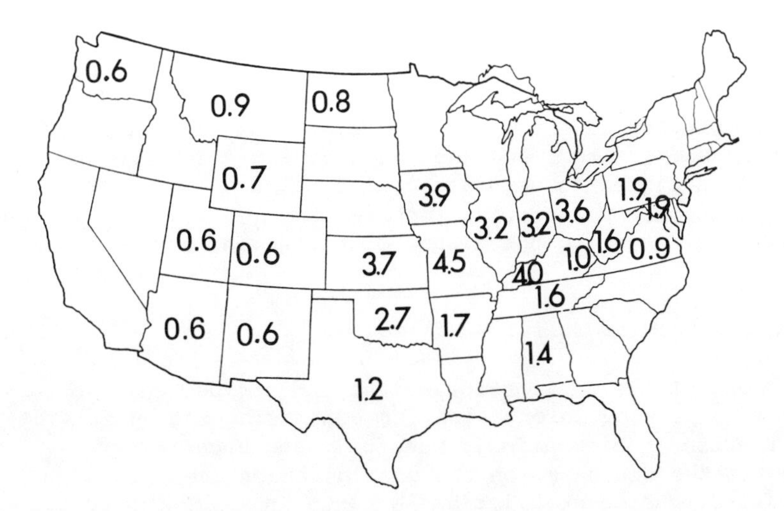

FIGURE 2.7 Average sulfur content of coal produced by state over the period from 1970 to 1978. SOURCE: Energy Information Administration 1981.

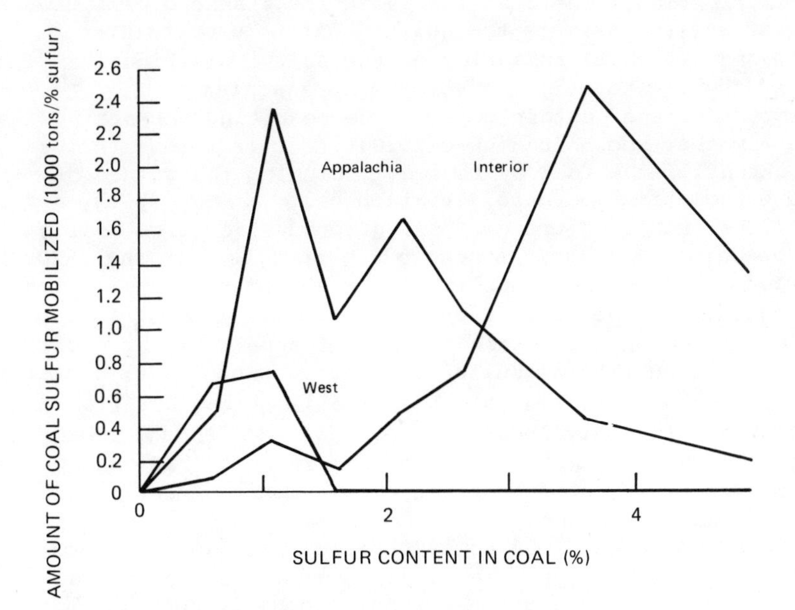

FIGURE 2.8 Coal sulfur distribution function for the three producing regions. The area under each curve is proportional to the amount of sulfur mobilized by each region. Data are averaged for the 9-year period 1970 to 1978. SOURCE: Energy Information Administration 1981.

to improve their handling and burning qualities. Sometimes coal is only crushed or screened, mainly to make it easier to handle. The quality of coal is upgraded the most when it is processed at preparation plants, where it is crushed, sized, and then cleaned and washed in special units designed to separate coal from rock and other impurities. Preparation improves the quality and heating value by lowering the amount of sulfur and ash.

Uncertainty of Sulfur Mobilization

One of the reasons for collecting the preceding data was to provide information for estimating the uncertainty of sulfur mobilization. The following uncertainty estimates are based on the consideration that each coal-producing district has a range in which its sulfur content will be confined and that sustained long-term coal production outside that range is not likely.

The distribution in sulfur content for each coal-producing region was approximated by a log normal distribution. The best-fit geometric standard deviation was used to estimate the sulfur-content uncertainty. Figure 2.9 shows summaries of the sulfur distributions and the log normal fits for the Appalachian, Interior, and Western coals. The resulting uncertainty estimates, shown in Figure 2.10, indicate the sulfur content of the coal by region, including the values of one geometric standard deviation below and above the mean.

The resulting trend of sulfur production from coal in the United States is shown in Figure 2.11. In the 1920s, most of the sulfur mobilization was from the Appalachian region. By the early 1980s, the mobilization of sulfur from Interior coal exceeded that of Appalachia by about 1.5 million tons of sulfur per year. Since 1970, Western coal has contributed to sulfur mobilization, but it accounts for only about a quarter of the tonnage mobilized from either the Appalachian or the Interior region and only about 12 percent of the total amount of coal sulfur mobilization. Hence, while U.S. coal production has increased in the 1970 to 1980 period from about 500 to 800 million tons/yr, the corresponding increase of the coal sulfur mobilization was only about 12 percent.

Based on the above production data, the average sulfur content of the three coal-producing regions is depicted in Figure 2.12. The sulfur content of Interior coal has remained higher than 3 to 3.5 percent. The sulfur in

Appalachian coal has dropped from 1.8 to 1.4 percent, primarily because of the increased production of the low-sulfur coal from eastern Kentucky. The sulfur content of Western coal has remained low at about 0.6 percent. The average sulfur content of coal for the entire United States has dropped significantly over the past 100 years from about 2.5 percent to about 1.6 percent. The significant reduction since 1970 is a result of the increased consumption of the low-sulfur Western coal, increased production of low-sulfur Interior coal, the closing of high-sulfur mines in Appalachia, and additional coal cleaning before combustion.

Surface Transfer Matrix

A key link in the flow of coal sulfur is its transfer from the producers to the consumers of coal. Railroads transport more than half of the total coal shipped each year. Unit trains provide high-speed shipments of large quantities of coal to electric power plants, often hauling more than 10,000 tons per train. Coal is also transported by trucks, barges, and ships, and relatively small amounts are delivered by conveyors and slurry pipeline. Some coal shipments are short-haul deliveries sent directly to mine-mouth electric power plants.

Here again we draw on extensive compilations since 1957 by the U.S. Bureau of Mines (1957-1977) and the Energy Information Administration (1977-1982) on the yearly surface transport of coal. In addition, the U.S. Geological Survey has provided similar compilations for the years 1917 (Lesher 1917) and 1927 (Tryon and Rogers 1927), which allow us to extrapolate to earlier time periods. The data sets indicate the amount of coal that has been shipped from a given producing district to the consuming state. In effect, the data bases yield the surface transfer matrices s_{ij} and show how the matrices have varied over time. As an example, a normalized surface transfer map for Appalachian coal is shown in Figure 2.13(a). The number in each state represents the percent of Appalachian coal consumed within that state. It appears that Pennsylvania, West Virginia, and Ohio account for almost half of the Appalachian coal consumption. The normalized transfer matrix for Interior coal shows that most of it is consumed within the interior region (Figure 2.13(b)). The shipments of Western coal (Figure 2.13(c)) show a rather broad spatial distribution covering the West as well as areas of the Midwest.

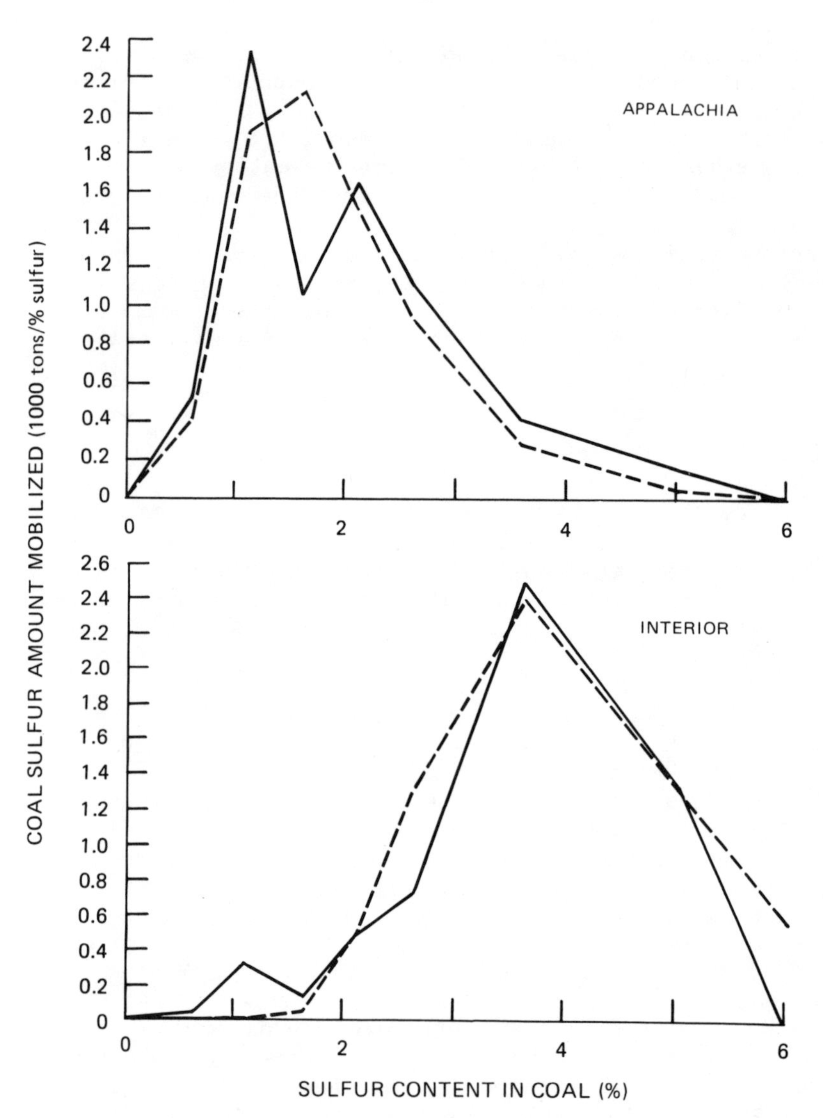

FIGURE 2.9 Log normal fit (dashed line) of sulfur distribution function (solid line) of coals shipped from the Appalachian, Interior, and Western coal mining regions. The data are averaged over the 9-year period 1970 to 1978. SOURCE: Energy Information Administration 1981.

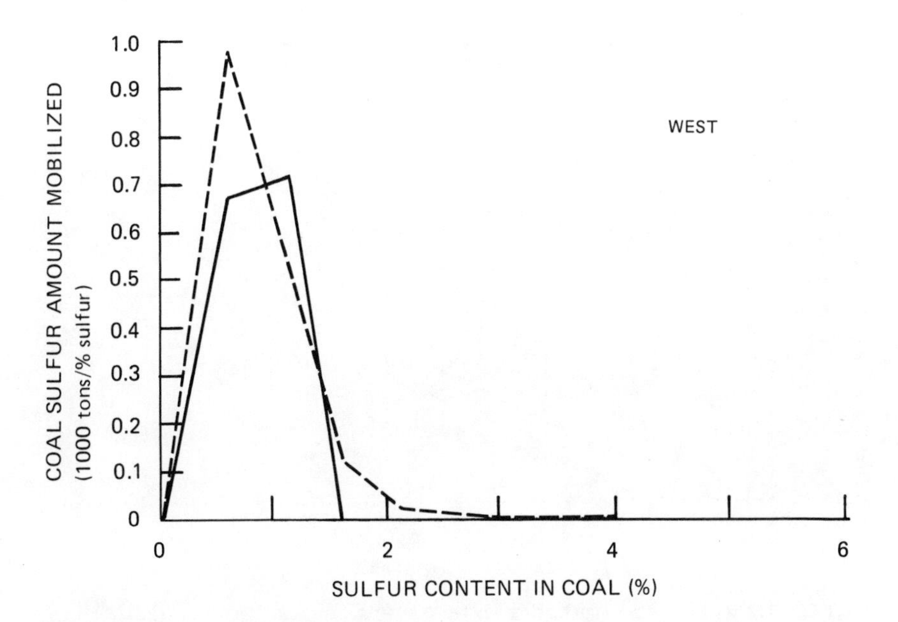

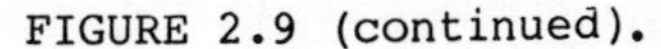

FIGURE 2.9 (continued).

The average sulfur content C_j of the coal consumed within each state j can be calculated from the expression

$$C_{ij} = \frac{\sum s_{ij} c_i P_i}{\sum s_{ij} P_i}$$

As an example, a map of the resulting consumed coal sulfur content data for 1978 is given in Figure 2.14. The northern and mid-Atlantic states consume coal with a content of about 1.4 percent sulfur. The midwestern states, as expected, consume coal with the highest sulfur content.

Coal Consumption

In the early 1900s coal was the nation's major fuel, supplying almost 90 percent of its energy needs. Later, coal's importance declined, mainly because of competition from petroleum and natural gas. Coal's share of the total energy consumed in the United States fell to about 38 percent in 1950, 23 percent in 1960, 19 percent in 1970,

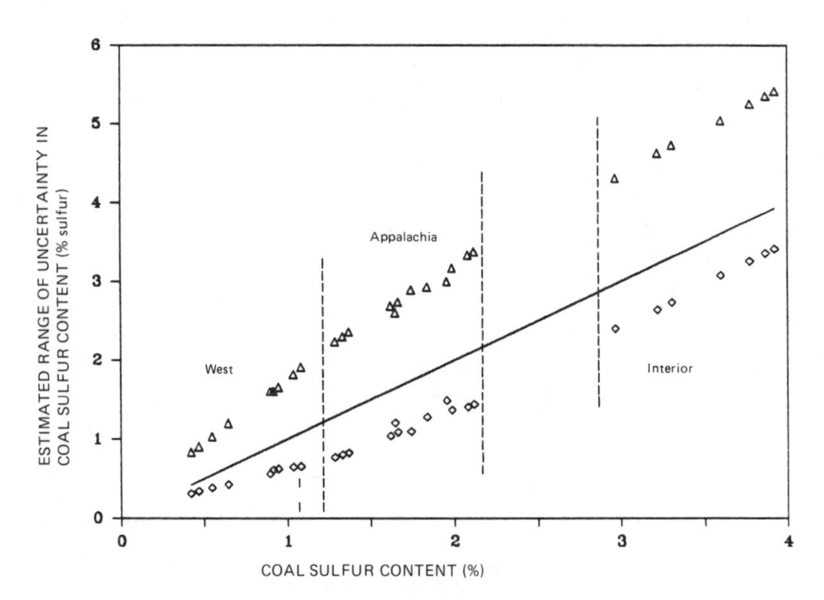

FIGURE 2.10 Estimated range of uncertainty of sulfur content for each coal-producing region. The solid line represents the best fit. The upper and lower estimates (designated by triangles and diamonds, respectively) are one geometric standard deviation from the geometric mean.

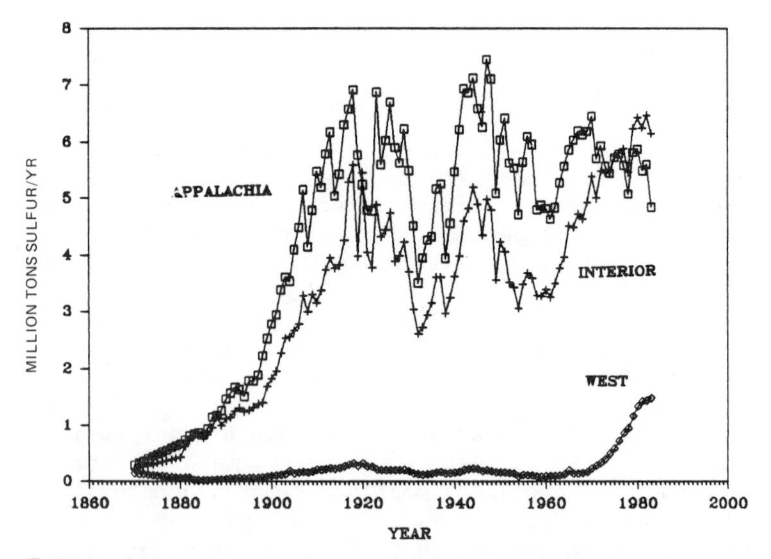

FIGURE 2.11 Trends in coal sulfur production for the Appalachian, Interior, and Western coal-producing regions.

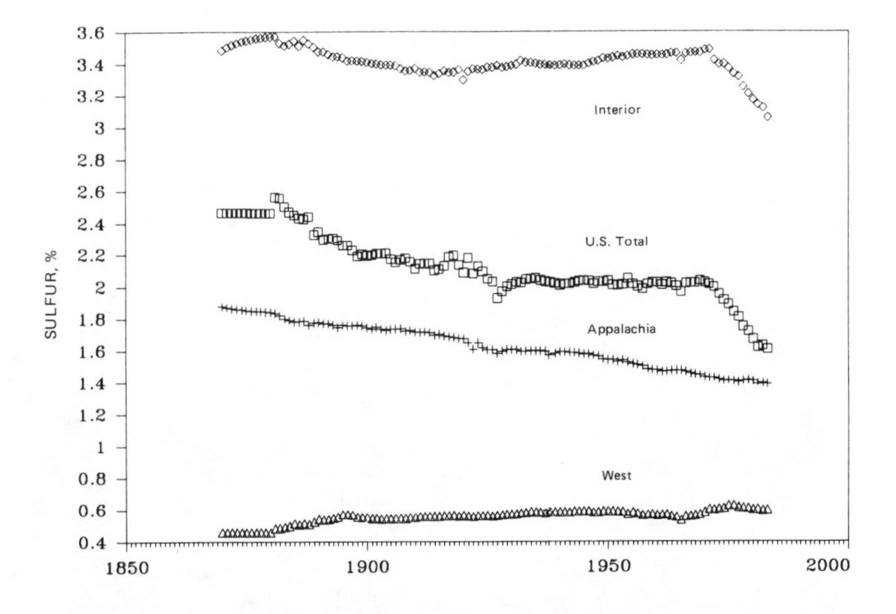

FIGURE 2.12 Trends in the average sulfur content for the Appalachian, Interior, and Western coal-producing regions and the production-weighted aggregate trend for the entire United States.

and less than 18 percent during parts of the 1970s. In 1980 and 1981, however, annual coal consumption increased to more than 700 million tons, exceeding the 500- to 680-million-ton range of the 1970s. As a result, coal's share of total energy consumption rose to 20 percent in 1980 and almost 22 percent in 1981.

The four leading coal-consuming states are Ohio, Pennsylvania, Indiana, and Illinois. Together, these states account for about one third of the coal consumed in the United States. In 1975, coal consumption was about 550 million tons/yr, roughly the same as around 1920 and 1943 (Figure 2.15). However, since the 1930s there has been a total transformation in the economic sectors that consume coal. Before 1945, coal consumption was divided among electric utilities, railroad, residential and commercial heating, oven coke, and other industrial processes. The railroad demand was particularly high during the war years of the early to mid-1940s. Within one decade, the 1950s, coal consumption by railroads and by the residential-commercial sector

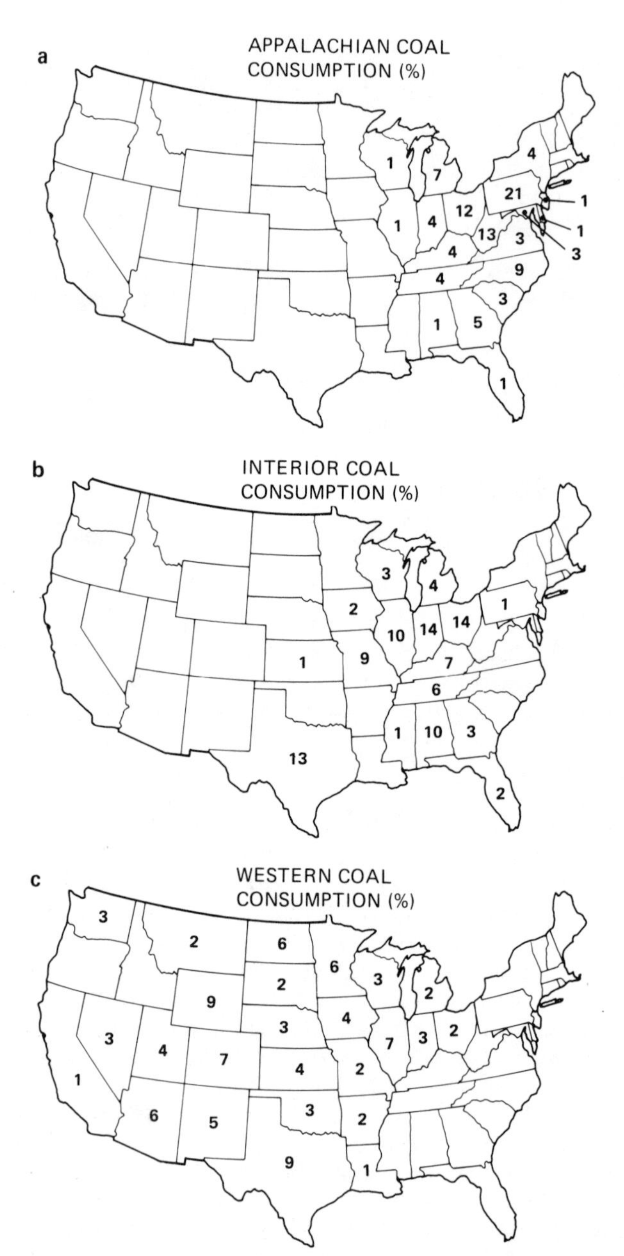

FIGURE 2.13 Normalized surface transfer maps of shipment to consumer states from coal producing regions. The numbers assigned to the states represent the percentages of total production consumed by the states: (a) Appalachia; (b) Interior; (c) the West.

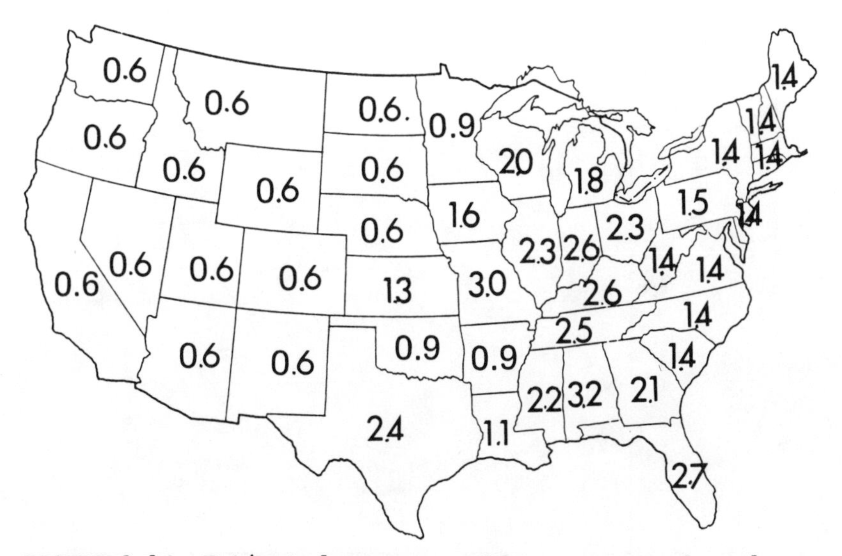

FIGURE 2.14 Estimated average sulfur content of coal consumed in the United States in 1978.

essentially vanished. Currently, electric utilities constitute the main coal-consuming sector, and the trend of total coal use in the United States since 1960 has been determined by the electric utility coal demand.

The changes of the consumer mix over the past decades have resulted in a change in seasonal coal consumption (Figure 2.16). In 1951, for instance, the total national coal consumption was maximum during the winter months of December, January, and February, mainly because of the increased residential-commercial coal demand. By 1974, the seasonal pattern of coal use was determined by the winter and summer peaks of utility coal use. The seasonal shift of coal sulfur emissions has undoubtedly resulted in significant changes in seasonal patterns in acid deposition.

According to the U.S. Bureau of the Census reports for 1889 and 1919, most of the coal consumption of the nation was confined to the states north of the Ohio River, stretching from Illinois to New York and New England. More recently, the amount consumed by the states south of the Ohio River and east of the Mississippi has been increasing rapidly. In the context of the current concern about acid rain, the spatial shift of coal consumption has major implications. In the Ohio River region (Indiana, Michigan, Ohio, Pennsylvania), coal

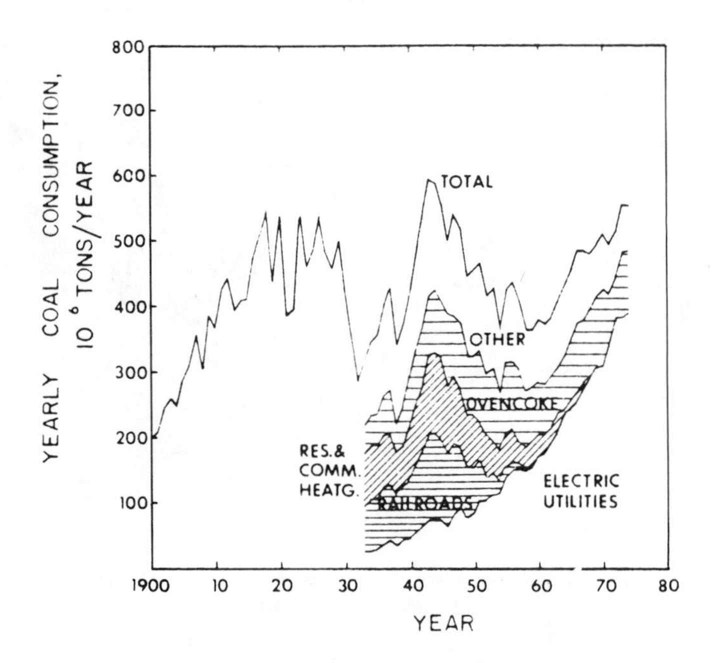

FIGURE 2.15 Trend of U.S. coal consumption by consuming sector. SOURCES: Mineral Resources of the United States, Yearbooks (U.S. Geological Survey 1900-1932); Mineral Yearbooks (U.S. Bureau of Mines 1933-1973).

consumption was high in the 1950s, increased sharply during the 1960s, and increased only slightly or stabilized during the 1970s (Figure 2.17). Coal consumption in New England, New York, Maryland, and Virginia has declined since about 1965, reflecting the fuel shift from coal to oil. The greatest increase in coal consumption has been recorded for Kentucky, West Virginia, North Carolina, South Carolina, and Tennessee. In the remainder of the southeastern U.S. coal consumption has been appreciable only in Georgia and Alabama. In Missouri and Illinois, there has been a moderate increase, again because of the rising demand for electric utility coal. The western states (Montana, Idaho, Wyoming, Nevada, Utah, Colorado, Arizona, and New Mexico) have undergone a rapid increase in coal consumption.

Figure 2.18 illustrates the changing geographical pattern of coal consumption. In 1917 the coal use density in Pennsylvania exceeded 400 g m^{-2} yr^{-1}, and in the industrial states east of Iowa it exceeded 100 g m^{-2} yr^{-1}.

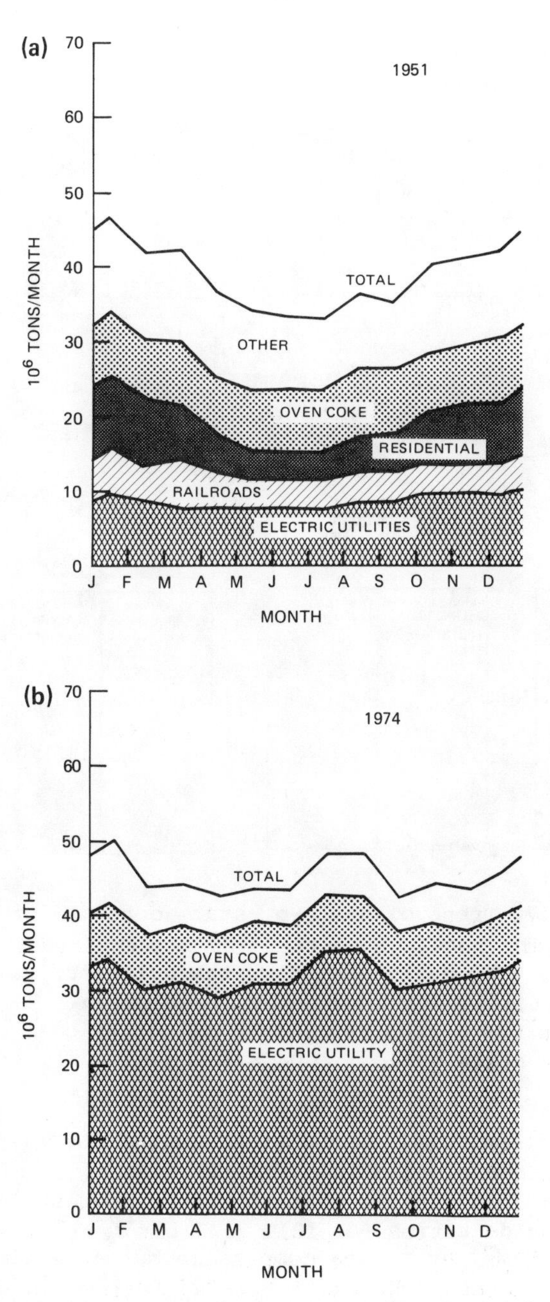

FIGURE 2.16 Seasonal patterns of U.S. coal demand: (a) 1951; (b) 1974. Data from Mineral Yearbooks (U.S. Bureau of Mines 1951, 1974).

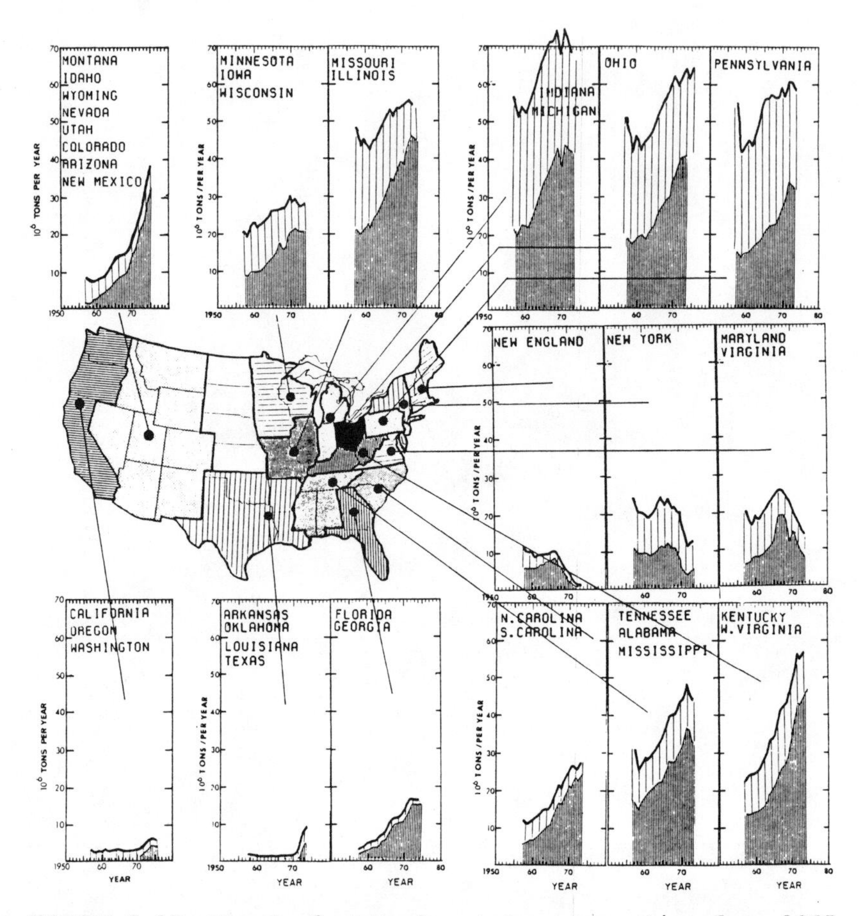

FIGURE 2.17 Trend of state-by-state consumption from 1957 to 1973. The shadings on the U.S. map represent groupings of states. The dark shading on the graphs represents electric utility consumption by state group. The light shading on the graphs represents all other coal-consuming sectors by state group. SOURCE: Husar et al. (1981).

For comparison, coal consumption density for 1975, calculated and plotted in a similar manner, is given in Figure 2.18(b). In the eastern United States average consumption densities for 1917 (147 g m^{-2} yr^{-1}) and for 1975 (155 g m^{-2} yr^{-1}) are comparable but show different spatial patterns. The most striking feature is the shift in the highest consumption densities westward and southward from the populous northeastern states. Coal consump-

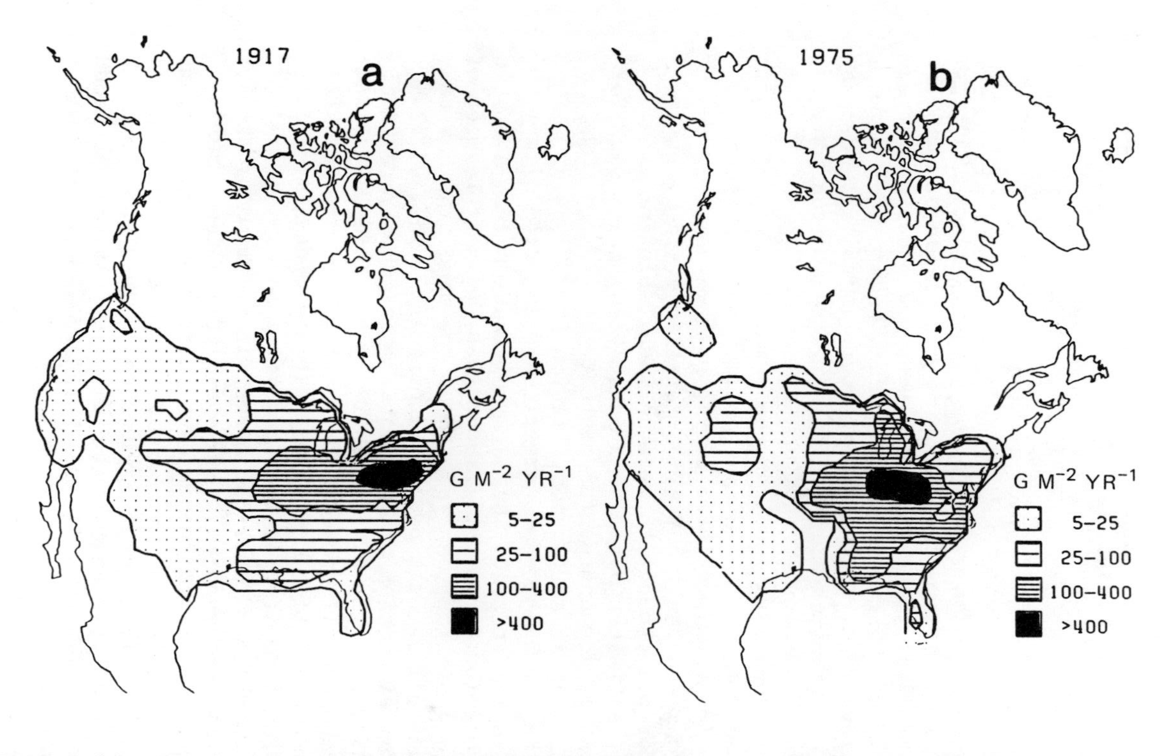

FIGURE 2.18 Total U.S. coal consumption density for (a) 1917; (b) 1975. Data for 1917 are from U.S. Bureau of Statistics (1917) and Lesher (1917); data for 1975 are from the Energy Information Administration (1983).

tion in New England, New York, and Pennsylvania was the highest around 1920 (approximately 140 millon tons/yr) and it has declined to 70-80 million tons/yr in the 1970s. In Pennsylvania alone, the coal consumption has declined from 80-90 million tons/yr around 1920 to 60 million tons/yr in the 1970s.

Oil Products

The consumption of petroleum products since 1950 is displayed in Figure 2.19, showing the trends of residual fuel oil, distillate oil, gasoline, and other products. Residual fuel oil consumption trends and spatial patterns are relevant for sulfur and nitrogen emissions, whereas those for gasoline are more relevant to hydrocarbon and nitrogen emissions.

Sulfur mobilization from the combustion of oil products can be estimated from either production or consumption data. In principle these two approaches should yield identical results. But since detailed state-by-state data for oil production were not available for this report, estimates were based on state-by-state data for oil consumption and sulfur content.

The trend in sulfur mobilization in the United States from the consumption of domestic and imported oils is shown in Figure 2.20. Sulfur mobilization from domestic crude oil increased until about 1960, when it leveled off at 3 to 4 million tons/yr. About the same time the role of crude oil imports became significant. The sulfur imported with other oil products, most notably residual fuel oil, also was significant. By the late 1970s, sulfur from imported oil exceeded the sulfur from domestically produced oil, but since 1978 there has been a significant reduction in sulfur in imported oil products.

As crude oil is refined a certain fraction of the sulfur is recovered as the by-product, sulfuric acid. The recovered fraction has been increasing steadily since 1950. According to U.S. Bureau of Mines Mineral Yearbooks the sulfur recovered at refineries in 1980 was about 4 million tons/yr. Hence, more than half of the estimated sulfur from crude oil is now retained and recycled at the refineries.

The emitted sulfur from oil products is calculated as crude oil sulfur content minus recycled sulfur. As seen in Figure 2.21, the oil sulfur emission estimated in this manner ranged between 3 and 4 million tons/yr for the

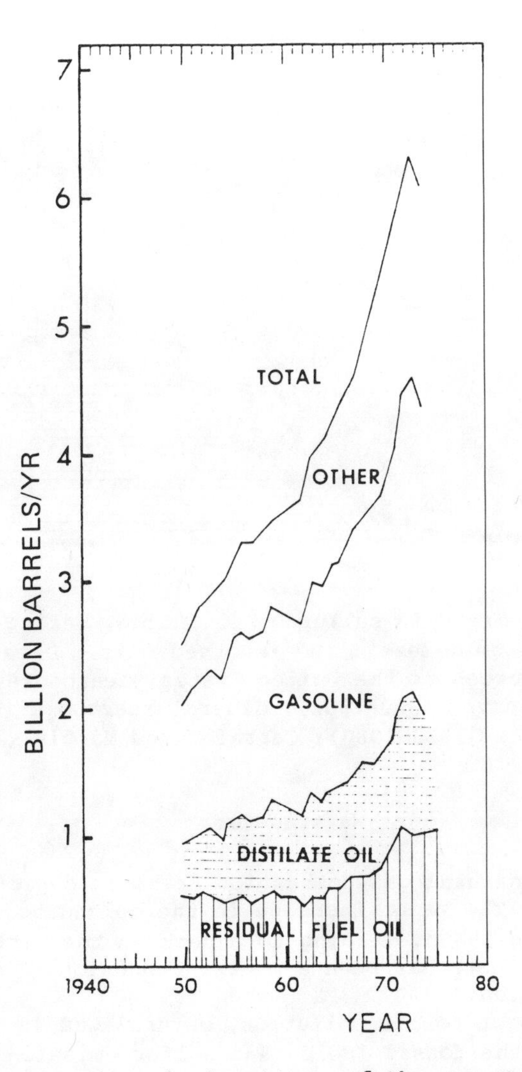

FIGURE 2.19 Historical pattern of the consumption of oil products. SOURCE: Mineral Yearbooks (U.S. Bureau of Mines 1950-1975).

period 1950 to 1978. Since then, there has been a significant decrease, caused primarily by declining imports and the increasing fraction of recycled sulfur. For 1982, emissions of sulfur from oil consumption were about 2 million tons/yr, which is less than 20 percent of the sulfur emissions from coal.

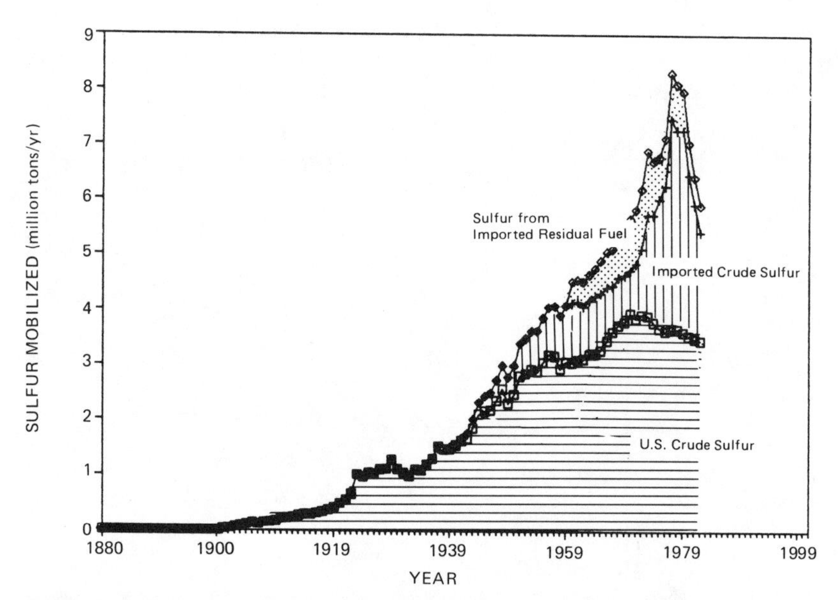

FIGURE 2.20 Trend in sulfur mobilization, before recycling, from domestic and imported oils. Data from Mineral Resources of the United States, Yearbooks (U.S. Geological Survey 1900-1932); Mineral Yearbooks (U.S. Bureau of Mines 1933-1980); Carrales and Martin (1975).

Natural Gas

The U.S. natural gas combustion trend is depicted in Figure 2.3. The major increase in the demand occurred in the 1950s and 1960s, leading to a peak in the early 1970s. Since then, there has been a modest reduction in natural gas consumption.

With respect to air pollution, natural gas is the cleanest of the fossil fuels, and sulfur emissions from combustion of natural gas are negligible.

Copper and Zinc Smelting

Significant production of copper began in the United States about 1895 and reached approximately 1 million tons annually by 1920. For the next 40 years copper production fluctuated at that level, with no significant trend. During the 1960s smelter copper production again increased, reaching a peak of over 2 million tons around

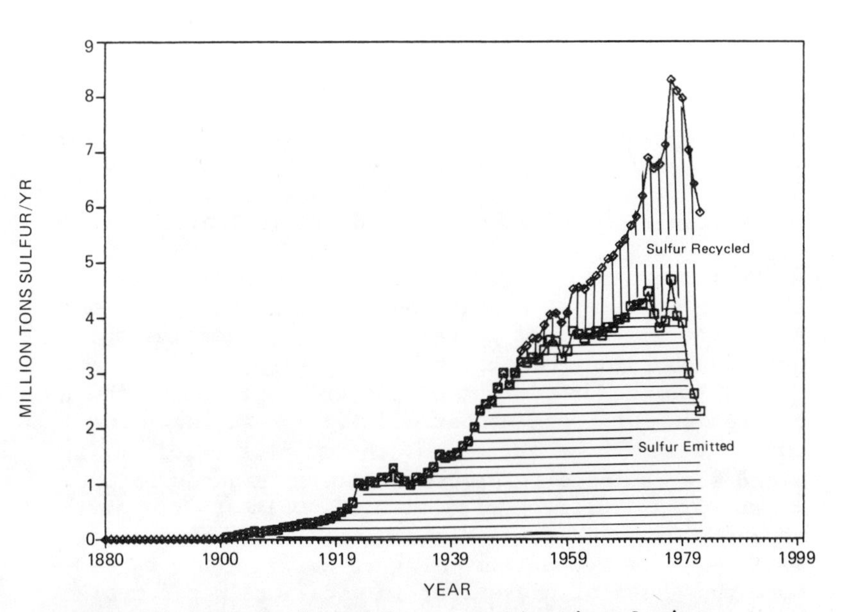

FIGURE 2.21 Trends in sulfur recyclization during petroleum processing and sulfur emissions from petroleum consumption in the United States. Data from Mineral Resources of the United States, Yearbooks (U.S. Geological Survey 1900-1932); Mineral Yearbooks (U.S. Bureau of Mines 1933-1980); Carrales and Martin (1975).

1970, followed by decline in 1970s. Arizona was the leading copper-producing state in the 1970s with 54 percent of the total national product, followed by Utah, New Mexico, Montana, Nevada, and Michigan. These six states accounted for 97 percent of 1970 mine production. Open-pit mines supplied 84 percent of the mine output. Most of the copper and zinc production in North America occurs in the Canadian province of Ontario.

Virtually all copper ore is treated at concentrators near the mines. Concentrates are further processed at sixteen smelters, eight in Arizona and one each in Utah, Michigan, Montana, Nevada, New Mexico, Tennessee, Texas, and Washington. In 1970, four companies accounted for 80 percent of the smelter capacity (U.S. Bureau of Mines 1971). Production of sulfuric acid is the main process for removing sulfur oxides from smelter gases. However, acid production is practical only from converter gases. With tightly hooded converters, 50 to 70 percent of the sulfur oxides can be removed; removal of additional sulfur oxides requires scrubbing.

Zinc smelting in 1960s and 1970s was about 800,000 tons/yr. Foreign imports of zinc ores constitute a significant fraction of zinc production.

Lead smelter production in the United States was about 700,000 tons/yr. However, sulfur emission from lead smelting is small compared with that from the smelting of copper and zinc and is not considered further in this report.

Sulfur emissions from metal smelting are estimated from the tonnage of sulfur mobilized by mining the ore minus the sulfur that is retained at smelters as sulfuric acid. According to data from the U.S. Bureau of Mines (1971) the sulfur content of copper ore is estimated as 1.2 ton sulfur/1 ton copper and 0.5 ton sulfur/1 ton zinc. Using these estimates, the quantities of sulfur mined are depicted in Figure 2.22(a). Figures on the amount of sulfur recovered at U.S. smelters were obtained from the U.S. Bureau of Mines Mineral Yearbooks (1944-1980). It is evident from Figure 2.22(a) that by 1980 more than half of the sulfur in metal ore was recycled. Hence, sulfur emissions (mobilized minus recycled) have fluctuated between 0.5 and 1.5 million tons/yr since the turn of the century. A particularly significant drop of emissions has occurred since 1970 (1.5 to 0.5 million tons/yr) as a result of both decline of smelter production and increased sulfur recovery (Figure 2.22(b)).

SULFUR EMISSION TRENDS

This section comprises a summary of the sulfur emission trends for eastern North America. These include sulfur emissions in the eastern United States and southeastern Canada from coal and oil combustion and copper and zinc smelting. The Canadian emission trends were taken from the United States-Canada Memorandum of Intent (1982) and extrapolated for pre-1955 years using smelter production data. Since there are substantial differences in the emission trends of smaller regions within eastern North America, the subcontinent was divided into five regions, labeled A, B, C, D, and E in Figure 2.23.

The regional and aggregate sulfur emission trends for eastern North America are shown in Figure 2.24; the shading represents a range of estimates based on the degree of uncertainty of sulfur content in coal. Aggregate sulfur emissions of all five regions (Figure 2.24(a)) exhibit a fluctuating trend since the early

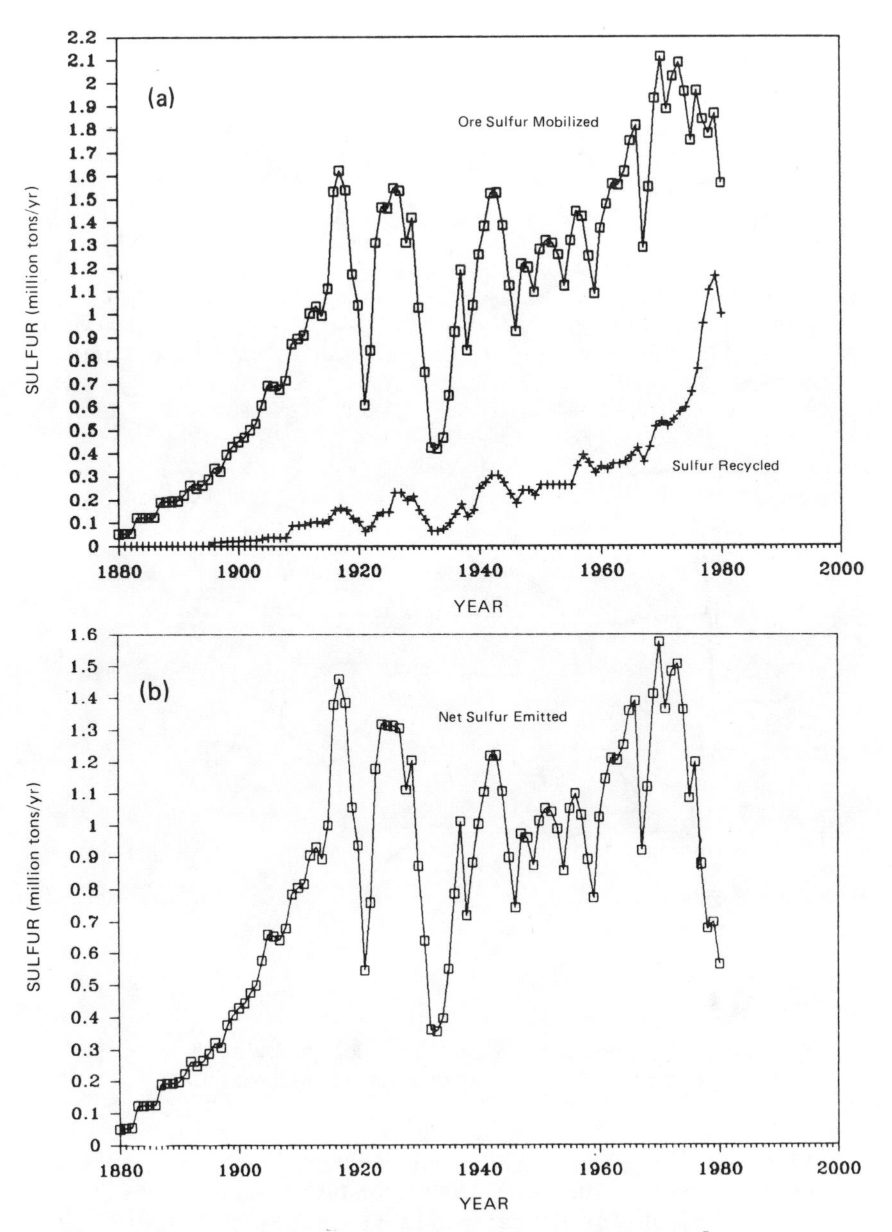

FIGURE 2.22 (a) Trend in sulfur mobilization and recyclization from copper and zinc smelting and processing; (b) trend in sulfur emissions from copper and zinc smelting. SOURCES: Mineral Resources of the United States, Yearbooks (U.S. Geological Survey 1880-1932); Mineral Yearbooks (U.S. Bureau of Mines 1933-1980).

FIGURE 2.23 Regions in eastern North America selected for the study of regional trends in emissions.

1900s, showing downward trends in the 1930s and 1950s and peaks in the 1920s, the 1940s, and the early 1970s. Currently, sulfur emissions in the subcontinent are estimated to range between 11 and 15 million tons/yr (22 to 30 million tons of sulfur dioxide/yr).

The emission trend for eastern Canada (Region A; Figure 2.24(b)) shows an increase since the turn of the century leading to a peak at about 3 million tons in the late 1960s, followed by a subsequent decline.

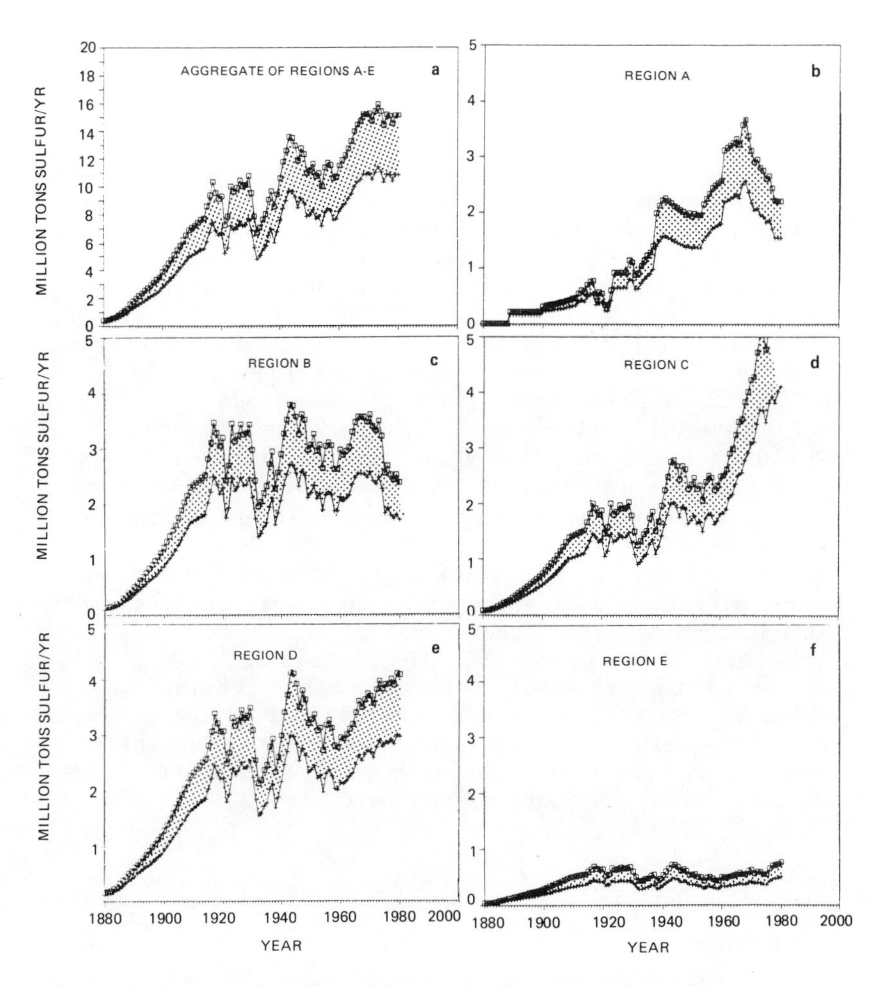

FIGURE 2.24 Regional sulfur emission trend estimates for (a) eastern North America (the aggregate of Regions A, B, C, D, and E); (b) Region A; (c) Region B; (d) Region C; (e) Region D; (f) Region E.

The sulfur emission trend for the northeastern states, including Pennsylvania (Region B) is given in Figure 2.24(c). An increase in emissions began in the 1880s, leading to a peak between 1910 and 1930 and followed by a decrease in the 1930s. Another increase occurred in the early 1940s, followed by a decrease in the late 1940s and 1950s, an increase in the 1960s, and a decrease in the 1970s. Current sulfur dioxide emissions are comparable with those in the 1930s and early 1900s but are substantially below those of the 1920s, 1940s, and 1960s.

The emission trend for the eastern states south of the Ohio River (Region C) is given in Figure 2.24(d). The emissions were relatively low in the southeastern United States until about 1960, when a strong increase occurred. This upward trend has persisted into 1980s.

The sulfur emission trend for the industrialized Midwest (Region D), including Illinois, Indiana, Michigan, Missouri, and Ohio, is given in Figure 2.24(e). The emission trend in this region shows a doubling since the turn of the century with peaks in the 1920s, 1940s, and 1970s and depressions in the 1930s and 1950s.

The sulfur emission trend for the upper Midwest (Region E)--Iowa, Minnesota, and Wisconsin--is given in Figure 2.24(f). The emissions in this region, currently less than 1 million tons/yr, have been consistently low over the last 100 years.

A comparison of sulfur emission densities north and south of the Ohio River is given in Figure 2.25, expressed as emission per unit area (g sulfur m^{-2} yr^{-1}). Emissions north of the Ohio River (Regions A, B, D, and E) have increased about 33 percent since the 1920s. In contrast, emissions south of the Ohio River (Region C) show a threefold increase since the 1930s. Currently, the sulfur emission densities are comparable for the regions north and south of the Ohio River.

REGIONAL TRENDS IN EMISSIONS OF NITROGEN OXIDES

Nitrogen oxides constitute the second major source of acidifying compounds. The overwhelming fraction of nitrogen oxide emissions arises from the combustion of fossil fuels; emissions from metal-processing plants are insignificant. Fuel consumption data constitute one of two important inputs needed for estimating nitrogen oxide emissions. The other is data for nitrogen oxide emission factors. For a given source of combustion, this factor is the quantity of nitrogen oxide emitted per unit of fuel consumed.

Estimating historical emission trends of nitrogen oxides is difficult because most of the nitrogen oxide is formed by the fixation of atmospheric nitrogen at high temperatures of combustion rather than by oxidation of the nitrogen contained in the fuel. The nitrogen oxide emissions depend primarily on the combustion temperature

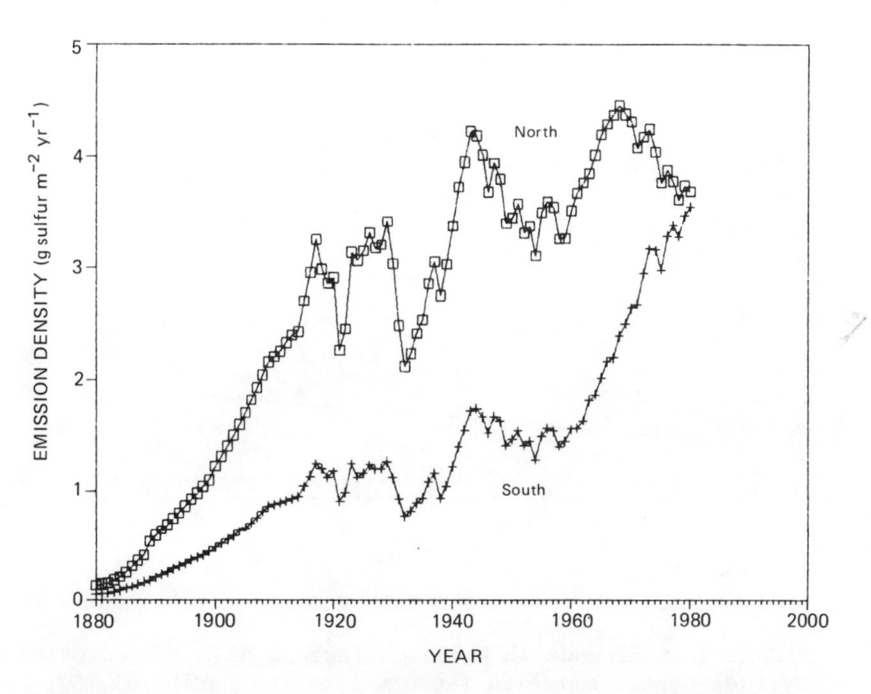

FIGURE 2.25 Sulfur emission densities for regions north (Regions A, B, D, E) and south (Region C) of the Ohio River.

and to a lesser degree on fuel properties. Since combustion processes in internal-combustion engines and boilers have undoubtedly changed since the turn of the century, it is likely that nitrogen oxide emission factors also have changed historically. Because combustion parameters can vary randomly over a wide range, and because information on historical combustion processes is generally lacking, assumptions concerning changes in emission factors over time constitute the major source of uncertainty in developing trends in nitrogen oxide emissions.

The emission factors for 1970 to 1980 used in this chapter were derived from extensive inventories that list nitrogen oxide emission factors according to source type of combustion (U.S. Environmental Protection Agency 1977, 1978). The numerous emission factors listed in these compilations were aggregated into four weighted-average emission factors by fuel type: coal, gasoline, natural gas, and other petroleum products. The emission factors

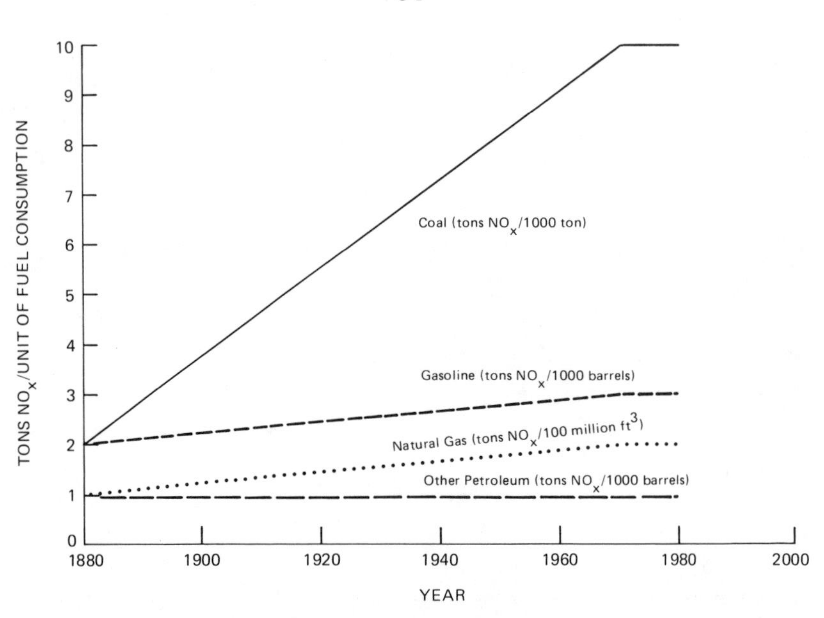

FIGURE 2.26 Trends in emission factors of nitrogen oxide by fuel type. Emission factors from the period of 1970 to 1980 were derived from data given from the U.S. Environmental Protection Agency (1977, 1978). For the period from 1880 to 1970, trends of historical emission factors were assumed to be linear, with slopes varying by fuel type.

before 1970 were estimated to reflect the fact that the average combustion temperature, and hence the production of nitrogen oxide per unit of fuel consumed, was lower, especially for coal combustion, over the past 100 years (Figure 2.26). A simple linear trend was assumed for all emission factors. For coal combustion the emission factor was assumed to increase fivefold from 1880 to 1970. For combustion of gasoline and natural gas the emission factors were assumed to increase by 50 and 100 percent, respectively, during the same period. The emission factor for other petroleum products was assumed to be constant over time.

Based on these estimates of emission factors and data on fuel consumption, regional emission trends were calculated (Figure 2.27). The shaded areas represent a range of uncertainty of ±30 percent, reflecting the large uncertainties inherently associated with the

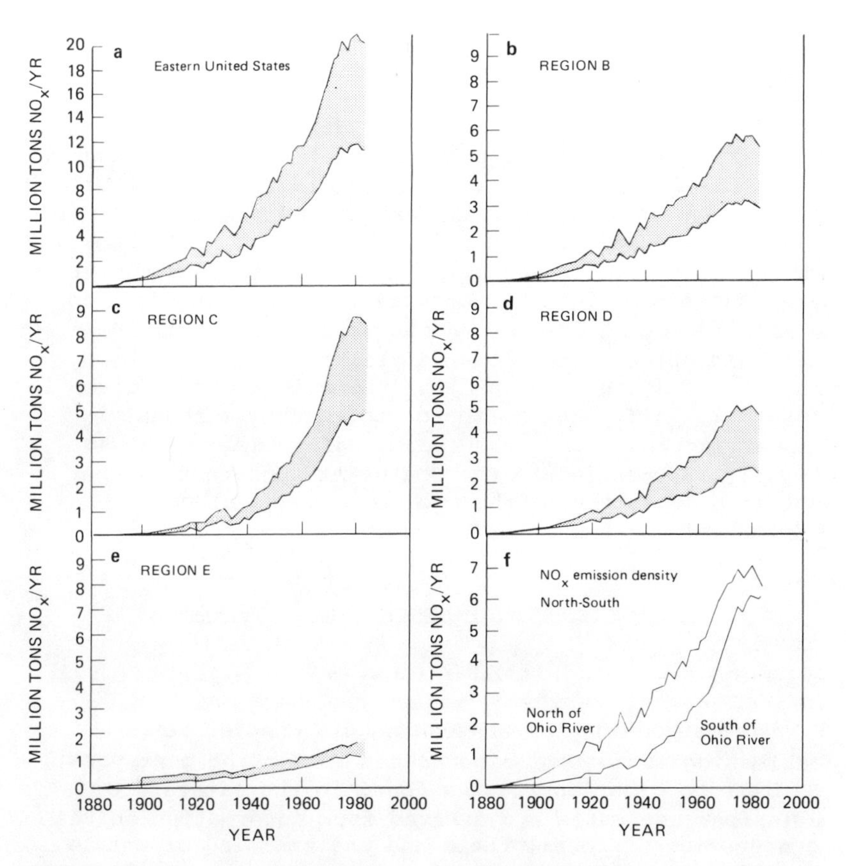

FIGURE 2.27 (a) Trends in emissions of nitrogen oxides in the eastern United States (the aggregate of Regions B, C, D, and E); (b) Region B; (c) Region C; (d) Region D; (e) Region E; (f) trends in emission densities of regions north (Regions B, D, and E) and south (Region C) of the Ohio River.

assumptions made in these calculations. For the eastern United States (the aggregate of Regions B, C, D, and E; Figure 2.27(a)), the estimates indicate a strong monotonic increase since the late 1800s. Evidently, there has been no significant change since the mid-1970s. The nitrogen oxide emission estimates for Region B are given in Figure 2.27(b). A steady increase is evident since the turn of the century, leading to a peak in the late 1970s. The emission trend for states south of the Ohio River (Region C; Figure 2.27(c)) shows a roughly exponential increase since the turn of the century. The nitrogen oxide emission trend for the industrialized midwestern states (Illinois, Indiana, Michigan, Missouri, and Ohio (Region D)) is shown in Figure 2.27(d). This trend resembles that of the northeastern states, showing a roughly linear increase since the 1920s. The upper midwestern states (Iowa, Minnesota, and Wisconsin (Region E)) show a steady upward trend (Figure 2.27(e)). However, the tonnage of nitrogen oxide emissions is substantially less than those for the northeastern, southeastern, and other midwestern states. A comparison of trends in nitrogen oxide emission densities north (Regions B, D, and E) and south (Region C) of the Ohio River is given in Figure 2.27(f). The northern states show a roughly linear increase since the 1920s, while the growth was roughly exponential in the Southeast. Currently, the emission densities in the two regions are almost equivalent.

COMPARISONS WITH OTHER TREND ESTIMATES

Sulfur Dioxide

The methodology presented in this chapter for estimating sulfur emission trends is for the most part similar to other approaches found in the literature. Emissions estimates are derived from information on the consumption of fossil fuels and the smelting of metals. The methodology differs from other reports by the way in which sulfur content of coal is estimated. The method described here uses mining data to determine the tonnage and distribution of the sulfur content in the coal at the mining site and information on the transport of the coal to the consumer to determine the emission source. From such an analysis it is possible to specify the uncertainty in sulfur emissions based on the sulfur distribution function of the mined coal.

Recently, an extensive compilation of emissions of sulfur dioxide and nitrogen oxides was developed by Gschwandtner et al. (1985). They report yearly state-by-state emissions estimates in the United States from 1900 to 1980. The difference between their analysis and the one presented here is in the method for calculating the sulfur content of coal in each state and estimating its uncertainty as a function of time.

A direct comparison of the results of the two estimates is shown in Figure 2.28. The shaded areas represent the estimated range of uncertainty in our analysis, and the diamonds represent the results of Gschwandtner et al. As shown in Figure 2.28(a) for the eastern United States (Regions B, C, D, and E), the estimates agree qualitatively over much of the time period since 1900. Both estimates show peaks in the 1920s and in the early to mid-1940s. Both show steeply rising emissions in the 1960s and dips in the 1930s and 1950s.

However, the estimates differ in some respects. Until about 1970 the estimate of Gschwandtner et al. for the eastern United States is consistently higher than our estimate. Furthermore, beginning around 1970 the trends appear to diverge. The estimate of Gschwandtner et al. shows a decrease in emissions in the eastern United States that continues through the 1970s. Our estimate suggests either no change or a possible increase, although the difference between the two estimates lies within the range of estimated uncertainty (see the shaded area, Figure 2.28(a)). A comparison of the emission trends is shown for three states in Figures 2.28(b)-2.28(d) to illustrate in more detail the extent of the differences in the two data sets. For Illinois, our estimate is substantially below that of Gschwandtner et al., except for the 1970s; the trend estimates for the state of Georgia are almost identical; for New York, the two estimates overlap since the 1930s but deviate before that time. These differences undoubtedly arise from the different assumptions employed in deriving the estimates.

Nitrogen Oxides

Gschwandtner et al. (1985) have also developed comprehensive estimates of emissions of nitrogen oxides. A comparison of our estimates with theirs is shown in Figure 2.29. Our analysis attempts to account for changes in emission factors that may have occurred over the past 100 years because of changes in average temperatures of combustion; the estimates of Gschwandtner et al. did not attempt such a correction. Nevertheless, as shown in the figure, after about 1930 the two estimates fall well within the range of estimated uncertainty, except for the lower estimates of Gschwandtner et al. for Region B in the 1970s.

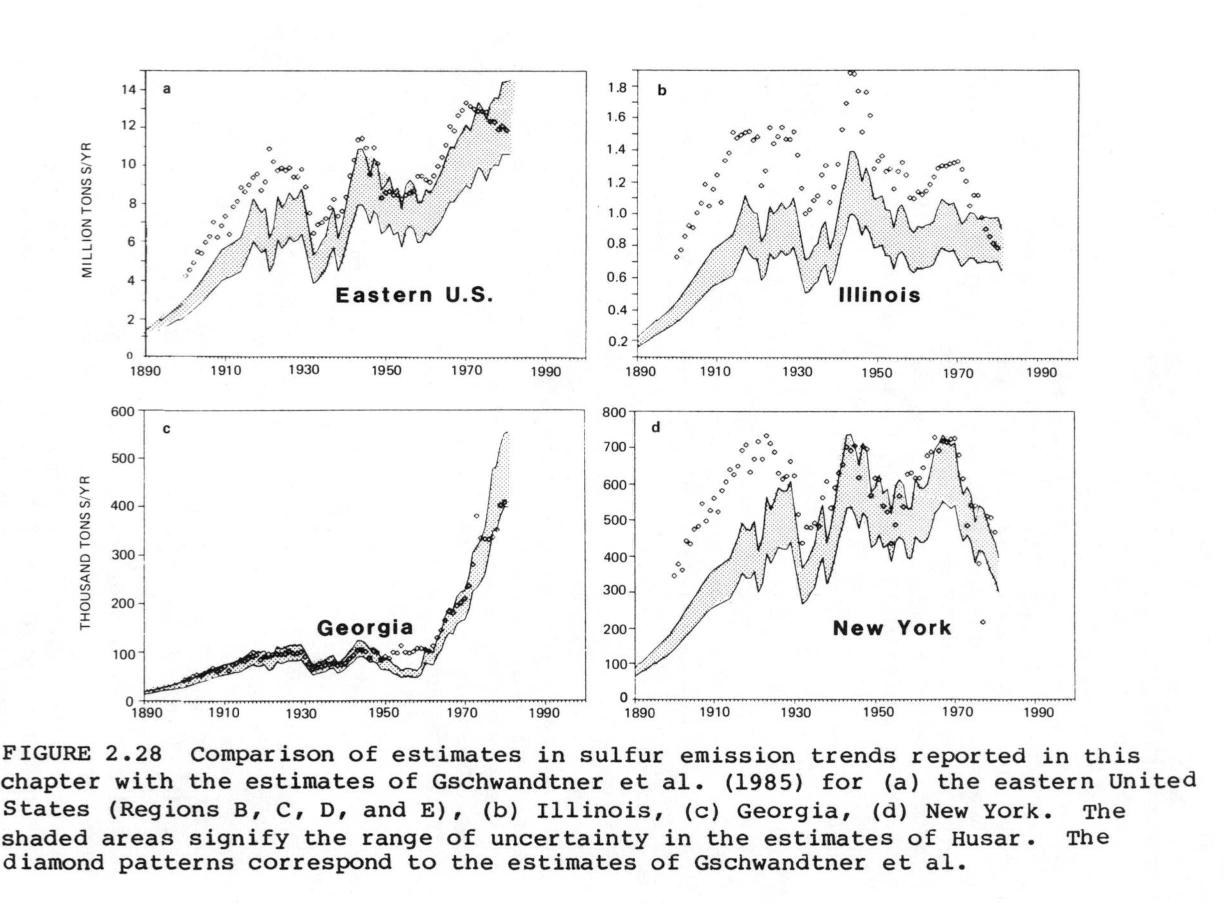

FIGURE 2.28 Comparison of estimates in sulfur emission trends reported in this chapter with the estimates of Gschwandtner et al. (1985) for (a) the eastern United States (Regions B, C, D, and E), (b) Illinois, (c) Georgia, (d) New York. The shaded areas signify the range of uncertainty in the estimates of Husar. The diamond patterns correspond to the estimates of Gschwandtner et al.

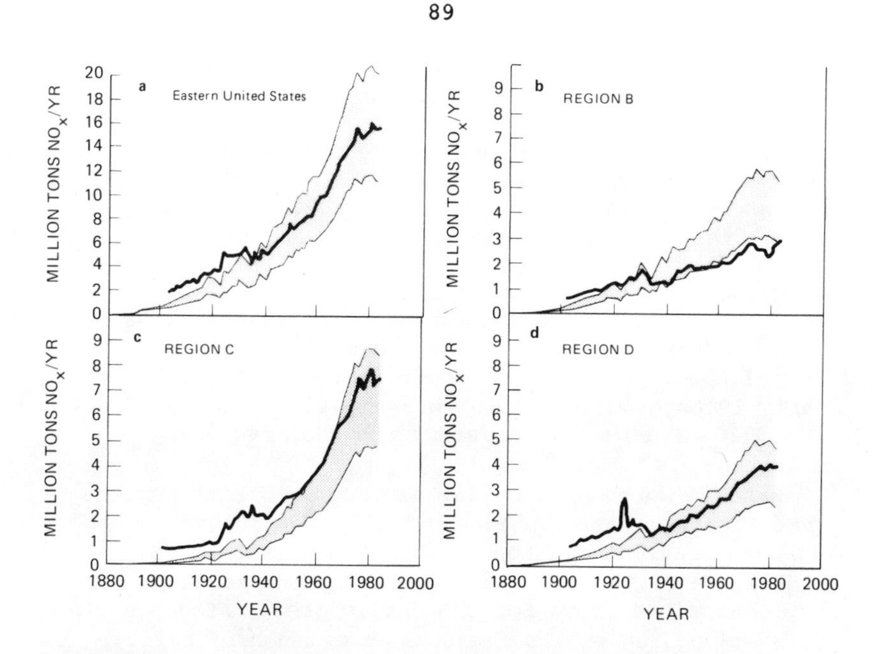

FIGURE 2.29 Comparison of estimates in nitrogen oxide emission trends reported in this chapter with the estimates of Gschwandtner et al. (1985) (dark lines) for (a) the eastern United States (Regions B, C, D, and E); (b) Region B; (c) Region C; (d) Region D.

SUMMARY

Sulfur emission trends for the eastern half of the United States and southeastern Canada are reconstructed for the past 100 years (1880-1980). The approach adopted uses a sulfur-flow accounting scheme from production in mines through distribution over land to emission sources. The average sulfur content of coal consumed in each state is estimated in this manner. The state-by-state sulfur emissions are calculated as the product of coal consumed and the average sulfur content. Coal combustion accounts for most of the current sulfur emissions, which are estimated to be between 11 and 15 million tons of sulfur/yr.

Sulfur mobilization by oil products increased from the 1940s until about the 1960s, when it leveled off at 3 to 4 million tons of sulfur/yr. Since 1978 there has been a significant reduction of oil sulfur emissions that is attributed to reduced oil imports and increased sulfur recycling at refineries.

The sulfur mobilization from copper and zinc smelting fluctuated from 0.5 to 1.5 million tons of sulfur/yr since the turn of the century. These emissions have dropped significantly since 1970 as a consequence of reduced sulfur production and increased sulfur recovery.

In eastern North America, sulfur dioxide emissions showed the most rapid rate of increase in the period from approximately 1880 to 1910. Since then sulfur emissions have increased overall by about 50 percent, but the increase has not been monotonic. Rather, the emissions have fluctuated between peaks and dips as a result of social, political, and economic factors. The 1920s, early 1940s, and late 1960s were "peak" periods, whereas the 1930s and the 1950s were periods of declining emissions.

Eastern North America was divided into five regions (see Figure 2.23). Until about 1970, emissions in each of these regions, although differing in magnitude, exhibited for the most part similar patterns of peaks and dips described above for the entire area. After about 1970 strong regional differences in trends of sulfur dioxide emissions emerged. The most distinctive differences were between the regions north and south of the Ohio River. Southeastern Canada (Region A) and the northeastern United States (Region B) show distinct downward trends. The southeastern United States (Region C) exhibits a steeply increasing trend. The midwestern United States (Region D) shows little change or perhaps a slightly increasing trend in sulfur dioxide emissions, and emissions in the north central United States (Region E) have remained low.

Nitrogen oxides are produced during combustion of fossil fuels and arise mainly through the fixation of atmospheric nitrogen at high temperatures. Thus, nitrogen oxide emissions depend primarily on the combustion process rather than on fuel properties. Hence, the nitrogen oxide emission trends have substantially higher uncertainties than those for sulfur dioxide. The regional nitrogen oxide emission trends indicate that in the states both north and south of the Ohio River, emissions have increased monotonically since the turn of the century. The northern states (Regions B, D, and E) show a roughly linear increase since that time, while the southern states (Region C) exhibit an exponential increase.

A comparison is made of trends in emissions of sulfur dioxide obtained in our analysis with those from a recent comprehensive report. Similarities and differences are examined. A similar comparison is presented for nitrogen oxide emissions.

ACKNOWLEDGMENTS

The development of the SO_x and NO_x emission inventories was supported by the Washington University School of Engineering and Applied Sciences and by the National Academy of Sciences. Janja Djukic Husar was instrumental in the acquisition of the data sets. Her help is greatly appreciated.

REFERENCES

Beaton, J. D., D. W. Bixby, S. L. Tisdale, and J. S. Platou. 1974. Fertilizer sulfur, status and potential. U.S. Technical Bulletin No. 21, The Sulfur Institute, Washington, D.C., and London.

Carrales M., Jr., and R. W. Martin. 1975. Sulfur content of crude oils. Information Circular 8676. U.S. Bureau of Mines, Department of Interior, Washington, D.C.

Energy Information Administration. 1977-1982. U.S. Department of Energy, Washington, D.C. Quarterly Reports, June 1977 to December 1982.

Energy Information Administration. 1981. Content in coal shipments, 1978. U.S. Department of Energy, Washington, D.C. DOE/EIA-0263(78).

Energy Information Administration. 1983. Coal distribution, January-December, 1982. U.S. Department of Energy, Washington, D.C. DOE/EIA-0125(82/4Q).

Gschwandtner, G., K. C. Gschwandtner, and K. Eldridge. 1985. Historic emissions of sulfur and nitrogen oxides in the United States from 1900 to 1980. Volume I. Results. U. S. Environmental Protection Agency. EPA-600/7-85-009a.

Hamilton, P. A., D. H. White, Jr., and T. K. Matson. 1975. The reserve base of U.S. coals by sulfur content. 2. The western states. Information Circular 8693, U.S. Bureau of Mines, Department of the Interior, Washington, D.C.

Husar, R. B., J. M. Holloway, D. E. Patterson, and W. E. Wilson. 1981. Spatial and temporal pattern of eastern U.S. haziness: a summary. Atmos. Environ. 15:1919-1928.

Lesher, C. E. 1917. Coal in 1917. U. S. Geological Survey. Part B. Mineral Resources of the United States. Part II, pp. 1908-1956.

Thompson, D. R., and H. F. York. 1975. The reserve base of U.S. coals by sulfur content. 1. The eastern states. Information Circular 8680, U.S. Bureau of Mines, Department of the Interior, Washington, D.C.

Tryon, F. G. and H. O. Rogers. 1927. Consumption of bituminous coal. U.S. Geological Survey, Mineral Resources of the United States. Part II, pp. 1908-1956.

United States-Canada Memorandum of Intent on Transboundary Air Pollution. Atmospheric Sciences and Analysis Work Group 2. 1982. Report No. 2F-M.

U.S. Bureau of the Census. 1889. Census of Manufacturing, Washington, D.C.

U.S. Bureau of the Census. 1919. Census of Manufacturing, Washington, D.C.

U.S. Bureau of the Census. 1975. Historical statistics of the United States, colonial times to 1970. U.S. Department of the Interior, Washington, D.C., pp. 587-588.

U.S. Bureau of Mines. Minerals Yearbook. U.S. Department of the Interior, Washington, D.C. Annual Publications, 1933-1980.

U.S. Bureau of Mines. Minerals Yearbook. U.S. Department of the Interior, Washington, D.C. Annual Publications, 1944-1980.

U.S. Bureau of Mines. Distribution of bituminous coal and lignite shipments. U.S. Department of the Interior, Washington, D.C. Quarterly Publications, 1957-1977.

U.S. Bureau of Mines. 1971. Control of sulfur oxides, emissions, in copper, lead, and zinc smelting (with list of references). Information Circular 8527, U.S. Department of the Interior, Washington, D.C.

U.S. Bureau of Statistics. 1917. Statistics of railways in the United States. Interstate Commerce Commission, Washington, D.C.

U.S. Environmental Protection Agency. 1977. Compilation of Air Pollutant Emission Factors. AP-42, 3rd ed. (NTIS PB-275525), Supplements 1-7 and 8-14. Springfield, Va.: National Technical Information Service.

U.S. Environmental Protection Agency. 1978. Mobile source emission factors. EPA-400/9-78-005 (NTIS PB295672/A17), Washington, D.C.

U.S. Geological Survey. 1880-1932. Mineral Resources of the United States. Yearbooks, U.S. Department of the Interior, Washington, D.C.

3
Uncertainties in Trends in Acid Deposition: The Role of Climatic Fluctuations

Raymond S. Bradley

INTRODUCTION

Acid deposition is the end product of a series of complex processes involving emission of precursors, chemical transformations in the atmosphere, physical transport of the pollutants through the atmosphere, and eventual dry or wet deposition. Day-to-day changes in weather play an important role in this sequence of events, since meteorological conditions at the time the pollutants are emitted determine the direction and the rate of both horizontal and vertical dispersal. Subsequent changes in these conditions determine whether precipitation will occur and to what extent the atmosphere will be cleansed of its pollutant load.

Considerable attention has focused on the influence of meteorological processes on acid deposition, with particular emphasis on source-receptor relationships to determine the origin and the route of transport of deposited materials (National Research Council 1983). In assessing changes in acid deposition over time scales longer than seasons, however, short-term meteorological variability is less significant than longer-term climatic variability. Climate is a statistical expression of daily weather events; but, over time periods of years and decades, changes in climate do occur, and these may be of considerable importance when evaluating long-term records of acid deposition. Clearly, temporal and spatial variations in atmospheric circulation over extended periods may directly influence patterns of deposition. Climatic fluctuations may also bring about changes in ecological conditions that could either accentuate or mask the effects of acid deposition. Indeed, climatic fluctuations could induce changes that might be confused with the effects of acid deposition.

Climatic variability is thus a pervasive factor in all aspects of the problem of detecting long-term trends in acid deposition. Consequently, any long-term records of acid deposition, whether they be direct or surrogate measures, must be evaluated in the context of climatic fluctuations over the same period. To this end, this chapter summarizes certain aspects of climatic variability in the eastern United States over the past 50 to 100 years that may be relevant to evaluating trends in acid deposition and its effects.

CYCLONE TRACKS

The occurrence of precipitation and precipitation-bearing weather systems is of prime interest to those studying acid deposition, particularly wet deposition. Much attention has centered on meteorological conditions preceding acid rain events, generally in an attempt to understand source-receptor relationships. This chapter focuses not on this aspect but rather on the general synoptic weather conditions that produce precipitation in the eastern United States.

Besides summertime convective storms, which are often dispersed spatially, the primary sources of precipitation in the eastern United States are organized frontal system associated with low-pressure centers (cyclones) traveling eastward across the region. Airflow into these systems carries polluted emissions far from the source area. It is the flow of air from different directions into the storm systems that carries polluted emissions away from the source area. Large interannual changes in the primar tracks that precipitation-producing systems follow, and hence changes in the flow of air into these systems, will affect interannual deposition patterns. Similarly, longer-term changes in the tracks of low-pressure systems and in the total number of cyclones per year will affect long-term airflow patterns and acid-deposition trends.

By studying the movement of cyclones over long periods of time, the primary tracks that they follow can be mappe (Figure 3.1). These maps, however, provide statistical summaries only and obscure the marked variability in storm tracks that characterize any one year. Not only may the actual tracks followed by individual storms and the associated airflows vary markedly from month to

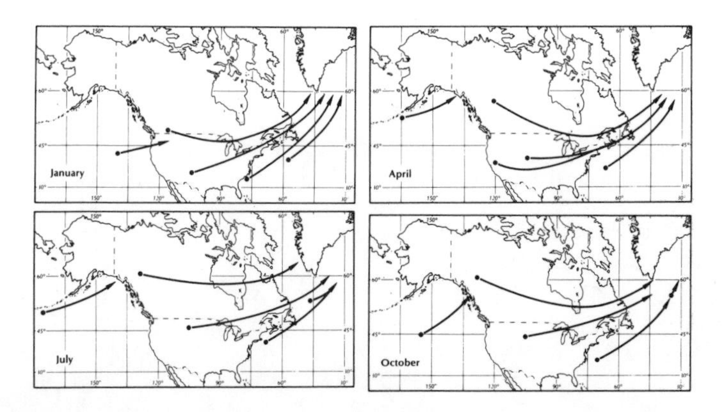

FIGURE 3.1 Primary depression tracks for four months of the year based on synoptic weather charts for 1951-1970. SOURCE: Reitan (1974).

month, but the total number of storms affecting the region per year may vary over time (Figure 3.2).

What is the significance of these changes for acid deposition? The processes by which material passes from source to receptor are complex and not fully understood. Every weather situation is unique and must be interpreted individually. Nevertheless, changes in airflow over a region that accompany long-term changes in cyclone tracks and associated frontal systems must play a fundamental role in producing temporal and spatial variations in acid deposition. This aspect of climatic variability deserves further study to quantify its potential significance in assessing trends in acid deposition.

PRECIPITATION AND DROUGHT

Since acid deposition results from both wet and dry processes, variations in the amount of precipitation and in the frequency and the size of precipitation events must play an important role in long-term trends in acid deposition. However, there has been surprisingly little research on this matter.

Precipitation records for the eastern United States are characterized by fairly large year-to-year variations of precipitation superimposed on long-term trends. In the Adirondack Mountains, for example, precipitation in some years is approximately 50 percent higher than in

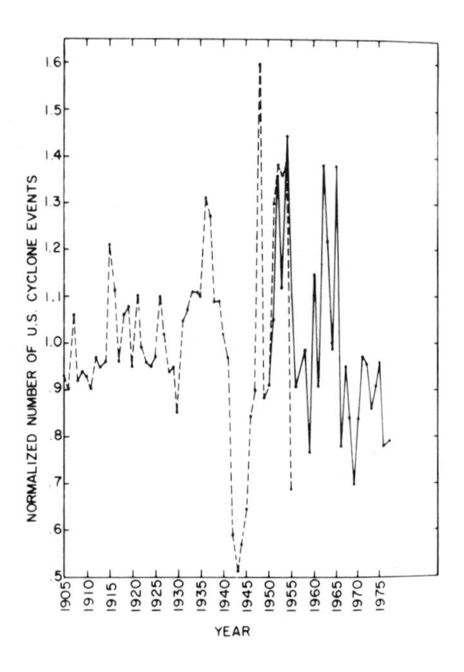

FIGURE 3.2 Normalized number of U.S. depression events, 1905-1977 (after Hosler and Gamage (1956) and Zishka and Smith (1980)).

others, and precipitation may be above or below the long-term average for several consecutive years (Figure 3.3). Furthermore, the _type_ of precipitation event may be significant to total acid deposition. In some areas, the frequency of days with heavy rain has increased markedly over the past 20 to 30 years with associated changes in cloudiness (Changnon 1983). Large precipitation events (greater than 1 in. (25 mm)) tend to have higher pH than smaller precipitation events but greater impact on total acid deposition (Likens et al. 1984). Thus, even if total annual precipitation were identical in any two years, differences in wet deposition could arise from differences in the size and the frequency of precipitation events. Unfortunately, little is known about the extent to which year-to-year variations in the number and size of precipitation events affect total acid deposition, principally because reliable long-term precipitation chemistry data sets are not available.

Changes in cloud cover (cloud type and extent) have not yet been studied in any detail. However, the frequency of convective clouds may be important in redistributing pollutants from low to high altitudes, thereby promoting their long-range transport. Further study of this aspect of climatic variability is warranted.

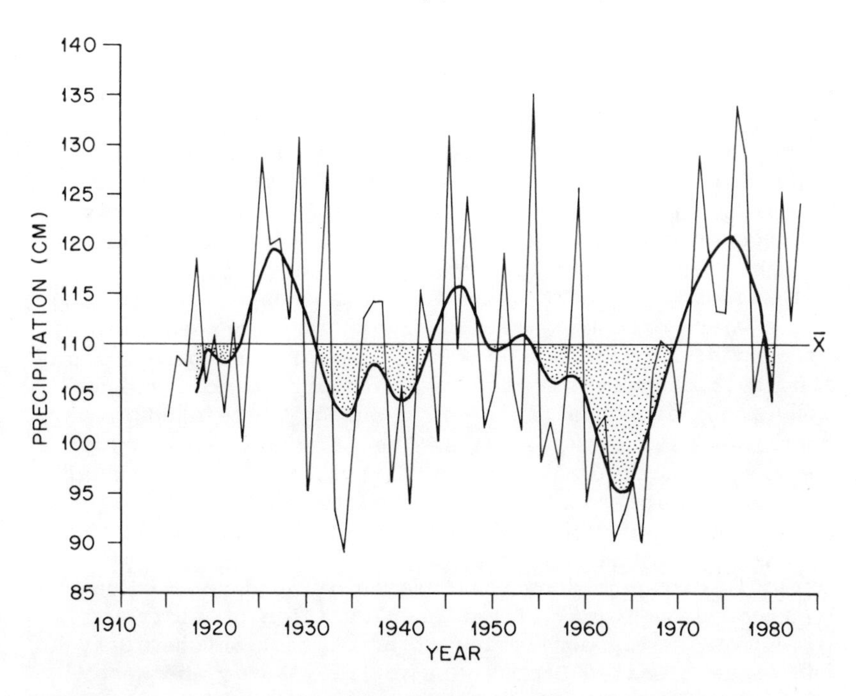

FIGURE 3.3 Annual precipitation in the Adirondack Mountains, New York. Horizontal line ($\overline{X}$) is the long-term average. Darker line shows low-frequency trends in the record (based on an 11-point binomial filter). Shaded areas indicate years of lower than average precipitation.

The occurrence of individual precipitation events with respect to the timing of sample collection may be of considerable importance in comparing historical data sets, particularly of stream-water chemistry. The acid shock phenomenon--low-pH water at the start of snowmelt--is fairly well known (National Research Council of Canada 1981), but even individual precipitation events during summer months can produce sharp reductions in pH similar to spring snowmelt changes (Kurtz et al. 1984). Thus, it is important to take into account the recent precipitation history of a site (i.e., the period before sampling) in comparing discrete stream-water samples.

Just as the amount of precipitation is important in evaluating trends in acid deposition, so is the absence of precipitation. Drought conditions, defined as periods of anomalously low precipitation (generally coupled with above average temperatures) are relatively common occurrences in some areas of the country, such as in the upper Midwest; in other areas, New England and New York for example, drought is a relatively uncommon event.

Wet deposition is reduced during droughts, but dry deposition may increase. In assessing trends in acid deposition and its effects, the ecological consequences of drought may be most significant. In those areas where droughts are relatively common, ecosystems capable of withstanding periodic water shortages have evolved. In areas where droughts are relatively rare, the ecological effects of drought are likely to be more profound and may result in changes and readjustments within ecosystems long after droughts per se have ended. Such changes may complicate attempts to assess the effects of acid deposition on ecosystems. In particular, studies of annual tree growth increments as a surrogate measure of the effects of acid deposition on biomass production may not differentiate unequivocally the long-term effects of periodic droughts from the effects of acid deposition.

Also, drought may affect precipitation chemistry directly. Early monitoring of precipitation chemistry in the eastern United States (Junge 1956, Junge and Werby 1958) provides an important data set with which to compare recent measurements. A direct comparison reveals significantly lower pH values over most of the East in the late 1970s compared with those approximately 25 years earlier. However, drought in the Midwest during the mid-1950s and consequent windblown dispersal of aerosols rich in calcium and magnesium may have resulted in overestimating pH values for this interval (Figures 3.4(a) and 3.4(b)). By adjusting the calculated pH values to take "excess" calcium and magnesium into account, changes in pH between the 1950s and late 1970s appear to be less than reported previously for most of the eastern United States (compare Figures 3.4(b) and 3.4(c)). Although one could take issue with the methods employed in making such adjustments and with the many assumptions necessary (Butler et al. 1984, Stensland and Semonin 1984), it is nevertheless clear that climatic variations over years and decades may have consequences that preclude directly comparing data for different time periods.

The severity of a drought is commonly expressed in terms of the Palmer Index (P.I.) (Palmer 1965). The occurrence of drought for at least two consecutive months with a P.I. of less than 3 is considered a severe event. The frequency of such events in the eastern United States

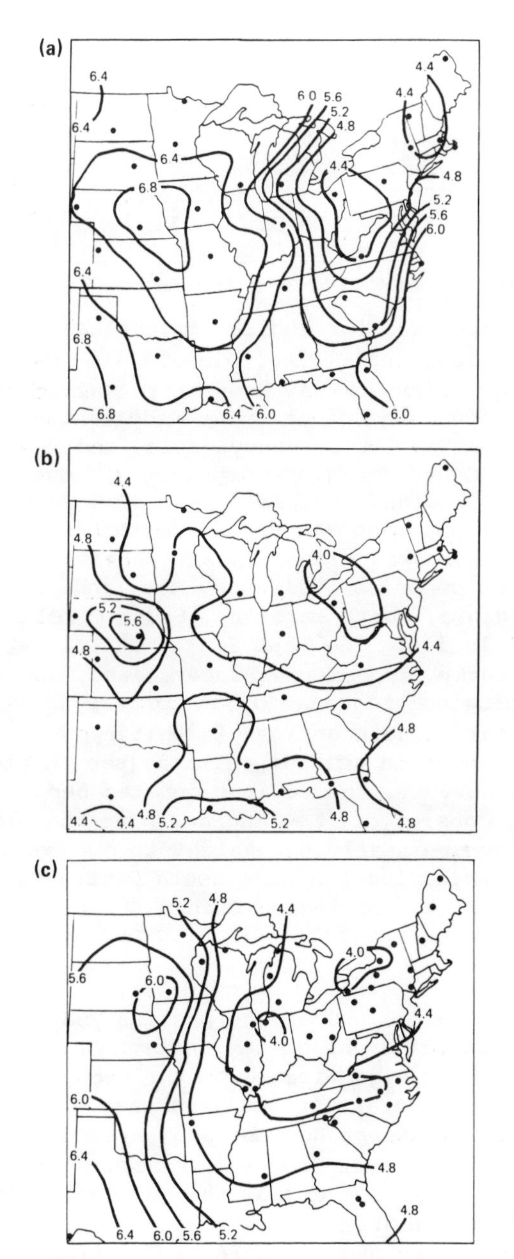

FIGURE 3.4 1955-56 pH distribution; (a) based on adjusted Junge data; (b) after correcting for assumed anomalously high concentrations of calcium and magnesium; and (c) September 1980 median pH distribution from National Acid Deposition Program network. See Chapter 5 for a detailed discussion of these data (Stensland and Semonin 1982).

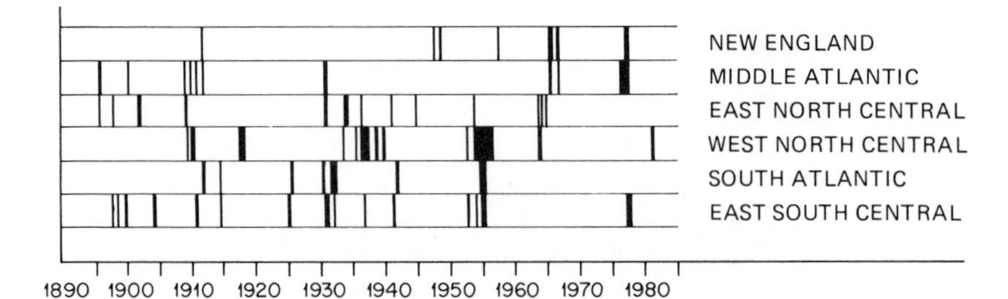

FIGURE 3.5 P.I.s (<3) for regions of the eastern United States, 1895-1981 (after Diaz 1983). New England comprises Maine, Vermont, New Hampshire, Massachusetts, Rhode Island, and Connecticut. The Middle Atlantic region comprises New York, Pennsylvania, and New Jersey. The East North Central region comprises Wisconsin, Michigan, Ohio, Indiana, and Illinois. The West North Central region comprises Minnesota, North Dakota, South Dakota, Iowa, Nebraska, Missouri, and Kansas. The South Atlantic region comprises Delaware, Maryland, West Virginia, Virginia, North Carolina, South Carolina, Georgia, and Florida. The East South Central region comprises Kentucky, Tennessee, Alabama, and Mississippi. These climatological regions do not correspond exactly to the regions that we have defined in this report to characterize trends in acid deposition (see Chapter 1, Figure 1.2). However, the regions denoted here are somewhat analogous; i.e., New England plus the Middle Atlantic States are nearly equivalent to our Region B, the South Atlantic plus the East South Central States are nearly equivalent to our Region C, and the East North Central States are somewhat similar to our Region D.

is shown in Figure 3.5. Over the past 90 years droughts have occurred in all areas of the eastern United States at some time. Most significant was the drought of the 1930s, which began to affect the area east of the Mississippi River, except New England, around 1930. The drought later became centered over the North Central States, lasting until the end of the decade in the West North Central region (see Figure 3.5). In the 1950s, a major but less severe drought affected an area from the South Atlantic seaboard to the upper Midwest; once again New England for the most part escaped its effects, as did the two other regions of the North (the Middle Atlantic and East North Central States). In the mid-1960s, however, an exceptionally severe drought affected New England and the Middle Atlantic States. Dendroclimatic studies have reconstructed drought severity in the Hudson River Valley since the seventeenth century (Figure 3.6), and from this work it appears that the 1960s drought was one of the most severe in the past 300 years (Cook and Jacoby 1979). Interestingly, this was a cool drought accompanied by below-average temperatures and anomalous northwesterly airflow. The most recent drought (1976-1977) also affected New England and the Middle Atlantic States but was not so extreme as the 1960s drought.

AIR STAGNATION EPISODES

It is well known that high air pollution levels are often associated with certain weather conditions, the most noteworthy of which are slow-moving anticyclones (high-pressure systems). Anticyclones are generally associated with subsiding air motion, clear skies, and strong nighttime radiative cooling, giving rise to pronounced temperature inversions that limit vertical dispersal of pollutants. Since pollutants are not readily dispersed during air stagnation episodes, relatively high levels of major contaminants, such as ozone, may occur for extended periods during these events. Thus, in evaluating the relationship between ecological indicators and acid deposition one should account for long-term changes in the frequency of air stagnation events in each region under study.

In addition, changes in air stagnation trends may play a role in changes in regional visibility over time. Visibility is also affected by variations in temperature and relative humidity, since hygroscopic sulfate aerosols, a primary ingredient in reducing visibility, absorb water from the air. Comparisons of long-term visibility data should thus account for variations in these parameters (Sloane 1983, 1984). This is discussed further in Chapter 4.

The frequency of air stagnation episodes in the eastern United States has been studied by Korshover (1976), who found the maximum frequency of such cases to be centered over Georgia and South Carolina (Figure 3.7). Seasonally, the fall (August-October) has the maximum number of air stagnation cases. An analysis of the number of stagnation cases each year in the eastern United States over the past 50 years reveals an increase

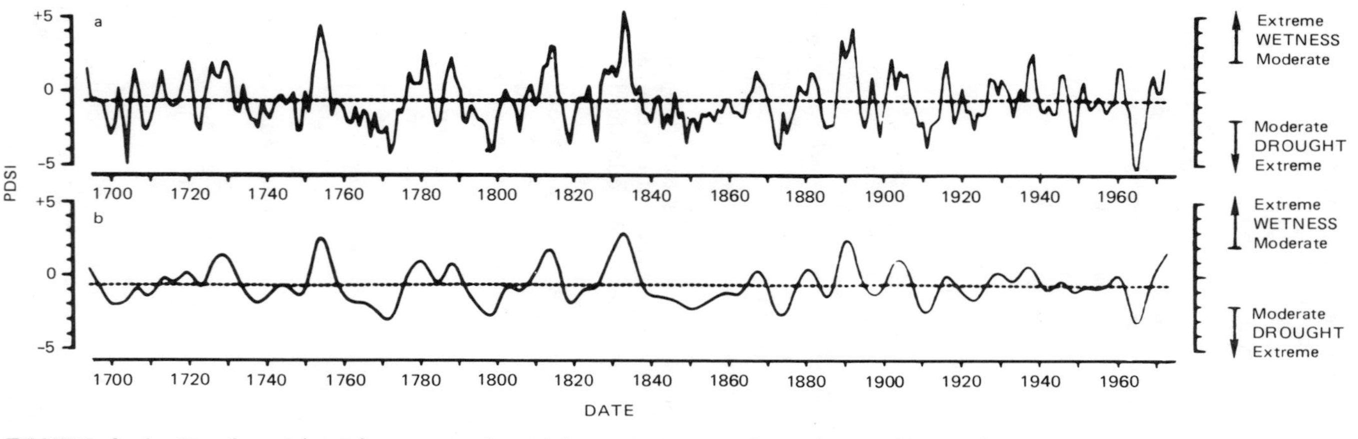

FIGURE 3.6 Dendroclimatic reconstruction of P.I.s for the Hudson River Valley area. The more negative the index the more severe the drought event; (a) raw data, (b) smoothed data. PDSI, means Palmer Drought Severity Index. SOURCE: Cook and Jacoby (1979).

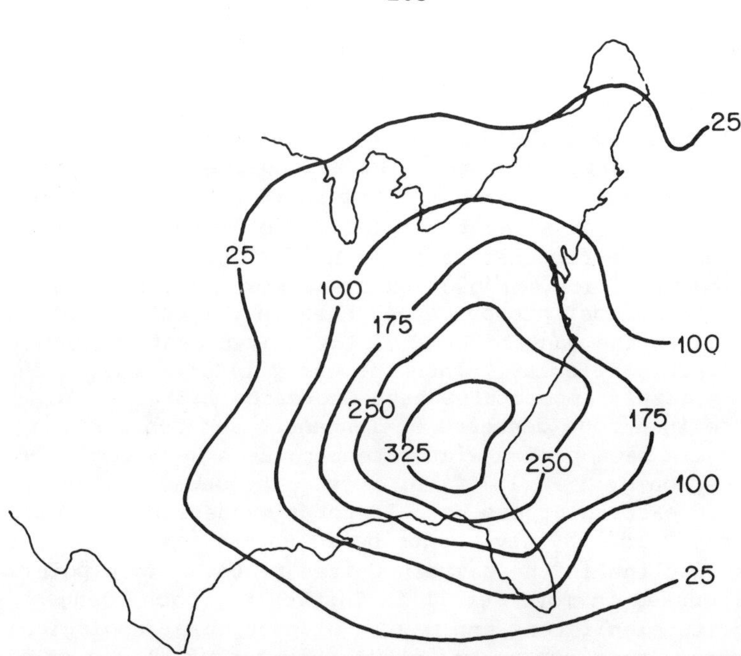

FIGURE 3.7 Geographical distribution of number of air stagnation days in the eastern United States over the period from 1936 to 1970 (after Korshover 1976).

from a low of 30 days/yr in 1936 to 1940 to a high of 57 days/yr in 1979 to 1983 (Figure 3.8). This does not imply that all parts of the country experienced an increase in stagnation days over the 50 years, but it does mean that the total number of cases within the entire eastern United States has increased. Nevertheless, it is clear that air stagnation conditions are variable in time and space and thus play a role in evaluating spatial and temporal trends in acid deposition and its effects.

TEMPERATURE

Changes in temperature over time are not of direct significance for trends in acid deposition. Nevertheless, they may affect a biological indicator, such as tree growth, and thereby confound efforts to isolate unequivocally any acid deposition signal in the tree ring records. (See Chapter 6.)

An analysis of long-term temperature data for the United States shows that the record since 1895 divides into three periods (Diaz and Quayle 1980) (Figure 3.9). The first period (1895 to 1920) was characterized by relatively low temperatures, the second period (1921 to 1954) by a warming trend, and the third period (1955 to 1979 or later) by a return to cooler conditions. When the changes in temperature between these periods are examined by region, it becomes clear that over the past 60 years the most pronounced changes in temperature have occurred in the eastern half of the United States (Figure 3.10). In particular, since the early 1920s summer (June-August) temperatures have decreased 1.5°F (0.8°C) or more in a zone centered on Tennessee and Kentucky. During the same period winter temperatures declined in this region by 3°F (1.7°C) or more. The decade of the 1960s in particular was notable for a succession of cold winters, which may have contributed to the onset of forest decline in the eastern United States. This point is discussed in more detail in Chapter 6. Such changes are quite significant and may be of particular ecological importance near the normal geographical range limit of a species or at higher elevations. It is also in these areas, where vegetation is under environmental stress, that increases in acid deposition might have the greatest impact. Thus, separating the climatic and acid deposition signals in vegetation can be complex and may produce equivocal results.

The changes in mean monthly or seasonal temperature that have been observed are only one isolated measure of climate. It is probable that these temperature changes reflect adjustments in large-scale circulation patterns of the eastern United States involving a multitude of other, more subtle changes in climate, such as length of the growing season, frequency of frosts, growing degree days, type and amount of cloudiness, vapor pressure, net radiation, and wind direction. Such changes may be significant ecologically, yet they are rarely subjected to careful long-term analysis because adequate data sets are lacking. Even with good long-term data and reliable models of the important interactions, it would be extremely difficult to isolate the multiplicity of effects that such changes could have on ecosystems from the additional influence of acid deposition. As it is, both data and models are generally inadequate as a means of resolving the unique effects of any particular factor.

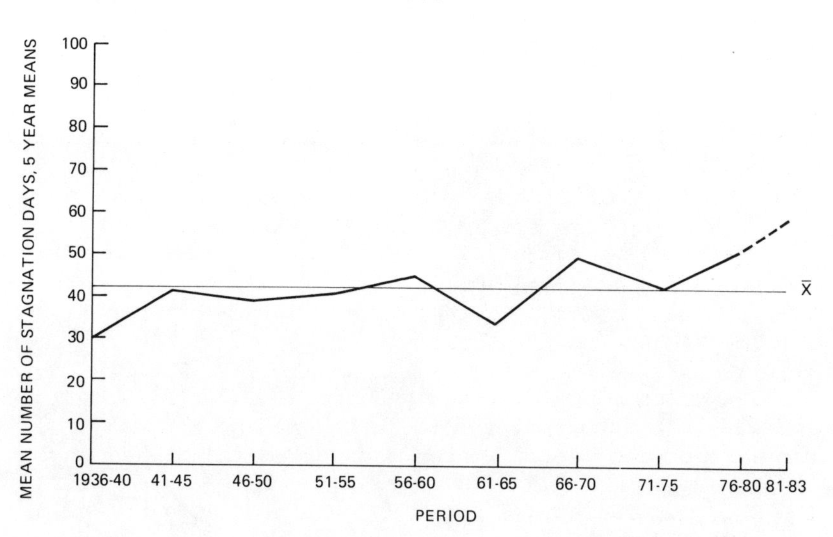

FIGURE 3.8 Air stagnation days in the eastern U.S., 1936-1983. Horizontal line ($\overline{X}$) is the long-term average. Dashed line signifies the 3-year mean from 1980 to 1983 (data courtesy of J. Korshover, Air Resources Laboratory, NOAA, Silver Spring, Md.).

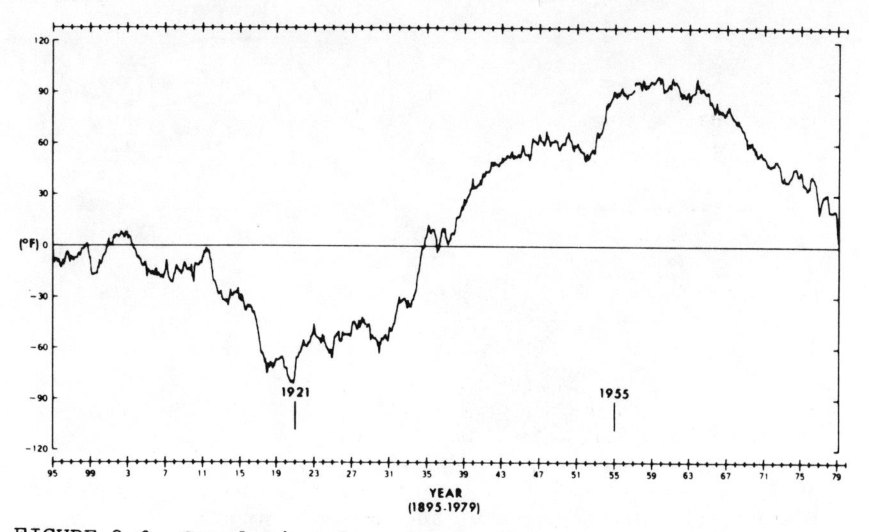

FIGURE 3.9 Cumulative departures of long-term mean monthly temperature in the contiguous United States; monthly means compiled from January 1895 to March 1979. SOURCE: Diaz and Quayle (1980).

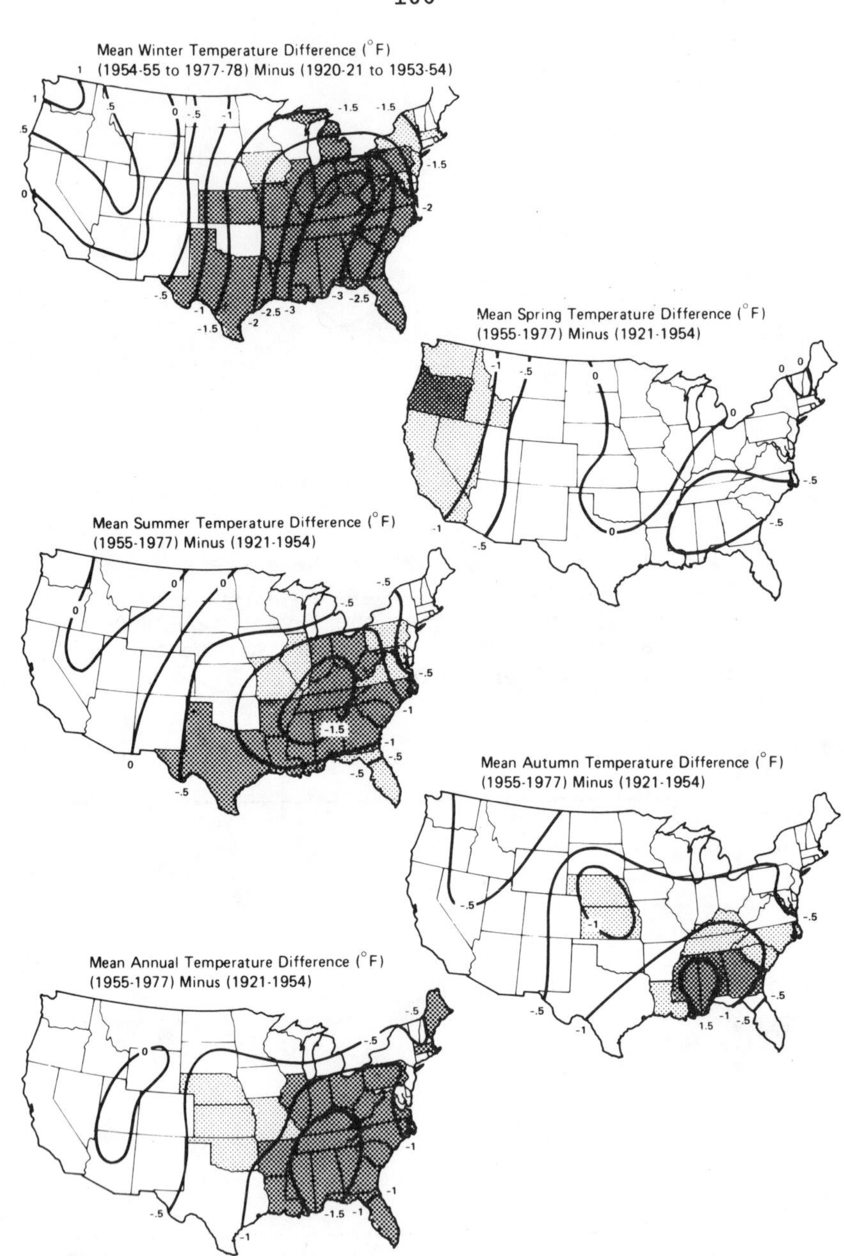

FIGURE 3.10 Mean seasonal and annual temperature differences (°F) for (1955-1977) minus (1921-1954). Shading indicates significance at the 99% (dark) and 95% (light) levels, respectively (after Diaz and Quayle 1980).

SUMMARY

Acid deposition is a consequence of atmospheric processes acting on air pollutants. As such, the phenomenon is subject to the variability of the atmospheric system on both short (meteorological) and long (climatological) time scales. In assessing trends in acid deposition, climatic trends must be an implicit part of the assessment. Furthermore, in trying to understand the effect that acid deposition may have had on various ecosystems it is equally important to have a clear understanding of the effects of climatic variations on the ecosystems. Because this type of analysis has rarely been carried out, research should focus on the following topics:

1. The effects that climatic variability may have on those systems thought to be affected by acid deposition, such as aquatic and forest ecosystems.
2. The implications of long-term climatic changes for trends in acid deposition both in the past and in the future.

REFERENCES

Butler, T. J., C. V. Cogbill, and G. E. Likens. 1984. Effect of climatology on precipitation acidity. Bull. Am. Meteorol. Soc. 65:639-640.

Changnon, S. 1983. Trends in floods and related climatic conditions in Illinois. Climatic Change 5:341-363.

Cook, E. R., and G. C. Jacoby, Jr. 1979. Evidence for quasi-periodic July drought in the Hudson Valley, New York. Nature 282:390-392.

Diaz, H. F. 1983. Some aspects of major wet and dry periods in the contiguous United States, 1895-1981. J. Climatol. Appl. Meteorol. 22:3-16.

Diaz, H. F., and R. G. Quayle. 1980. The climate of the United States since 1895: spatial and temporal changes. Mon. Weather Rev. 108:249-266.

Hosler, C. L., and L. A. Gamage. 1956. Cyclone frequencies of the United States for the period 1905-1954. Mon. Weather Rev. 84:388-390.

Junge, C. E. 1958. The distribution of ammonia and nitrate in rainwater over the United States. Trans. Am. Geophys. Union 39:241-248.

Junge, C. E., and R. T. Werby. 1958. The distribution of chloride, sodium, potassium, calcium and sulfate in rainwater over the United States. J. Meteorol. 15:417-425.

Korshover, J. 1976. Climatology of stagnating anticyclones east of the Rocky Mountains, 1936-1975. NOAA Tech. Mem. ERL/ARL-55. National Oceanic and Atmospheric Administration. 26 pp.

Kurtz, J., A. J. S. Tang, R. W. Kirk, and W. H. Chan. 1984. Analysis of an acidic deposition episode at Dorset, Ontario. Atmos. Environ. 18:387-394.

Likens, G. E., F. H. Bormann, R. S. Pierce, J. S. Easton, and R. E. Munn. 1984. Long-term trends in precipitation chemistry at Hubbard Brook, New Hampshire. Atmos. Environ. 18:2641-2647.

National Research Council. 1983. Acid Deposition: Atmospheric Processes in Eastern North America. Washington, D.C.: National Academy Press. 375 pp.

National Research Council of Canada. 1981. Acidification in the Canadian Aquatic Environment: Scientific Criteria for Assessing the Effects of Acid Deposition on Aquatic Ecosystems. National Research Council Publication No. 18475 196.

Palmer, W. C. 1965. Meteorological drought. U.S. Weather Bureau Res. Paper No. 45. U.S. Department of Commerce. 58 pp.

Reitan, C. H. 1974. Frequencies of cyclones and anticyclones for North America 1951-1970. Mon. Weather Rev. 102:861-686.

Sloane, C. S. 1983. Summertime visibility declines: meteorological influences. Atmos. Environ. 17:763-774.

Sloane, C. S. 1984. Meteorologically adjusted air quality trends: visibility. Atmos. Environ. 18:1217-1229.

Stensland, G. J., and R. G. Semonin. 1982. Another interpretation of the pH trend in the United States. Bull. Am. Meteorol. Soc. 63:1277-1284.

Stensland, G. J., and R. G. Semonin. 1984. Response to: "Comments on effect of climatology on precipitation acidity." Bull. Am. Meteorol. Soc. 65:640-643.

Zishka, K. M., and P. J. Smith. 1980. The climatology of cyclones and anticyclones over North America and surrounding ocean environs for January and July 1950-1977. Mon. Weather Rev. 108:387-401.

4

Patterns and Trends in Data for Atmospheric Sulfates and Visibility

John Trijonis

INTRODUCTION

In this chapter, we examine the geographical patterns and historical trends in atmospheric sulfate concentrations and visibility. The relationship between particulate sulfate and acid deposition is a close one. Sulfur dioxide emissions, which contribute to acid deposition, also produce particulate sulfate. In fact, the sulfate particles are a fundamental component of acid deposition.

In the eastern United States, sulfate particles are also the dominant contributors to visibility reduction (light extinction or light attenuation). Thus, visibility in the East is indirectly related to acid deposition. Particulate sulfate is an important component of visibility reduction for two reasons. First, sulfate particles tend to form in the 0.1- to 1.0-micron size range, by far the most efficient size range for scattering visible light. Second, since sulfates are hygroscopic, they attract a substantial volume of water into the particulate phase, which further increases the mass of light-scattering aerosol. It is now well established that, on average, sulfates and associated water account for about 50 percent of visibility reduction in the East, slightly less in urban areas but somewhat greater in nonurban areas (Weiss et al. 1977; Trijonis and Yuan 1978; Leaderer and Stolwijk 1979; Pierson et al. 1980; Ferman et al. 1981; Wolff et al. 1982; Trijonis 1983, 1984). In fact, the most definitive program conducted at a nonurban site during the summertime concluded that sulfates and associated water accounted for three-fourths of the total light extinction during the study period (Ferman et al. 1981).

Some have even suggested that visibility observations, i.e., light-extinction levels, can serve as a quantitative surrogate for sulfates in the eastern United States on a day-to-day basis (Leaderer et al. 1979). More recent studies, however, indicate that visibility is not necessarily an accurate daily predictor of sulfates (Trijonis 1984). Nevertheless, the relationship still appears to hold for longer-term averages; geographical patterns and historical trends in average light-extinction levels in the East often reflect corresponding changes in average sulfate concentrations.

DESCRIPTION OF DATA BASES

Three types of nationwide data are examined in this chapter: visibility observations at a large number of airports, turbidity measurements at a few monitoring stations, and sulfate aerosol measurements at a few nonurban Environmental Protection Agency (EPA) sites.

The visibility data consist of visual range estimates made by trained observers at airport weather stations. To help to ensure the quality of these data, we have followed several steps discussed and recommended in the literature, such as using daytime measurements only, plotting cumulative percentiles according to appropriate methods, prescreening airports for adequate visibility markers, and checking patterns for consistency among various stations in a region (Trijonis and Yuan 1978; Husar et al. 1979; Sloane 1982). Some questions remain, however, regarding possible inconsistencies in the data because of variations in reporting practices, observer detection thresholds, and visibility markers.

Because spatial patterns in visibility usually involve much greater differences than historical trends, airport visibility data are considered to be of better quality for detecting spatial patterns than for detecting historical changes. Nevertheless, airport visibility data are generally considered to be fairly good for historical trend analysis, especially if care is taken to check for a general consistency for trends at similiar locations.

In this chapter, we often convert the visual range data (V) into estimates of light-extinction coefficient (B), using the Koschmeider formula, $B = (-\ln C_0)/V$. This relationship holds for a uniform atmosphere with an observer able to detect a minimal contrast of C_0. We have assumed a value of $C_0 = 0.02$, the so-called standard observer, although several studies have shown that visibility data from airport observers more closely correspond to $C_0 = 0.05$ (Aerovironment, Inc. 1983; Malm 1979; Middleton 1952; Douglas and Young 1945). The extinction levels reported here can be corrected to the latter value by dividing by 1.3.

The turbidity data were obtained from the EPA/NOAA network of Volz sunphotometers. Sunphotometers measure variations in the amount of direct sunlight lost before reaching the Earth's surface. Because the measurement is made only with an unobstructed line of site to the Sun, data are missing whenever clouds obscure the Sun, so that the number of observations varies with location and season. The turbidity data are similar to the visibility data, except that the sight path is vertical in the former rather than horizontal. Thus, turbidity depends not only on atmospheric extinction but also on the vertical depth of the extinction layer.

The sulfate data come from the EPA's National Air Surveillance Network (NASN), which not only collects particulate (Hi-Vol) filter samples at nonurban stations but also performs chemical analysis of those samples for sulfate ion and other constituents. A potential problem with the sulfate data concerns an artifact caused by sulfate formed from SO_2 absorbed on glass-fiber filters. The error introduced by this factor should not be of great concern in this study, however, for two reasons. First, the data are from nonurban sites, where gaseous SO_2 concentrations should be relatively low. Second, in terms of acid deposition, it is total sulfur loadings that are of interest and not just sulfate aerosol per se. It is not necessarily a liability to include SO_2 in the measurements unless the SO_2 collection introduces significant biases in either the regional distributions of sulfate or the trend of sulfate over time.

Because the results of these analyses are compared in Chapter 1 with other measures of overall acid deposition patterns in eastern North America, the data bases cited here pertain, whenever possible, to rural locations (as defined in World Aeronautical Charts, 1976). Since metropolitan areas can produce significant changes in sulfate concentrations and sometimes in visibility, it seems best to avoid large urban centers when trying to characterize overall regional patterns and trends.

Many of the data presented in this chapter represent gross annual or seasonal statistics with no presorting for the influence of meteorological factors. There are questions concerning the effect of climatic variations on geographical patterns and, especially, on historical trends. For example, the work of Sloane (1983, 1984) suggests that shifts in relative humidity and to some extent in temperature over the past 30 years may have affected visibility trends in the eastern United States substantially. Husar and co-workers, in providing the visibility data examined in this chapter, sought to minimize the influence of humidity by eliminating observations with fog or precipitation and normalizing for relative humidity. Nevertheless, as is noted in Chapter 3, questions remain regarding the potential effect on visibility trends of changes in stagnation episodes and temperature. Unfortunately, we cannot definitively resolve the issue of climatological trends and visibility with our present state of knowledge.

For the most part, this chapter relies on published data and analyses. The main exception is that Husar and coworkers have updated and revised some of their studies for this report. Their new work extends the analysis to 1983 and restricts it to nonurban sites.

GEOGRAPHICAL DISTRIBUTION

Figures 4.1-4.3 summarize the geographical distributions for visibility, atmospheric turbidity, and sulfate concentrations in the continental United States. Figure 4.1 is based on median values of visibility at 100 rural airports. Figure 4.2 presents annual averages from 26 stations of the EPA/NOAA atmospheric turbidity network. Figure 4.3 is based on annual averages for sulfate concentrations at 22 nonurban EPA monitoring sites.

The main spatial features of all three figures are similar. The best air quality occurs in the mountain/desert West, with fairly sharp gradients to worse air quality toward the east and the west. The highest sulfate concentrations and atmospheric extinction (lowest visibility) occur east of the Mississippi River and south of the Great Lakes. The location of worst air quality varies somewhat from figure to figure, but the Ohio Valley generally exhibits nearly the worst, if not the worst, air quality.

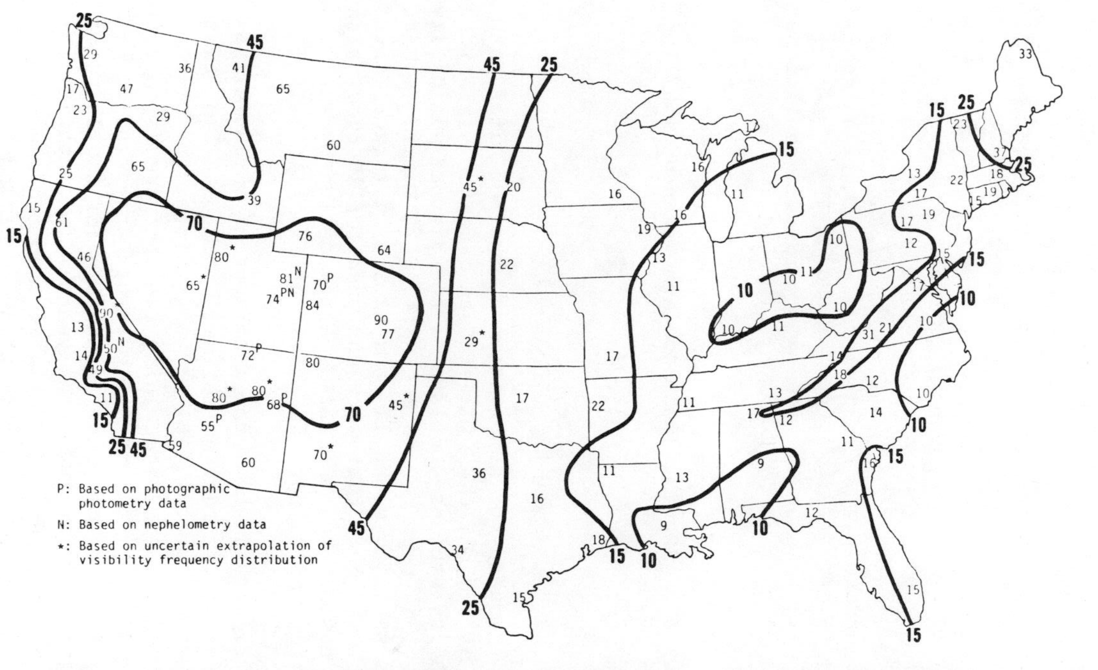

FIGURE 4.1 Median annual visual range (in miles) at suburban/nonurban locations in the United States (1974-1976). SOURCE: Trijonis (1982).

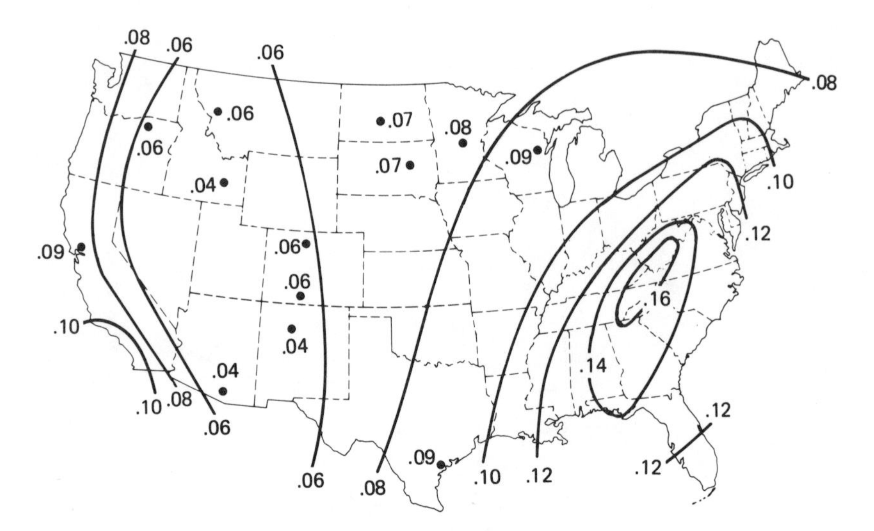

FIGURE 4.2 Annual average background turbidity (1967-1976). SOURCE: Robinson and Valente (1982).

HISTORICAL TRENDS

For frequency of observation, number of sites, and length of record, the richest data base for historical air quality studies consists of the airport visibility observations reported to the National Climatic Center. In this section, we shall look at the historical trends in that data base that were analyzed for 35 rural sites in the eastern United States, supplemented by visibility, turbidity, and sulfate data from a few other eastern locations.

Figure 4.4(a) shows the locations of the 35 sites according to the regions defined in Chapter 1 of this report, and Figure 4.4(b) shows the locations of the 15 NASN sites used in determining temporal trends in sulfate concentrations. Figure 4.5 presents the trend in 3-year running averages of light extinction for the 35 rural sites from 1949 to 1983 (the years with computerized historical records). The figure also includes an extrapolation of the light-extinction trend to 1942-1944 based on an analysis of hard-copy data at eight eastern rural locations and historical trends for nonurban sulfate concentrations at 15 NASN sites for 1965-1978.

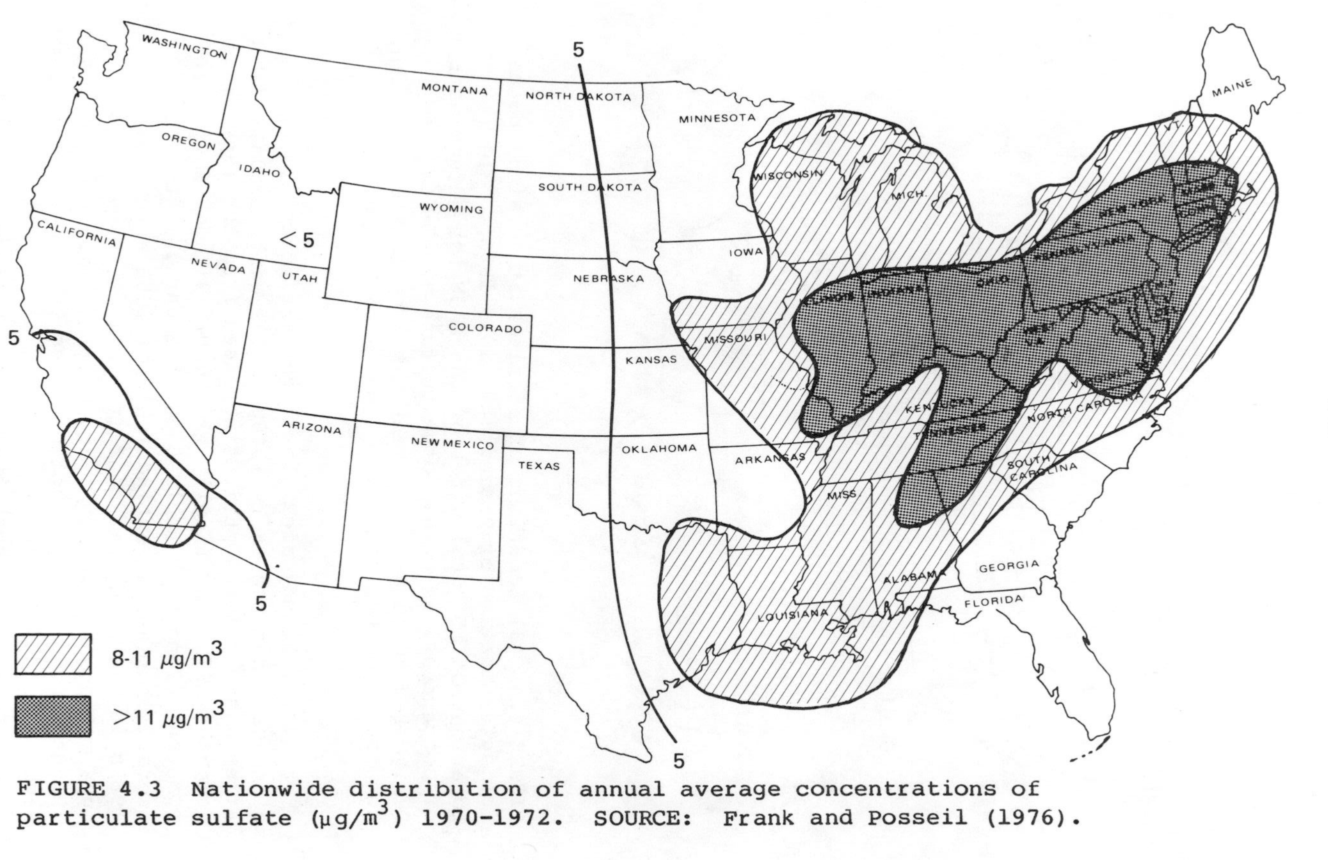

FIGURE 4.3 Nationwide distribution of annual average concentrations of particulate sulfate (μg/m^3) 1970-1972. SOURCE: Frank and Posseil (1976).

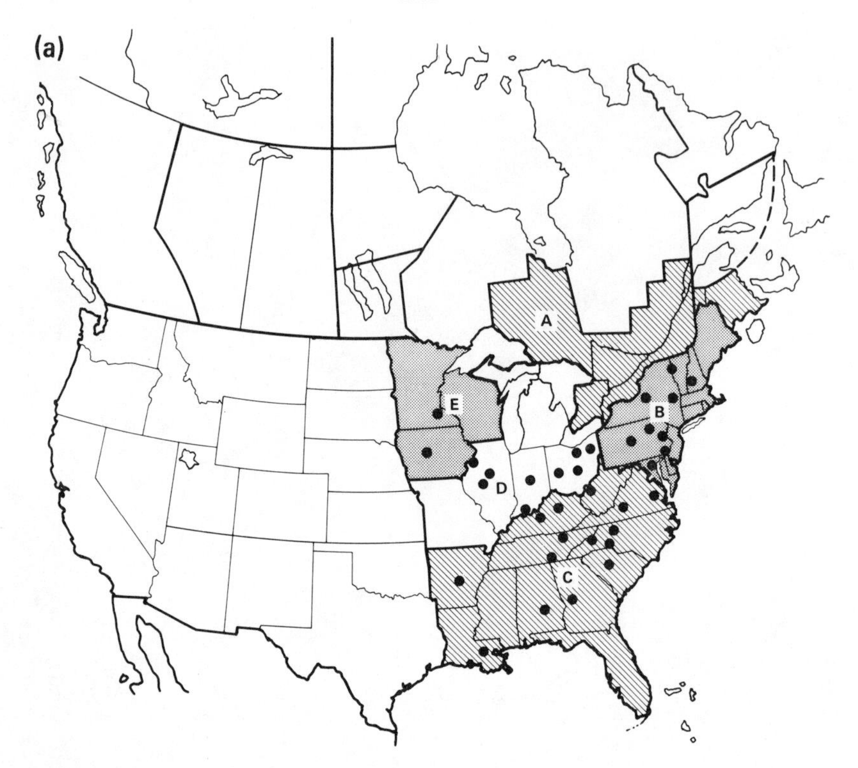

FIGURE 4.4 (a) Locations of 35 sites at rural airports for determination of visibility (1949-1983). (b) Locations of 15 rural sites of the NASN for determination of concentrations of airborne sulfate (1965-1978). Designated areas A-E represent the 5 regions defined in Chapter 1 of this report.

Figure 4.5 indicates that light-extinction levels decreased from the early 1940s to the early 1950s, then after a short level period increased from the late 1950s to about 1970, and finally fluctuated without any apparent trend between 1970 and the early 1980s. The overall increase in extinction from 1950 to the early 1980s is highly significant ($t = 6.1$). However, the apparent changes in slope after 1950, e.g., the seeming change around 1970, do not pass the 95 percent confidence level.

Figure 4.5 also shows that rural sulfate concentrations increased from the 1960s to the 1970s. The percentage increase in sulfates, however, was greater

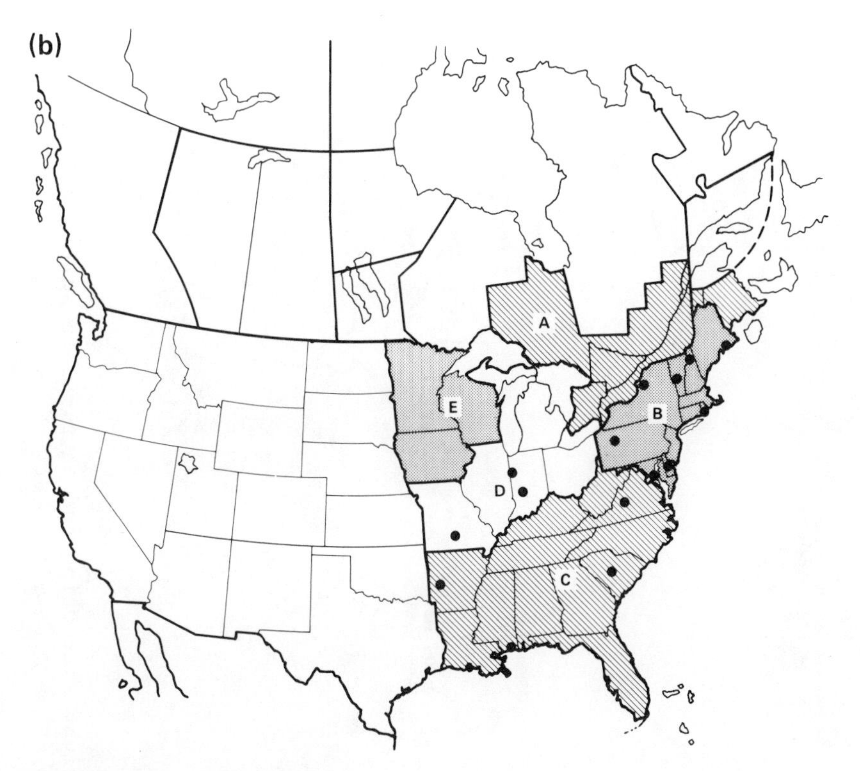

FIGURE 4.4 (continued).

than the percentage increase in light extinction. Recognizing that sulfates account for only about half of the total extinction in the East, this suggests that the sulfates may have been the dominant particulate component producing the light-extinction increase.

Figure 4.6 presents lengthy historical records that Husar et al. (1981) have compiled for two sites. One of these records--visibility observations at Blue Hill, Massachusetts--covers the years 1889 to 1958 with a recent addition in 1981-1983. The other record--turbidity at Madison, Wisconsin--covers the period 1916 to 1977. Comparing the two records reveals three notable features. First, the trends at the two locations are dissimilar between 1916 and the early 1940s. Second, there is fairly good agreement from the early 1940s to the late 1950s. Third, although the two trends seem divergent from the late 1950s to the present, it is hard to interpret a trend from the single Blue Hill data point for 1981-1983.

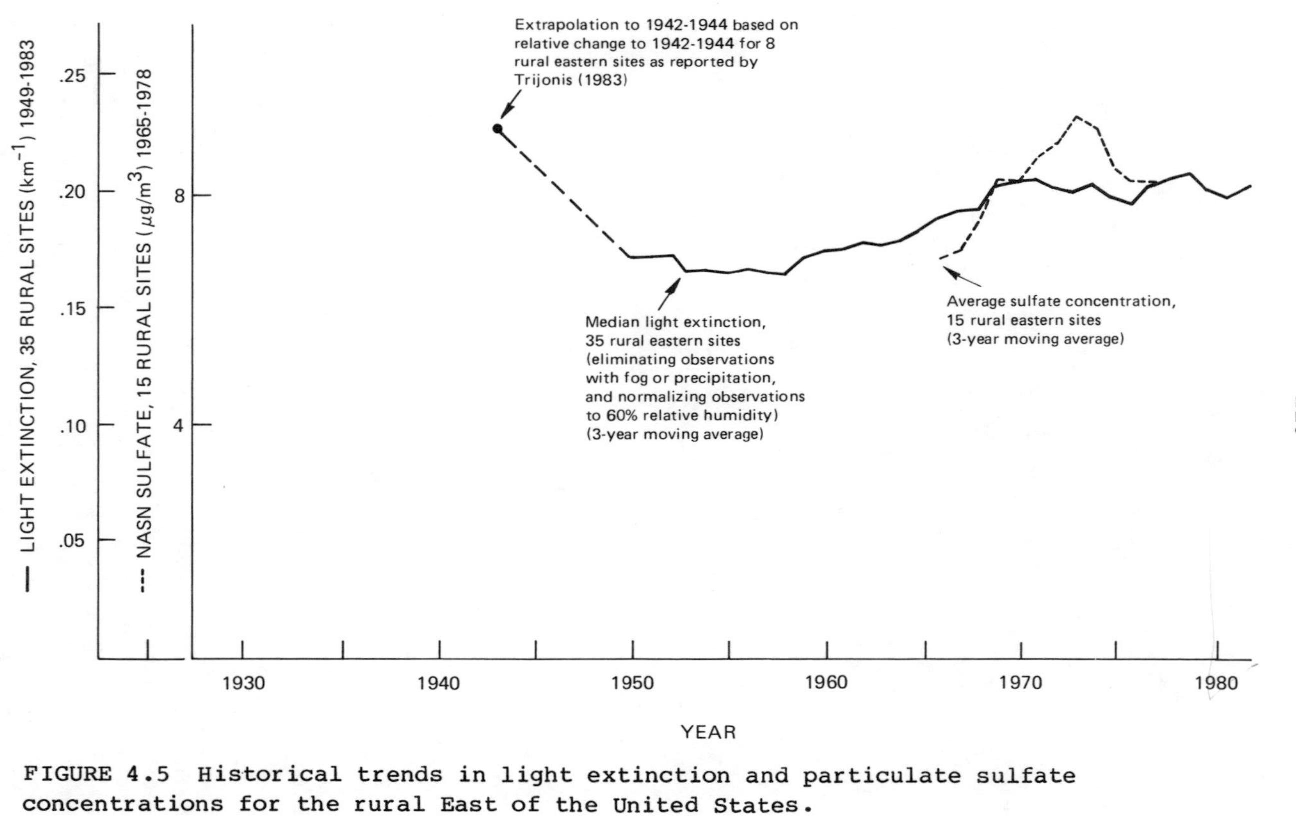

FIGURE 4.5 Historical trends in light extinction and particulate sulfate concentrations for the rural East of the United States.

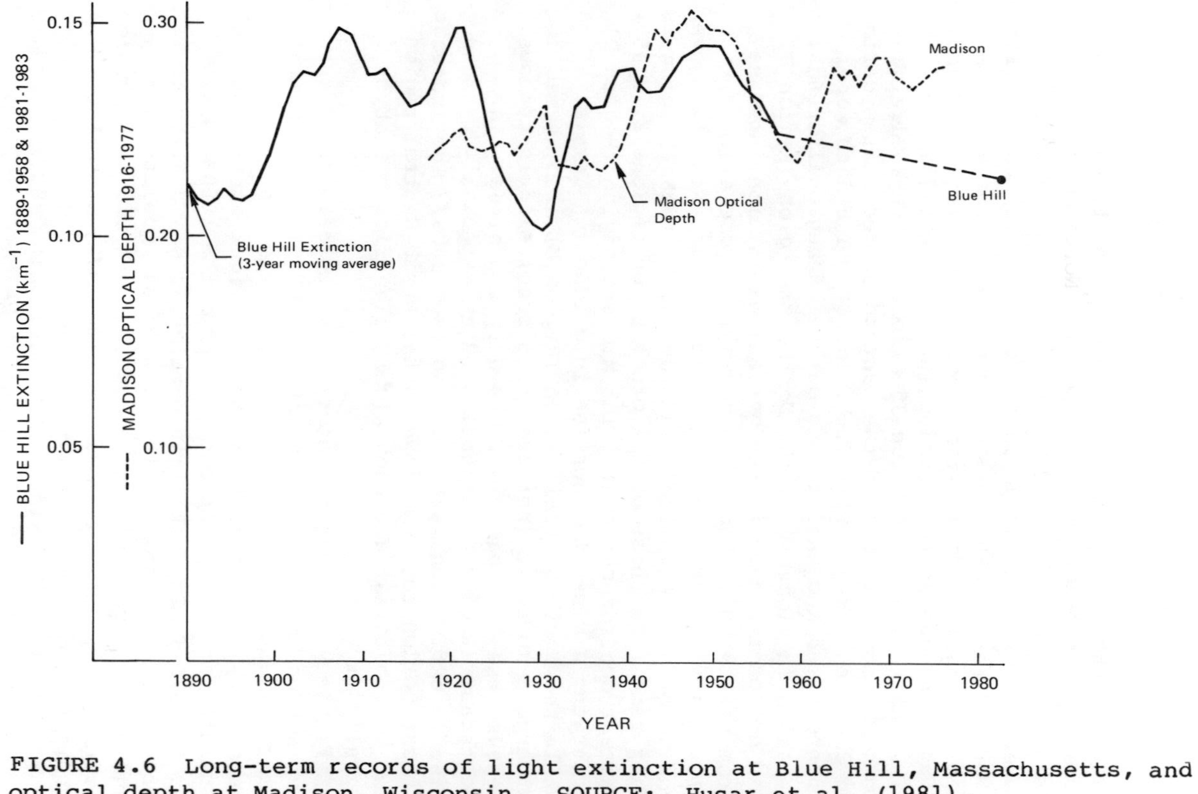

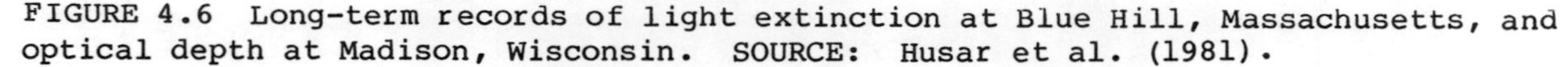

FIGURE 4.6 Long-term records of light extinction at Blue Hill, Massachusetts, and optical depth at Madison, Wisconsin. SOURCE: Husar et al. (1981).

Part of the disagreement in the trends might be explained by noting that the two sites are widely separated geographically and that a single site may not be representative of its own region. The trend at Madison appears to reflect the trend for the North Central region (Region E) from the 1950s to the 1970s (see Figure 4.7). However, the apparent decrease at Blue Hill from the late 1950s to the present does not seem consistent with the nearly constant overall trend for the Northeast region (Region B) during that period (see Figure 4.7).

Visibility data from the 35 airports were subdivided into the four regions of the eastern United States defined in Chapter 1 (Regions B, C, D, and E). The data were then plotted and analyzed to determine regional trends. As shown in Figure 4.7, the Northeast region (Region B) did not participate in the overall extinction increase from the 1950s to the 1970s shown in Figure 4.5 but apparently underwent a slight decrease in extinction over the three decades. The trend for the Northeast region is different from those of the other three regions at a 95 percent confidence level. The Midwest region (Region D) underwent a moderate increase in extinction levels. While the North Central region (Region E) also seemed to undergo a moderate increase, this conclusion is tenuous because the data base includes only two sites. The Southeast (Region C) demonstrated an especially pronounced rise in extinction levels during the 1960s. The increase in extinction for the Midwest, the North Central region, and the Southeast are each highly significant statistically ($t > 6$).

Figure 4.8 presents historical trend data for light extinction and sulfate concentrations during the summer quarter (July-September). It is apparent that an especially sharp rise in summertime light extinction and sulfate concentrations occurred from the 1950s to the early 1970s. In fact, this particular seasonal trend is one of the most salient features of air quality history and has been pointed out by most workers who have studied trends in visibility (Miller et al. 1972, Munn 1973, Trijonis and Yuan 1978, National Research Council 1979, Husar et al. 1979, Sloane 1982, Trijonis 1982), turbidity (Flowers et al. 1969, Robinson and Valente 1982), or sulfates (Trijonis 1975, Environmental Protection Agency 1975, Altshuller 1976, Frank and Possiel 1976, National Research Council 1979) in eastern North America. The summer was a season of relatively good visibility during the 1950s but became the season of distinctly lowest

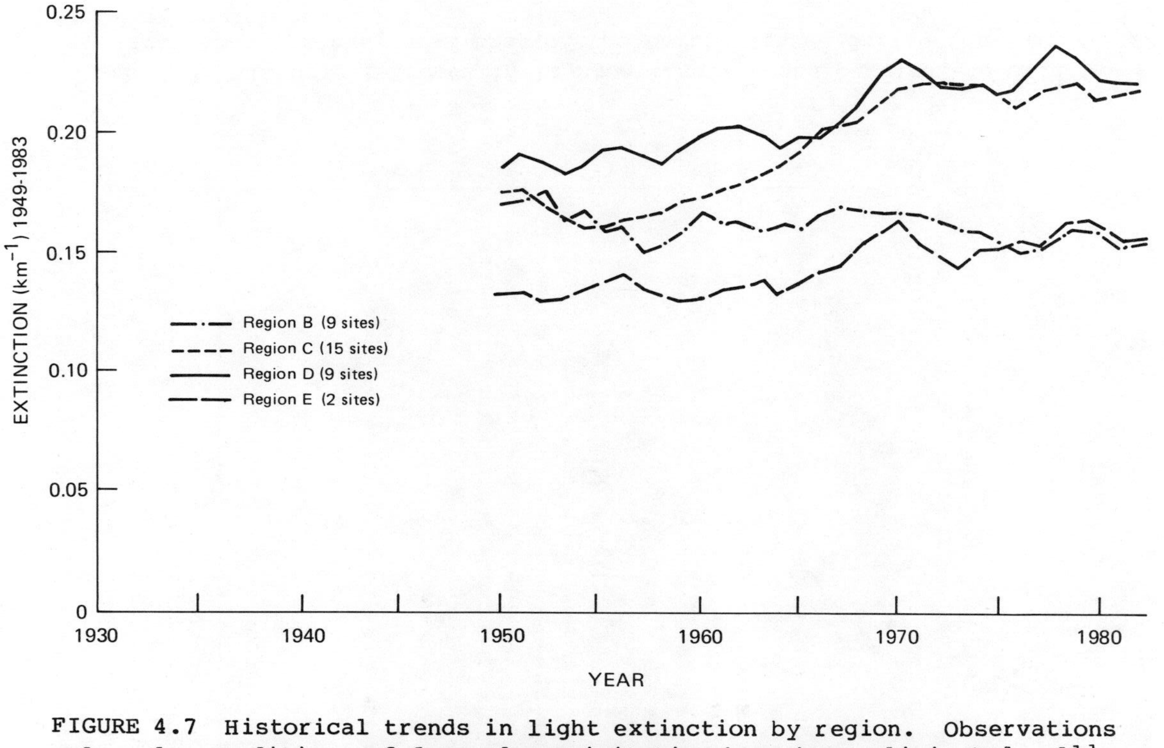

FIGURE 4.7 Historical trends in light extinction by region. Observations made under conditions of fog and precipitation have been eliminated. All values are normalized to 60-percent relative humidity.

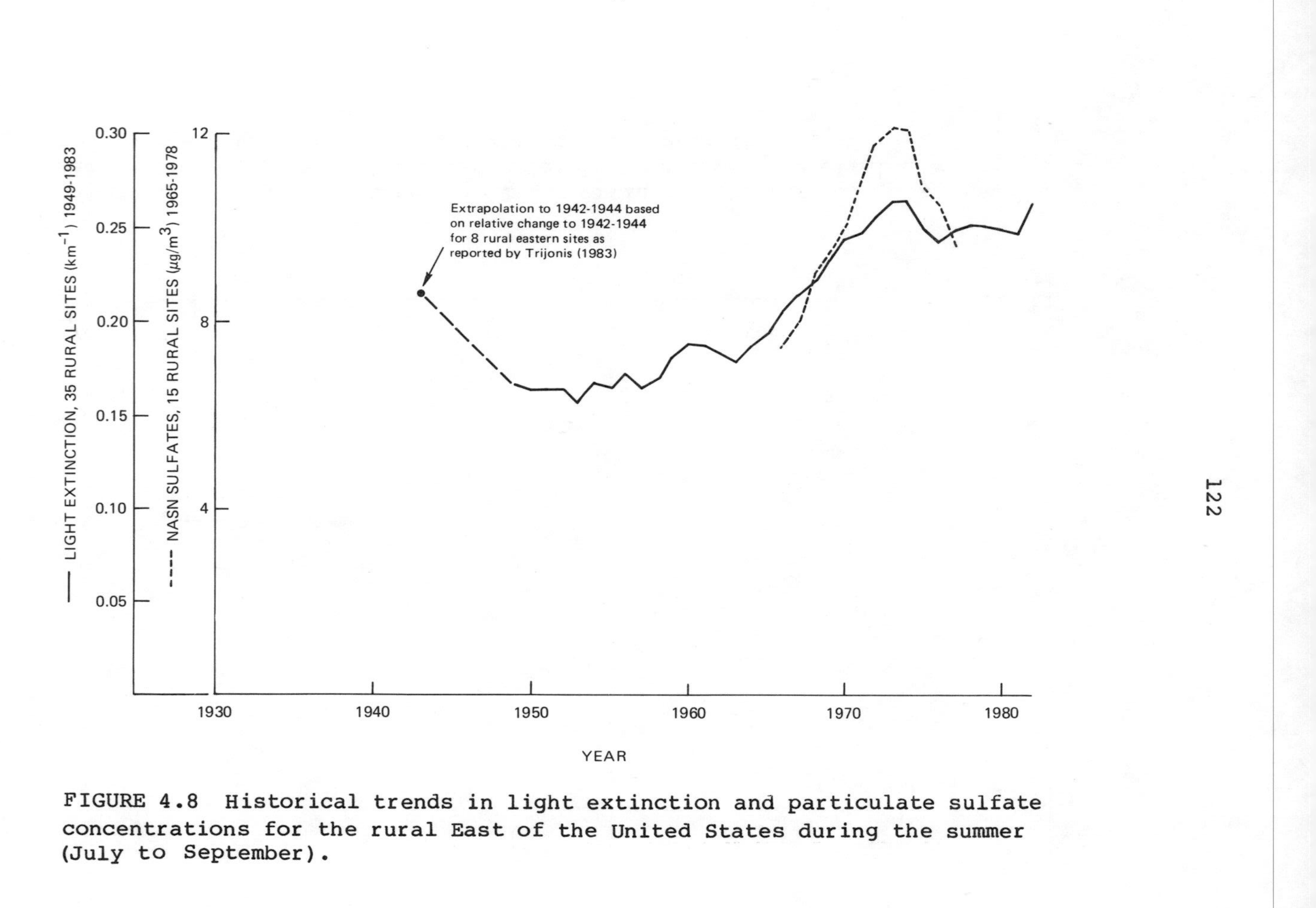

FIGURE 4.8 Historical trends in light extinction and particulate sulfate concentrations for the rural East of the United States during the summer (July to September).

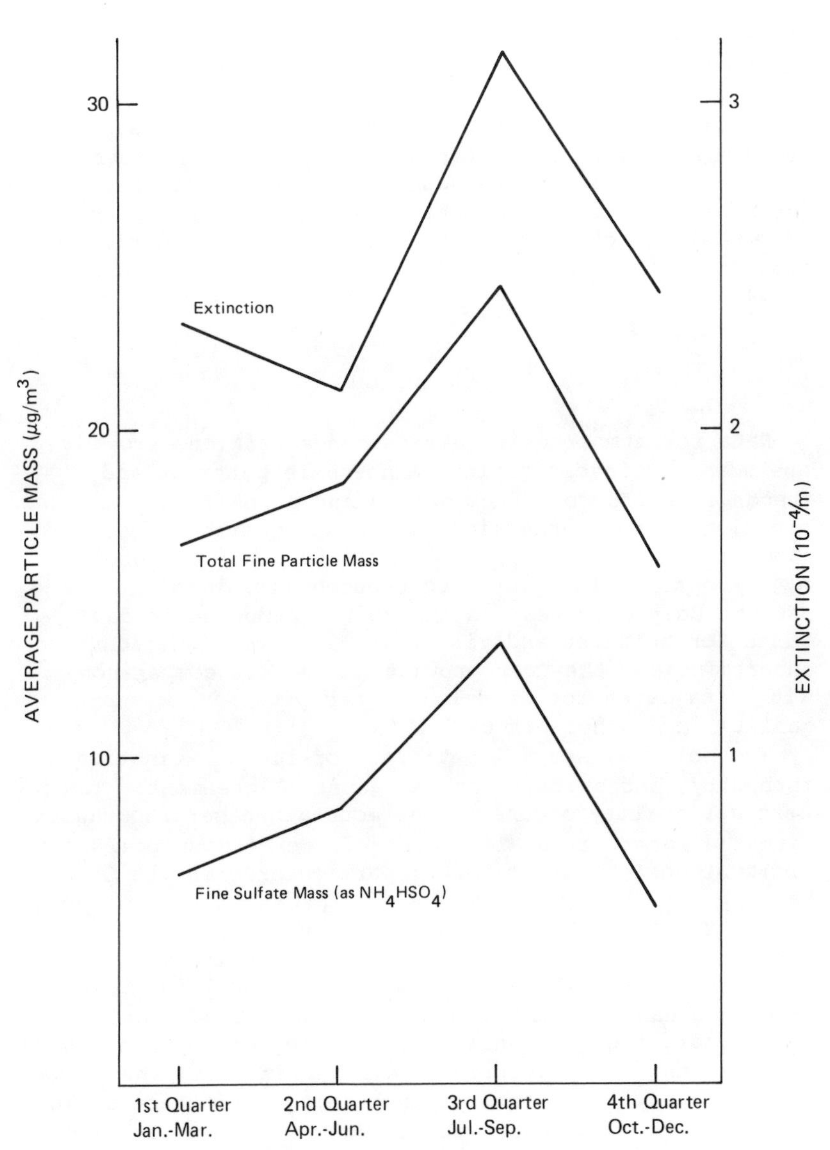

FIGURE 4.9 The seasonal patterns of fine-sulfate mass per unit volume, total fine-particle mass per unit volume, and light extinction for rural areas of the eastern United States. Fine-particle mass and fine-sulfate mass data are averages over five rural EPA dichotomous sampler sites (Will County, Illinois; Jersey County, Illinois; Monroe County, Illinois; Erie County, New York; and Durham County, North Carolina). The light-extinction data are averages over eight rural airports. SOURCES: Trijonis 1982, Trijonis and Yuan 1978.

visibility and highest sulfate by the early 1970s (see Figure 4.9). We are quite certain of the observed seasonal trend because artifacts from measuring inconsistencies are canceled out when trends for one season are compared with those for other seasons. The summertime deterioration of air quality has been attributed to the rapid growth of summertime sulfur oxide emissions from coal-fired power plants caused by air conditioning demands (Holland et al. 1977, Husar et al. 1979, Sloane, 1981).

CONCLUSIONS

Data for atmospheric sulfate concentrations provide one means of investigating large-scale patterns and trends in sulfur oxide emissions and deposition. Another useful measure is the rich historical data base for atmospheric visibility, since sulfate particles are a major contributing factor to reduced visibility in the eastern United States. Although the atmospheric data bases for sulfates and visibility involve significant uncertainties, they can provide productive comparisons with data bases for emissions, precipitation, water quality, and other effects.

The national spatial patterns for light extinction, turbidity, and sulfates are in general agreement. The best air quality occurs in the mountain/desert Southwest, with the worst air quality east of the Mississippi River and south of the Great Lakes, particularly in the Ohio Valley.

Light extinction (a measure of inverse visibility) in nonurban areas of the East decreased substantially from the 1940s to the 1950s, increased from the 1950s to about 1970, and has been about constant since then. The southeast quadrant of the United States (Region C) experienced the greatest deterioration of air quality from the 1950s to the 1970s. The trend of increasing extinction during the 1960s is consistent with trends in ambient sulfate. An especially marked increase in extinction during the summer season occurred from the 1950s to the 1970s.

REFERENCES

Aerovironment, Inc. 1983. Comparison of visibility measurement techniques: eastern United States. EPRI-EA 3292 Research Project 862-15. Electrical Power Research Institute, Palo Alto, Calif.

Altshuller, A. P. 1979. Regional transport and transformation of sulfur dioxide to sulfates in the United States. J. Air Pollut. Assoc. 26:318-324.

Douglas, C. A., and L. L. Young. 1945. Development of transmissometer for determining visual range. U.S. Civil Aeronautics Administration. Tech. Development Report No. 47.

Environmental Protection Agency. 1975. Position on regulation of atmospheric sulfates. EPA 450/2-75-007. Environmental Protection Agency, Washington, D.C.

Ferman, M. A., G. T. Wolff, and N. A. Kelly. 1981. The nature and source of haze in the Shenandoah Valley/Blue Ridge Mountains area. J. Air Pollut. Control Assoc. 31:1074-1082.

Flowers, E., R. McCormick, and R. Kurfis. 1969. Atmospheric turbidity over the United States 1961-68. Appl. Meteorol. 8:955-962.

Frank, N., and N. Possiel. 1976. Seasonality and regional trends in atmospheric sulfates. Paper presented at the meeting of the American Chemical Society, San Francisco, Calif., August 30 to September 3.

Holland, T. C., R. L. Short, and A. H. Wehe. 1977. SO_2 emissions from power plants over the 1969-1975 period. Environmental Protection Agency, Washington, D.C.

Husar, R. B., D. E. Patterson, and J. M. Holloway. 1979. Trends of eastern United States haziness since 1948. Paper presented at the Fourth Symposium on Atmospheric Turbulence, Diffusion, and Air Pollution, Reno, Nevada, January 15-18.

Husar, R. B., J. M. Holloway, and D. E. Patterson. 1981. Spatial and temporal patterns of eastern U.S. haziness: a summary. Atmos. Environ. 15:1919-1928.

Leaderer, B. P., and J. A. Stolwijk. 1979. Optical properties of urban aerosol and their relation to chemical composition. Ann. N.Y. Acad. Sci. 338:70-85.

Leaderer, B. P., T. R. Holford, and J. A. Stolwijk. 1979. Relationship between sulfate aerosols and visibility. J. Air Pollut. Control Assoc. 29:154-157.

Malm, W. C. 1979. Visibility: a physical perspective. Paper presented at the Workshop on Visibility Values. Fort Collins, Colorado, January 28-February 1.

Middleton, W. E. K. 1952. Vision Through the Atmosphere. Toronto: Toronto University Press.
Miller, M. E., N. L. Canfield, T. A. Ritter, and C. R. Weaver. 1972. Visibility changes in Ohio, Kentucky, and Tennessee from 1962 to 1969. Mon. Weather Rev. 100:67-71.
Munn, R. E. 1973. Secular increases in summer haziness in the Atlantic provinces. Atmosphere 11:156-161.
National Research Council. 1979. Controlling Airborne Particles. Washington, D.C.: National Academy Press.
Pierson, W. R., W. W. Brachaczek, T. J. Truex, J. W. Butler, and T. J. Korinski. 1980. Ambient sulfate measurements on Allegheny Mountains and the question of atmospheric sulfate in the northeastern United States. Ann. N.Y. Acad. Sci. 338:114-173.
Robinson, E., and R. J. Valente. 1982. Atmospheric turbidity over the United States from 1967 to 1976. EPA 600/S3-82-076. Environmental Protection Agency, Washington, D.C.
Sloane, C. S. 1982. Visibility trends--II: mideastern United States, 1948-1978. Atmos. Environ. 16:2309-2321.
Sloane, C. S. 1983. Summertime visibility declines: meteorological influences. Atmos. Environ. 17:763-774.
Sloane, C. S. 1984. Meteorologically adjusted air quality trends: visibility. Atmos. Environ. 18:1217-1230.
Trijonis, J. 1975. The relationship of sulfur oxide emissions to sulfur dioxide and sulfate air quality. Pp. 233-275 in Air Quality and Stationary Source Emission Control. Prepared for the Committee on Public Works, United States Senate. Serial no. 94-4. Washington, D.C.: U.S. Government Printing Office.
Trijonis, J. 1982. Existing and natural background levels of visibility and fine particles in the rural East. Atmos. Environ. 16:2431-2445.
Trijonis, J. 1984. Analysis of particulate concentrations and visibility in the eastern United States. Prepared under contract 68-02-3578 for Environmental Protection Agency, Washington, D.C.
Trijonis, J., and K. Yuan. 1978. Visibility in the Northeast: long-term visibility trends and visibility/pollutant relationships. EPA 600/3-78-075. Environmental Protection Agency, Washington, D.C.
Weiss, R. E., A. P. Waggonner, R. J. Charlson, and N. C. Ahlquist. 1977. Sulfate aerosol: its geographical extent in the midwestern and southern United States. Science 197:977-981.

Wolff, G. T., N. A. Kelly, and M. A. Ferman. 1982. Source regions of summertime ozone and haze episodes in the eastern United States. Water Air Soil Pollut. 18:65-82.

World Aeronautical Charts. 1976. U.S. Department of Defense, Federal Aviation Administration and U.S. Department of Commerce, Washington, D.C.

5

Precipitation Chemistry

Gary J. Stensland, Douglas M. Whelpdale, and *Gary Oehlert*

INTRODUCTION

The Role of Precipitation Composition in This Study

The chemical composition of wet deposition has been central to the question of the effects of acid deposition and the long-range transport of air pollution since the issue of "acid rain" rose to prominence in Europe two decades ago. Early evidence linked acidic substances in precipitation to harmful effects on surface waters and fisheries in Scandinavia, and as a consequence interest has since focused on precipitation acidity and acid deposition rather than on the more general case of deposition of air pollutants. This focus can be attributed to several factors: Changes in precipitation composition could actually be detected, and techniques were available to make measurements; this was not true for dry deposition. In addition, precipitation composition was starting to be used as an indicator of atmospheric quality, so researchers were aware of the value of such measurements. Finally, the correlation between adverse effects and wet deposition could be made more directly than that between effects and emissions.

The chemistry of precipitation has been studied both in the expectation that it will provide a measure of the consequences of emissions in time and space and also as an indicator of effects on the environment. Unfortunately, problems exist both with the quality of past data and with the interpretation of the data. These difficulties have stimulated the examination of other historical data in efforts to improve our understanding of the phenomenon of acid deposition and to corroborate the available precipitation-chemistry records.

Wet and Dry Deposition

Atmospheric constituents, and thus pollutant emissions, reach the surface of the Earth by a variety of processes: those in which precipitation is involved contribute to wet deposition; those that (primarily during dry periods) involve sedimentation, turbulent and molecular diffusion, impaction, and interception contribute to dry deposition; and those that include cloud and fog impaction, dew, and frost are sometimes termed occult deposition. Wet deposition and bulk deposition (i.e., the sum of wet, dry, and occult, obtained with the collector always open) have been measured with varying degrees of reliability for decades; dry deposition has been measured sporadically in a research mode but not routinely; and measurements of the third type of deposition are just beginning. Even historical air-concentration data, from which dry deposition might have been estimated, are sparse and incomplete.

Out of necessity, therefore, we focus on wet deposition data. From the point of view of determining historical trends, this is not necessarily a serious limitation on the regional-to-continental scale, where the wet deposition is expected to reflect total deposition to a reasonable extent. At a specific location and at a specific time, however, wet deposition may not be representative of total deposition.

In addition, omitting dry and occult deposition is certainly a serious constraint on determining mass balance and interpreting effects. Over eastern North America the total wet and total dry deposition are thought to be of approximately equal magnitude. Dry deposition is, in general, greater than wet near emission source regions, where ambient concentrations of pollutants are higher. An additional complexity for dry deposition is that the chemical species behave differently: in general, gaseous nitric acid, sulfur dioxide, and ammonia are removed more efficiently than their particulate forms, nitrate, sulfate, and ammonium.

In high-elevation ecosystems the direct input from cloud and fog water is likely to exceed that from either of the other pathways. Interpreting observations of surface ecosystems must allow for these other inputs, even though the majority of past deposition data is for wet or bulk deposition.

Constituents of Interest

We limit this discussion of wet deposition to the major soluble species in precipitation that account for most of the measured conductance of the samples. The species include the following ions: hydrogen (H^+), bicarbonate (HCO_3^-), calcium (Ca^{2+}), magnesium (Mg^{2+}), sodium (Na^+), potassium (K^+), sulfate (SO_4^{2-}), nitrate (NO_3^-), chloride (Cl^-), and ammonium (NH_4^+). Experience has shown that a pH value can be calculated from measurements of the last eight ions in the list, and the calculated value is usually in good agreement with the measured pH value. The fact that we can often successfully calculate the pH of precipitation samples indicates that the small list of measured ions is probably sufficient for studies of wet deposition that emphasize the acid precipitation phenomenon. However, there are exceptions. For example, samples from remote locations can be strongly affected by organic acids (Galloway et al. 1982).

The data in Table 5.1 demonstrate the relative importance of these ions at three monitoring sites of the National Acid Deposition Program (NADP) in the United

TABLE 5.1 Median Ion Concentrations for 1979 for Three NADP Sites (μeq/L)

Ion	GA (42 Samples)[a]	MN (37 Samples)[b]	NY (49 Samples)[c]
SO_4^{2-}	38.9	45.8	44.8
NO_3^-	11.6	24.2	25.0
Cl^-	8.2	4.2	4.2
HCO_3^- (calculated)	0.3	10.3	0.1
Anions	59.0	84.5	74.1
NH_4^+	5.5	37.7	8.3
Ca^{2+}	5.0	28.9	6.5
Mg^{2+}	2.4	6.1	1.9
K^+	0.7	2.0	0.4
Na^+	17.6	13.7	4.9
H^+	17.8	0.5	45.7
Cations	49.0	88.9	67.7
Median pH	4.75	6.31	4.34

NOTE: See Georgia, Minnesota, and New York in Figure 5.23 for location of these sites.
[a]The Georgia Station site in west central Georgia.
[b]The Lamberton site in southwestern Minnesota.
[c]The Huntington Wildlife site in northeastern New York.

States. Concentrations are expressed in microequivalents per liter (μeq/L) to permit a direct evaluation of each ion's contribution to the anion or cation sum. If all ions were measured and there were no analytical uncertainty, then the anion sum would equal the cation sum. The values for hydrogen ion concentration were calculated from the measured median pH value, and the values for bicarbonate were calculated by assuming that the sample was in equilibrium with atmospheric carbon dioxide. Although the sulfate and nitrate levels shown are similar at the Minnesota and New York sites, the pH differs greatly owing to the higher levels of ammonium, calcium, magnesium, sodium, and potassium ions at the Minnesota site. These ions are frequently associated with basic compounds. The data in Table 5.1 suggest, therefore, that the concentrations of all the major ions must be considered if we are to understand the time and space patterns of pH. The data also indicate that sites in different regions of the United States exhibit large differences in ionic concentrations.

Currently, sites in states such as Ohio, New York, Pennsylvania, and West Virginia have the feature shown for the New York site in Table 5.1, for which hydrogen, sulfate, and nitrate are the dominant ions. For the New York site, 98 percent of the acidity could be accounted for if all the sulfate were sulfuric acid, whereas nitrate, as nitric acid, could have accounted for about 55 percent of the acidity. Bowersox and de Pena (1980) have concluded, by applying multiple linear-regression analysis for a central Pennsylvania site, that on the average, sulfuric acid is the principal contributor to hydrogen-ion concentration in rain, but the acidity in snow is principally from nitric acid.

Variability in Acid Deposition: The Role of Meteorology

Any discussion of short- and long-term trends in acid deposition cannot ignore the role of meteorological and climatological variability. Even if pollutant emissions were absolutely constant every hour of the day and every day of the year, variations in atmospheric conditions would bring about uneven deposition patterns in both time and space. For example, both variability throughout the course of a year and long-term trends in precipitation records reflect changes in storm tracks and in the frequency of precipitation-producing circulation patterns that affect a given region each year. Differences in precipitation amount from year to year affect total wet deposition, but even if the precipitation amount were constant every year, wet deposition of sulfate, nitrate, and other common ions would not necessarily be, because of the dependence of the concentrations of these ions on the size of the precipitation events and the time interval between the events.

Atmospheric processes vary continuously across a wide temporal and spatial range; at whatever scale one selects, atmospheric variability plays a fundamental role in the acid deposition process. Unless we understand this role thoroughly, the meaning of trends in acid deposition will be difficult to resolve unequivocally. (See Chapter 3 for a more detailed discussion.)

DATA EVALUATION

Given that the object of our evaluation is to determine trends in concentrations (or deposition) of chemicals in precipitation, our first task is to identify approximate data sets and determine their quality. Since adequate methods to determine dry deposition rates for most chemical species do not exist, wet-only deposition data collected with automatic samplers are generally the most reliable and useful. However, owing to the sparsity of such data we must also consider data bases in which less desirable sampling methods were used. Such methods include bulk sampling and manual wet-only sampling in which dust leakage into the sampler and water evaporation and snow sublimation from the collection vessels have apparently introduced serious bias into the data sets. The same biases may occur even in an automatic wet-only sampling device when mechanical failure causes exposure of the sample to the open atmosphere. In this chapter we emphasize trends in concentration rather than in deposition to avoid the introduction of spatial variability caused by variability in the amount of precipitation.

A primary tool in evaluating possible biases in monitoring techniques is comparison of different data sets, such as bulk data with concurrent wet-only data. This approach has not always been used in previous reports but may be one of the most important evaluation techniques. However, since local sources can affect bulk data through high levels of dry deposition, the only satisfactory comparison must involve concurrence in space and time. Unfortunately, only in a few cases are the two sampling methodologies concurrent spatially and temporally, so that such comparisons are usually far from ideal.

Several different evaluations are described in this chapter, with emphasis on evaluating those United States data sets quoted most frequently in trend studies.

Comparing Three-Month Regional Bulk Data Concurrently with Wet-Only Data

Most of the published comparisons between bulk and wet-only ion concentrations are for eastern sites. The difference between these two sampling methods would be expected to be greater in locations where local sources may be influential, for example, where sampling occurs near tilled fields with an abundance of windblown dust.

Peters amd Bonelli (1982) have reported the results of a bulk sampling program for the period from about December 1, 1980, to March 1, 1981. They indicated that considerable effort in siting was made to avoid the occurrence of local contamination of regional deposition. States from Minnesota and Illinois eastward to Maryland and Maine were included in the study. The NADP network had a large number of stations operating in the same area for the same time period. Therefore we are able to present a simple comparison of the results for the wet-only NADP concentration data versus the bulk concentration data and examine geographic variations in the differences between the two methods.

The wet-only and bulk concentration data were plotted and examined visually to select five contiguous geographical areas in which there was some degree of uniformity in values. The areas selected are shown in Figure 5.1. The bulk sample at each site was a single time-integrated sample collected with a continuously open cylindrically shaped sampler. The concentration ratios of the ions (mean of bulk values divided by mean of wet-only values) are shown in Table 5.2 for each area. One may observe that in all cases the value of the ratio is greater than or equal to 1.0 and that there is considerable variation in the ratio by area and by chemical parameter. The ratios for calcium, magnesium, and potassium are greater than 2 (and as high as 30) for all areas, and bulk sulfate concentrations exceeded

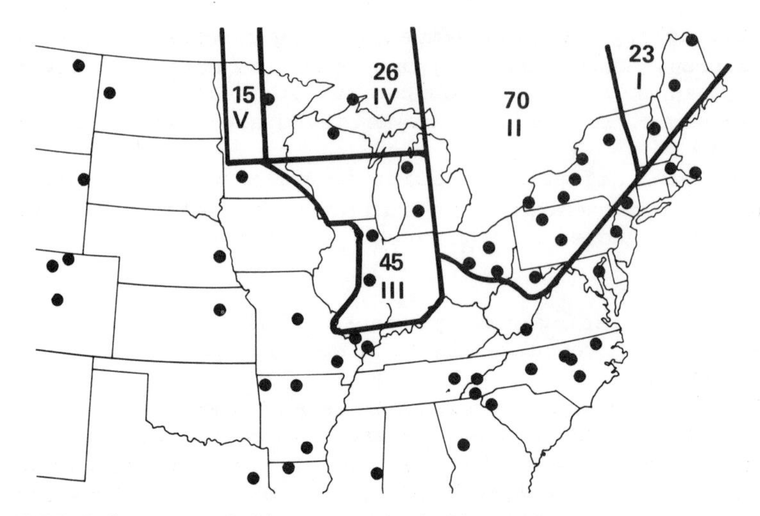

FIGURE 5.1 Map of five areas (indicated by roman numerals) used for comparing concentrations of various ions in wet-only deposition with those in bulk samples. The arabic numbers indicate the number of bulk sampling sites within each area. The NADP sites are indicated by circles. Bulk-measuring sites are distributed uniformly about the areas but are not co-located with NADP sites. See also Table 5.2.

wet-only concentrations by factors of 2.0 and 8.1 in areas III and V, respectively. All ion concentrations were elevated by more than 90 percent in areas III and V, with a median ratio value of 3.7 for these two areas. Peters and Bonelli (1982) reported both sample volume and an estimated sample volume derived from the data from the nearest National Weather Service precipitation gauge site. Using these data, we calculated the medians of the ratios of the measured divided by estimated volumes to be 0.88, 0.79, 0.81, 0.69, and 0.45 for areas I, II, III, IV, and V, respectively. These ratios suggest that bulk samples may be concentrated up to a factor of 2 by evaporation. Thus, the many ratios greater than 2.0 in Table 5.2 suggest that differences between concentrations of ions in bulk samples and in wet-only samples cannot by fully explained by evaporation.

Even when an average bulk to wet-only ion concentration ratio for an area in Table 5.2 has a value close to 1, an individual bulk site in that area may still differ widely in concentration from the wet-only data. Each site needs

TABLE 5.2 Ion-Concentration Ratios (Mean of Bulk Values Divided by Mean of Wet-Only Values) for All Sites in Areas I to V for the Period December 1980-February 1981

	Ion									
Area	Ca^{2+}	Mg^{2+}	K^+	Na^+	Cl^-	NO_3^-	NH_4^+	SO_4^{2-}	N_B[a]	N_W[b]
I	2.5	5.6	2.7	1.2	1.0	1.1	1.1	1.0	23	3
II	3.7	4.8	5.9	2.9	2.0	1.5	2.0	1.3	70	11
III	7.2	3.5	11.2	2.7	2.1	1.9	3.5	2.0	45	4
IV	2.7	2.3	3.1	1.9	1.6	1.1	1.3	1.3	26	3
V[c]	30.5	19.2	6.0	2.8	6.6	3.9	2.1	8.1	15	–

NOTE: Areas are depicted in Figure 5.1.
[a]Number of bulk sampling sites used to determine the means.
[b]Number of wet-only sampling sites used to determine the means.
[c]Since there were no NADP sites in area V, computer-generated contour maps (for weighted-average concentration) for all data available for the period 1978 - mid-1983 were used to estimate average NADP ion concentrations for this area.
SOURCES: NADP data were provided by Illinois State Water Survey. Bulk data were obtained from Peters and Bonelli (1982).

to be evaluated separately for each chemical parameter to determine if bulk concentrations are likely to be similar to wet-only concentrations.

Comparing Bulk and Wet-Only Data at Hubbard Brook, New Hampshire

A comparison of bulk and wet-only data from the Hubbard Brook site shows good agreement for most major ions, in contrast to other bulk versus wet-only comparisons (cf. Tables 5.3 and 5.4 (below)). Weighted-average concentrations of the colocated Hubbard Brook bulk samples and NADP wet-only samples were determined for the period from June 1979 to May 1982. Table 5.3 presents the results as 3-year averages except for pH, for which the averages of three 12-month periods are given.

Bulk concentrations of sulfate plus nitrate were 4.8 µeq/L (7 percent) higher than for the wet-only values. The sum of calcium plus magnesium, an indicator of dust input, was actually lower for the bulk than for the wet-only data. Apparently, dry deposition of sulfate and nitrate to the continuously exposed bulk collector, either through gaseous precursors or in association with dust particles, was not an important factor. Another

TABLE 5.3 Comparison of Bulk and Wet-Only Weighted-Average Concentrations (μeq/L) at Hubbard Brook, New Hampshire

Ion	Hubbard Brook (bulk)[a]	NADP (wet only)[b]
SO_4^{2-}	49.9	48.5
NO_3^-	27.3	23.9
Cl^-	6.6	6.2
NH_4^+	11.0	9.4
Ca^{2+}	5.6	5.8
Mg^{2+}	2.2	3.0
K^+	1.0	0.4
Na^+	5.3	8.4
H^+	60.3	47.9
pH	4.24, 4.19, 4.22	4.39, 4.27, 4.32

[a] Weighted-average concentrations for the period June 1979-May 1982. Data provided by John Eaton, Cornell University, personal communication 1985.
[b] Weighted-average concentration for wet-only samples from the NADP collector for the period June 1979-May 1982.

typical problem with bulk data, concentration effects owing to sample evaporation, was probably not a major factor given the close agreement of bulk ion concentrations with the wet-only concentrations. This finding suggests that the funnel and bottle design of the Hubbard Brook bulk sampler effectively prevented this problem for long-term averages at this site. The three (12-month) weighted averages of pH for the bulk samples are lower than the corresponding values for the wet-only samples. The 3-year weighted-average concentration of hydrogen ions in bulk samples is 26 percent larger than in wet-only samples. Such a difference is likely to be significant for these data.

In addition to different sampling devices and different laboratories for sample analyses of the Hubbard Brook data, different screening and sampling procedures were used for bulk data and the wet-only data. More bulk data than wet-only data were deleted from the data base for the 3-year period. Since the two sampling devices were generally serviced on different days of the week, it was not possible to calculate the data as exactly matched pairs. It would be useful to undertake a more detailed study involving a week-by-week comparison--both to learn more about the long-term record at this important site and to help to relate trends in bulk measurements to those in wet-only measurements.

Comparing U.S. Geological Survey Bulk Data with National Acid Deposition Program Wet-Only Data

Nine U.S. Geological Survey (USGS) bulk collection sites have operated in New York state and Pennsylvania since 1965 (Barnes et al. 1982), and several investigators have summarized and analyzed these data. (See later discussion.) Here, we compare these data with NADP wet-only data (Table 5.4).

The first column of the table gives the mean and the standard deviation of all New York NADP data available through the end of 1982; the seven NADP sites began operation at different times between November 1978 and

TABLE 5.4 Ion Concentrations (μeq/L) and pH for All New York NADP Sites and for Two New York USGS Bulk Sites

Ion	Mean and Standard Deviation for Seven NADP Sites[a]	Weighted Average		Time-Adjusted[d] Weighted Average	
		Hinckley[b]	Mays Point[c]	Hinckley	Mays Point
N[e]		23	24	23	24
Ca^{2+}	8.4 ± 2.0	30.2	74.2	26.3	71.6
Mg^{2+}	3.5 ± 0.8	7.7	17.1	6.7	16.5
K^{+}	0.8 ± 0.6	2.1	2.9	1.8	2.8
Na^{+}	6.2 ± 3.5	5.4	10.4	4.7	10.0
NH_4^{+}	16.5 ± 5.5	31.2	22.7	37.3	27.5
Sum[f]	35.4	76.6	127.3	76.8	128.4
NO_3^{-}	30.2 ± 5.4	41.7	46.5	62.1	54.7
SO_4^{2-}	64.8 ± 13.4	72.5	82.2	65.9	76.4
Cl^{-}	5.8 ± 3.3	8.9	15.2	8.9	15.2
pH	4.22 (4.15-4.33)[g]	3.99	3.99	3.90	3.86
H^{+}	60.4 ± 10.3	102.3[h]	102.3[h]	126.4[h]	138.3[h]

[a] Volume-weighted concentrations for 1978-1982.
[b] Hinckley USGS bulk data for 8/31/75 to 9/15/77.
[c] Mays Point USGS bulk data for 8/31/75 to 8/31/77.
[d] Time adjustments made using time-trend results for 1965-1978 period as determined by Kendall's test (Barnes et al. 1982). An adjustment factor for a 4.5-year time shift was multiplied by the USGS bulk concentrations: Ca^{2+}, Mg^{2+}, K^{+}, and Na^{+} concentrations were multiplied by 0.87 and 0.965 for Hinckley and Mays Point, respectively; SO_4^{2-} concentration by 0.909 and 0.93; NO_3^{-} concentration by 1.489 and 1.176; NH_4^{+} concentration by 1.194 and 1.211; and H^{+} concentration by 1.236 and 1.352.
[e] N is the number of samples at USGS sites.
[f] "Sum" gives the total microequivalents per liter potentially available for neutralizing acids in precipitation.
[g] Range.
[h] USGS weighted-average H^{+} concentrations are likely influenced by a few outliers. Median concentrations are ~50% less.

June 1980. The sample volume weighted-average concentrations were calculated for each NADP site, and the means of these values are shown in Table 5.4.

The table also presents weighted-average concentrations of data from Hinckley and Mays Point, two of the USGS bulk sampling sites in New York; later in this chapter these two sites are listed with three others for consideration of regional time-trend analysis. The Hinckley and Mays Point sites were chosen for inclusion in Table 5.4 because these sites tend to represent the extremes in concentrations found at the USGS stations, with the Hinckley site tending to have the lower concentrations and the Mays Point site the higher concentrations (Barnes et al. 1982).

Because the two data sets do not overlap in time (1978-1982 for NADP and 1975-1977 for USGS) the USGS data are presented in two ways. The two middle columns in Table 5.4 present the data from 1975-1977. The two right-hand columns present the USGS data adjusted to account for the difference in time periods. Temporal trends of the concentrations of the ions listed in the table were determined by Barnes et al. (1982) for the period from 1965 to 1978. We estimated the data for the 1978-1982 period by multiplying the USGS data (1975-1977) by a factor derived from the calculated trend from 1965-1978 (see footnote d in Table 5.4 for details).

Regardless of whether the actual or adjusted USGS values are used, large differences exist between the NADP and USGS data sets. Calcium concentrations are three to nine times higher in the bulk data. Magnesium, potassium, ammonium, and nitrate concentrations are also higher in the bulk data (factors of 1.4 to 4.9). Sodium concentrations are about the same, and the hydrogen-ion concentration is much higher in the bulk data, although one might anticipate the reverse. The bulk sulfate values are somewhat higher than the wet-only values at Mays Point but are about the same as Hinckley values.

The overall conclusion is that the USGS bulk concentration data are significantly different from NADP wet-only data for many ions. The sum of the basic cations is larger at the bulk sites than at the wet-only sites by more than twofold.

Comparing Two Wet-Only Networks at Three Sites

In this section wet-only data from the Multistate Atmospheric Power Production Pollution Study (MAP3S) network (event samples) and the NADP network (weekly samples) are compared for three locations. Data were selected for the same time interval (March or April 1979 to December 1981) for each location in each network.

Median concentrations and pH values for Bondville, Illinois, are shown in Table 5.5; the two collectors at this site are separated by about 10 m. The data in Table 5.6 are for sites near State College, Pennsylvania, and the two collectors are separated by about 25 km. The data in Table 5.7 are for sites near Ithaca, New York, where the two collectors are about 40 km apart.

The best agreement of data might be expected for the Bondville site since the two collection devices are located so close to each other. At Bondville the NADP sulfate and nitrate data are similar to those from MAP3S; pH agrees within 0.02 pH unit; calcium, magnesium, and sodium are higher in NADP data; and potassium, ammonium, and chloride are higher for MAP3S data. A similar pattern is found for State College even though the two sites are about 25 km apart at this location and thus not exactly the same precipitation events were sampled.

The average calcium, magnesium, and sodium NADP and MAP3S values in Table 5.7 for the Ithaca site are less similar than those in Tables 5.5 and 5.6. The NADP values are also less similar than the MAP3S values for sulfate and nitrate. Therefore, we suggest that the environment may be less pristine at the Ithaca NADP site than at the MAP3S site, resulting in higher ion concentrations in samples at the former. Despite this, the pH values for the NADP and MAP3S sites are essentially the same for this 32-month period, indicating that pH comparisons should be done in conjunction with comparisons of the other major ions.

These data show that the 32-month median concentrations of ions for NADP and _nearby_ MAP3S sites may differ by more than 10 percent for the various ions. On _average_, however, the pH agreement is excellent, and nitrate and sulfate data agree within 10 percent for two of the locations.

It is likely that some precipitation constituents at the Ithaca NADP site are more affected by local sources than those at the MAP3S site. This illustrates another caution to be observed in carrying out a time-trend

TABLE 5.5 Median Concentrations (mg/L) in Precipitation at Bondville, Illinois

Ion	MAP3S	NADP	NADP ÷ MAP3S	MAP3S ÷ NADP
Ca^{2+}	0.23	0.29	1.26	
Mg^{2+}	0.029	0.041	1.41	
Na^+	0.061	0.073	1.20	
K^+	0.038	0.027		1.41
NH_4^+	0.43	0.41		1.05
SO_4^{2-}	3.07	3.28	1.07	
NO_3^-	1.74	1.85	1.06	
Cl^-	0.19	0.18		1.06
Median pH	4.32	4.30		
N[a]	140	109		

[a]N is number of samples measured.

TABLE 5.6 Median Concentrations (mg/L) in Precipitation for Two Sites Near State College, Pennsylvania

Ion	MAP3S	NADP	NADP ÷ MAP3S	MAP3S ÷ NADP
Ca^{2+}	0.22	0.24	1.09	
Mg^{2+}	0.034	0.048	1.41	
Na^+	0.090	0.132	1.47	
K^+	0.063	0.036		1.75
NH_4^+	0.40	0.32		1.25
SO_4^{2-}	3.75	3.71		1.01
NO_3^-	2.60	2.49		1.04
Cl^-	0.32	0.24		1.33
Median pH	4.11	4.15		
N[a]	202	109		

[a]N is number of samples measured.

TABLE 5.7 Median Concentrations (mg/L) in Precipitation for Two Sites Near Ithaca, New York

Ion	MAP3S	NADP	NADP ÷ MAP3S	MAP3S ÷ NADP
Ca^{2+}	0.12	0.22	1.83	
Mg^{2+}	0.020	0.044	2.20	
Na^+	0.041	0.080	1.95	
K^+	0.028	0.023		1.22
NH_4^+	0.34	0.44	1.29	
SO_4^{2-}	3.07	4.28	1.39	
NO_3^-	2.23	2.55	1.14	
Cl^-	0.19	0.20	1.05	
Median pH	4.15	4.16		
N[a]	187	116		

[a]N is number of samples measured.

analysis of constituents in precipitation. Changes in time in the local sources may be quite different from the changes in time of the regional sources affecting a site. The problem would generally be expected to be more severe if bulk data were being evaluated.

Concluding Remarks

The paucity of wet-only precipitation data has made essential the use of bulk data for trend analysis. We recognize that bulk concentration data can be affected by evaporation and by dry deposition from distant and local sources. Few data sets are available with which to examine precisely the differences between bulk and wet-only data and the influences of these factors on deposition. Even when bulk sampling sites are located carefully, bulk concentrations usually exceed wet-only values. The differences clearly depend on sampling location, time period, and chemical species.

The USGS network data show higher concentrations than do data from wet-only sampling stations in the same region. Thus, the data have limitations for trend analysis despite the length of the record available from this network.

In at least one case, the site at Hubbard Brook, New Hampshire, the sulfate and nitrate concentrations in bulk and wet-only samples agree well, and the difference in the sums of the basic cations is relatively small. (See Table 5.3.) Thus one may have reasonable confidence in applying the bulk data for trend analysis at this site. The sampling procedure at Hubbard Brook eliminated the evaporation problem, and evidently siting has minimized dry deposition effects, probably because of the long distance of this site from major sources of SO_x emissions. Thus, Hubbard Brook seems to be the exception regarding the general lack of compatibility between bulk and wet-only data.

Comparing NADP and MAP3S data for three locations suggests that even the wet-only data may differ because of local effects. Differences in data from these two networks at nearby (i.e., 40-km) stations in New York are probably caused by differences in local emission sources or perhaps by different sampling protocols. (See Tables 5.5 to 5.7.)

GEOGRAPHICAL DISTRIBUTION

Current Concentration and Deposition Patterns

The geographical distribution of major ionic constituents of precipitation and of their wet deposition over North America is generally well known. The many pH, concentration, and deposition maps prepared using various combinations of the 1978-1980 data from several networks show, essentially, the same features; maps for the year 1980 are reproduced below from Barrie and Hales (1984). These data were prepared during the Canada-USA Memorandum of Intent work on the basis of observations from the Canadian Network for Sampling Precipitation (CANSAP), the Air and Precipitation Network (APN), and the MAP3S and NADP networks and are from the first year when a compatible merger of data from these four federal networks was possible.

Figures 5.2 to 5.6 show the annual precipitation amount weighted averages for hydrogen-ion concentration, pH, and concentrations of sulfate, nitrate, and ammonium. Figures 5.2 and 5.3 show hydrogen-ion concentrations up to 60-80 μmol/L (pH 4.2-4.1) over a region of some 10^6 km^2 stretching from Illinois and Kentucky northeastward through New York and southern Ontario. Figure 5.4 shows a background sulfate concentration of less than 10 μmol/L in northern Canada and 3-15 μmol/L in the western United States, and values reaching greater than 40 μmol/L in Ohio and southern Ontario. Nitrate in Figure 5.5 has a background level of 2-6 μmol/L and rises to 40-50 μmol/L over the lower Great Lakes. The patterns of hydrogen, sulfate, and nitrate concentration in precipitation are similar. In contrast, Figure 5.6 shows that ammonium concentration reaches a maximum of greater than 40 μmol/L in and downwind of the northern plains states of the United States.

Annual average calcium concentrations were not reported by Barrie and Hales (1984) but are shown for U.S. sites in a report of the U.S. Environmental Protection Agency (1984). Values are lowest for sites in states along the East Coast and in the Northwest, with concentrations generally below 5 μmol/L. Sites in the Upper Great Plains in an area from eastern Wyoming to Lake Michigan and in the Southwest from western Texas to southern New Mexico and Arizona had the highest values, about 12 to 25 μmol/L. The patterns for ammonium and calcium ions identify those areas with the greatest acid-neutralizing potential.

Wet deposition is determined as the product of precipitation-amount-weighted concentration (Figures 5.2 to 5.6) and annual precipitation amount. Precipitation-amount fields have greater spatial variability than the concentration fields and in addition may be a source of large year-to-year differences. Figure 5.7 shows that precipitation during 1980 was not highly anomalous in the Northeast, and it indicates the high variability of the field. The wet deposition fields, shown in Figures 5.8 to 5.11 for hydrogen, sulfate, nitrate, and ammonium, are similar to the regional patterns of the concentration fields. Wet deposition of hydrogen ion reaches 80 $mmol/m^2$, sulfate and nitrate 40 $mmol/m^2$, and ammonium 30 $mmol/m^2$ in the eastern portion of the continent.

Barrie and Hales (1984) present 5 years of observations from three MAP3S stations to demonstrate how representative the year 1980 was of the 5-year normal for the high-deposition region in the Northeast. Table 5.8 shows the deviations (D) of the 1980 values for precipitation amount and concentration from 5-year (1977-1981) means. The 1980 precipitation amount was 8-16 percent below the 5-year (1977-1981) average, similar to the difference ir Figure 5.7. Despite this, the deviations of the ionic concentrations from the 5-year means were variable, ranging from -9 percent for hydrogen ion at Pennsylvania State University to +25 percent for sulfate and ammonium at Whiteface Mountain, New York. Longer data records are required in order to establish whether such deviations from long-term means are typical and acceptable.

Seasonal variations in the concentrations of particular ions, such as sulfate and hydrogen, have been discussed frequently. For example, Herman and Gorham (1957) reported that in Nova Scotia snow sampled in the early 1950s contained lower sulfur and nitrogen concentrations than did rain sampled during the same period. In the late 1960s, Fisher et al. (1968) observed lower concentrations of precipitation sulfate in the cold season. Bowersox and de Pena (1980), Pack and Pack (1980), and Pack (1982) reported strong seasonal variations in sulfate in precipitation at MAP3S sites in New York, Pennsylvania, and Virginia.

Bowersox and de Pena (1980) found only slightly higher nitrate in precipitation in the winter than they did in other seasons at the MAP3S site in Pennsylvania. Hydrogen ion had a maximum in the warm months, and sulfate was the principal anion affecting acidity. Nitrate, at concentrations similar to those of sulfate, did not correlate

(text continues on page 157)

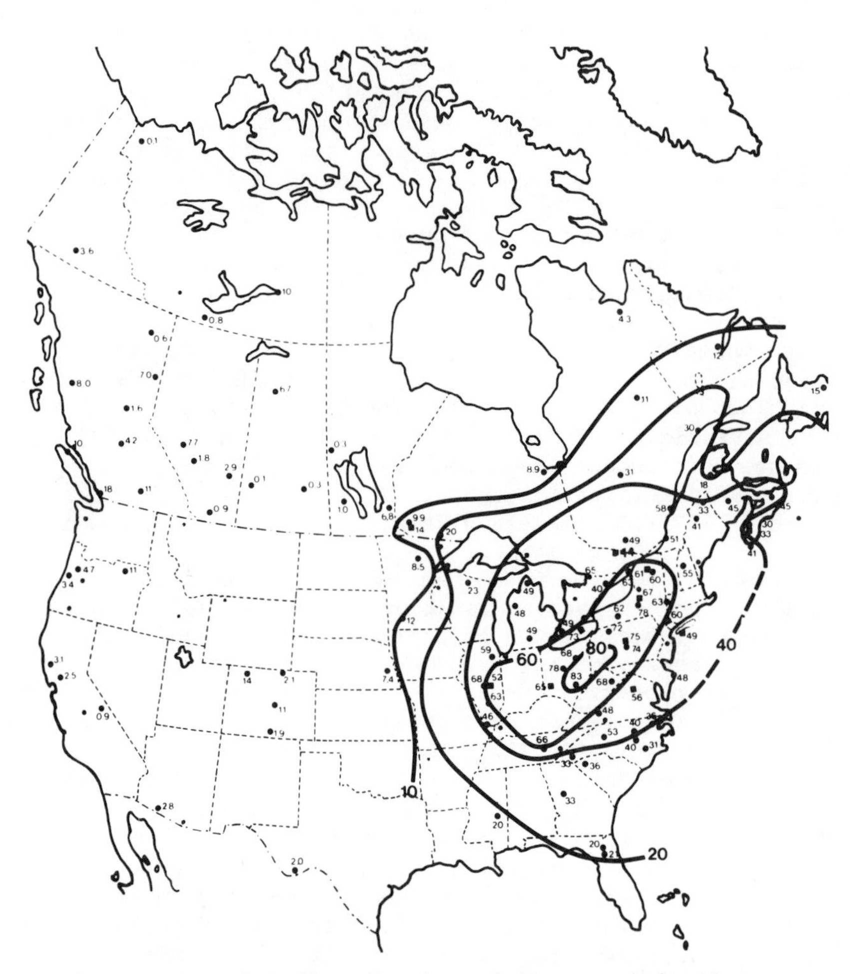

FIGURE 5.2 Spatial distribution of the precipitation-amount-weighted annual mean hydrogen-ion concentration (micromoles per liter) in North America in 1980. Data from four networks are plotted: Canada, CANSAP (circles) and APN (squares); United States, NADP (circles) and MAP3S (squares). SOURCE: Barrie and Hales (1984).

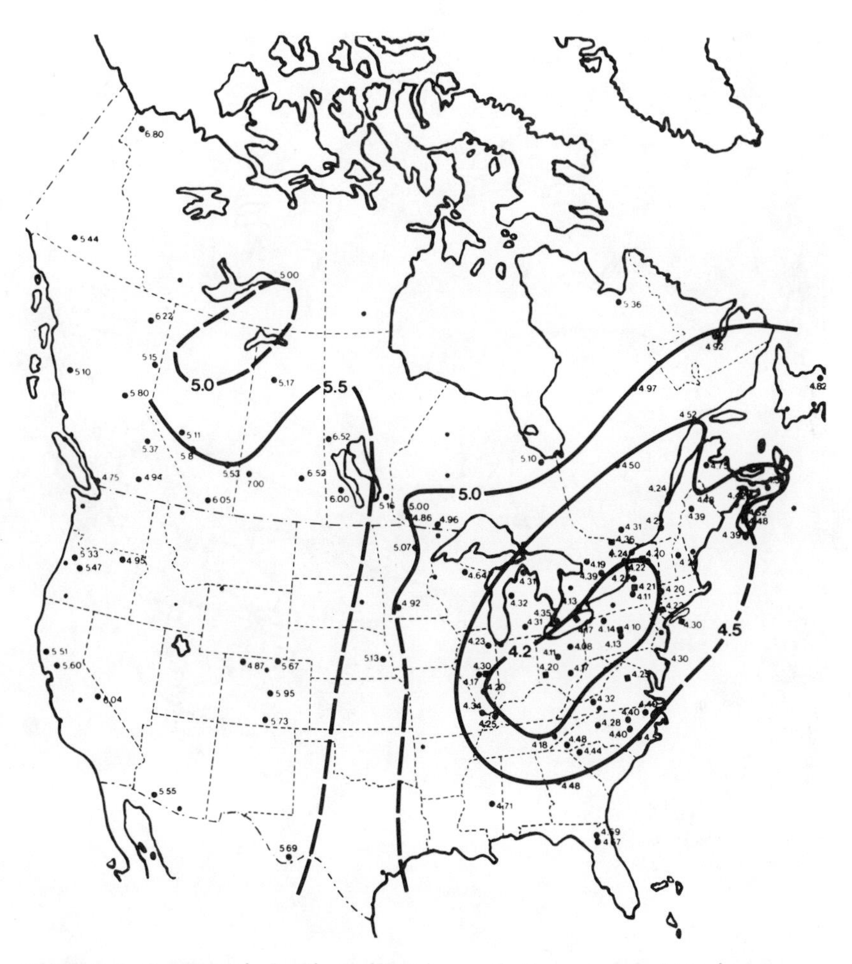

FIGURE 5.3 Spatial distribution of the precipitation-amount-weighted annual mean hydrogen-ion concentration (expressed as pH) in North America in 1980. Data from four networks are plotted: Canada, CANSAP (circles) and APN (squares); United States, NADP (circles) and MAP3S (squares). SOURCE: Barrie and Hales (1984).

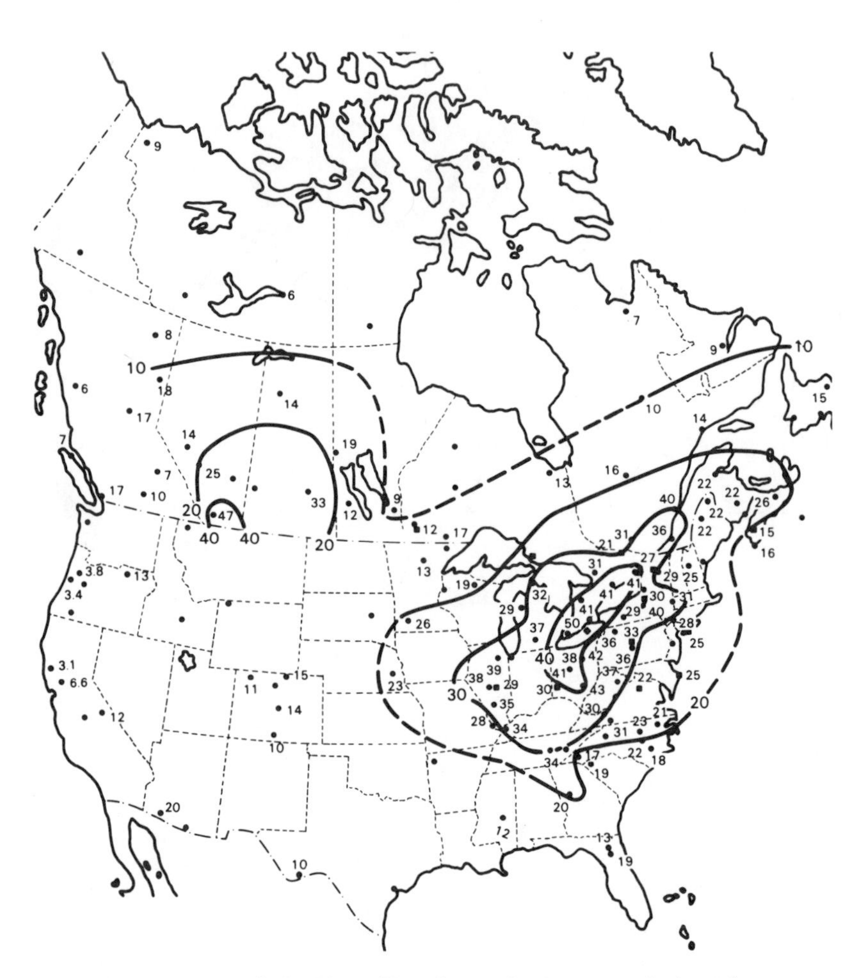

FIGURE 5.4 Spatial distribution of the precipitation-amount-weighted annual mean sulfate-ion concentration (micromoles per liter) in North America in 1980. Data from four networks are plotted: Canada, CANSAP (circles) and APN (squares); United States, NADP (circles) and MAP3S (squares). SOURCE: Barrie and Hales (1984).

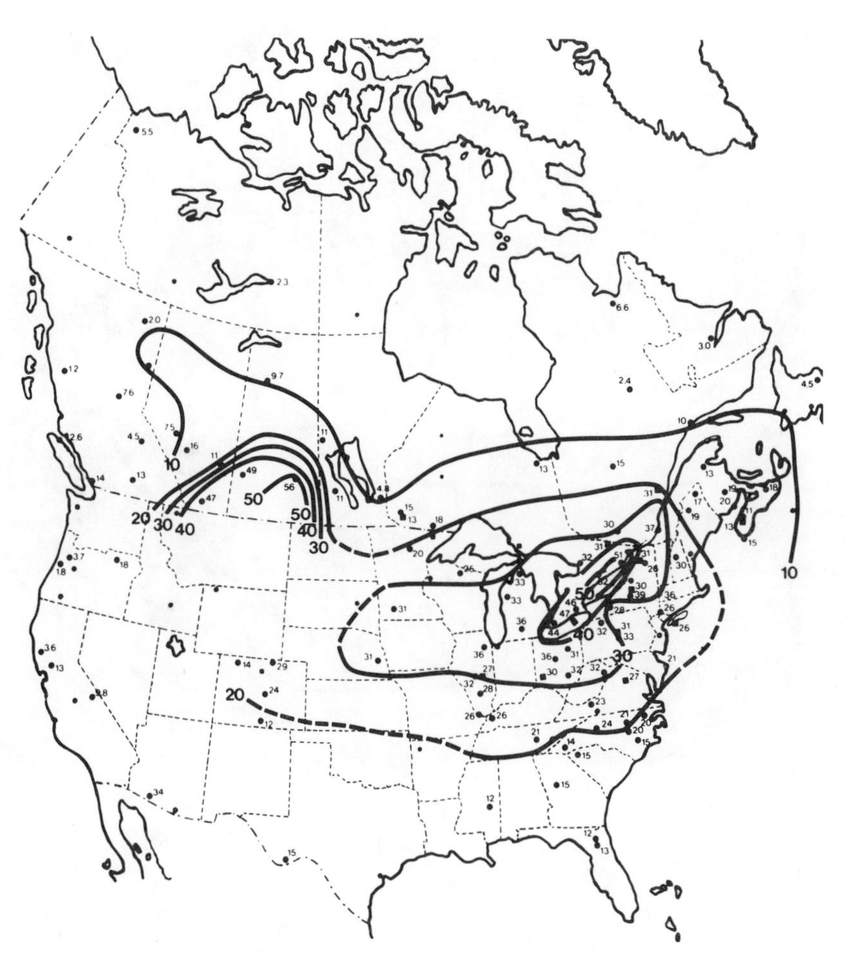

FIGURE 5.5 Spatial distribution of the precipitation-amount-weighted annual mean nitrate-ion concentration (micromoles per liter) in North America in 1980. Data from four networks are plotted: Canada, CANSAP (circles) and APN (squares); United States, NADP (circles) and MAP3S (squares). SOURCE: Barrie and Hales (1984).

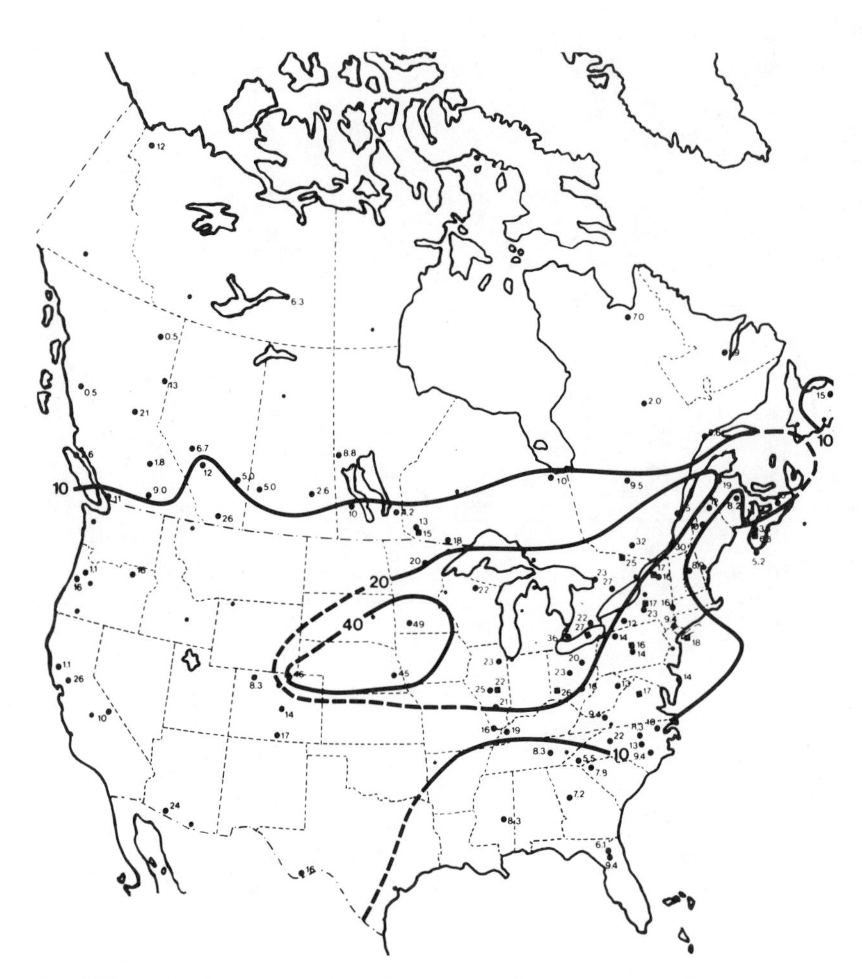

FIGURE 5.6 Spatial distribution of the precipitation-amount-weighted annual mean ammonium-ion concentration (micromoles per liter) in North America in 1980. Data from four networks are plotted: Canada, CANSAP (circles) and APN (squares); United States, NADP (circles) and MAP3S (squares). SOURCE: Barrie and Hales (1984).

FIGURE 5.7 Precipitation amount in 1980 expressed as a percentage of the long-term mean amount of precipitation in North America. SOURCE: Barrie and Hales (1984).

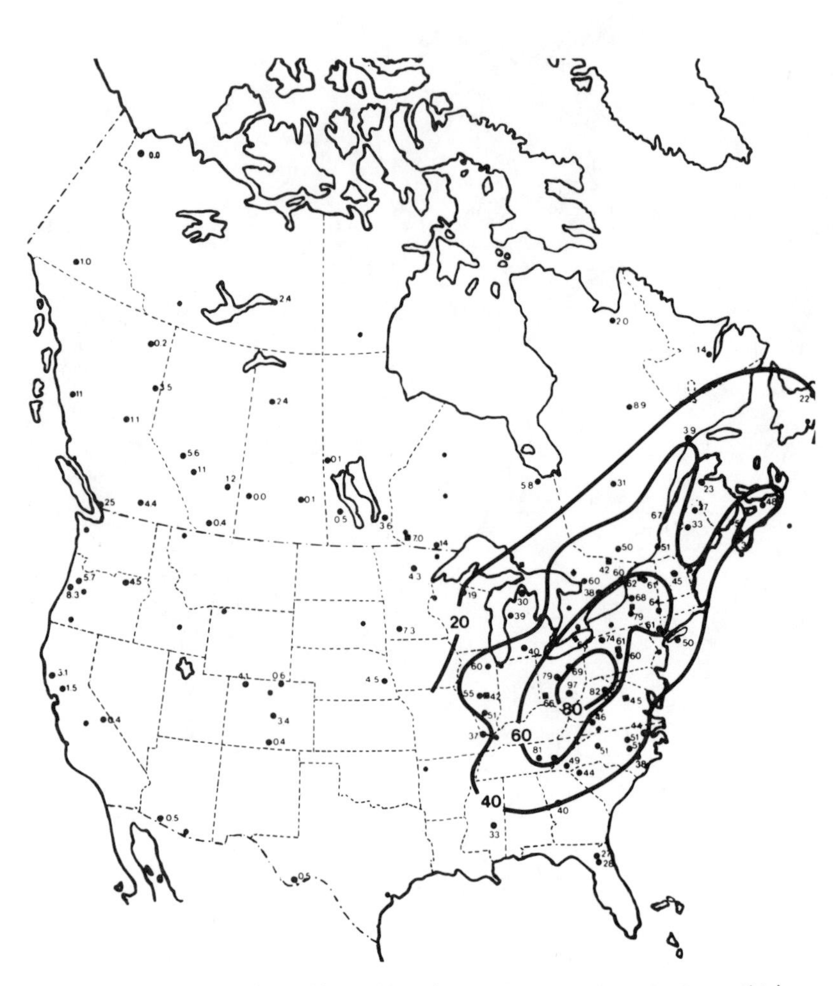

FIGURE 5.8 Spatial distribution of annual wet deposition of hydrogen ions (millimoles per square meter) in North America in 1980. Data from four networks are plotted: Canada, CANSAP (circles) and APN (squares); United States, NADP (circles) and MAP3S (squares). SOURCE: Barrie and Hales (1984).

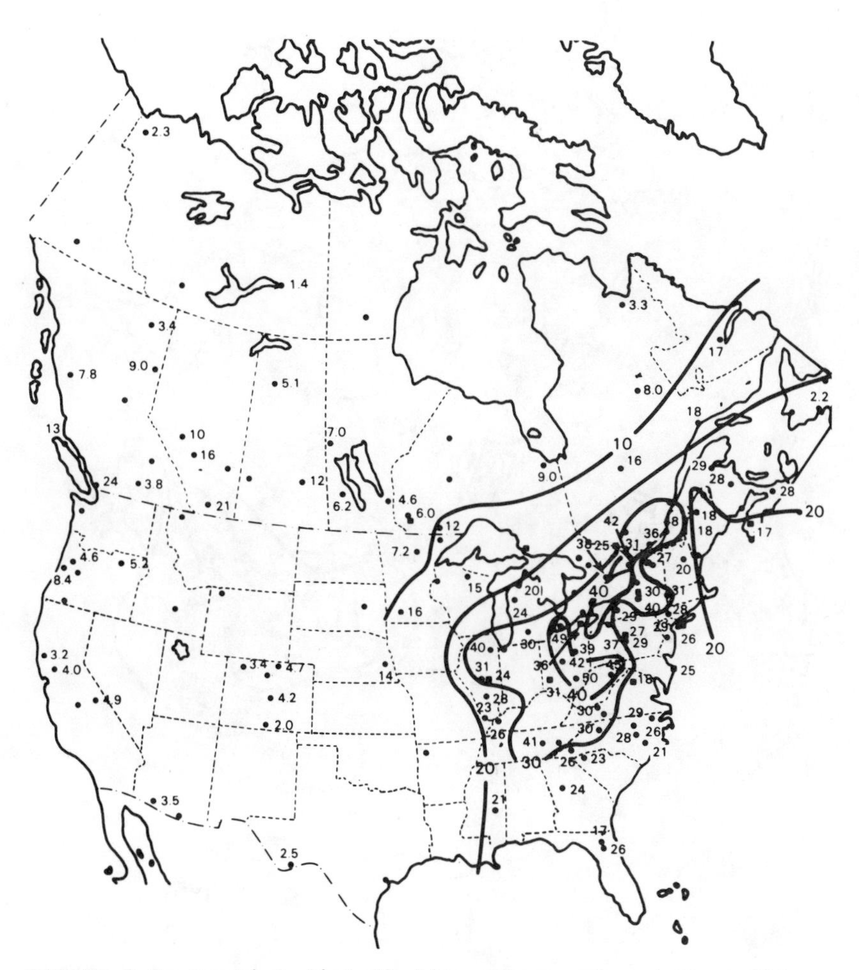

FIGURE 5.9 Spatial distribution of annual wet deposition of sulfate ion (millimoles per square meter) in North America in 1980. Data from four networks are plotted: Canada, CANSAP (circles) and APN (squares); United States, NADP (circles) and MAP3S (squares). SOURCE: Barrie and Hales (1984).

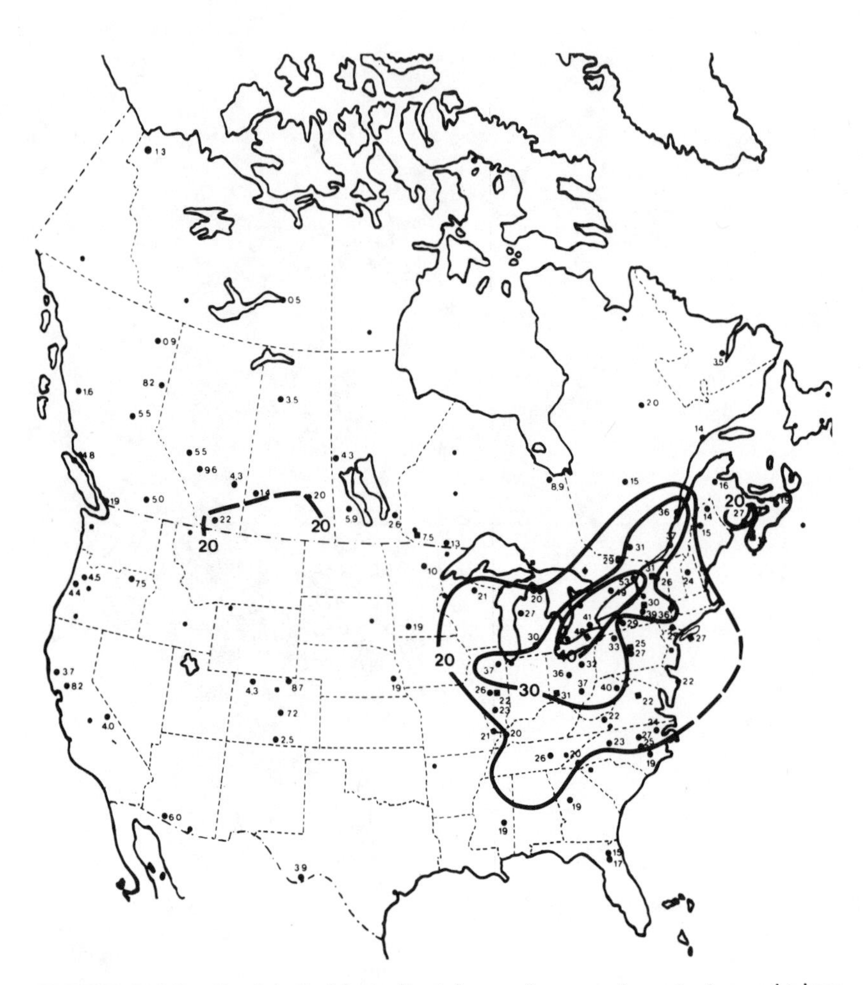

FIGURE 5.10 Spatial distribution of annual wet deposition of nitrate ion (millimoles per square meter) in North America in 1980. Data from four networks are plotted: Canada, CANSAP (circles) and APN (squares); United States, NADP (circles) and MAP3S (squares). SOURCE: Barrie and Hales (1984).

FIGURE 5.11 Spatial distribution of annual wet deposition of ammonium ion (millimoles per square meter) in North America in 1980. Data from four networks are plotted: Canada, CANSAP (circles) and APN (squares); United States, NAPD (circles) and MAP3S (squares). SOURCE: Barrie and Hales (1984).

TABLE 5.8 Annual Mean Precipitation Amount (P) and Concentration of Major Ions (μeq/L) at Three Sites in the MAP3S Network in the Northeastern United States from 1977 to 1981

		Ion			
Year	P (cm)	H^+	SO_4^{2-}	NO_3^-	NH_4^+
Site: Whiteface Mountain, New York (73°51′W, 44°23′N)					
1977	116	46	22	22	11
1978	80	60	29	26	17
1979	107	50	20	19	11
1980	95	61	27	26	17
1981	124	37	21	18	12
D(%)[a]	−9	20	25	17	25
Site: Ithaca, New York (76°43′W, 42°23′N)					
1977	99	69	30	29	13
1978	78	84	34	31	16
1979	102	70	28	25	15
1980	91	67	30	30	17
1981	124	70	37	31	17
D(%)	−8	−7	−6	3	9
Site: Pennsylvania State University (77°57′W, 40°47′N)					
1977	105	73	33	31	15
1978	94	78	21	31	13
1979	132	58	26	26	15
1980	88	75	33	31	16
1981	104	59	35	30	25
D(%)	−16	−9	7	4	−4

[a]D is the deviation of 1980 observations from the 5-year mean.
SOURCE: Barrie and Hales (1984).

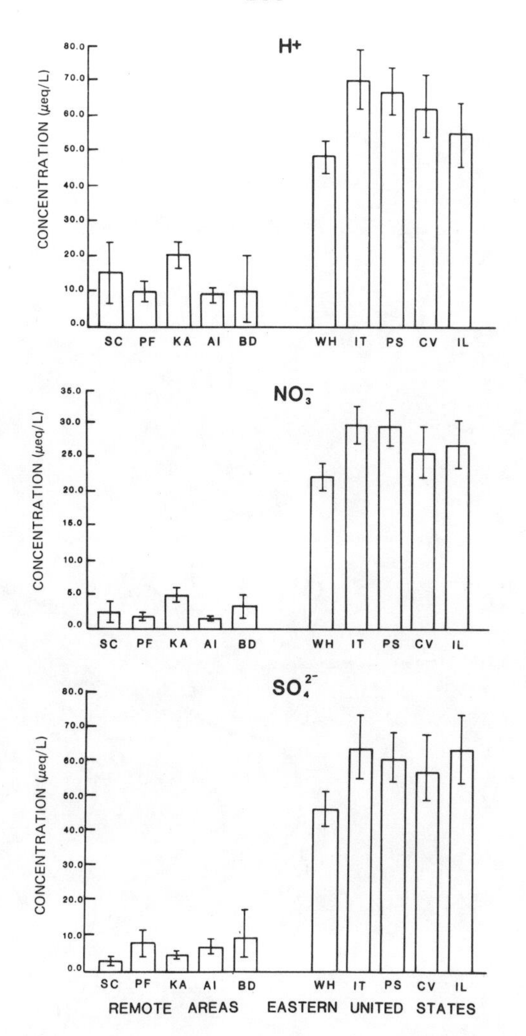

FIGURE 5.12 Comparison of the volume-weighted-mean concentration and the upper and lower error estimates among five sites in remote areas and five sites in the eastern United States. Remote sites: San Carlos, Venezuela (SC); Poker Flat, Alaska (PF); Katherine, Australia (KA); Amsterdam Island, Indian Ocean (AI); and Bermuda, Atlantic Ocean (BD). Eastern United States sites: Whiteface Mountain, New York (WH); Ithaca, New York (IT); State College, Pennsylvania (PS); Charlottesville, Virginia (CV); and Champaign, Illinois (IL). SOURCE: Galloway et al. (1984).

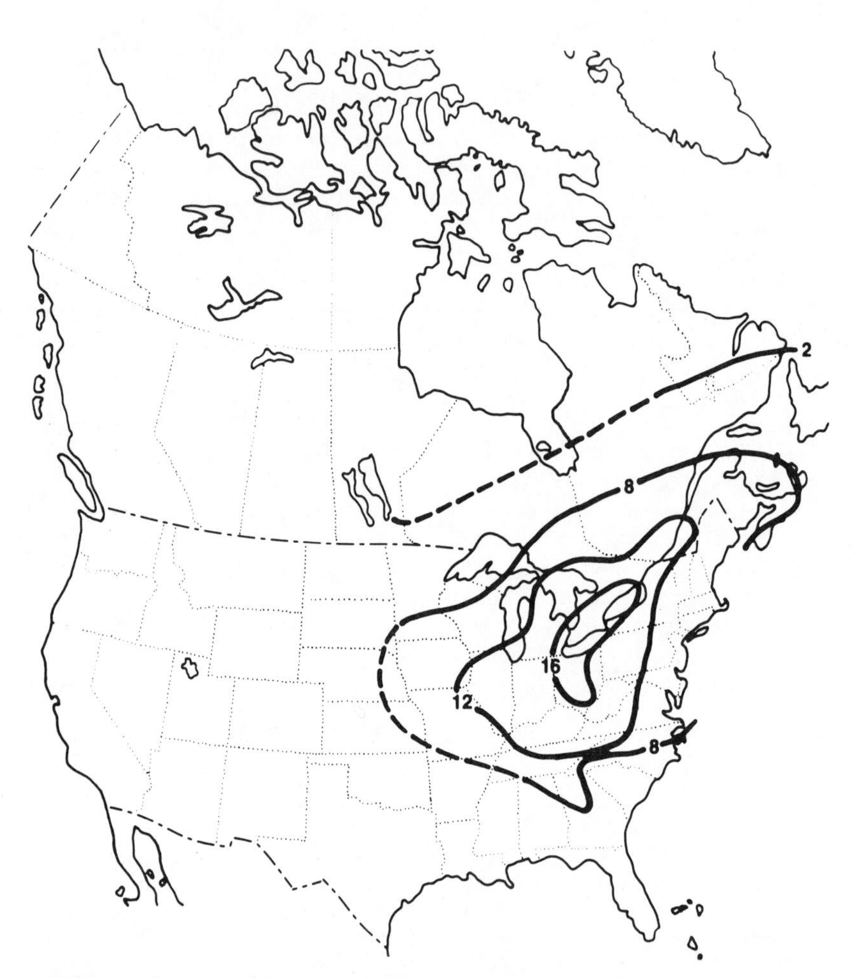

FIGURE 5.13 Enrichment factors for sulfate concentrations in North American precipitation. Numbers signify the ratio of sulfate concentration in precipitation in eastern North America to sulfate concentration in precipitation in remote areas of the world. SOURCE: Galloway et al. (1984).

well with hydrogen ion in liquid precipitation but did in snow and frozen precipitation.

Bowersox and Stensland (1981) analyzed NADP data for seasonal variations in sulfate, nitrate, and hydrogen-ion concentrations. The sampling sites were grouped into five regions in the eastern United States, each containing from two to seven sites. The data for each region were averaged for the cold season (November to March) and the warm season (May to September). The resulting warm-to-cold period ratios for sulfate varied from a maximum of about 2.0 in the New England region to a minimum of 1.25 in the Illinois and Michigan regions. In terms of seasonal (three-month interval) trends, maximum sulfate concentrations occurred in the summer. The investigators noted that aerosol sulfate has a similar seasonal variation but that sulfur oxide emissions for the Northeast have a relatively small seasonal variation. For nitrate, Bowersox and Stensland (1981) found a maximum warm-to-cold period ratio of 1.5 for the region in the Southeast and a minimum ratio of 0.9 for the Michigan region. Thus, the nitrate and sulfate seasonality ratios had different geographical patterns. The free acidity of the precipitation was greater in the warm period for all the regions and reflected the mixture of the patterns for sulfate and nitrate. The free-acidity ratio of warm-to-cold season values was a maximum of 1.5 for the Pennsylvania region and a minimum of 1.2 for the Illinois region.

The seasonal pattern of precipitation sulfate concentration is different for western Europe from that for the eastern United States. Granat (1978) averaged the data for many European sites and reported a maximum sulfate concentration in the spring, that was 1.6 times greater than the minimum value observed in the fall. The sulfur emissions in the region are at maximum in the winter (Ottar 1978).

Comparing Global Background and Eastern North American Precipitation Composition

Comparing precipitation composition and wet deposition in heavily populated or industrialized regions with those in remote regions of the world shows clearly the higher levels caused by anthropogenic emissions. The elevated concentrations of acid-related species in precipitation in eastern North America, Europe, and Japan attest to this.

The results of Barrie and Hales (1984) show that within North America the concentrations of hydrogen, sulfate, and nitrate ions in the Northeast may exceed the continental background levels found in the West and the North by some 20, 5, and 10 times, respectively. On the global scale, Galloway et al. (1984) have compared data from five sites in remote areas of the world--San Carlos, Venezuela; Poker Flat, Alaska; Katherine, Australia; Amsterdam Island, Indian Ocean; Bermuda, Atlantic Ocean--with those from five wet-only sites in the eastern United States--Whiteface Mountain, New York; Ithaca, New York; State College, Pennsylvania; Charlottesville, Virginia; and Champaign, Illinois (Figure 5.12). Values are precipitation-weighted means of 14 to 125 samples from remote locations and 186 to 373 samples from eastern North America. Figure 5.12 shows that hydrogen-ion concentrations at remote locations are significantly lower than those in eastern North America and that remote sulfate and nitrate concentrations are consistently lower by more than a factor of 5. To illustrate the enrichment of North American concentrations of sulfate over global background, Figure 5.13 (Galloway et al. 1984) shows the ratio of 1980 concentrations in North America to global background concentrations.

Although up to a fivefold enrichment of hydrogen ion occurs, Galloway et al. (1984) point out that such a comparison is somewhat misleading because the hydrogen-ion concentration in remote areas may be determined by an entirely different mix of substances, primarily organic acids, from that in eastern North America, where strong mineral acids are the main contributors. For nitrate and sulfate, variability from the different remote sites is small, and the North American levels are from 2 to 10 and 2 to 16 times greater, respectively.

From the beginning of the acid rain debate, precipitation pH has been used as the central variable, or indicator, of the phenomenon. As Galloway et al. (1984) showed, this has not been a particularly useful choice. At remote continental locations, organic acids may lower the pH to levels below 5.0 (Galloway et al. 1982). In addition, Charlson and Rodhe (1982) have shown that "clean" rain may contain naturally occurring acids (notably sulfuric acid in the natural portion of the sulfur cycle) in high enough concentrations, in the absence of basic materials, to produce pH values in the 4.5 to 5.6 range. Thus, pH values less than 5.0 may be found in remote areas in the absence of anthropogenic sulfur and nitrogen. In such circumstances, the concentrations of sulfur and nitrogen are much lower than in industrialized areas.

Comparing precipitation concentrations of acid species in remote and industrialized regions of the world shows the effect of anthropogenic emissions in the latter. In the absence of lengthy time-trend records, concentrations in remote areas serve to establish baseline values in regions currently affected.

TIME-TREND INFORMATION

Data Available

The optimum data set for analyzing trends would consist of a long unbroken series of measurements, made using the same methods, at sites known to be representative of the phenomenon being examined. Unfortunately, no precipitation chemistry data set exists in North America that has all these attributes.

The desired record of data would be long enough to cover the time period in which the phenomenon of interest occurred and extend some time before and after. These extended time periods would help to keep erroneous conclusions from being drawn based on shorter-term trends or drifts in the data. Changes in protocols or methods of data collection and analysis introduce uncertainties into the record, unless the changes have been thoroughly documented and there has been a period of overlap when both old and new procedures have been used.

With continuous records available from so few locations, it is tempting to assume that trends in such locations are indicative of what has happened over a larger area. This may or may not be the case. Hidy et al. (1984) discuss the problem of site representativeness in this context in their analysis of the USGS data. (See the section below on trends for Hubbard Brook and USGS bulk data.)

Figure 5.14 summarizes the types of information available on which an evaluation of trends must be made. Before Junge's measurements in 1955-56, only sporadic measurements were made, beginning about 1910. These have been reviewed by Hidy et al. (1984) and Cogbill et al. (1984) and are discussed below.

It is possible to compare the various "snapshots in time" provided by measurements by Junge in the 1950s and

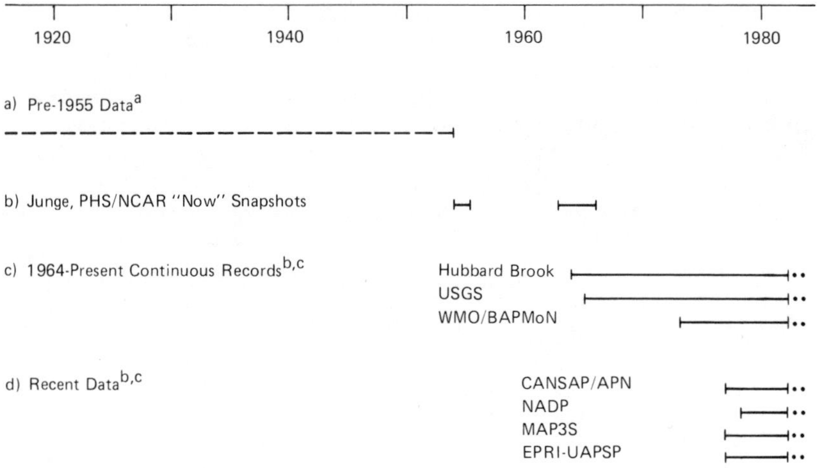

1920 1940 1960 1980

a) Pre-1955 Data[a]

b) Junge, PHS/NCAR "Now" Snapshots

c) 1964-Present Continuous Records[b,c]

Hubbard Brook
USGS
WMO/BAPMoN

d) Recent Data[b,c]

CANSAP/APN
NADP
MAP3S
EPRI-UAPSP

[a]Dashed line indicates period of sporadic measurements.
[b]Dotted lines indicate that measurements are continuing.
[c]Solid lines indicate time period over which data were available to authors of this report.

FIGURE 5.14 Summary of available precipitation chemistry data bases.

by the Public Health Service/National Center for Atmospheric Research (PHS/NCAR) in the 1960s with present-day measurements. Beginning in 1963, a continuous and valuable series of measurements has been made at Hubbard Brook, New Hampshire. Soon afterward the USGS network was established, and in the early 1970s the World Meteorological Organization (WMO) network was established in cooperation with the governments of the United States and Canada as part of a more extensive global program. (This network is sometimes referred to as WHO/NOAA/EPA in the United States after the National Oceanic and Atmospheric Administration and the U.S. Environmental Protection Agency, the two agencies responsible for administering the monitoring program). More recently, since approximately 1978, several extensive sampling networks have come into operation. NADP and CANSAP/APN (Canadian Network for Sampling Precipitation/ Air and Precipitation Monitoring Network) are discussed in subsequent sections. A detailed description of the MAP3S and EPRI-UAPSP (Electric Power Research Institute-Utility Acid Precipitation Sampling Program) is presented in Wisniewski and Kinsman (1982). They also discuss several other monitoring activities in North America not covered in this chapter.

Pre-1955 Data

Most precipitation chemistry data available before approximately 1955 were obtained by bulk sampling. In most cases, measurements were taken to characterize nutrient input to agricultural soils for purposes of soil management. The reliability of many of these data is questionable because not only were they based on bulk samples but there were substantial differences in collector materials, sampler designs, chemical analysis methods, and site location. Hidy et al. (1984) have reviewed many of these historical data and have tabulated times, locations, and methods of collection.

Figure 5.15, adapted from Hidy et al. (1984), shows sulfate concentrations in New York and Connecticut derived from bulk samples for the period 1910 to 1954. The data were from stations that used stainless steel, aluminum, or glass samplers located outside towns or cities or that reported the lowest deposition in an area for a given year. (Hidy et al. assumed that areas of lowest deposition provided the cleanest samples since they were less likely to be contaminated by dust or other impurities.) These data provide a qualitative picture of the levels encountered during the first half of the present century.

In the Northeast, ion concentration and deposition are relatively constant over the period: annual concentration averages for sulfate are in the range of 10 to 40 mg/L, and deposition ranges from 20 to 80 kg ha^{-1} yr^{-1}. The few available data for the Midwest are comparable to those for the Northeast, whereas those for the Southeast show lower concentrations and depositions. As we will show below, concentration and deposition values after 1955 are generally much lower. We cannot draw conclusions about the cause of the higher values before 1950; they may be the result of sampling and analysis practices, siting, higher ambient concentrations, or some combination thereof. These data are not indicative of regional wet-only concentration patterns and cannot be compared with modern wet-only data to establish time trends.

Trends from Mid-1950s to Present

As shown in Figure 5.14, data from four relatively brief periods of time have been used to investigate changes in precipitation composition, primarily pH, in

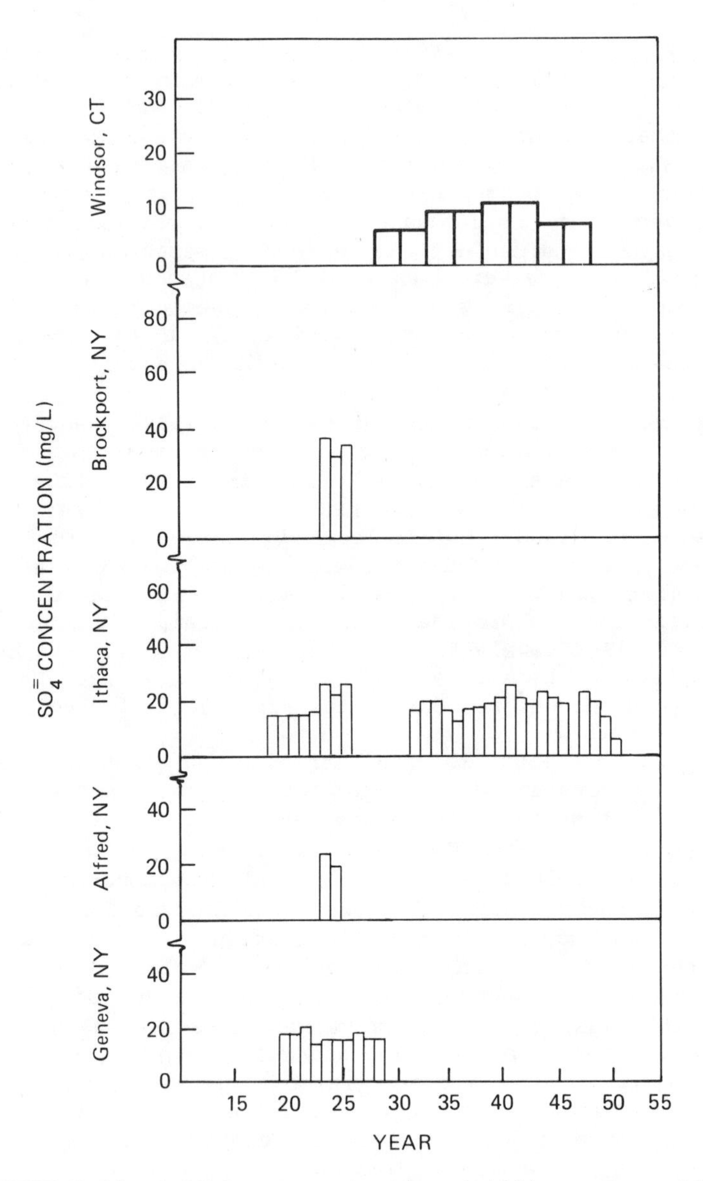

FIGURE 5.15 Sulfate concentration (milligrams per liter) in bulk precipitation samples for locations in the northeastern United States. The width of the bar designates the period of sampling. Adapted from Hidy et al. (1984). Data sources are listed in Table 1 of Hidy et al.

the period since the mid-1950s. The major sources of these data are Junge and Werby (1958) for the mid-1950s, Lodge et al. (1968) (also referred to as the PHS/NCAR data base) for the early 1960s, a variety of sources for the mid-1970s (see Likens and Butler 1981), and the NADP and CANSAP networks in recent years. Analysis of these data sets and the conclusions drawn from them have given rise to considerable debate in the published scientific literature over the past few years.

On the basis of their analyses Likens and Butler (1981) and Cogbill et al. (1984) concluded that a large spread and probable intensification of acid precipitation (pH < 5.6) in eastern North America have occurred during the past 25 years and that acid precipitation may not have been widespread in the eastern United States before 1930. More specifically, they concluded that precipitation has become more acidic, especially in the Midwest and Southeast, between the 1950s and the present and that many areas of the East show an increase in acidity of greater than 22 μeq/L during the past three decades.

The major points of concern about using these data for trend analysis involve (1) the amount of uncertainty and bias introduced by differences in sampling, analytical, and calculational methods and (2) the influence of differing climatological conditions, particularly drought, during the periods of data collection.

Hansen and Hidy (1982) reviewed the available post-1955 data and assessed the biases and uncertainties associated with (1) use of different sampling, analytical, and calculational methods to deduce precipitation acidity, (2) use of data from different times and conditions, (3) failure to collect brief or light rainfall and initial precipitation in an event, (4) incomplete chemical analyses, and (5) selection of data for analysis. The outcome of this assessment was to determine a series of uncertainty values in pH by decade and geographical region. Cogbill et al. (1984) dispute several of these estimates. Hansen and Hidy (1982) conclude as follows: "The results show that the historical data are of insufficient quality and quantity to support any long-term trends in precipitation acidity change in the eastern United States." Somewhat contradictorily, however, they do allow that ". . . it now seems that the precipitation acidity on the average, was lower in the mid-1950s than in the mid-1960s and thereafter in parts of the eastern United States"; they also conclude that despite limited observations in the Southeast it appears that a region of elevated acidity (pH < 5.0) in rainfall from Alabama through Florida exists that was not evident in the mid-1950s.

Stensland and Semonin (1982) evaluated the potential role of elevated levels of calcium and magnesium from crustal sources in the composition of precipitation in the mid-1950s. Drought conditions in the mid-1950s are likely to have caused increased concentrations of wind-raised soil components in precipitation samples, possibly resulting in calculated acidity values lower than would have been typical of the period. Stensland and Semonin (1984) also point out that data from the 1960s (PHS/NCAR) and the 1970s (WMO/NOAA/EPA) show generally higher ion concentrations than do current NADP data. Figures 5.16 and 5.17 show a comparison between older and more recent data for calcium and sulfate ions. The analysis suggests that the earlier measurements--1950s, 1960s, 1970s--were perhaps affected by evaporation and dry deposition (dust leakage) and may have been more like bulk than wet-only samples in many cases. On the basis of a simple pH model Stensland and Semonin (1982) recalculated the 1955-1956 pH distribution and found a smaller pH change between the mid-1950s and the present than do the calculations of Likens and Butler (1981) (see Chapter 3, Figure 3.4). Stensland and Semonin (1982) point out the crucial importance of (1) measuring for all major ions in samples to be used for trend analysis and (2) analyzing concurrently meteorological/ climatological factors when investigating large-scale temporal and spatial changes.

Investigators have not reached a consensus on the subject of precipitation acidity changes since the 1950s; the quality and the quantity of the earlier data make a quantitative evaluation of the change difficult. Nevertheless, workers agree that precipitation acidity has increased since the 1950s in parts of the eastern United States despite disagreement on the amount of the change and the mechanism. More specifically, there is agreement that the acidity of precipitation has increased in the Southeast (e.g., Barrie and Hales 1984) since the 1950s. However the reason for the change has not been thoroughly explored. A comparison with the Junge data shows that the concentrations of sulfate in the southeastern states have generally increased since the 1950s (see Figure 5.18). A similar trend is indicated for nitrate and hydrogen ions (Barrie and Hales 1984). However, concentrations of calcium and calculated magnesium have

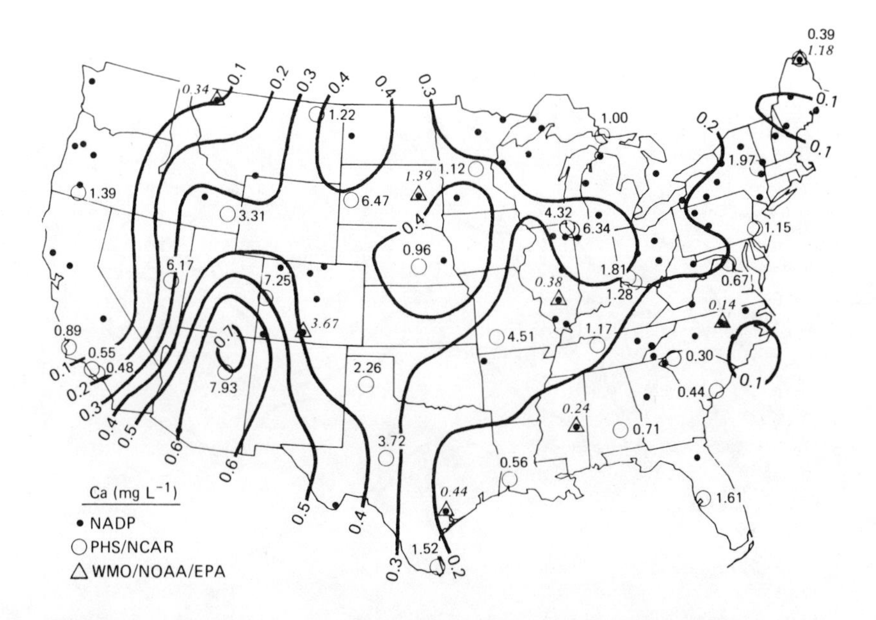

FIGURE 5.16 Weighted average calcium concentration in precipitation (milligrams per liter). The computer-generated contours are for the NADP wet deposition data for 1978-1981 for sites with at least 20 valid samples. The other sites on the map with numerical data show the calcium concentrations determined in earlier years by PHS/NCAR (1960-1966) (bold numbers) and WMO/NOAA/EPA (1972-1973) (italic numbers). SOURCE: Stensland and Semonin (1984).

decreased significantly. We do not know the relative importance of strong acids contributing hydrogen ions versus decreased soil components resulting in less neutralization.

For the northeastern states, Figure 5.18 indicates that sulfate concentrations have risen in some areas (Massachusetts, New York, Ohio, and Indiana) although generally not to the same degree as the Southeast. The other northeastern states either exhibit no change (Pennsylvania and Michigan) or show decreases (Maine, Massachusetts, Vermont, and Maryland).

The very large negative differences in sulfate concentration (Figure 5.18) at many sites in the Great Plains and the West raise the question of comparability of the 1955-1956 sulfate data with the recent wet-only

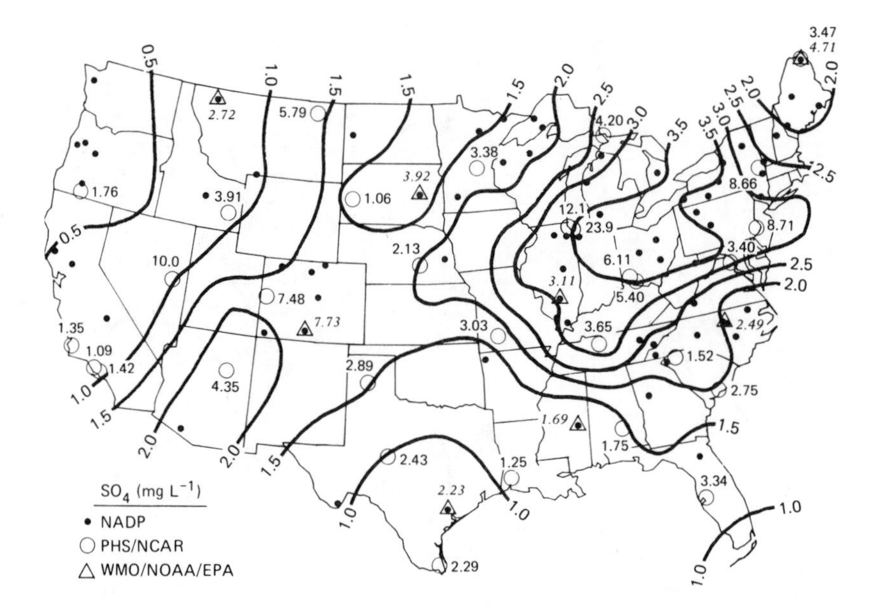

FIGURE 5.17 Weighted-average sulfate concentration in precipitation (milligrams per liter). The computer-generated contours are for the NADP wet deposition data for 1978-1981 for sites with at least 20 valid samples. The other sites on the map with numerical data show the sulfate concentrations determined in earlier years by PHS/NCAR (1960-1966) (bold numbers) and WMO/NOAA/EPA (1972-1973) (italic numbers). SOURCE: Stensland and Semonin (1984).

data. It seems likely that dry deposition and evaporation may have biased the 1950s sulfate data upward for many western sites.

The need to have a continuing data base with which to evaluate trends is clearly evident. Although the need for quality-assured data is widespread, such data are crucial in the case of trend analysis since the signal-to-noise ratio may be small. In addition to the particular species of concern in trend analysis of acid deposition, we need to examine other atmospheric chemical components as well as meteorological/climatological parameters.

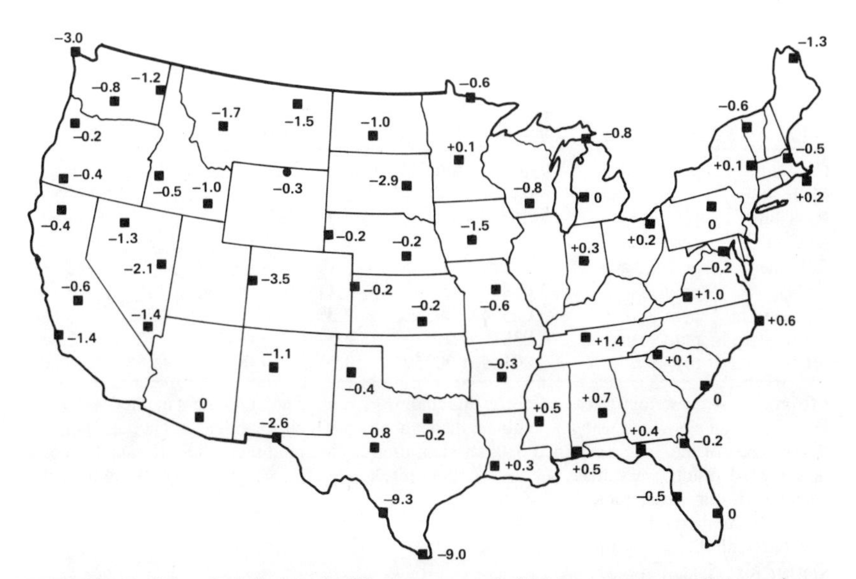

FIGURE 5.18 Sulfate differences (milligrams per liter) determined by subtracting 1955/1956 precipitation sulfate concentrations (Junge 1958) from NADP sulfate concentrations.

Trends for Hubbard Brook and USGS Bulk Data

The Hubbard Brook Data

In this section we describe a trend analysis undertaken on the Hubbard Brook weekly bulk-precipitation chemistry data (Oehlert 1984a). The data consist of analyses of the more than 500 bulk samples collected between 1964 and 1979. The trend analysis was done for the following ions: hydrogen, sulfate, nitrate, chloride, ammonium, calcium, magnesium, potassium, and sodium; chloride data were not available until June 1967. Data were made available through the courtesy of G. Likens and J. Eaton (Cornell University, personal communication, 1984).

Monthly volume-weighted average concentrations and monthly depositions were calculated from the raw data. We analyzed the data for chloride, calcium, magnesium, potassium, and sodium on the logarithmic scale. Seasonal influences were eliminated from the variables by subtracting from each monthly value the mean of the time series for that month.

TABLE 5.9 Precipitation-Chemistry Trends at Hubbard Brook

		Observed Data			Smoothed Data[a]		
Ion	Scale	Trend	se[b]	t[c]	Trend	se[b]	t[c]
Calicium	percent/year	−9.49	1.06	−8.95	−9.28	1.07	−8.67
Magnesium	percent/year	−6.96	1.94	−3.59	−7.13	1.84	−3.87
Potassium	percent/year	−7.01	1.87	−3.75	−6.48	1.61	−4.02
Sodium	percent/year	−6.25	1.39	−4.50	−5.82	1.18	−4.93
Chloride	percent/year	−9.19	4.20	−2.19	−8.95	3.34	−2.68
Sulfate	ppm/year	−0.0383	0.0159	−2.41	−0.0467	0.0147	−3.18
Ammonium	ppm/year	−0.00554	0.00236	−2.35	−0.00651	0.00247	−2.63
Nitrate	ppm/year	0.0381	0.0161	2.37	0.0367	0.0152	2.41
pH	pH/year	0.00143	0.00259	0.55	0.00204	0.00256	0.80

[a] To prevent undue weighting of outliers on this analysis, a rough outlier resistant method was used. Data were first smoothed using a nonlinear data smoother, which smoothed single outliers more than the rest of the data. Smoothed data are then used in the trend regression. If smoothed and unsmoothed results are similar, the unsmoothed result is preferable; if they are dissimilar, the smoothed result is preferable.

[b] se is the standard error.

[c] t is the trend divided by the standard error.

SOURCE: Oehlert (1984a).

The statistical methods for the trend analysis are those described in Oehlert (1984b). Basically, trends are determined by ordinary least-squares regression, but the standard errors of the trends are adjusted for the presence of autocorrelation in the residuals. Nominally, a t value greater than 2 in absolute value would be considered statistically significant. Here the cut-off value of 2 is only approximate.

The linear trends observed in the data over the period of record are shown in Table 5.9. The results show that some strong trends exist in the Hubbard Brook data. In particular, concentrations of all ions but hydrogen and nitrate are decreasing. pH has remained fairly constant. Nitrate has increased significantly over the period 1964 to 1979, although the result is somewhat dependent on statistical technique and the time period of analysis. Inclusion of data more recent than 1979 tends to indicate no change in nitrate concentrations over the entire time span. Figure 5.19 shows the deseasonalized monthly values for pH, sulfate, nitrate, and calcium.

One feature of the sulfate trend is that it varies seasonally. With summer defined as June, July, and August, and the other seasons defined accordingly, the trends for summer, fall, winter, and spring are -0.025, -0.023, -0.061, and -0.055 ppm/yr. Thus, spring and

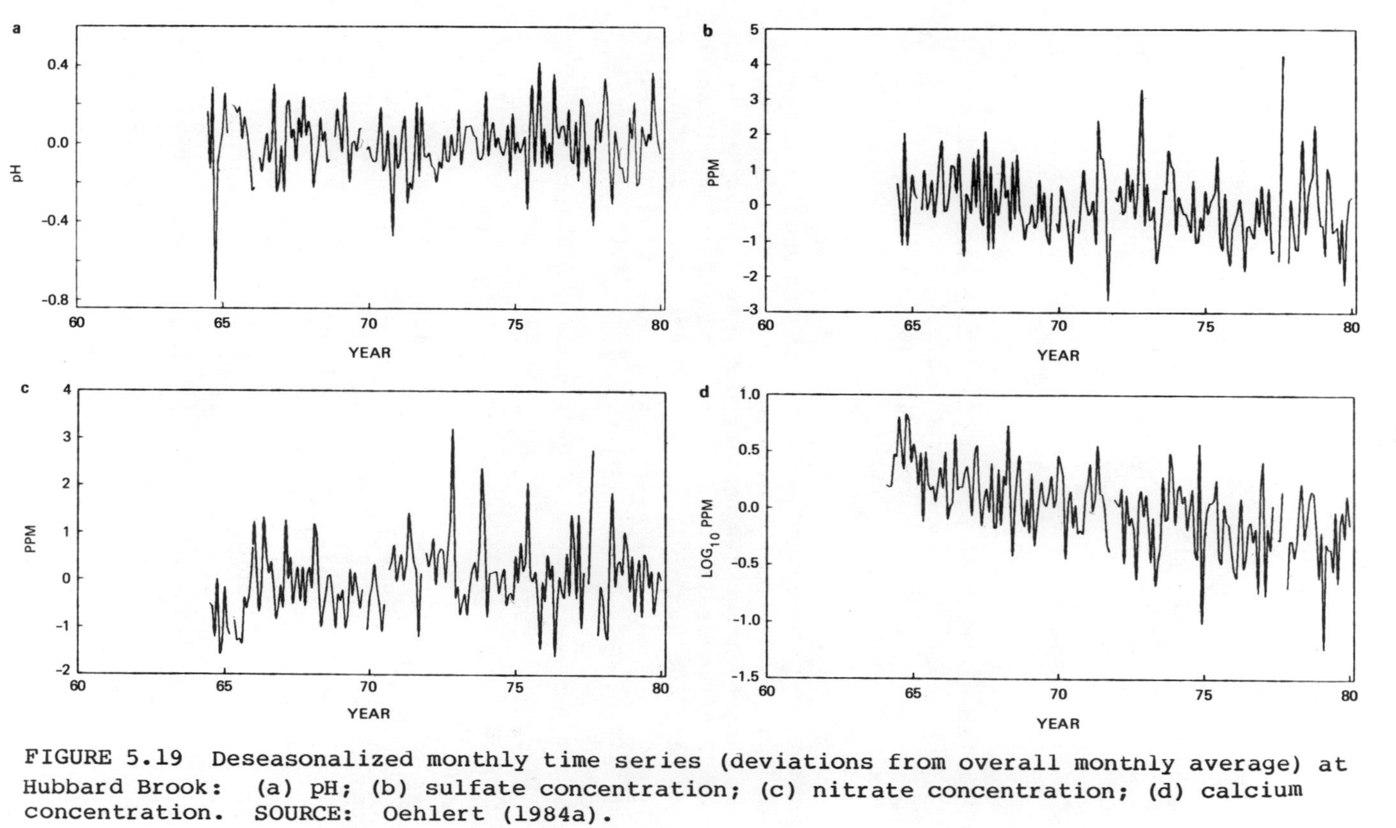

FIGURE 5.19 Deseasonalized monthly time series (deviations from overall monthly average) at Hubbard Brook: (a) pH; (b) sulfate concentration; (c) nitrate concentration; (d) calcium concentration. SOURCE: Oehlert (1984a).

TABLE 5.10 Trends in Ratios of Sulfate, Nitrogen Species, and Total Nitrogen at Hubbard Brook

		Observed Data			Smoothed Data		
Ion	Scale	Trend	se	t	Trend	se	t
SO_4^{2-}/NO_3^-	percent/year	−6.19	1.62	−3.83	−5.81	1.42	−4.10
NH_4^+/NO_3^-	percent/year	−7.52	1.41	−5.34	−7.52	1.34	−5.62
Total N	ppm/year	0.00431	0.00473	0.91	0.00301	0.00467	0.64
SO_4^{2-}/total N	ppm/year	−3.24	1.23	−2.64	−3.12	1.05	−2.98

SOURCE: Oehlert (1984a).

winter concentrations have been falling more rapidly than the summer and fall concentrations.

Ratios of species' concentrations in the data have been used to analyze atmospheric processes (National Research Council 1983) and trends. The data indicate that the ratio of sulfate to nitrate concentrations in Hubbard Brook data has been declining at about 6 percent per year, while the ammonium to nitrate ratio has been declining by about 7.5 percent per year. There has been, however, no overall trend in total nitrogen concentration (nitrate plus ammonium). There is still a decreasing trend in the ratio of sulfate to total nitrogen caused by the decreasing trend in sulfate. Table 5.10 summarizes these trends.

We are aware of no plausible explanation for the decreasing trends in calcium-, magnesium-, potassium-, sodium-, and chloride-ion concentrations. The trends are not related to systematic bias introduced by changes in sampling procedures since these methods have not changed over the period of record (G. Likens, Institute of Ecosystem Studies, New York Botanical Garden, personal communication, 1985). That the ions exhibit this change would suggest changes in local source strengths since these substances are usually found in particulate matter that is not transported large distances from sources. As one possible scenario, the decrease in sulfate (about 20 μeq/L for the 18-year period) and the decrease in calcium plus magnesium (about 18 μeq/L) could result, for example, from a decrease in the input of calcium sulfate, either as dry deposition (sulfur dioxide plus calcium on the funnel surface) or as wet deposition (calcium sulfate in precipitation). Then the portion of the sulfate transported from distant sources would be much less. We have no way to interpret the sulfate trend unequivocally.

Several authors using different statistical methods have examined the Hubbard Brook precipitation chemistry data for trends. Table 5.11 summarizes the methods and results from recent publications on trend analysis of the Hubbard Brook data. A summary of the trends since 1964 follows:

Hydrogen ion: Record is variable with no significant trend; slight tendency to lower concentrations for parts of the period.

Sulfate: Significant decrease of approximately 2%/yr superimposed upon periods of drift; decrease in winter and spring is greater than for summer and fall.

Nitrate: Trend over entire record is variously described as "not significant" (1963-1982) and "increasing" (1964-1979); an apparent change in slope in about 1971 may be from chance.

Ammonium: Three authors describe the trend as "no change" and two others say "decreasing (1.9 and 6.4%/yr)"

Sodium, chloride, calcium, magnesium, potassium aluminum: Agreement among three authors of a strong decrease in all these constituents.

ratios: Sulfate/nitrate, ammonium/nitrate, and sulfate/hydrogen show a consistent, significant decrease; nitrate/hydrogen shows a significant increase.

Several features of this review stand out. Results from the various authors are generally consistent for hydrogen; sulfate; the group including sodium, chloride, calcium, magnesium, potassium; and aluminum and their ratios. For nitrate and aluminum there is qualitative agreement on the behavior of the time-series segments but some disagreement on the trend in the overall record.

The different statistical techniques may introduce some differences in the results, but the length of record used in different studies is more important. The end point of the time series varies among studies between 1977 and 1982. In addition, several authors have noted the differences in trends that may occur with an analysis of selected portions of the time series. Finally, the differences in estimated rates of change obtained in the various analyses point to the need for caution if quantitative comparisons are to be made between trends in precipitation composition and, for example, emissions.

TABLE 5.11 Summary of Trends in Hubbard Brook Precipitation Composition

	% change per Year (Based on First-Year Concentration Value)
1. Likens et al. (1980)	
• Annual weighted-mean pH (1964-1977) has been variable, but no statistically consistent trend	
• Annual weighted-mean SO_4^{2-} has decreased; regression-line slope of -0.07 mg L^{-1} yr^{-1} with highly significant correlation coefficient of 0.71	SO_4^{2-}: -2.2
• Annual weighted-mean NO_3^- increased rather markedly after 1964 but has not increased much since 1970-1971	
• Annual weighted-mean NH_4^+ changed very little	
2. Galloway and Likens (1981)	
• Summer and winter changes (1964-1979) were not significant for H^+	
• SO_4^{2-} showed a significant ($p < 0.05$) decrease in winter, but summer changes were not significant	
• NO_3^- showed a significant ($p < 0.01$) increase in summer, but winter changes were not significant	
• NO_3^-/SO_4^{2-} increased significantly ($p < 0.001$) in both summer and winter, dominated by NO_3^- increase in summer and SO_4^{2-} decrease in winter	
• NO_3^-/H^+ increased significantly in summer ($p < 0.01$) and winter ($p < 0.02$)	
• SO_4^{2-}/H^+ decreased significantly in summer ($p < 0.02$) and winter ($p < 0.03$) but scatter is large	
3. Munn et al. (1982)	
• On the basis of ridit analysis (1964-1979) no trend was found in pH	
• SO_4^{2-} showed a slight fall for annual, summer, and winter	
• NO_3^- showed no change for annual, summer, and winter	
• NH_4^+ showed no change for annual, summer, and winter	

• Na^+, Cl^-, Ca^{2+}, Mg^{2+}, K^+, and Al^{3+} all showed a fall for annual, summer, and winter • SO_4^{2-}/NO_3^- show a fall for annual, summer, and winter • Analysis of time series before and after 1971-1979, at which time a discontinuity appeared in the record, showed some differing trends for each constituent	
4. National Academy of Sciences (1983)	
• SO_4^{2-} declined 33 ± 18% (95% confidence limits) between 1965-1966 and 1979-1980; regression line had a slope of −0.074 (standard error 0.016) and an intercept of 3.186 (standard error 0.138) • NO_3^- data suggest an erratic trend toward a maximum around 1970, followed by a leveling off or a slight decrease	SO_4^{2-}: 2.2
5. Hidy et al. (1984)	
• 1964-1965 through 1980-1981 data qualitatively suggest a shift to lowered H^+ at least through 1976; the linear-regression slope for H^+ is −0.648 μeq L^{-1} yr^{-1} • Record shows a general decrease in SO_4^{2-} superimposed on periods of drift, particularly during 1968-1971 and a linear decrease of about 34% since 1964-1965; regression equation is SO_4^{2-} = −0.0743 (water year) + 3.186, with sigma for the coefficient = ±0.01571 and sigma for the constant = ±0.138 • NO_3^- showed initial trend of increase, with a leveling out after 1971 • NH_4^+ has remained roughly constant throughout the period	SO_4^{2-}: −2.0
6. Likens et al. (1984)	
• 1963-1982 data show no statistically significant trend in annual volume-weighted H^+; since 1970-1971 H^+ showed a significant decline of about 28% ($p < 0.025$, $r = -0.69$) • SO_4^{2-} declined approximately 34%; peak summer concentrations and minimum winter values have also decreased • NO_3^- showed no statistically significant trend; NO_3^- increased from 1964-1965 to about 1970-1971 but has fluctuated since without showing trend	SO_4^{2-}: −1.9

TABLE 5.11 (continued)

	%change per Year (Based on First-Year Concentration Value)
• NH_4^+ showed a 34% decrease	NH_4^+: −1.9
• Cl^- showed a 63% decrease	Cl^-: −3.5
• K^+ showed a statistically significant decrease; this was influenced by high concentration in 1963-1964 and 1964-1965, two drought years	
• Ca^{2+} showed an 86% decrease	Ca^{2+}: −4.8
• Mg^{2+} showed a 79% decrease	Mg^{2+}: −4.4
7. Oehlert (1984a)	
• Used monthly volume-weighted averages; variables were deseasonalized; least-squares regression with adjustment for autocorrelation; 1964-1979	
• pH has remained fairly constant; the calculated trend is not significant	
• SO_4^{2-} is strongly decreasing, −0.0467 ppm/yr (standard error = 0.0147); spring and winter concentrations are falling more rapidly than summer and autumn ones (summer: −0.025 ppm/yr; autumn: −0.023 ppm/yr; winter: −0.061 ppm/yr; spring: −0.055 ppm/yr)	SO_4^{2-}: −1.2
• NO_3^- is increasing, 0.0367 ppm/yr (standard error = 0.0152)	NO_3^-: 3.6
• NH_4^+ is decreasing, −0.00651 ppm/yr (standard error = 0.00247)	NH_4^+: −6.4
• Cl^- is strongly decreasing, −9.19%/yr (standard error = 4.20)	Cl^-: −9.2
• Na^+ is stongly decreasing, −6.25%/yr (standard error = 1.39)	Na^+: −6.3
• K^+ is strongly decreasing, −7.01%/yr (standard error = 1.87)	$K+$: −7.0
• Ca^{2+} is strongly decreasing, −9.49%/yr (standard error = 1.06)	Ca^{2+}: −9.5
• Mg^{2+} is strongly decreasing, −6.96%/yr (standard error = 1.94)	Mg^{2+}: −7.0
• SO_4^{2-}/NO_3^- is strongly decreasing, −6.19%/yr (standard error = 1.62)	SO_4^{2-}/NO_3^-: −6.2
• NH_4^+/NO_3^- is strongly decreasing, −7.52%/yr (standard error = 1.41)	NH_4^+/NO_3^-: −7.5

Changes in sulfate deposition at a site are caused by changes in natural and anthropogenic emissions of sulfur compounds and by changes in the transport, conversion, and deposition processes that deliver the sulfur to the receptor. As a rule, a source close to a receptor will affect the receptor more strongly than a source of the same size farther away. Thus, changes in local source strengths are a potential concern when trying to infer regional trends from site data.

Recently, Coffey (1984) speculated about the possible influence of a local paper mill on the Hubbard Brook precipitation-sulfate data, although data on actual emissions from the plant were lacking. John Pinkerton (National Council of the Paper Industry of Air and Stream Improvement, personal communication, 1984) estimated sulfur dioxide emissions from the paper mill from 1963 to 1980 based on available information on strikes and closings at the plant, pulp and paper production rates, energy consumption, and knowledge of the boilers and processes involved. No direct measures of emissions were made at the plant.

The sulfur dioxide emissions from the plant were roughly constant at about 4200 tons/yr from 1963 until a shutdown of the plant in August of 1970. After that, the plant operated sporadically with emissions during operational periods at about one third of the pre-1970 rate (Figure 5.20). The longest period of total shutdown was from July 1976 to November 1978, during which there were no sulfur dioxide emissions from the mill.

No large step changes of the type illustrated in Figure 5.20 are evident in the monthly data for sulfate concentration in precipitation at Hubbard Brook. Qualitatively, the Hubbard Brook sulfate data show short-term increases and decreases superimposed upon a linear decrease from 1964 through 1980. The large step decrease in emissions in 1970 occurs at a time when the Hubbard Brook sulfate record is at the end of a short-term (4-year) decrease and the beginning of a short-term (3-year) increase. Quantitatively, linear regression of the deseasonalized monthly precipitation sulfate data on the deseasonalized emissions data from the mill is not statistically significant, having a t value of 1.22, while the regression of deseasonalized precipitation sulfate on a time trend has a t value of -2.41. Thus, we would accept the null hypothesis of no association between sulfate in precipitation at Hubbard Brook and sulfur dioxide emission from the paper mill, while we

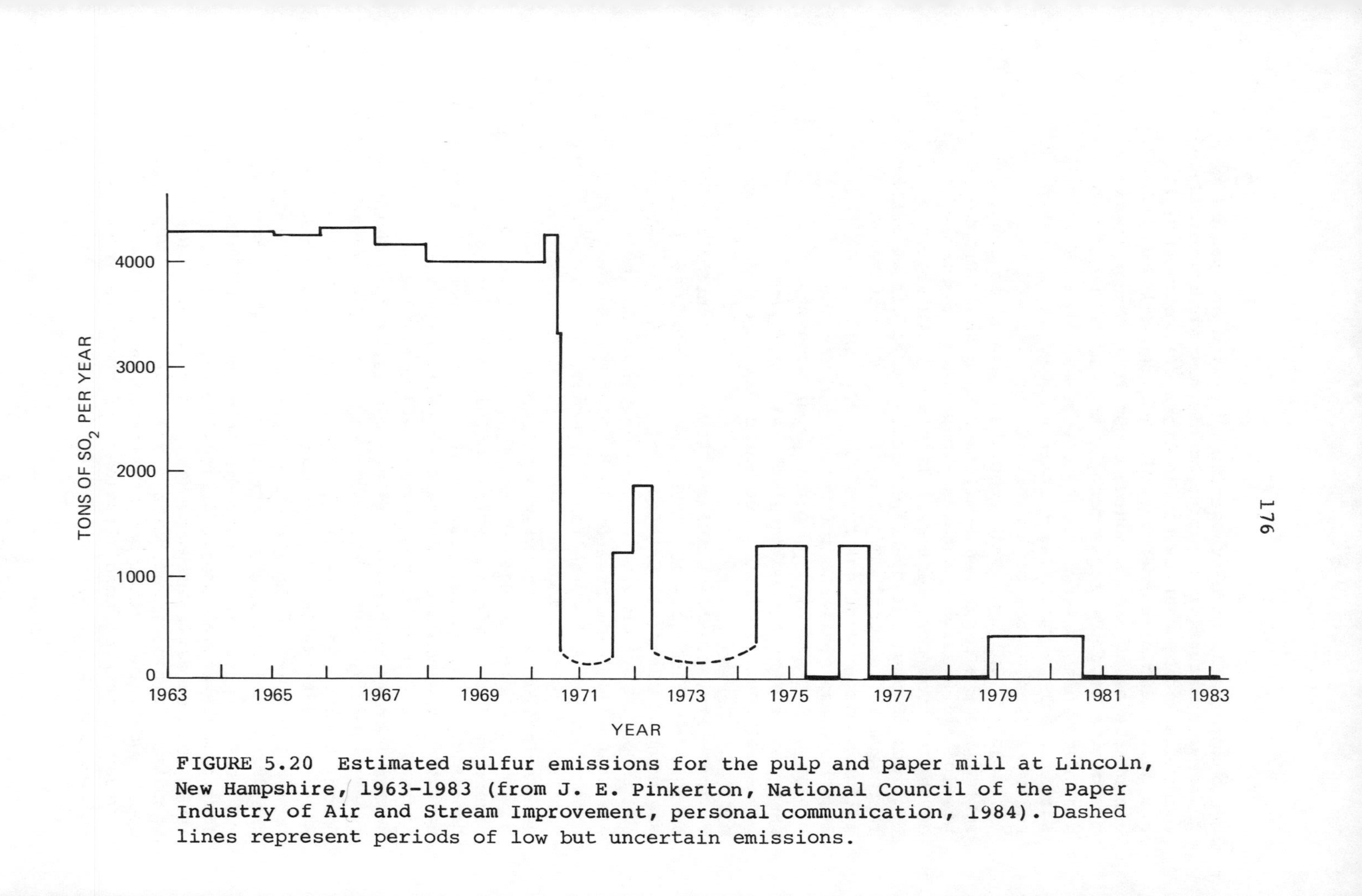

FIGURE 5.20 Estimated sulfur emissions for the pulp and paper mill at Lincoln, New Hampshire, 1963-1983 (from J. E. Pinkerton, National Council of the Paper Industry of Air and Stream Improvement, personal communication, 1984). Dashed lines represent periods of low but uncertain emissions.

would reject the null hypothesis of no linear trend. A linear time trend is a better predictor than the estimated emissions from the paper mill. In the general case, the deposition at Hubbard Brook is likely to be the result of a combination of regional and local influences, so it is reasonable to predict the sulfate concentrations using linear regression on both time and emissions from the mill. Using both variables, the t value for a linear effect over time is -3.62 and the t value for the emissions effect is -2.17. Thus, allowing for some regional linear trend, any effect of the shape of the paper mill emissions curve is statistically estimated to be significantly negative. From this nonphysical result we conclude that the trend present in the Hubbard Brook sulfate data is essentially uninfluenced by the paper mill emissions.

The U.S. Geological Survey Data

The USGS network was established in 1965 to provide data and continuous records on atmospheric contributions for use in hydrologic and geochemical studies (Peters et al. 1982). The network comprises nine stations, eight in New York and one in northern Pennsylvania; six sites are inland rural, two are urban, and two are coastal (one urban/coastal). Samples are monthly bulk samples.

Because this data set provides one of the longest continuous records of precipitation composition and deposition in North America, it has been examined by several authors. Virtually all these authors have commented on problems associated with the data (in particular, see Miller and Everett 1981, Peters et al. 1982, Miles and Yost 1982, Bilonick and Nichols 1983, Hidy et al. 1984, Bilonick 1984, Cogbill et al. 1984, Peters 1984). Potential problems include changes in sites, collectors, analytical methods, and laboratories. Authors differ in their assessments of the importance and implications of these factors for trend analysis; however, the various analyses have been instructive and have produced some valuable trend information.

A time-trend regression analysis by Miller and Everett (1981) for pH data showed no significant trend for any of the nine sites but a slight downward trend for the pooled data. Analysis of the nitrate data showed slight increases for individual stations and a greater increase for pooled data.

TABLE 5.12 Time Trends for Five New York Sites in the USGS Bulk Data Set

Ion	Site	% per year		
		a	b	c
SO_4^{2-}	Canton	−0.7	−1.6[d]	−1.2
	Hinckley	−1.8[d]	−3.4[e]	−2.6[f]
	Mays Point	−0.8	−2.6[e]	−2.0
	Salamanca	−0.8	−2.7[d]	0.8
	Rock Hill	−0.4	−2.0[d]	−1.1
NO_3^- (after April, 1969)	Canton	−2.3	7.6[d]	9.4
	Hinckley	4.5[e]	10.9[f]	13.9[f]
	Mays Point	2.4	6.3[e]	5.0[f]
	Salamanca	−3.7	−1.7	−2.5
	Rock Hill	−2.5	3.0	1.5
NH_4^+	Canton	2.3[f]	2.2	4.6[f]
	Hinckley	5.2[e]	−3.6[e]	5.5[e]
	Mays Point	7.6[e]	−1.0	6.0[e]
	Salamanca	2.9	−3.7[e]	0.2
	Rock Hill	6.1[d]	3.4[d]	4.4
Sum of Ca^{2+}, Mg^{2+}, Na^+, and K^+	Canton	2.9[d]	2.2	3.7[d]
	Hinckley	−4.7[f]	−3.6[e]	−3.7[d]
	Mays Point	−0.8	−1.0	−1.0
	Salamanca	−2.7	−3.7[e]	−2.3
	Rock Hill	6.5[f]	3.4[d]	6.1[f]
H^+	Canton	−1.2	−6.8[e]	−0.5
	Hinckley	3.4[f]	8.4[e]	6.7[f]
	Mays Point	4.5[e]	4.3	10.0[f]
	Salamanca	8.6[f]	10.2[e]	8.7[f]
	Rock Hill	−2.6	−8.9[f]	6.3[f]

[a]Trend by simple linear regression between measured ion concentration and date of collection.
[b]Trend by linear regression between monthly normalized ion concentrations and date of collection.
[c]Trend by seasonal Kendall's test.
[d]$p < 0.10$.
[e]$p < 0.05$.
[f]$p < 0.01$.
SOURCE: Barnes et al. (1982).

Barnes et al. (1982) analyzed the USGS data for time trends at all nine sites with monthly bulk data since 1965. Results for five of the sites are shown in Table 5.12; data for the other four sites are not included as they are viewed as less likely to be regionally representative; i.e., urban sites (Albany and Mineola, New

York), coastal sites (Upton and Mineola, New York), and sites affected by local sources. The site at Athens, Pennsylvania, was characterized by high levels of nitrate and ammonium, which were perhaps due to a nearby dairy farm. The time-trend values were calculated with three different methods as described in Table 5.12. The time trend by simple linear regression is commonly used, while the seasonal Kendall test is described as the most robust of the three. The data in Table 5.12 show that the magnitude of a time trend varies by as much as threefold depending on which of the three methods is used (for example, note the figures on nitrate and hydrogen ions at Hinckley, New York, as extreme cases). The time trend can change sign as the method is changed (for example, note those for ammonium at Hinckley and hydrogen ion at Rock Hill). Also, the magnitude of the time trend can change by more than a factor of 2 among the five sites for the same method (for example, note the change in sulfate for method (b)). The data in Table 5.12 suggest that quantitative time-trend results from the USGS bulk data are subject to imprecision. Qualitatively, the most consistent result from Table 5.12 is that sulfate has decreased by about 1 to 3 percent/yr at the five nonurban, noncoastal sites. Nitrate increases by 4 to 14 percent/yr for significant values ($p < 0.10$) and decreases by up to 4 percent/yr for some nonsignificant values. The significant values of ammonium-ion trends vary from -4 to +8 percent. Similarly, values of cation-sum (calcium plus magnesium plus sodium plus potassium) trends vary from -5 to +7 percent, and the significant hydrogen ion trends vary from -9 to +10 percent. The large variability in trends observed at the five USGS bulk sampling sites may be caused by local effects masking the regional patterns.

Bilonick and Nichols (1983) applied the Box-Jenkins time series analysis method to "cleaned," pooled, monthly deposition data for 1965-1979. They found that the long-term mean levels of hydrogen, sulfate, nitrate, and calcium ions in bulk samples were essentially constant.

Hidy et al. (1984) attempted to establish the reliability of USGS bulk data by comparing time series from the Hinckley and Mays Point stations with the event data from the MAP3S stations at Whiteface Mountain, New York and Ithaca, New York, respectively. They concluded that although the USGS data displayed a systematic positive bias in sulfate concentrations on a monthly basis, these data composed a self-consistent, reliable long-term record of sulfate deposition that was adequate to reflect drifts and trends. However, the long-term patterns of hydrogen, nitrate, and ammonium in the USGS data were less reliable and difficult to interpret for long-term trends.

Hidy et al. (1984) also examined the spatial representativeness of a number of stations in the Northeast. From an examination of intersite correlations for sulfate among the Hinckley, Mays Point, Canton, Salamanca, and Athens USGS stations they found that only the first three tended to have similar and simultaneous month-to-month variations in sulfate. This points out the difficulty of expecting that a single station or a grouping of a few stations will necessarily provide data that will represent regional or subregional patterns without having established the fact from supplementary information beforehand.

Figure 5.21 shows, as an example, box plots of the analysis of Hidy et al. (1984) for the Hinckley, New York, site. Median sulfate concentrations showed an overall downward trend, while median nitrate concentrations increased in the 1960s and fluctuated thereafter. Median ammonium concentrations appear to have increased after 1975, and median pH showed a weak downward trend. Results obtained from the box plot analyses of the five USGS stations and analysis of the data from Hubbard Brook, New Hampshire, are summarized in Table 5.13. (We note that there are some contradictions between Tables 5.12 and 5.13, presumably a result of different statistical analyses.) Two central New York sites, Hinckley and Canton, and Hubbard Brook show a downward trend in sulfate concentration. Nearby but farther west, Mays Point, Salamanca, and Athens do not. Two of the central New York sites, Hinckley and Mays Point, along with Athens, Pennsylvania, show a positive nitrate trend and the site at Hubbard Brook shows a slightly positive trend; Canton and Salamanca show no trend. Ammonium appears to have increased steadily at all USGS sites except Salamanca, and the site at Hubbard Brook shows a decreased trend. pH trends, although less reliable, are either neutral or downward over the period of record.

Hidy et al. (1984) also applied an autoregressive integrated moving-average model to the sulfate data from Hinckley and Mays Point, the two sites where the data record was of sufficient length. Results showed a statistically significant decrease in sulfate concentration of about 2 percent/yr between 1965 and 1980 at both sites.

In summary, despite the uncertainties inherent in the USGS data and the different statistical analysis techniques used, some general statements can be made about trends in the data. Although slight downward trends in pH are evident at some stations for some portions of the record, pH generally shows no trend. The sulfate records at Hinckley and Mays Point show a downward trend, consistent with the record at Hubbard Brook. Because of problems with the nitrate data the various analyses are equivocal--some show increases, some decreases, some no change. Ammonium generally appears to be increasing.

As important as the trend estimates themselves are, several points stand out from the published analyses. Large variability is evident between stations and parameters; even methods of analysis and data preprocessing affect the results. Stations within a limited geographical area may be affected by different processes or climates and therefore may not be assumed to be representative of a region. The trends in sulfate at the two stations examined in detail by Hidy et al. (1984) appear to be most certain because of their self-consistent, reliable, long-term record. The fact that the trends are similar to that at Hubbard Brook gives some confidence to proposing a regional trend in precipitation sulfate concentration.

The World Meteorological Organization/ Background Air Pollution Monitoring Network Data

In the early 1970s the United States and Canada established, respectively, ten and eight stations as part of the WMO Background Air Pollution Monitoring Network (BAPMoN) to measure wet-only precipitation composition on a monthly basis. These stations have undergone several changes over the years, including station relocation, changes in collectors, sample handling protocols, analytical methods, and sampling frequency with the result that the data are difficult to interpret. Preliminary analysis of both the U.S. and Canadian data indicates that sample contamination in the early years of the network and the subsequent changes in operation have made the data unsuitable for trend analysis without applying more rigorous data screening and correction criteria than have been used thus far. However, within the last 3 to 4 years, these BAPMoN stations have been incorporated into

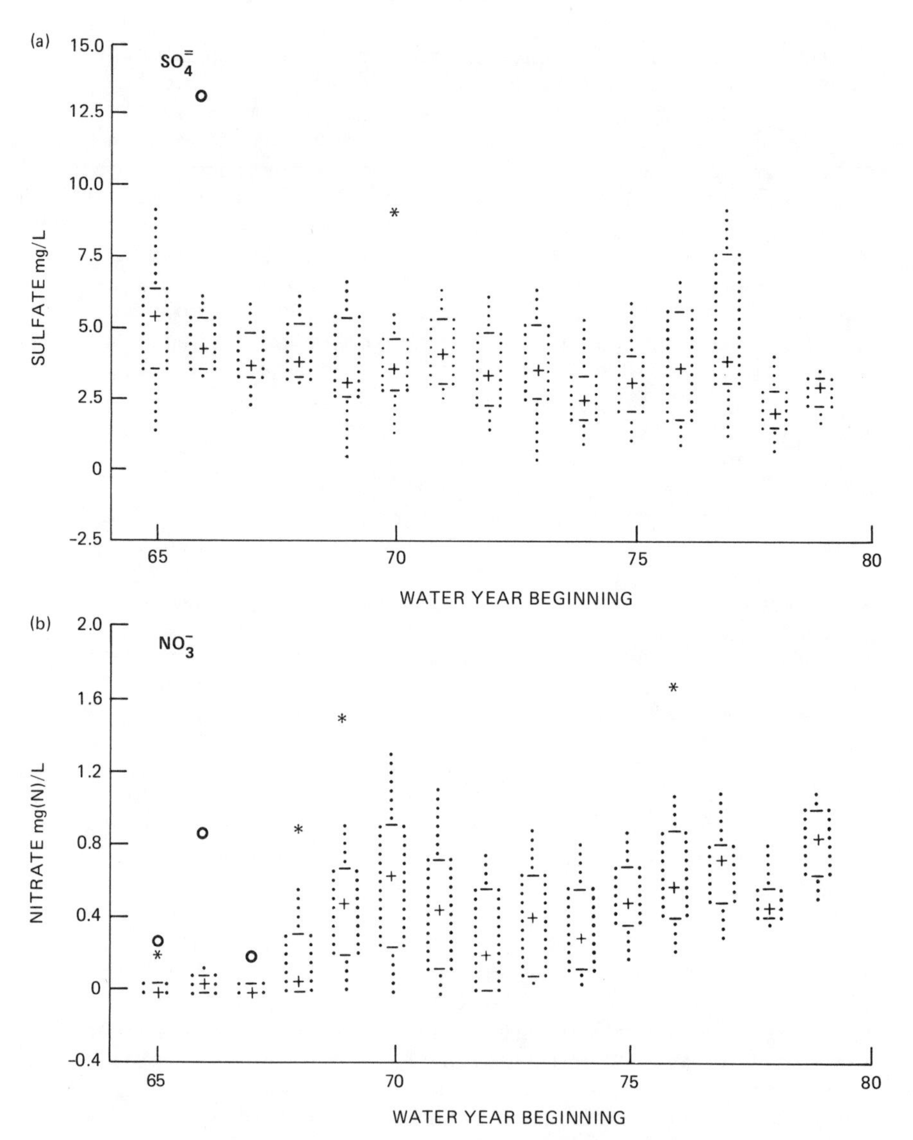

FIGURE 5.21 Box plots for precipitation chemistry at the Hinckley, New York, USGS site. (a) sulfate concentration, (b) nitrate concentration, (c) ammonium concentration, (c) pH. Data are from monthly bulk-deposition samples. The box encloses the middle 50 percent of the data for the year. The median value is denoted by a +. The dotted lines extending beyond the box give the range of all data for the year. The * and o symbols denote possible and probable outliers, respectively. SOURCE: Hidy et al. (1984).

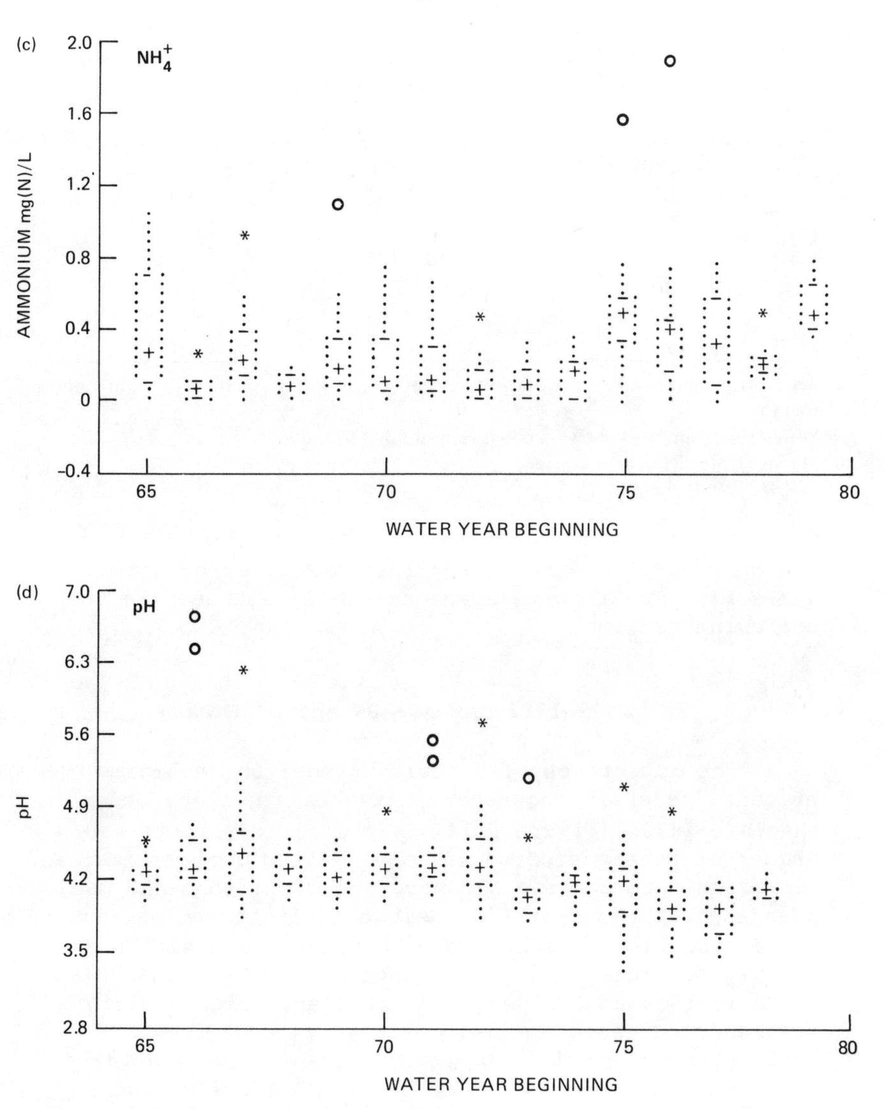

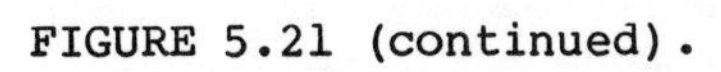
FIGURE 5.21 (continued).

TABLE 5.13 Summary of Apparent Trends[a] Based on Boxplot Analysis of Annual Median Precipitation Chemistry Data From the USGS Sites and the Hubbard Brook Site (1965-1980)

Parameter	Hubbard Brook	Hinckley N.Y.	Canton N.Y.	Mays Point N.Y.	Salamanca N.Y.	Athens Pa.
Precipitation	+	0	0	0	0	0
SO_4^{2-}	−	−	−	0[b]	0	0+
NO_3^-	0+	+	0	+	0	+
NH_4^+	−[c]	+	+	+	0	+
pH	0[c]	−	0	0	−	0

[a] + is an upward trend, − is a downward trend, 0 is no trend, and 0+ is a slightly upward trend.
[b] Time-series analysis shows a downward trend at this site.
[c] From Oehlert (1984a).
SOURCE: Hidy et al. (1984).

the NADP and Canadian Air Pollution Monitoring Network (CAPMoN). Their most recent data are discussed in the following section.

The 1979-1982 CANSAP-APN and NADP Data

As the importance of deposition monitoring became evident, several longer-term networks began operating in the late 1970s (Figure 5.14). In addition, there are a number of other networks of more limited geographical and jurisdictional scope. All have added stations and made changes in their operations since their inception, but of major importance has been the increased emphasis on quality control/quality assurance and on data analysis.

In most cases the period of available high-quality data and increased spatial coverage extends from 1979 or 1980, although MAP3S began providing high-quality data in late 1976. This is too brief a period for an analysis of trends. However, this period may have been characterized by rather large changes in emissions, and it is of great interest to determine whether these changes can be detected in the precipitation data. In what follows, recent results from two networks, CANSAP-APN and NADP, are examined in the context of emission changes over the past 3 to 5 years.

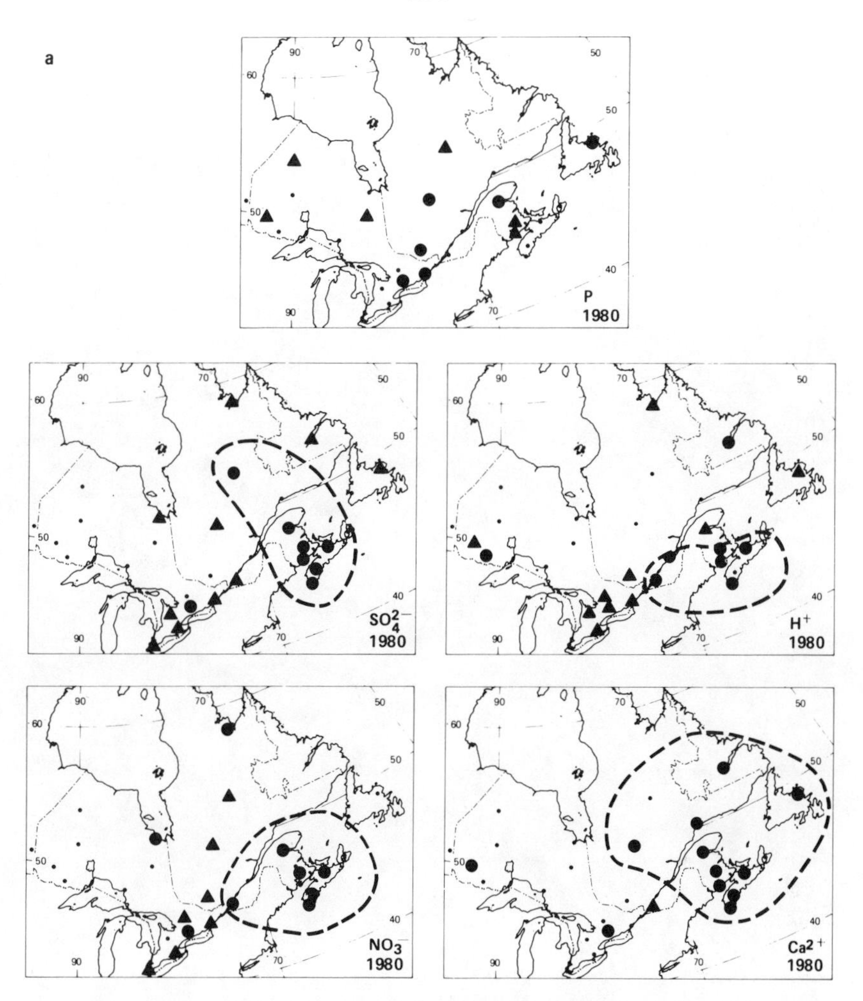

FIGURE 5.22 Comparison of the annual precipitation-weighted-mean concentrations of sulfate, hydrogen, nitrate, calcium, and precipitation amount (designated P in top panels in (a) to (c)) for (a) 1980; (b) 1981; (c) 1982 at eastern CANSAP stations. Filled circles denote maximum of three annual values; filled triangles denotes minimum.

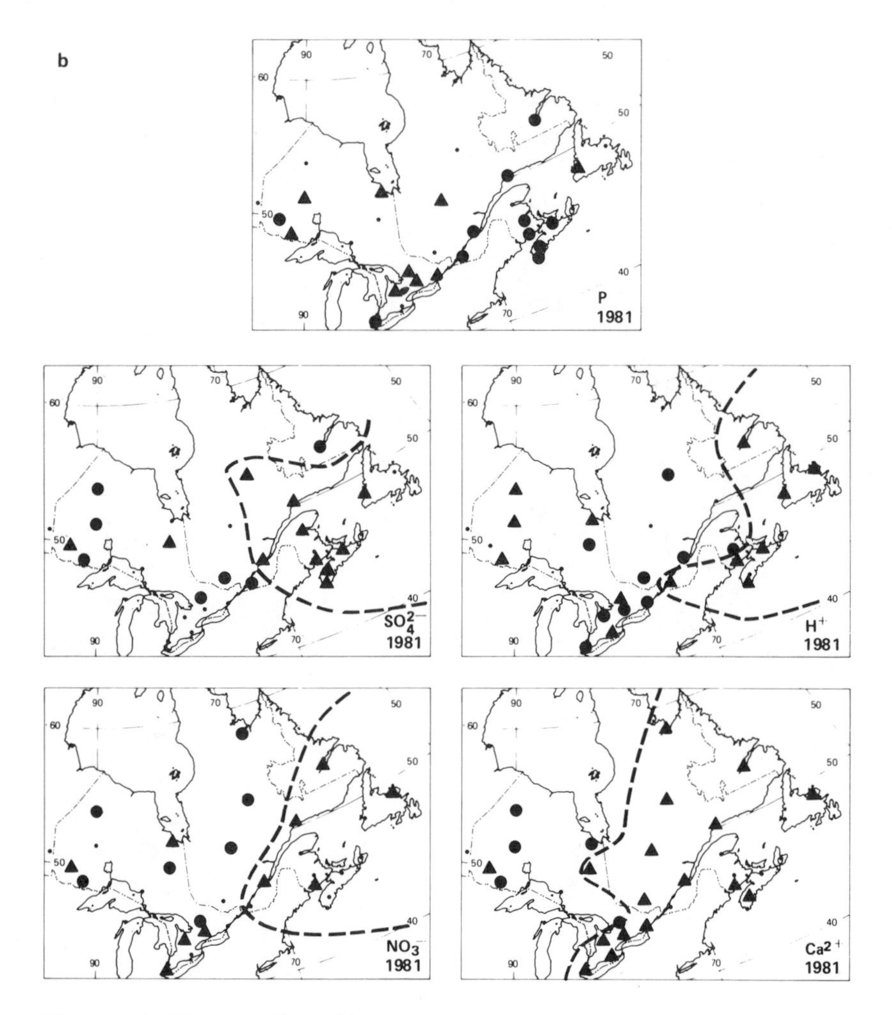

FIGURE 5.22 (continued).

CANSAP-APN Data

Data from the CANSAP-APN (now CAPMoN) network were examined with particular interest in the period of reduced emissions from 1980 to 1982. Between the beginning of 1980 and late 1983 sampling was carried out in a monthly composite mode, and these data are considered to be more reliable than earlier data (Barrie and Sirois 1982). Two features have emerged from the preliminary analysis of the data: (1) data from stations within the same

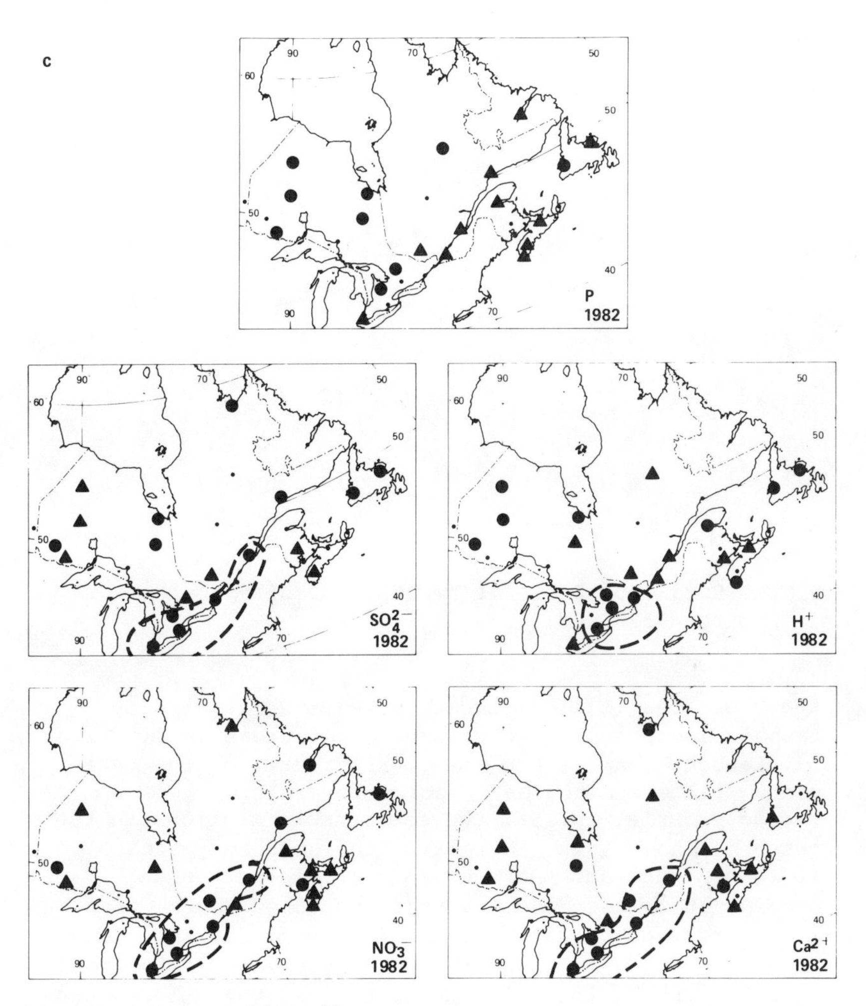

FIGURE 5.22 (continued).

geographical region of eastern Canada often showed similar behavior and (2) ions of different origin also tended to show similar behavior within regions.

Figures 5.22(a), 5.22(b), and 5.22(c) show these features for the ions sulfate, nitrate, hydrogen, and calcium during the years 1980, 1981, and 1982, respectively; precipitation amount (P) is also shown. In 1980, for example, several stations in the Atlantic Provinces and the upper St. Lawrence Valley had the highest annual precipitation-weighted-mean values of the three years for

FIGURE 5.23 Locations of twelve NADP sites referred to in Figures 5.24 and 5.25.

the four ions (filled circles, region designated by dashed lines). Similarly, several stations in the lower St. Lawrence Valley had the lowest values of the three years (filled triangles), except for calcium. As a second example, in 1981 the easternmost stations in the network, those in the Atlantic Provinces and parts of eastern Quebec and Labrador, had the lowest annual precipitation-weighted means of the three years. (See areas enclosed by dashed lines in Figure 5.22(b).) In 1982, the maximum values for the three-year period tended to occur in the lower St. Lawrence Valley and lower Great Lakes region. (See dashed lines in Figure 5.22(c).)

Even this descriptive approach does point out the strong role that factors other than emissions have in determining the composition of precipitation at a given station. Sulfate and nitrate tend to behave in a similar way at many stations, but calcium, not primarily an anthropogenic constituent, also shows similar behavior in many instances. The fact that calcium is often similar to sulfate and nitrate suggests that meteorology, both precipitation and wind regime, plays a major role in the spatial pattern of precipitation composition.

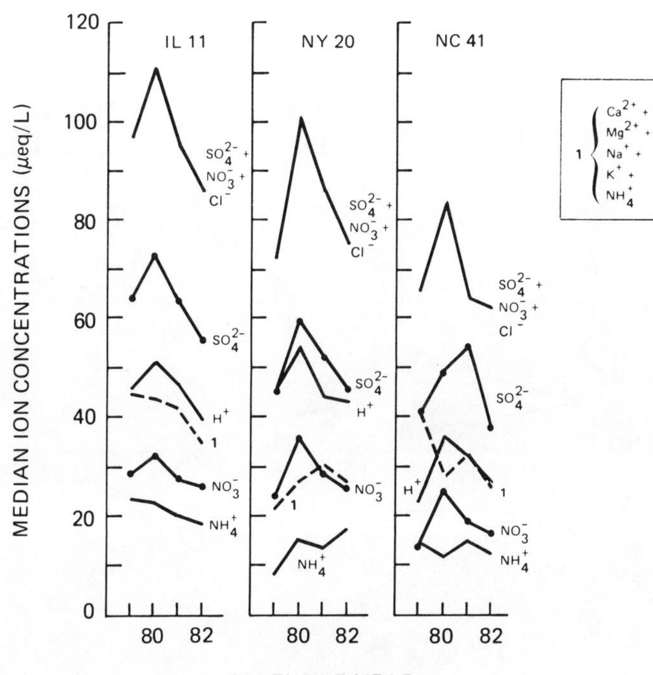

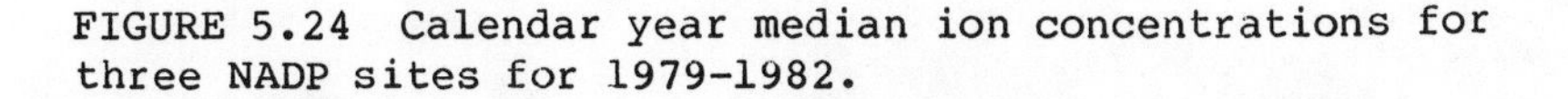

FIGURE 5.24 Calendar year median ion concentrations for three NADP sites for 1979-1982.

NADP Data

Because the data have only recently become available, no studies have been published that examine the time pattern for NADP stations (in particular for the 1979-1982 time period). Here an initial examination is presented.

When the NADP began in 1978 most of the stations were located in the northeastern quadrant of the United States, although there were fewer stations in other areas. These stations have compiled relatively complete records for the 1979-1982 period. Samples were collected for each week that had precipitation and problems of contamination or equipment malfunction were infrequent. Results are presented for 12 sites whose locations are shown in Figure 5.23.

The median annual concentrations for major ions and combinations of ions are shown in Figure 5.24 for three of the sites (IL11, NY20, and NC41), and volume-weighted

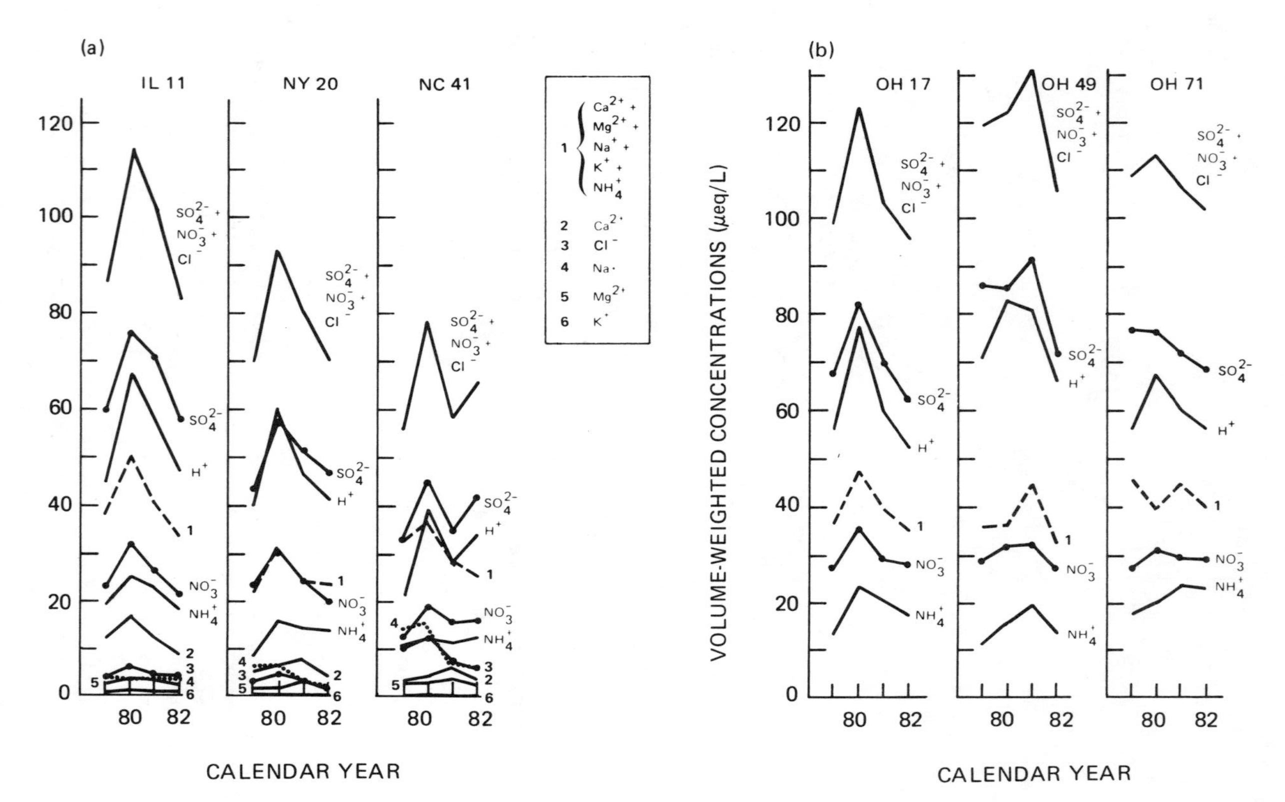
(a)
IL 11
NY 20
NC 41
VOLUME-WEIGHTED CONCENTRATIONS (μeq/L)
CALENDAR YEAR
SO4 2− + NO3 − + Cl −
SO4 2−
H +
NO3 −
NH4 +
1 { Ca 2+ + Mg 2+ + Na + + K + + NH4 + }
2 Ca 2+
3 Cl −
4 Na +
5 Mg 2+
6 K +
(b)
OH 17
OH 49
OH 71
VOLUME-WEIGHTED CONCENTRATIONS (μeq/L)
CALENDAR YEAR

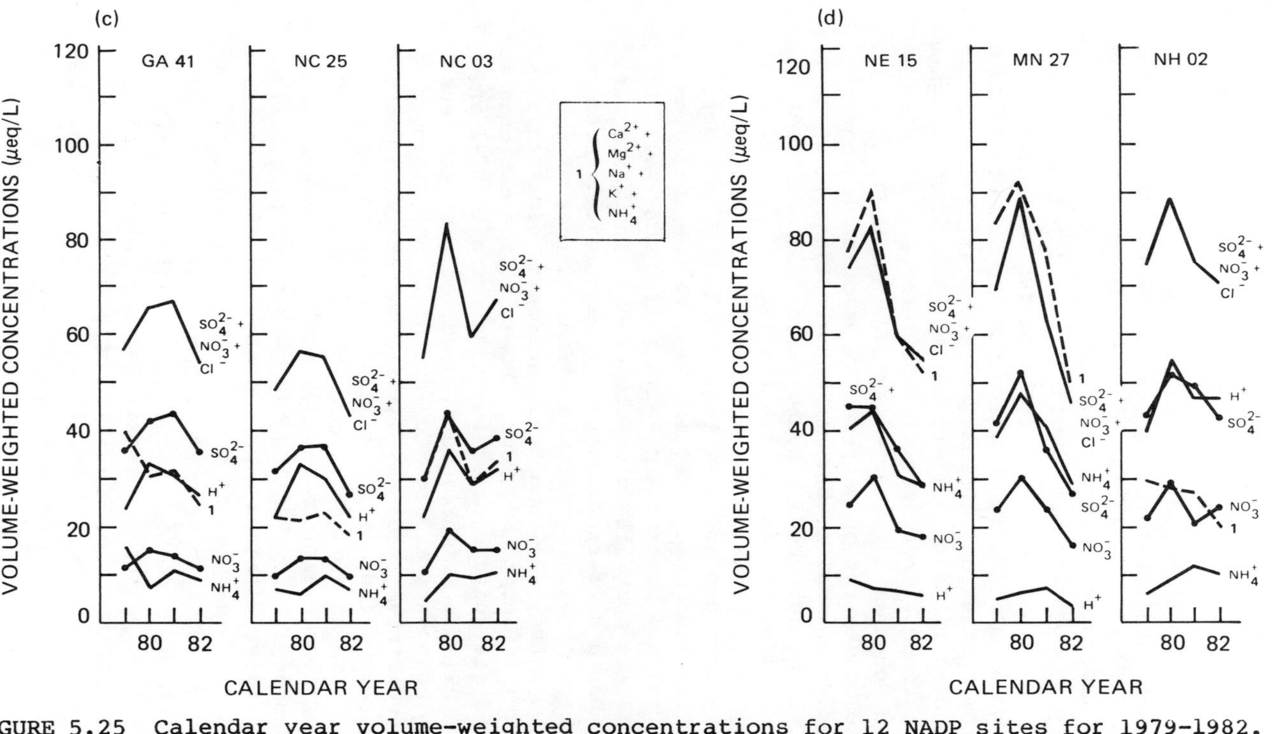

FIGURE 5.25 Calendar year volume-weighted concentrations for 12 NADP sites for 1979-1982.

average concentrations (to be referred to as weighted concentrations) are shown for all twelve sites in Figure 5.25.

The results shown in Figures 5.24 and 5.25(a) can be compared to evaluate differences in the two data presentations. For site IL11 the anions and hydrogen ion have similar median and weighted concentration patterns, although hydrogen ion is lowest in 1979 for the weighted concentration and in 1982 for the median concentration. For ammonium and the base cation sum (the sum of calcium, magnesium, potassium, sodium, and ammonium) the year of maximum concentration differs for the median and weighted concentrations. There are also differences in median and weighted concentrations at sites NY20 and NC41. For site NY20 the concentration patterns of ammonium ion and the base-cation sum differ between the two figures, while for site NC41 the sulfate and hydrogen ion concentration patterns differ in the two presentations.

To compare the 12 sites, volume-weighted average concentrations are shown in Figure 5.25. From Figure 5.25(b) for the three sites in Ohio, we see that almost every curve has some site-to-site differences even though the sites are spatially close together. In particular, sulfate has a different pattern for each site. From Figure 5.25(c) we see that the various curves of the first two sites, GA41 and NC25, are quite similar in shape but different from NC03. However, the patterns for site NC03 are quite similar to those for site NC41 in Figure 5.25(a). From Figure 5.23 we see that GA41 and NC25 are adjacent, as are NC03 and NC41, suggesting a spatial consistency for the annual concentration patterns.

In summary, the most persistent pattern for all ions and all sites is higher concentrations in 1980 and 1981 and lower concentrations in 1979 and 1982. However, not all sites or all ions follow the general pattern. Even multiple sites within a single state have different time patterns (for example, in North Carolina and Ohio). It is not uncommon to have most of the major conservative ions (sulfate, nitrate, ammonium, and calcium) following the same time pattern (for example, see IL11). This suggests that for the 1979-1982 period meteorology, as opposed to the temporal changes in source emissions, was important in determining time trends of ion concentrations.

CONCLUSIONS AND RECOMMENDATIONS

Conclusions

- The eastern half of the United States and southeastern Canada south of James Bay experience concentrations of sulfate and nitrate in precipitation that are, in general, greater by at least a factor of 5 than those in remote areas of the world, indicating that levels have increased by this amount in northeastern North America since sometime before the 1950s.
- The northeastern quadrant of the United States, consisting of Region B, the northernmost states of Region C, and most of Region D, is the area most heavily affected by acidic species (the ions of hydrogen, sulfate, and nitrate) in precipitation.
- Spatial distributions of annual (precipitation-weighted-mean) concentrations and wet deposition for the major ionic species in precipitation are adequately known on the regional to continental scale in North America to afford comparison with spatial patterns of emissions and other possible indicators of acid deposition, such as visibility and the quality of surface waters. The 1980 concentration and deposition fields are acceptably representative of the broad-scale patterns present in the 7-year period 1977-1983.
- Data on the chemistry of precipitation before 1955 should not be used for trend analysis.
- Precipitation is currently more acidic in parts of the eastern United States than it was in the mid-1950s or mid-1960s; however, the amount of change and its mechanism are in dispute. Changes in natural dust sources, anthropogenic acid sources, and sampling methods are all major factors that need to be considered in interpreting the changes in precipitation.
- Precipitation sulfate concentrations and possibly acidity have increased in the southeastern United States (Region C) since the mid-1950s.
- For Hubbard Brook in New England, general agreement exists among several reports on the magnitude and direction of trends since 1964 in some species: hydrogen ion shows no overall significant trend; sulfate has decreased at approximately 2 percent/yr; sodium, chloride, calcium, magnesium, and potassium have shown strong decreases with time; nitrate appears to have increased until about 1970-1971 and subsequently leveled off; the analyses for the ammonium record are contradictory.

• On the basis of past analyses of the USGS bulk data from New York and Pennsylvania since the mid-1960s a consistent network-wide trend does not emerge: results differ from station to station (with a few exceptions). Thus the USGS data base should be screened carefully when general conclusions about regional-scale trends are drawn. The USGS stations at Mays Point and Hinckley, New York, show sulfate trends based on time-series analysis to be consistent with those at Hubbard Brook, making it more likely that a regional trend exists in this parameter.

• On the basis of preliminary analyses of CANSAP-APN and NADP data for the eastern half of the continent, changes in anthropogenic emissions during the period 1979-1982 are not readily apparent; concentration changes from year to year occur concurrently in both anthropogenic and natural constituents and are thought to be more strongly affected by meteorological rather than emissions changes during this period.

• Because data from the WMO BAPMoN network have not yet been rigorously quality assured, trend analyses have not yet been carried out.

• The uncertainties associated with available time-trend estimates of precipitation chemistry make it unwise to draw quantitative conclusions on the basis of direct comparisons between changes in concentrations or deposition and changes in emissions. In cases in which artifacts may be introduced in a precipitation-chemistry data set, for example, by changes in analytical methods over time, even qualitative conclusions related to emission changes may not be possible.

• In general, individual sites or groups of a few neighboring sites cannot be assumed a priori to provide regionally representative information; regional representativeness must be demonstrated on a site-by-site basis.

• General conclusions drawn from merging bulk precipitation data sets with wet-only precipitation data sets are not valid unless detailed site-by-site analyses justify the approach. Similar evaluations should be made when different wet-only precipitation data sets are merged.

• Conclusions regarding time trends depend to some extent on the statistical analysis technique chosen (see Table 5.12). They also depend on the period of the record used for analysis, and this point is frequently ignored.

- Trends for particular ions in precipitation (including hydrogen) should be interpreted only in relation to trends of all the major ions present.

Recommendations

- Expand the scope of deposition monitoring to include dry and occult deposition.
- Strive for improved understanding of the characteristics of all forms of deposition, particularly as a function of altitude and species.
- Continue long-term, wet-only national monitoring, preferably under the auspices of a single jurisdiction.
- Use measures other than pH or hydrogen-ion concentration as a measure or an indicator of the acid deposition problem.
- Investigate the role of organic acids in precipitation acidity.
- Discourage bulk sampling except in specific circumstances in which it can be demonstrated that bulk and wet-only samples are providing comparable data.
- Use other major ionic species (in addition to hydrogen and sulfate) in future trend analyses; also use ancillary information such as meteorological/climatological data.
- Investigate the role of meteorological and climatological variability in temporal and spatial variations of acid deposition.
- Monitor air concentrations in closely located sites.
- Take great care in all aspects of selecting and maintaining a station, particularly its siting and operation, if it is to be used in the future for trend analysis.
- Devote more effort to the analysis of available data.
- Emphasize network design to ensure that networks are capable of detecting changes, particularly those changes that may result from strategic emissions reductions.
- Establish a mechanism and a procedure to compile data, to perform quality control and assurance, to analyze routinely, to synthesize results, and to publish multinetwork data.

REFERENCES

Barnes, C. R., R. A. Schroeder, and N. E. Peters. 1982. Changes in the chemistry of bulk precipitation in New York State, 1965-1978. Northeast. Environ. Sci. 1(3-4):187-197.

Barrie, L. A., and J. M. Hales. 1984. The spatial distributions of precipitation acidity and major ion wet deposition in North America during 1980. Tellus 36B:333-355.

Barrie, L. A., and A. Sirois. 1982. An analysis and assessment of precipitation chemistry measurements made by CANSAP: 1977-1980. Internal Report AQRB-82-003-T, Atmospheric Environment Service, Downsview, Ontario, 163 pp.

Bilonick, R. A. 1984. Discussion of Miles and Yost, 1982. Atmos. Environ. 18:479.

Bilonick, R. A., and D. G. Nichols. 1983. Temporal variations in acid precipitation over New York state--what the 1965-1979 USGS data reveal. Atmos. Environ. 17:1063-1072.

Bowersox, V. C., and R. G. de Pena. 1980. Analysis of precipitation chemistry at a central Pennsylvania site. J. Geophys. Res. 85:5614-5620.

Bowersox, V. C., and G. J. Stensland. 1981. Seasonal patterns of sulfate and nitrate in precipitation in the United States. Paper No. 81-6.1 presented at the 74th Annual Meeting, Air Pollution Control Association, Philadelphia, Pa., June 21-26.

Charlson, R. J., and H. Rodhe. 1982. Factors controlling the acidity of natural rainwater. Nature 295:683-685.

Coffey, P. E. 1984. Source-receptor relationships for acid deposition: pure and simple? Critical Review Discussion Papers. J. Air Pollut. Control Assoc. 34:905-906.

Cogbill, C. V., G. E. Likens, and T. A. Butler. 1984. Uncertainties in historical aspects of acid precipitation: getting it straight. Atmos. Environ. 18:2261-2270.

Environmental Protection Agency. 1984. Deposition Monitoring, Chapter A-8 in Critical Assessment Review Papers, 1984. Volume I. EPA-600/8-83-016AF.

Fisher, D., A. Gambell, G. Likens, and F. Bormann. 1968. Atmospheric contributions to water quality of streams in the Hubbard Brook Experimental Forest, New Hampshire. Water Resour. Res. 4:1115-1126.

Galloway, J. N., G. E. Likens, and M. E. Hawley. 1984. Acid precipitation: natural versus anthropogenic components. Science 226:829-831.

Galloway, J. N., G. E. Likens, W. C. Keene, and J. M. Miller. 1982. The composition of precipitation in remote areas of the world. J. Geophys. Res. 87:8771-8776.

Galloway, J. N., and G. E. Likens. 1981. Acid precipitation: the importance of nitric acid. Atmos. Environ. 15:1081-1085.

Granat, L. 1978. Sulfate in precipitation as observed by the European atmospheric chemistry network. Atmos. Environ. 12:413-424.

Hansen, D. A., and G. M. Hidy. 1982. Review of questions regarding rain acidity data. Atmos. Environ. 16:2107-2126.

Herman, F., and F. Gorhman. 1957. Total mineral material, acidity, sulfur and nitrogen in rain and snow at Kentville, Nova Scotia. Tellus 9:180-183.

Hidy, G. M., D. A. Hansen, R. C. Henry, K. Ganesan, and J. Collins. 1984. Trends in historical and precursor emissions and their airborne and precipitation products. J. Air Pollut. Control Assoc. 31:333-354.

Junge, C. E., and R. T. Werby. 1958. The concentration of chloride, sodium, potassium, calcium, and sulfate in rainwater over the United States. J. Meteorol. 15:417-425.

Likens, G. E., F. H. Bormann, R. S. Pierce, J. S. Eaton, and R. E. Munn. 1984. Long-term trends in precipitation chemistry at Hubbard Brook, New Hampshire. Atmos. Environ. 18:2641-2647.

Likens, G. E., and T. J. Butler. 1981. Recent acidification of precipitation in North America. Atmos. Environ. 15:1103-1109.

Likens, G. E., F. H. Bormann, and J. S. Eaton. 1980. Variations in precipitation and streamwater chemistry at the Hubbard Brook Experimental Forest during 1964-1977. In Effects of Acid Precipitation on Terrestrial Ecosystems, T. C. Hutchinson and M. Havas, eds., New York: Plenum Press, pp. 443-464.

Lodge, J. P., J. B. Pate, W. Basbergill, G. S. Swanson, K. C. Hill, E. Lorange, and A. L. Lazrus. 1968. Chemistry of United States precipitation. Final report on the National Precipitation Sampling Network. Boulder, Colo., National Center for Atmospheric Research. 66 pp.

Miles, L. J., and K. J. Yost. 1982. Quality analysis of USGS precipitation chemistry data for New York. Atmos. Environ. 16:2889-2898.

Miller, M. L., and A. G. Everett. 1981. History and trends of atmospheric nitrate deposition in the eastern U.S.A. In Formation and Fate of Atmospheric Nitrates. Workshop Proceedings, EPA-600/9-81-025, 162-178. Environmental Protection Agency, Washington, D.C.

Munn, R. E., G. E. Likens, B. Weissman, J. Hornbeck, C. W. Martin, F. H. Bormann, G. W. Oehlert, and R. Bloxam. 1982. A climatological analysis of the Hubbard Brook (New Hampshire) precipitation chemistry data. Final report to the U.S. Department of Energy, Grant DE-FG01-SOEV-10455, 221 pp.

National Research Council. 1983. Acid Deposition: Atmospheric Processes in Eastern North America. Washington, D.C.: National Academy Press. 375 pp.

Oehlert, G. W. 1984a. A statistical analysis of the trends in the Hubbard Brook bulk precipitation chemistry data base. Technical Report No. 260, Series 2, Department of Statistics, Princeton University, 64 pp.

Oehlert, G. W. 1984b. Basic statistical methods for trend analysis. Unpublished manuscript, Department of Statistics, Princeton University.

Ottar, B. 1978. An assessment of the OECD study on long range transport of air pollutants (LFAP). Atmos. Env. 12:445-454.

Pack, D. H. 1982. Precipitation chemistry probability--the shape of things to come. Atmos. Environ. 16:1145-1157.

Pack, D. H., and D. W. Pack. 1980. Seasonal and annual behavior of different ions in acidic precipitation. World Meteorological Organization Special Environmental Report No. 14, pp. 303-313.

Peters, N. E. 1984. Discussion of Miles and Yost, 1982. Atmos. Environ. 18:1041-1042.

Peters, N. E., and J. E. Bonelli. 1982. Chemical composition of bulk precipitation in the North-Central and Northeastern United States, December 1980 through February 1981. Circular 874, U.S. Geological Survey, 63 pp.

Peters, N. E., R. A. Schroeder, and D. E. Troutman. 1982. Temporal trends in the acidity of precipitation and surface waters of New York. Water Supply Paper 2188, U.S. Geological Survey, 35 pp.

Stensland, G. J., and R. G. Semonin. 1982. Another interpretation of the pH trend in the United States. Bull. Am. Meteorol. Soc. 63:1277-1284.
Stensland, G. J., and R. G. Semonin. 1984. Response to comment of T. S. Butler, C. V. Cogbill, and G. E. Likens. Bull. Am. Meteorol. Soc. 65:640-643.
Wisniewski, J., and J. D. Kinsman. 1982. An overview of acid rain monitoring activities in North America. Bull. Am. Meteorol. Soc. 63:598-618.

6
The Nature and Timing of the Deterioration of Red Spruce in the Northern Appalachian Mountains

Arthur H. Johnson and *Samuel B. McLaughlin*

INTRODUCTION

Over the past decade, researchers have postulated a number of possible effects of environmental pollutants on forest ecosystems. Theoretical and experimental investigations have shown that acidic substances might affect many aspects of plant and forest function. In addition, some researchers are now beginning to study how acidic substances interact with other pollutants and natural stresses to affect forest ecosystems. Thus, there is a reasonable amount of evidence suggesting how acid deposition might affect forests, but to date no studies have shown the postulated effects to be present in the field.

The finding of widespread mortality of red spruce (Picea rubens Sarg.) in the northern Appalachians (Siccama et al. 1982, Johnson and Siccama 1983, Carey et al. 1984, Foster and Reiners 1983, Scott et al. 1984) sparked research and speculation about whether acid deposition is a cause. Because of the lack of documented mechanisms, only circumstantial evidence supports claims that acid rain has caused damage to spruce in eastern North America. The suspicion that atmospheric deposition may play a role in the decline of red spruce is supported mainly by the facts that large-scale changes in the forest have taken place in high-elevation areas receiving airborne heavy metals and acidic substances at rates that appear to be greater than those experienced by almost all other forested areas in North America and that no obvious natural cause has been documented.

Red spruce have died in uncharacteristic numbers near the top of their elevational range where they are subject to extreme climatic conditions as well as to pollutant input. We find evidence that the current episode of mortality could have been caused by climatic changes and that there are other possible natural contributors that remain unevaluated at the present time. To some extent, understanding the nature and timing of this forest disturbance puts constraints on the range of natural and anthropogenic factors that might be involved, but there is no way now to determine rigorously whether the deposited materials are assimilated benignly by the forest or if the airborne pollutants have altered life-sustaining processes enough to cause significant mortality.

HIGH-ELEVATION CONIFEROUS FORESTS OF THE APPALACHIANS

Researchers have long studied the zones of vegetation types that exist on the mountains of the eastern United States where forest composition varies along an altitudinal gradient. Subalpine stands dominated by fir and/or spruce occur from tree line down to 750 m in the White Mountains (New Hampshire), 850 m in the Green Mountains (Vermont), and 900 m in the Adirondack Mountains (New York). To the south, spruce-fir forests continue as patches on the highest peaks of the Appalachians. The lower boundary varies from about 1060 m in the Catskill Mountains (New York) to 1525 m in the Great Smoky Mountains (Siccama 1974, Oosting and Billings 1951).

Gradients in flora and environmental conditions in the eastern montane forests have been described by numerous investigators (i.e., Foster and Reiners 1983, Siccama 1974, Myers and Bormann 1963, Harries 1966, Adams et al. 1920, Holway et al. 1969, Scott and Holway 1969, McIntosh and Hurley 1964, Whittaker 1956). Figure 6.1 summarizes altitudinal gradients on Camels Hump (Vermont), a site that has figured prominently in recent research. In the Green Mountains (Vermont), sugar maple is the most important canopy species (by density and basal area) below 750 m, and balsam fir is most important above 850 m. A rather narrow transition zone lies between. Red spruce is a minor species in the hardwood forest (<750 m) and above 1150 m. It is most important in stands of the transition and lower boreal zones (760 to 1150 m). Vegetation patterns are similar, but somewhat more complex, in the Catskill, Adirondack, and White mountains because they are irregular massifs rather than linear ridges like the Green Mountains (Siccama 1974, Holway et al. 1969, McIntosh and Hurley 1964).

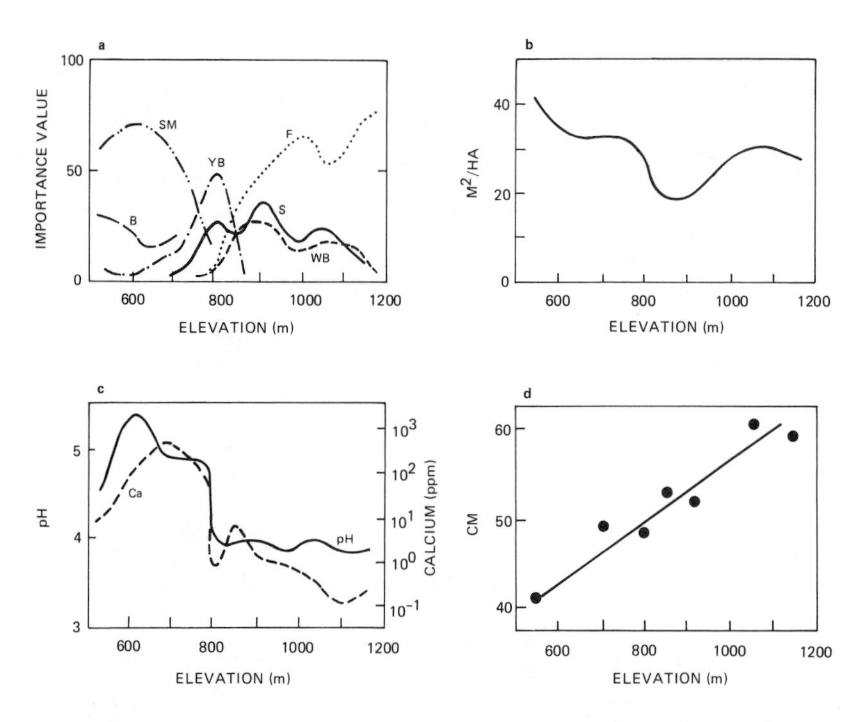

FIGURE 6.1 Altitudinal gradients in tree species and environmental characteristics at Camels Hump, Vermont. Curves are based on measurements at 11 elevations (except Figure 6.1(d)) by Siccama (1974) in 1964-1966. (a) Species importance value (based on basal area, density, and frequency) for sugar maple (Acer saccharum Marsh) (SM); yellow birch (Betula alleghaniensis Britt.) (YB); beech (Fagus grandifolia Ehrh.) (B); red spruce (Picea rubens Sarg.) (S); balsam fir (Abies balsamea (L.) Mill) (F); white birch (Betula papyrifera var cordifolia (Marsh) Regel) (WB). (b) Basal area of all live woody species of >2-cm dbh. (c) Soil pH (H_2O) and exchangeable calcium (NH_4OAc extraction) in the B horizon. (d) Growing season throughfall collected by three to five rain gauges under the canopy at seven elevations. Amounts were measured weekly during the ice-free period during 1964-1966. (e) Maximum and minimum air temperature under the canopy. (f) Average number of frost-free days during 1964-1966.

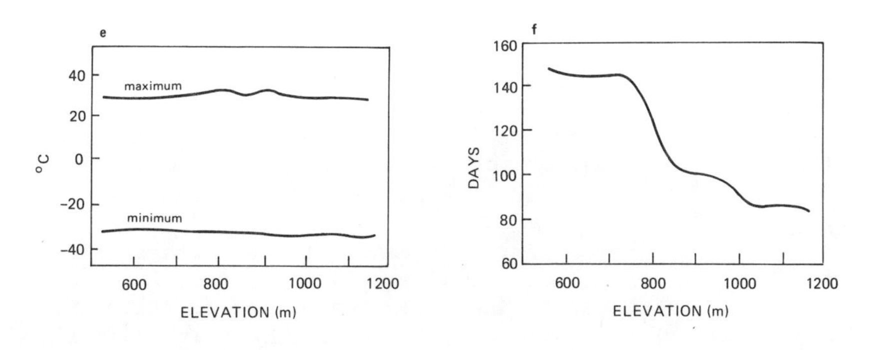

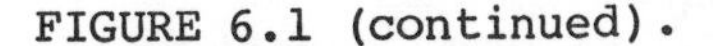
FIGURE 6.1 (continued).

In addition to the elevational gradients, patterns of vegetation are influenced by natural disturbance and disturbance caused by human activity. Logging, fire, landslides, and windthrow have affected the composition and structure of the coniferous forests that have been studied in relation to red spruce decline.

There is evidence that red spruce was a much more important component of the coniferous and hardwood forests of the northern Appalachians prior to the period of extensive timber harvesting in the late 1800s and early 1900s (Siccama 1971). Extensive harvesting of red spruce coupled with unfavorable conditions for regeneration after harvesting are generally thought to be the main reasons (Pielke 1979).

DEPOSITION OF ACIDIC SUBSTANCES AND HEAVY METALS IN NORTHERN APPALACHIAN FORESTS

The reasons that cause many to suspect pollutant stress as a factor contributing to mortality in red spruce are the following: (1) Pollutant deposition in subalpine forests appears to be very high relative to deposition in nearly all other extensive forested areas of North America. (2) The pollutants in sufficient doses are probably capable of altering many of the life-sustaining processes key to maintaining forests. (3) There have been no obvious natural causes offered to explain the observed mortality. The following factors hinder a satisfactory scientific assessment of the role of atmospheric deposition: (1) The variety of pollutants and their delivery rates are not defined precisely. (2)

Testing the impact of the array of known pollutants on the many life-sustaining processes is a complicated task. (3) Any change caused by pollutant substances is superimposed upon naturally occurring changes that alter forest composition, structure, and function, making it difficult to distinguish between natural and anthropogenic effects.

The subalpine coniferous forests of the northern Appalachians are above cloud base for considerable portions of the year. The estimates of the duration of cloud cover vary from approximately 200 h/yr at 600 m in the Green Mountains to about 40 percent of the year at 1220 m in the White Mountains (Siccama 1974, Lovett et al. 1982). Cloud moisture tends to be much more acidic than precipitation, with average H^+ concentration during the growing season reported to be 288 ± 193 µmol/L (Lovett et al. 1982). Based on a model of cloud droplet capture and on measurements of the chemistry of 10 cloud events, Lovett et al. (1982) estimated cloud water inputs of major ions to a subalpine balsam fir forest at 1220 m on Mount Moosilauke (New Hampshire) to be considerably greater than bulk precipitation inputs (Table 6.1). Excluding dry deposition, the deposition of acidic substances (H^+, NH_4^+, SO_4^{2-}, NO_3^-) in the subalpine balsam fir forest is estimated to be three to six times greater than in the lower lying hardwood forest

TABLE 6.1 Annual Deposition in Bulk Precipitation to a Northern Hardwood Forest at Hubbard Brook, New Hampshire, Compared with Estimated Annual Deposition by Bulk Precipitation and Cloud Droplet Capture in a Subalpine Balsam Fir Stand on Mt. Moosilauke, New Hampshire

	Hubbard Brook (Northern Hardwood Forest) (1963-1974) (kg ha^{-1} yr^{-1})[a]	Mt. Moosilauke (Subalpine Balsam Fir) (1980-1981) (kg ha^{-1} yr^{-1})[b]		
Ion	Bulk Precipitation	Bulk Precipitation	Cloud Water	Sum
H^+	1.0	1.5	2.4	3.9
NH_4^+	2.9	4.2	16.3	20.5
Na^+	1.6	1.7	5.8	7.5
K^+	0.9	2.1	3.3	5.4
SO_4^{2-}	38.4	64.8	137.9	202.7
NO_3^-	19.7	23.4	101.5	124.9

[a]Data from Likens et al. (1977).
[b]Data from Lovett et al. (1982).

TABLE 6.2 Mean (x), Standard Deviation (s), and Number (n) of Samples for Ionic Concentrations in Bulk Precipitation at the Huntington Forest, Newcomb, New York

	Bulk Precipitation (μeq/L)		
Ion	x	s	n
NO_3^-	31	22	67
SO_4^{2-}	66	37	66
H^+	88	56	64
K^+	6	15	70
Ca^{2+}	13	19	70
Mg^{2+}	5	9	70
Na^+	4	8	70

NOTE: Elevation of forest approximately 500 m.
SOURCE: Mollitor and Raynal (1983) in Burgess et al. (1984).

at nearby Hubbard Brook, New Hampshire (Likens et al. 1977). Values for major ion deposition at high elevation in the northern Green Mountains estimated from rime ice chemistry, cloud chemistry, and droplet capture by artificial collectors (Scherbatskoy and Bliss 1984) are in rather close agreement with the estimates of Lovett et al. (1982). The estimates of acid deposition at high elevation exceed by severalfold the values calculated from low-elevation data collected by the major monitoring networks (see Chapter 5), but it is clear that the cloud-water input estimates are subject to uncertainty owing to the small number of samples, the variability in cloud chemistry, and the difficulties associated with determining cloud-water capture by the forest canopy.

Roman and Raynal (1980) have reported recent alterations in patterns of tree rings of red spruce in mixed conifer/hardwood stands in the Huntington Forest (Adirondack Mountains). Precipitation in that area had a mean H^+ ion concentration of 88 ± 56 μmol/L; overall precipitation chemistry is presented in Table 6.2.

As in the case of lake sediments (see Chapter 9), lead has been shown to be a useful indicator of atmospheric deposition to forest soils (Reiners et al. 1975, Siccama et al. 1980, Johnson et al. 1982, Friedland et al. 1984a,b). High concentrations of lead have been observed in the subalpine forest floor in New England (Reiners et al. 1975, Johnson et al. 1982, Friedland et al. 1984 a,b, Hanson 1980). The amount of lead in the forest floor increases with elevation (Figure 6.2)--amounts at 1000 m

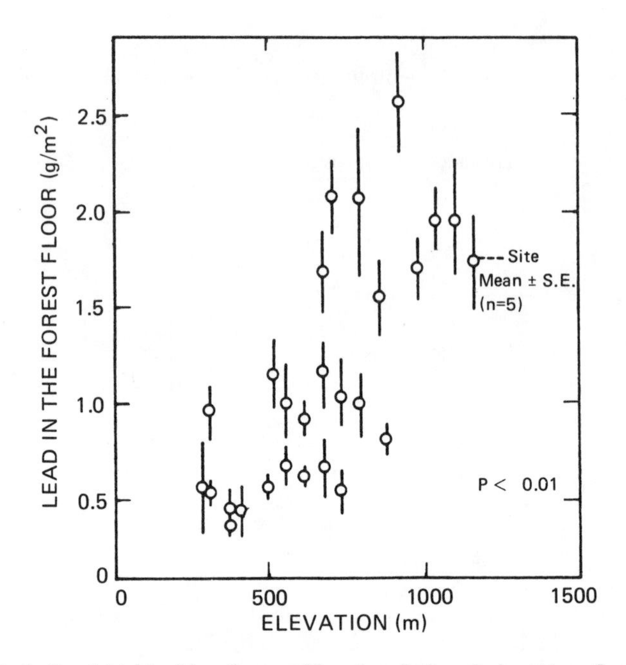

FIGURE 6.2 Altitudinal gradient of lead in the forest floor (Oi + Oe + Oa horizons) for 5 random samples collected at 27 sites in Massachusetts, Vermont, and New Hampshire. After Johnson et al. (1982).

are approximately three to four times higher than those at 500 m. A similar elevational trend in the concentrations and amount of lead in the forest floor was observed at Camels Hump (Table 6.3), but no trends in the concentrations of other metals were noted (Friedland et al. 1984a). Lead concentration in the forest floor at Camels Hump peaks in the Oe horizon (2 to 8 cm beneath the surface), reaching a mean of 302 ± 14 μg/g (Friedland et al. 1984b). Between 1966 and 1981, lead concentrations in the forest floor at Camels Hump increased by 57 percent in the northern hardwood (548 to 731 m), 124 percent in the transition (762 to 853 m), and 95 percent in the boreal (883 to 158 m) zones (Friedland et al. 1984a). Copper and zinc concentrations also increased at somewhat lower rates during the 15-year period.

Converting the 15-year increases in metal concentrations to annual rates, Friedland et al. (1984a) determined that the increased concentrations of lead, copper, and zinc in the forest floor were about equal to regional

TABLE 6.3 Trace-Metal Concentrations (μg/g) and Amounts (mg/m^2), Organic Matter (O.M.) and Organic Carbon (O.C.) Percents and Amounts (kg/m^2), and Dry Weights for Forest Floor (F.F.) Collected in Three Different Forest Vegetation Zones on Camels Hump Mountain

		Metal							
Forest Type	n[a]	Pb	Cu	Zn	Ni	Cd	O.M.	O.C.	F.F. Dry Wt. (kg/m^2)
		μg/g					%		
Northern hardwood	20	99.8 (5.8)	15.0 (0.86)	93.5 (8.3)	16 (4.6)	1.9 (0.2)	51.3 (4.5)	21.4 (1.5)	
Transition	10	151 (25)	14.0 (1.1)	87.1 (5.7)	5.1 (0.7)	1.6 (0.2)	74.3 (4.8)	29.5 (1.4)	
Boreal	25	197 (16)	15.0 (0.72)	77.9 (7.3)	7.7 (0.9)	1.5 (0.1)	72.6 (3.9)	28.8 (1.2)	
		mg/m^2					kg/m^2		
Northern hardwood	20	843 (86)	131 (15.5)	858 (145)	140 (23)	16 (2.1)	4.92 (0.46)	1.8 (0.19)	9.60 (1.45)
Transition	10	1280 (148)	143 (23.8)	785 (57.2)	47 (8.6)	16 (2.9)	7.09 (0.75)	2.78 (0.29)	9.39 (0.83)
Boreal	25	2000 (117)	161 (12.4)	818 (63.0)	85 (10)	15 (1.1)	8.41 (0.39)	3.14 (0.20)	11.6 (1.02)

NOTE: Numbers in parentheses are standard errors of the means.

[a] n is the number of samples.

SOURCE: Friedland et al. (1984a).

estimates of atmospheric deposition of those metals (Groet 1976, Siccama et al. 1980, Smith and Siccama 1981, Scherbatskoy and Bliss 1984). These data are consistent with the contention that subalpine coniferous forests receive considerably higher rates of atmospheric deposition than the adjacent lower-altitude hardwood forests.

To date, the estimates of atmospheric inputs are imprecise because there are few data from short periods of record that show large variations in chemical composition. Largely because of the effect of cloud-droplet interception, the subalpine coniferous forests apparently receive acidic substances and heavy metals at rates that appear to be greater than the rates in neighboring low-elevation forests. With the exception of areas near large sources of sulfur or metal emissions no other forests in eastern North America receive trace metals and acidic substances at rates as great as those estimated for subalpine coniferous forests of Vermont and New Hampshire. There is, however, little reliable evidence that we can use to assess whether we should expect chronic effects on biota resulting from the past and current atmospheric inputs of acidic substances.

CHARACTERISTICS OF RED SPRUCE IN HIGH-ELEVATION CONIFEROUS FORESTS

Red spruce are abundant in cool, moist climates of the high elevations of the southern and northern Appalachians, in coastal regions from Maine to Nova Scotia, and in interior areas of southern Quebec, northern Vermont, northern New Hampshire, Maine, New Brunswick, and Nova Scotia; in the montane forests, sites with a high capacity to hold moisture are particularly favorable (McIntosh and Hurley 1964). Red spruce older than 300 years are common in the northern Appalachians (Siccama 1974, Burgess et al. 1984), and tree ring patterns indicate that some individual trees remain suppressed beneath the canopy for 80 years or more, attesting to the shade tolerance of the species. These characteristics allow it to be competitive in high-elevation forests with its more vigorous but shorter-lived competitors, balsam fir and white birch.

At present, red spruce in the understory or in the canopy may show symptoms of disease or injury in the montane forests of New York, Vermont, and New Hampshire (Johnson and Siccama 1983, Burgess et al. 1984, Friedland et al. 1984c). One set of symptoms is common to all of the northern Appalachian sites discussed below. Trees show needle discoloration and death, crown thinning, an altered branching pattern, and the eventual death of branches and individuals. Friedland et al. (1984c) observed that in all size classes browning of the newest needles was widespread after the winters of 1980-1981 and 1983-1984. They suggest that the most common observation of foliar loss--that it is most noticeable at the outer tips of the branches--might be accounted for by repeated winter damage to new foliage. Typically (but not uniformly) the crowns of mature, declining red spruce appear to die back from the top down and from the outside inward (Johnson and Siccama 1983), and severely declining trees generally have five or fewer year classes of needles.

Studies in the northern Appalachians of root fungi indicate the presence of pathogens that play a secondary role. Early studies (Hadfield 1968, Mook and Eno 1956) found infection by *Cytospora kunzii*, *Fomes pini*, and *Armillaria mellea*. These studies also noted the absence of insects or pathogens that might be a primary cause of disease. A recent study of 288 red spruce in hardwood, transition, and boreal stands in the White, Green, Adirondack, and Catskill mountains showed that declining spruce were infected with *Armillaria mellea* (Carey et al. 1984). *Armillaria* infected approximately 4 percent of the severely declining trees in the boreal zone, 25 percent in the transition forests, and approximately 45 percent in the hardwood zone, indicating that *Armillaria* is an agent related to deteriorating red spruce. But owing to its absence in many declining trees it is not the main cause of deterioration.

Decline, Disease, and Injury

Burgess et al. (1984) have offered an important perspective on the current status of red spruce in the northern Appalachians. Most commonly, "decline" has been the term used to describe the current deterioration of red spruce (e.g., Johnson and Siccama 1983, Carey et al. 1984, Siccama et al. 1982, Scott et al. 1984). Declines differ from most diseases in that they are not the result of a single causal agent. Manion (1981) has suggested a model for declines whereby three categories of stress may act to produce mortality. Environmental factors (e.g., poor site quality, airborne pollutants, nutrient deficiencies) may predispose a tree to greater than normal damage from a short-term initiating stress (i.e., drought, frost damage, insect defoliation, severe air pollution event), which is followed by contributing stresses, which are often lethal attacks from biotic agents (e.g., root pathogens, bark beetles). Because there are several possible stress factors in each of the three categories, declines can be very difficult to resolve into their components. Historically, several tree declines dating back to the 1930s have been documented in North America (Manion 1981, Houston 1981, Mueller-Dombois 1983), often without satisfactory resolution of the causes. Many incidences of decline show a strong relationship to drought conditions (Burgess et al. 1984, Houston 1981), with several other factors presumably determining the ultimate fate of the affected trees, since not all droughts trigger declines. It is also important to note that widespread and synchronized episodes of canopy diebacks are occurring and have occurred in areas not subject to high levels of air pollution (e.g., Hawaii, Mueller-Dombois 1983).

The characteristic of slowly advancing dieback in red spruce (defined here as loss of foliage and branch death) and the presence of secondary pathogens are consistent with the concept of decline. On the other hand, declines are most commonly associated with mature trees (Manion 1981), and the observation of mortality and symptoms in smaller size classes is unlike most other known declines (Burgess et al. 1984). The observation of needle damage and dieback in red spruce in many size classes indicates that injury (short-term effects of a stress) may be involved at some stage in spruce deterioration. There is no evidence to date that red spruce are deteriorating solely because of a biotic disease (defined here as a long-term interaction of a host with a pathogen), but the understanding of biotic agents associated with deteriorating red spruce is incomplete because no systematic studies of the occurrence of pathogens have been reported.

Changes in Red Spruce Populations

Johnson and Siccama (1983) reported on the status of red spruce crown dieback in northern hardwood, transition, and boreal stands in the Appalachians. Methods and data from that survey are given in Appendix A and Figure 6.3. The percentage of spruce that are dead or in an advanced

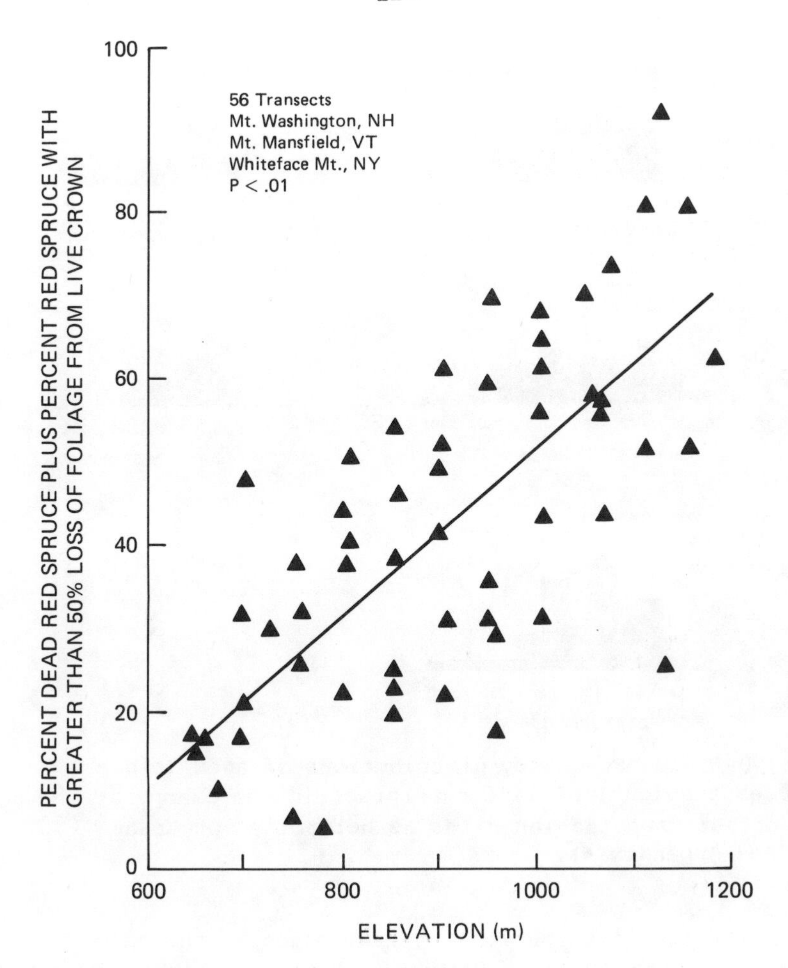

FIGURE 6.3 Populations of severely declining (>50 percent foliar loss from the live crown) and dead red spruce at systematically sampled sites in the northern Appalachian Mountains. Only trees of >10-cm dbh (diameter at breast height) reaching into the canopy were rated. Data correspond to spruce in categories 3 and 4 (Table A.1). Methods are given in Appendix A.

state of deterioration shows a positive correlation with elevation at the northern sites. Mortality and crown dieback at sites south of the Adirondacks were substantially lower.

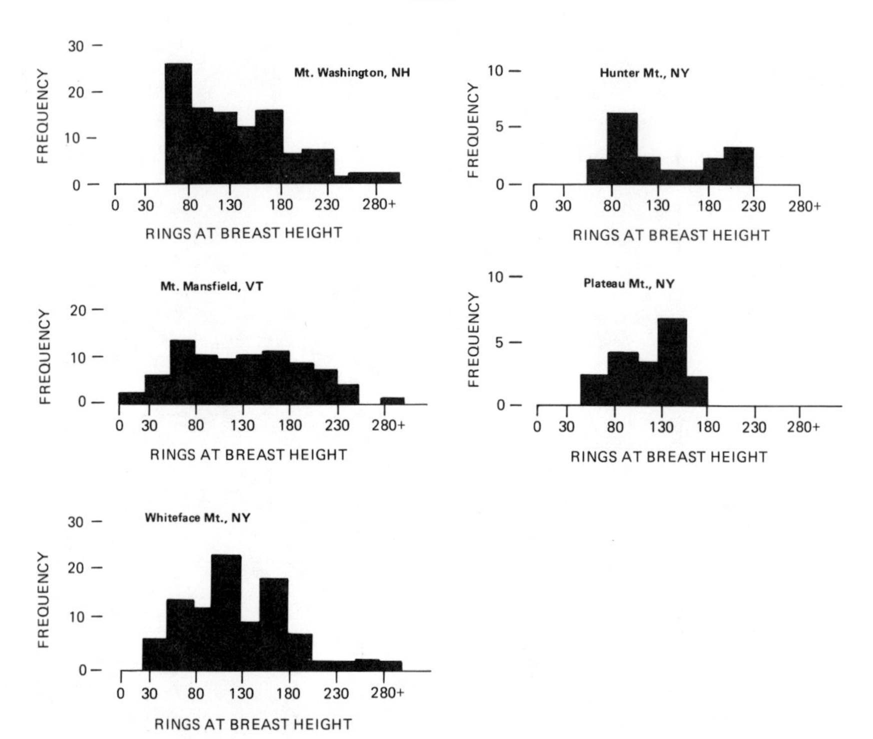

FIGURE 6.4 Frequency distributions of ages (rings at breast height) of live red spruce in the canopy at the northern Appalachian sites as determined on transects (see Appendix A).

The age distributions of red spruce in the canopy greater than 10 cm in diameter at breast height (dbh) are shown in Figure 6.4. The proportion of standing dead spruce (all dead stems at least 1.5 m tall) in six diameter classes is shown in Figure 6.5. These data indicate that the canopy of the northern stands is multiaged and that dead red spruce composed a similar proportion of all diameter classes with some increase in the larger size classes. The degree of dieback was largely independent of stand basal area at the time of sampling (Figure 6.6).

Table 6.4 and Figure 6.7 summarize data on the deterioration of spruce populations during the past two decades from quantitative surveys at Camels Hump and Whiteface Mountain. Those data indicate a 40- to 70-percent decrease in live spruce since the mid-1960s.

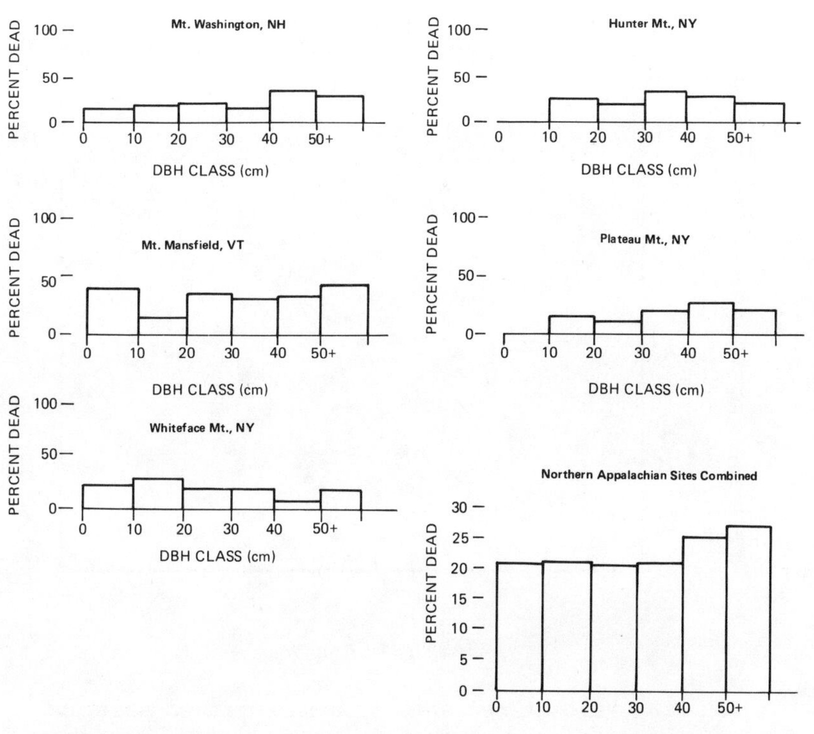

FIGURE 6.5 Percent dead red spruce in six diameter classes (DBH; diameter at breast height) at northern Appalachian sites. Total included 1199 red spruce.

Consistent with the data on standing dead spruce stems in 1982, decreases of substantial proportions occurred in both the 2- to 10-cm dbh and the greater-than-10-cm dbh size classes. Deteriorating or dead red spruce can occur as scattered individuals in conifer or hardwood stands where spruce is a minor component. Most commonly, patches of dead and deteriorating red spruce are observed in the subalpine coniferous forests where balsam fir (*Abies balsamea* (L.) *Mill.*) and white birch seedlings or saplings are now abundant. In some cases, the patches have enlarged to a hectare or more in size, probably because of blowdown of both spruce and fir. This secondary disturbance effect probably accounts for some of the decrease in spruce and most of the decrease in balsam fir at high elevations (Scott et al. 1984).

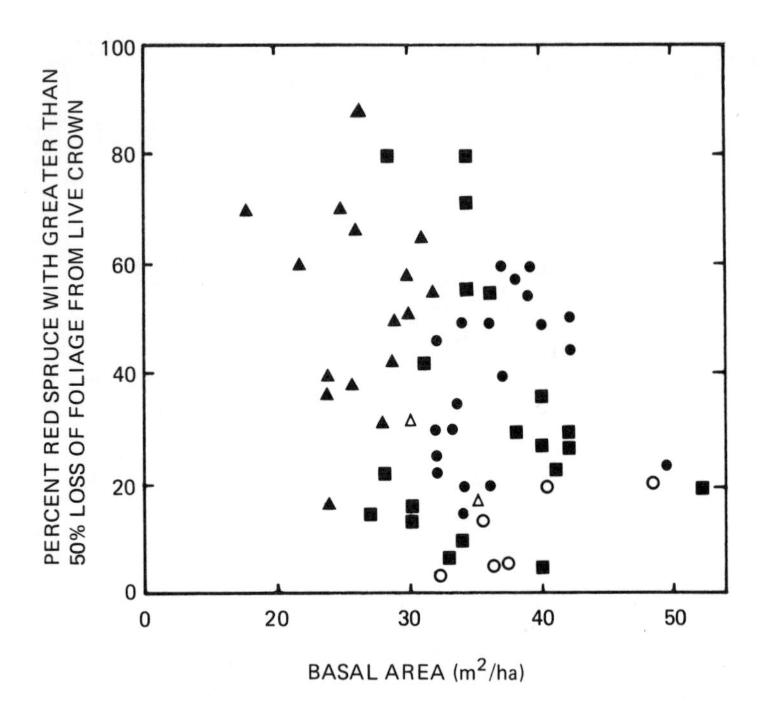

FIGURE 6.6 Percent red spruce in an advanced state of deterioration versus stand basal area. Data correspond to spruce in categories 3 and 4 (Table A.1). Filled circles, Mount Washington, New Hampshire; filled triangles, Mount Mansfield, Vermont; filled squares, Whiteface Mountain, New York; open triangles, Catskill Mountains, New York; open circles, southern Appalachian sites.

Initiation of Red Spruce Deterioration

The current period of rapid deterioration of red spruce in the northern Appalachians appears to have begun sometime between the late 1950s and mid-1960s. This conclusion is based on documentation of patches of dying spruce in Vermont and New Hampshire by the early 1960s and by examination of tree rings from several sites in the Northeast.

Wheeler (1965), Kelso (1965), and Tegethoff (1964) observed a general dying of spruce at elevations of 700 to 1070 m in New Hampshire with symptoms similar to those recently described. Patches of dying red spruce were

TABLE 6.4 Comparison of Density and Basal Area of Red Spruce (*Picea rubens* Sarg.) in Forests of the Green Mountains, Vermont, in 1965 and 1979

	Density (Stems/ha)				Basal Area (m^2/ha)			
Size (cm) Class	1965[a]	1979	%Change	Dead[b]	1965	1979	%Change	Dead
	Boreal Forest (> 880 m)							
	Camels Hump							
< 2	4211	2866	−32	–	–	–	–	–
2-9	206	151	−27	39	0.52	0.30	−42	0.09
≥ 10	125	60	−52	73	6.32	3.51	−44	7.00
	Other mountains (pooled data)[c]							
2-9	256	73	−71	73	0.48	0.15	−69	0.17
≥ 10	145	76	−48	66	3.17	3.22	+2	1.98
	Transition Forest (760-880 m)							
	Camels Hump							
< 2	8748	4333	−50	–	–	–	–	–
2-9	495	145	−70	102	0.89	0.14	−84	0.18
≥ 10	91	54	−41	65	5.73	3.42	−40	8.55
	Other mountains (pooled data)							
2-9	352	45	−87	54	0.79	0.08	−90	0.11
≥ 10	124	9	−93	9	2.03	0.26	−87	0.10

[a] Camels Hump studied in 1965, other mountains in 1964.
[b] Dead stems measured in 1979 only.
[c] Jay Peak, Bolton Mt., and Mt. Abraham; stems < 2 cm dbh were not tallied.
SOURCE: After Siccama et al. (1982).

reported in 1962 and were observed to have increased in 1964 and 1965. Browning of foliage, red spruce in all stages of deterioration, and browning and mortality in young spruce less than 0.6 m tall were noted. The early observers considered the causes to be severe winter conditions combined with summer drought. Hadfield (1968) and Mook and Eno (1956) studied areas of dead and dying spruce in Vermont and observed several secondary agents but no primary pathogens or insects. Siccama (1974) noted that dead and dying red spruce were abundant on Camels Hump during the period 1964-1966, and that mortality in all size classes was particularly conspicuous below 792 m.

Qualitative examination of annual rings of red spruce in the northern Appalachians showed a series of very narrow rings beginning during the early 1960s and often continuing to 1982, when the trees were cored (Johnson and Siccama 1983). Mean ring widths and ring width indices are shown in Figures 6.8 to 6.10 for representative areas of the northern study sites. Those data show a marked reduction in ring width sometime during the

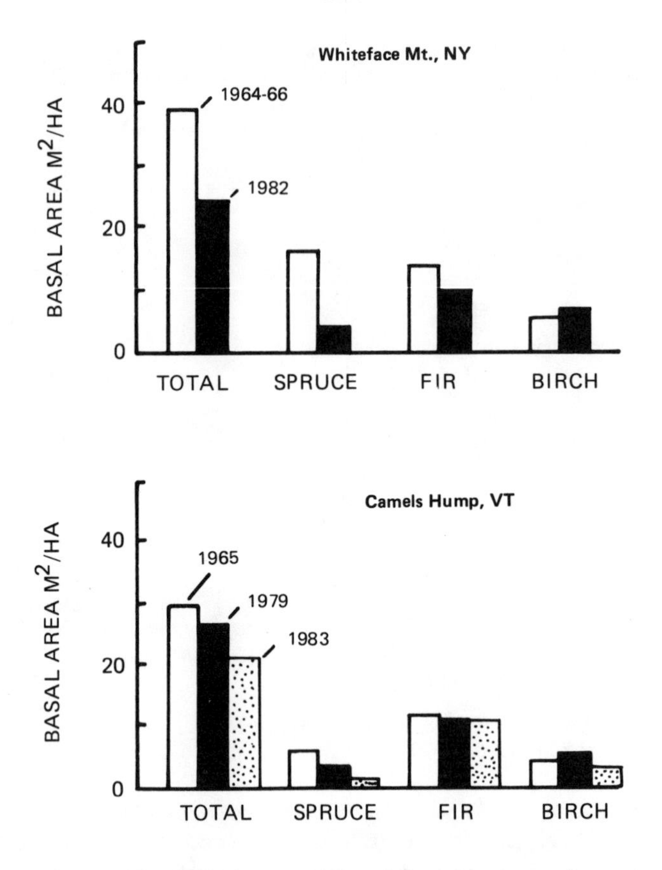

FIGURE 6.7 Basal area changes in major species at Whiteface Mountain, New York, and Camels Hump, Vermont. Basal area values include live trees of >10-cm dbh at Whiteface Mountain and of >2-cm dbh at Camels Hump. Data are for stands above 900 m at Whiteface and above 880 m at Camels Hump. (After Siccama et al. 1982 and Scott et al. 1984). 1983 Camels Hump data are from Ph.D. thesis of H. Whitney, Department of Botany, University of Vermont (unpublished).

early to mid-1960s. The start of the change in ring width appears to have begun in the period 1960-1965. The same pattern was found for ring width indices for red spruce in lower-elevation hardwood-spruce stands in the central Adirondacks (Roman and Raynal 1980), and for old growth, high-elevation red spruce at Lake Arnold, New York (Cook 1985).

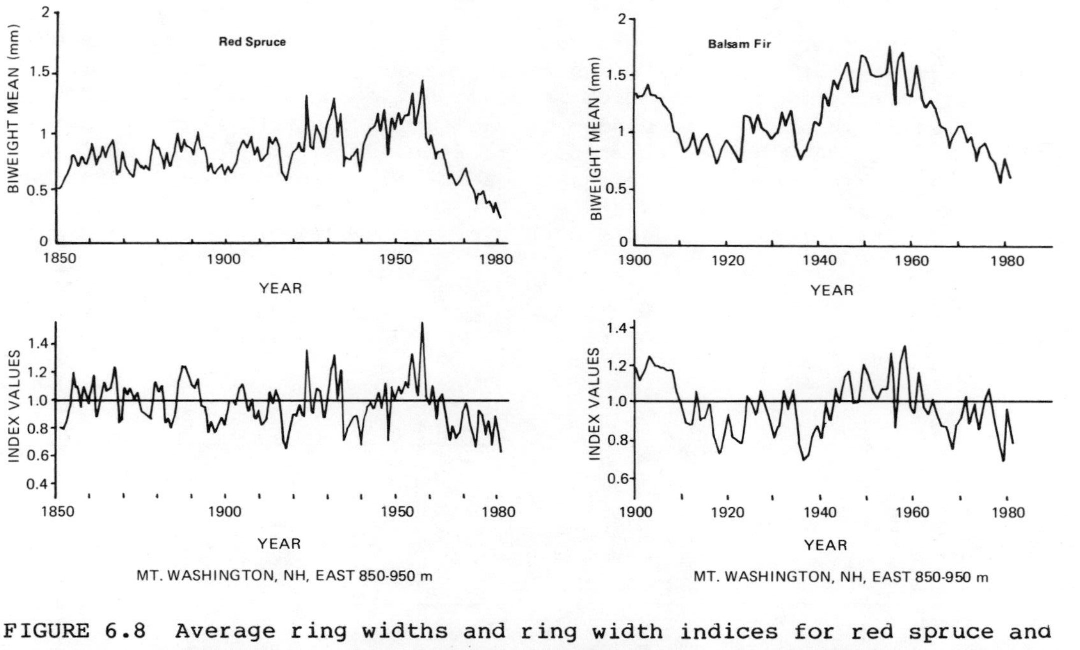

FIGURE 6.8 Average ring widths and ring width indices for red spruce and balsam fir on the east slope of Mount Washington, New Hampshire. Data based on two cores from each of 15 dominant trees at intermediate elevation (850 to 950 m). Procedures are given in Appendix A. The lengths of the chronologies ranged from approximately 80 to 245 years (mean = 144 years). Biweight means are weighted means of raw ring widths, which minimize the importance of outliers (as described in Mosteller and Tukey 1977).

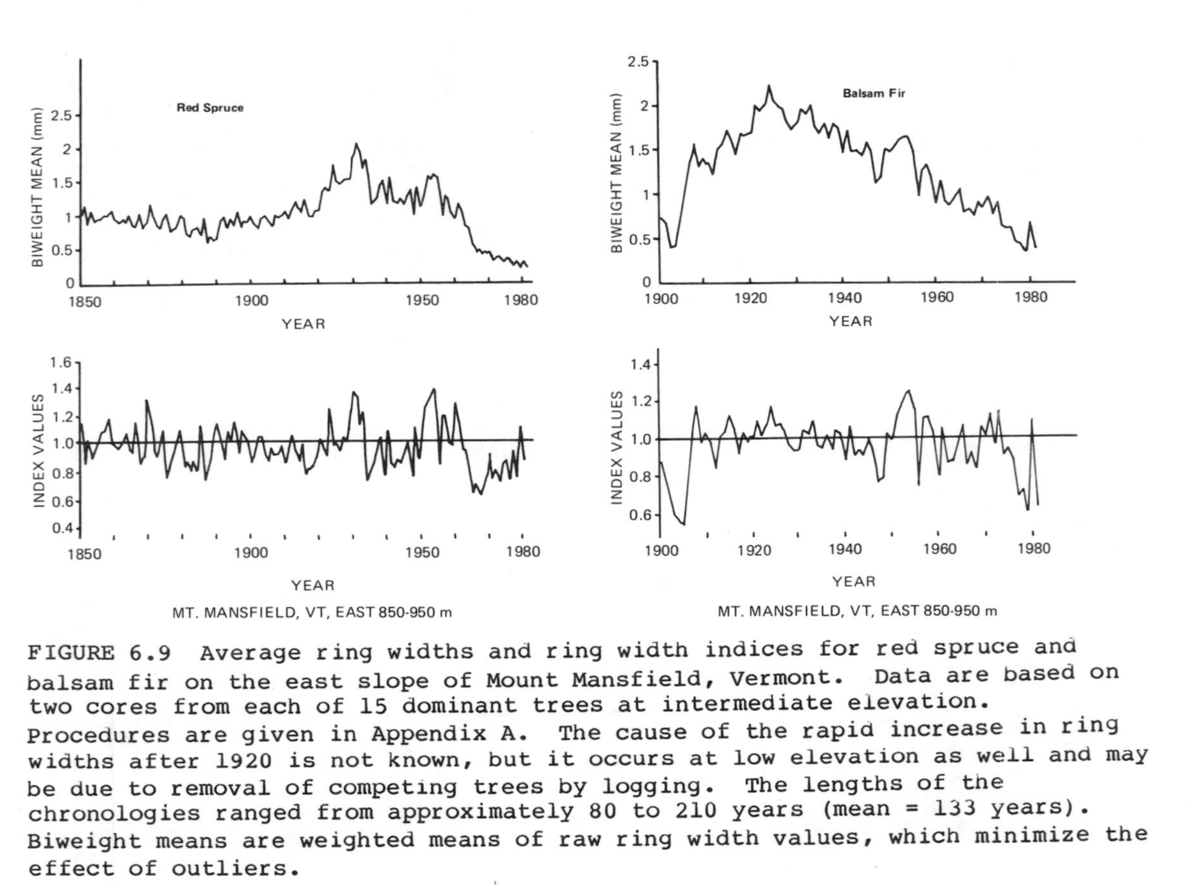

FIGURE 6.9 Average ring widths and ring width indices for red spruce and balsam fir on the east slope of Mount Mansfield, Vermont. Data are based on two cores from each of 15 dominant trees at intermediate elevation. Procedures are given in Appendix A. The cause of the rapid increase in ring widths after 1920 is not known, but it occurs at low elevation as well and may be due to removal of competing trees by logging. The lengths of the chronologies ranged from approximately 80 to 210 years (mean = 133 years). Biweight means are weighted means of raw ring width values, which minimize the effect of outliers.

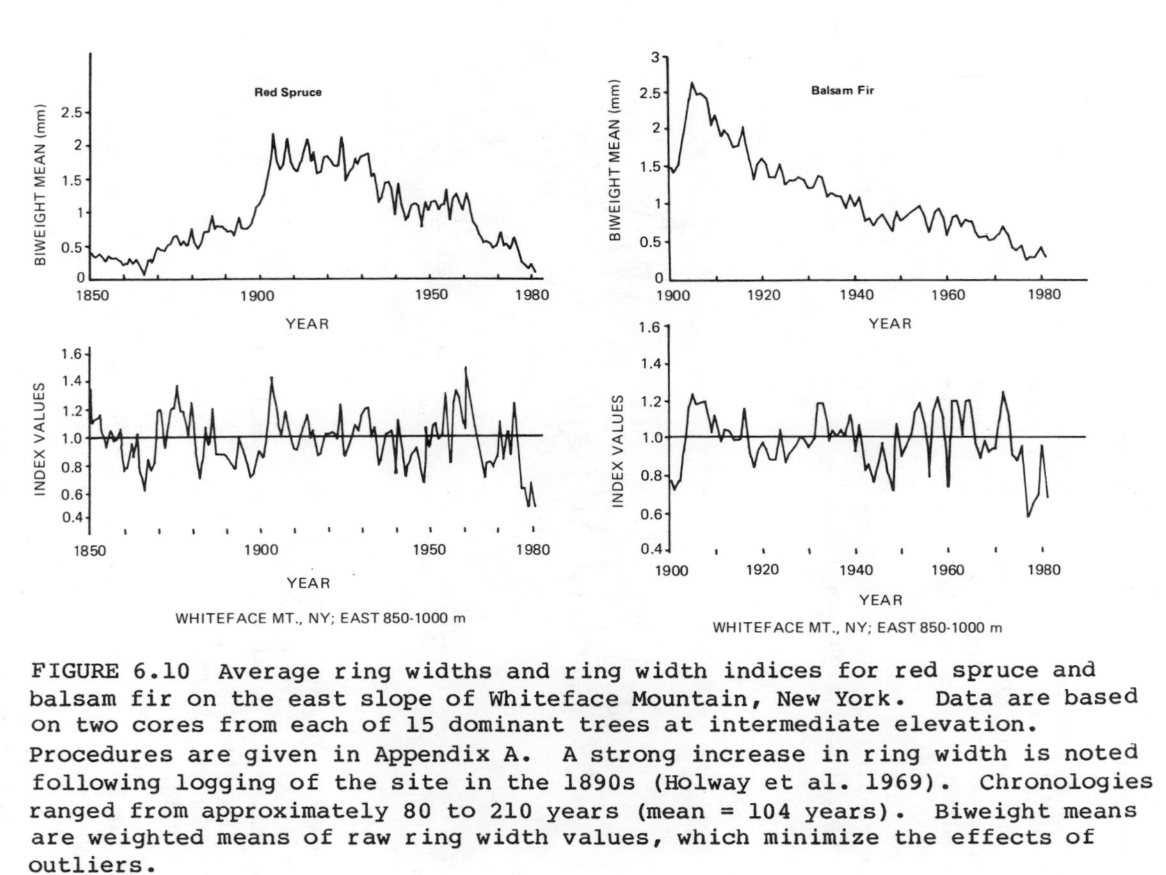

FIGURE 6.10 Average ring widths and ring width indices for red spruce and balsam fir on the east slope of Whiteface Mountain, New York. Data are based on two cores from each of 15 dominant trees at intermediate elevation. Procedures are given in Appendix A. A strong increase in ring width is noted following logging of the site in the 1890s (Holway et al. 1969). Chronologies ranged from approximately 80 to 210 years (mean = 104 years). Biweight means are weighted means of raw ring width values, which minimize the effects of outliers.

Multiple-regression analyses of red spruce chronologies from old-growth, high-elevation stands in Maine (Conkey 1979) and in New York (Cook 1985) indicate that spruce ring width is related to temperature, particularly in the winter and spring. Cook analyzed an old-growth red spruce chronology from a stand at Lake Arnold in the High Peaks region of the Adirondacks. Models based on average monthly temperatures provided a good fit (adj r^2 = 0.436 to 0.487) to annual index values for the calibration period 1890-1950 (Figure 6.11). The models predicted index values reasonably well from 1951 to 1967. After 1967, ring width was considerably less than predicted by the models, suggesting that the trees were responding to factors other than, or in addition to, temperature (Figure 6.11).

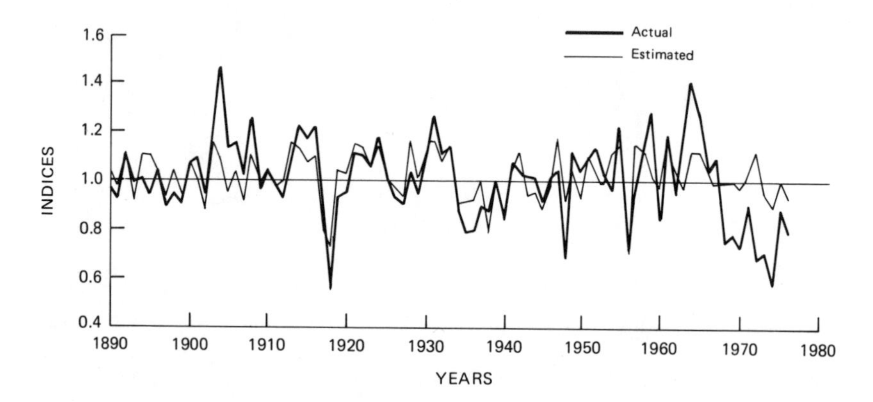

FIGURE 6.11 Actual and estimated red spruce tree ring indices at Lake Arnold in the Adirondack Mountains (elevation 1150 m) based on a temperature response model. The model was developed using mean daily temperatures for the Northern Plateau Climatic Division of New York. The model was calibrated using the 1890-1950 period, and model predictions run from 1951 to 1976. The dark line is the actual indice record, the light line is the estimated indice record. The climate model was Index = -0.261 Jp - 0.300 Ap + 0.180 Op + 0.202 Np + 0.291 Dp + 0.155 Mc + 0.209 Sc. R^2 = 0.502; R^a_{adj} = 0.436, DF = 53. (Jp, previous July temperature; A, August; N, November; D, December; Mc, current May temperature; S, September). (From Cook 1985). The number of rings at breast height ranged from 142 to 297, with most in the range 200 to 250.

Conkey (1979, 1984a,b, 1985) showed an abrupt and sustained change in red spruce maximum latewood density in high-elevation, old-growth red spruce in Maine after 1962 that she tentatively attributed to the effects of shortened growing seasons caused by a series of cool spring temperatures. The interpretation of latewood-density chronologies is a promising new tool in assessing the response of trees to their environment, but it is difficult to tell at present what factors are most important in determining density patterns. Conkey's findings are another signal of changes in tree ring properties probably related to temperature beginning in the 1960s.

The tree ring evidence is consistent with the contention that red spruce have been deteriorating for approximately 20 to 25 years and that the deterioration was probably synchronous across the northern Appalachians. One acceptable interpretation of the available evidence is that, sometime around 1960-1965, the balance of stand conditions and environmental factors stressed red spruce beyond its ability to recover completely and that a decline ensued. Another is that continual or increasing stress to which red spruce is particularly susceptible has caused widespread injury, mortality, and collapse of the canopy in high-elevation stands. We are unaware of any clear evidence that balsam fir are deteriorating except as a secondary consequence of gap formation and the attendant increase in microclimatic stress (i.e., increased susceptibility to wind and winter damage). Tree ring records for balsam fir generally do not show the abrupt change in the 1960s (Figures 6.8 to 6.10) observed in red spruce except at the front of naturally occurring fir waves at elevations above the spruce-fir zone (Marchand 1984).

SUMMARY OF POSSIBLE CAUSES

Dieback symptoms are most pronounced above 900 m, which is an environment subject to natural stresses such as wind, cold winter temperatures, and nutrient-poor soils and to high levels of pollutants such as acidic and heavy metal deposition. Sorting out the factors that have actually contributed to the large-scale deterioration of red spruce is complicated by the lack of information on pathogens, the range of known natural stresses and potential pollution-related stress factors, the background of mortality caused by normal stand dynamics, and the secondary effects of stand destabilization. The breakup of the high-elevation stands in some areas appears to have proceeded as a positive-feedback system--the gaps increase climatic stress and susceptibility to wind, then blowdown enlarges the disturbance patches and exposes more trees. It appears that the breakup of the deteriorating stands was triggered by rather extensive mortality in spruce, and the observation that red spruce in all size classes now show the development and progression of dieback argues that at least some of the factors causing spruce mortality are still operating. Although the formation of disturbance patches (particularly by windthrow) is an important determinant of vegetative patterns, the synchronized spruce mortality across a broad region, and mortality in all size classes, are different from the disturbance phenomena currently recognized as contributing to present-day vegetative patterns is subalpine spurce-fir forests (Foster and Reiners 1982). Of interest in a historical context are reports of widespread mortality in red spruce between 1871 and 1890 in eastern North America (Hopkins 1891, 1901). These suggest that large-scale mortality in red spruce populations can be unrelated to pollutant deposition. The episodes summarized by Hopkins (1891, 1901) occurred at roughly the same time in West Virginia, New York, Vermont, New Hampshire, Maine, and New Brunswick and were tentatively attributed to a spruce beetle (called Dendroctonous piceaperda by Hopkins (1901)) acting as a secondary agent following some other stress. A rigorous investigation of the nature, extent, and symptoms of previous episodes of spurce mortality might provide useful information for comparison with the recent conditions of spruce decline.

In evaluating the possible causes of the current red spruce deterioration we recognize the following possibilities at the present time. Further research and evaluation of each is warranted.

1. Stand dynamics resulting from aging or successional characteristics.
2. Drought.
3. Biotic disease.
4. Repeated or severe winter damage due to climatic change.
5. Pollutant effects.
6. Combinations of pollutant and natural stresses.

Considering the available data, aging or competitive pressures seem unlikely to be primary factors causing mortality of red spruce because the observed patterns argue against them: widespread and synchronous mortality; mortality in all age classes; mortality in stands of widely differing basal area, age, composition; and disturbance history. Although most of the deteriorating northern stands are multiaged and contain old spruce (more than 200 years old), few have basal area values characteristic of virgin, old-growth spruce stands, reported to be 45 to 69 m^2/ha in New Hampshire and 41 to 59 m^2/ha in Maine (Foster and Reiners 1983).

Some stands (see Figure 6.10) were subject to logging in the late nineteenth and early twentieth centuries. To some extent the regrowth of some stands after cutting in the late 1800s may have resulted in critical competitive pressure and sharply decreasing ring width by 1960 (Meyer 1929). On the other hand, other stands show no signs of release caused by logging or other major disturbance (e.g., Figure 6.8) but show the same sharp ring width decrease over the past 20 years. The subalpine forest on the west side of Mount Washington is a virgin forest, and it shows dieback, mortality, and decreased ring width in the same way as the rest of the high-elevation sites. (See also Foster and Reiners 1983.) The deterioration of stands harvested intensively less than a century ago argues against the contention that the observed dead spruce result only from the breakup of old-growth stands. Mortality in all size and age classes and in stands with different disturbance histories argues against the idea of "cohort senescence" (mortality of an even-aged canopy) that has been applied to other synchronous episodes of canopy tree death (Mueller-Dombois 1983). However, the previous episode of regionwide spruce mortality and periods of logging could have served as synchronizing events that led to a multiaged cohort, which grew into a highly competitive situation by 1960.

Widespread drought in the northeastern United States in the mid-1960s (see Chapter 3) was prolonged and severe (Cook and Jacoby 1977), and as droughts are often associated with declines of other species (Manion 1981, Houston 1981, Burgess et al. 1984) the 1960s drought has been proposed as a possible cause of the current spruce deterioration (Kelso 1965, Tegethoff 1964, Hadfield 1968, Siccama 1974, Johnson and Siccama 1983). However, it is difficult to assess the extent of drought stress in high-elevation forests. Examination of Palmer Drought

Severity Index values (which are, at best, a modest representation of high-elevation drought) shows that the peak drought years were 1964 and 1965 (Johnson and Siccama 1983). The data of Siccama (1974), reproduced in Figure 6.1(d), provide direct evidence that it was not dry in absolute terms at Camels Hump and that growing season precipitation amounts increased substantially at higher elevation. Even though 60 cm of warm season precipitation might be a critically low amount for trees adapted to higher precipitation, the increase in precipitation with increasing elevation argues against drought as a principal cause of deterioration, since the degree of deterioration increases at higher elevation. The persistence of dieback symptoms in smaller size classes (Friedland et al. 1984c) also argues against a drought 20 years past being a principal cause of the continuing deterioration. Finally, the observations of patches of dead and dying spruce in 1962 (Kelso 1965, Tegethoff 1964, Wheeler 1965) indicate that some mortality was occurring before the period of most-intense drought.

Based on the reported observations of the past 20 years it appears unlikely that red spruce are deteriorating because of a biotic agent. Although no primary pathogens have been reported, no systematic studies have been conducted so this possibility cannot be confidently rejected at present, and its definitive evaluation awaits further research.

More likely as a key factor in spruce decline was a second climatic anomaly that occurred during the 1960s and that could have had a region-wide, adverse influence on red spruce. In Chapter 3 we have described the occurrence of anonamously cold winters across the eastern United States from 1962 to 1971 (e.g., Namias 1970), and several observations suggest that such an occurrence could be of primary importance. Multiple regression analyses of tree ring chronologies (Cook 1985, Conkey 1979) suggest an adverse effect of cold winter temperatures and cool spring temperatures on annual ring width, although no physiological connection has been established. Numerous reports of severe winter damage, particularly during the past 20 years, suggest that red spruce is susceptible to damage from winter conditions, particularly after mid-winter thaw (Curry and Church 1952, Pomerleau 1962, Friedland et al. 1984c, Burgess et al. 1984). Early observers of the recent deterioration of spruce in New Hampshire observed severe winter damage in stands with dying spruce and suggested that severe winter

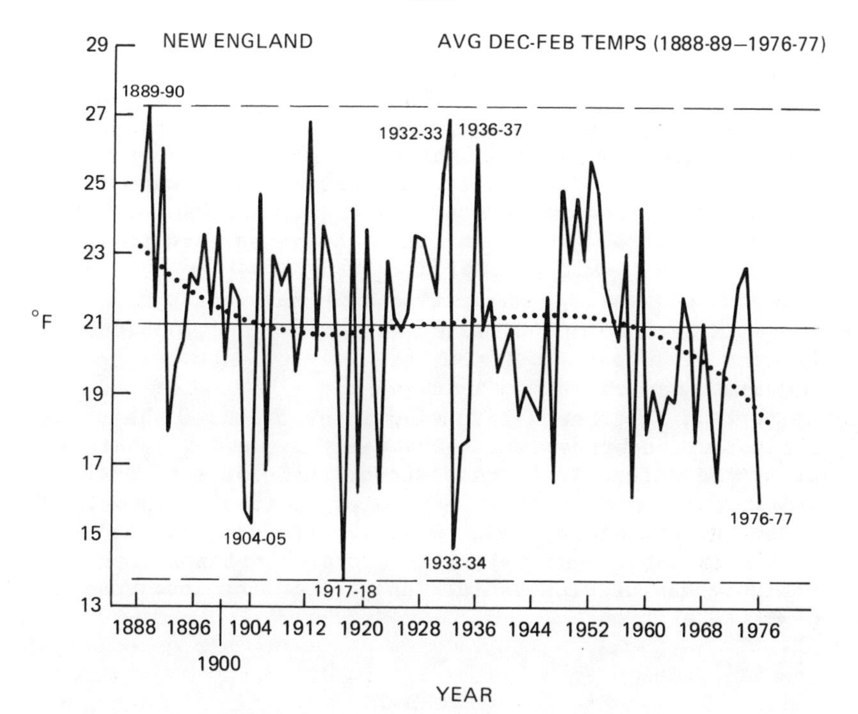

FIGURE 6.12 Trend in winter temperatures in New England (after Diaz and Quayle 1978). The smoothed curve is a third-order polynomial least squares fit.

conditions may have contributed to the mortality (Kelso 1965, Wheeler 1965, Tegethoff 1964). Trends in New England winter temperatures (Diaz and Quayle 1978) show that the past 25 years have been characterized by a trend toward colder winters (Figure 6.12). Diaz and Namias (1983) and Conkey (1985) have shown evidence of cool spring temperatures and shortened growing seasons in the 1960s and 1970s. The effects of cool springs and cold winters may be of particular importance to spruce growing near the top of its elevational range where the growing season length may be marginal under average conditions.

Acidic substances, heavy metals, and gaseous pollutants alone or in combination could alter life-maintaining processes in forests. Whether they can do so to the extent necessary to cause the mortality reported in the above discussion cannot be rigorously evaluated. There is no indication now that acidic deposition is an important factor in the red spruce decline. Ring width anomalies occur in the 1960s in areas of relatively low acidic deposition (Roman and Raynal 1980) and in higher elevation studies where deposition is higher, and where trees are in contact with acidic cloud water for a substantial part of the year. The abrupt and synchronous changes in ring width and wood density patterns across such a wide area seem more likely to be related to climate than to air pollution. Acid deposition or other airborne chemicals might play a role in predisposing the trees to the factors that are injurious, and when those factors are clearly defined, the role of airborne chemicals can be better assessed.

Testing hypotheses involving interactions of natural and pollution stress will eventually be warranted since it may be difficult to reconstruct the effect of climatic anomalies or biotic disease in a way that will show that those factors alone could have triggered the recent deterioration. Finally, refined procedures for tree ring analysis may provide useful insights on the involvement of climate.

REFERENCES

Adams, C. C., G. P. Burns, T. L. Hankinson, B. Moore, and N. Taylor. 1920. Plants and animals of Mt. Marcy, New York. Parts I, II, III. Ecology 1:71-94, 204-233, 274-288.

Burgess, R. L., M. B. David, P. D. Manion, M. J. Mitchell, V. A. Mohnen, D. J. Raynal, M. Schaedle, and E. H. White. 1984. Effects of Acidic Deposition on Forest Ecosystems in the Northeastern United States: An Evaluation of Current Evidence. New York State College of Environmental Science and Forestry, Syracuse, N.Y.

Carey, A. C., E. A. Miller, G. T. Geballe, P. M. Wargo, W. H. Smith, and T. G. Siccama. 1984. Armillaria mellea and decline of red spruce. Plant Dis. 68:794-795.

Conkey, L. E. 1979. Response of tree ring density to climate in Maine, U.S.A. Tree Ring Bull. 19:29-38.

Conkey, L. E. 1984a. X-ray densitometry: wood density as a measure of forest productivity and disturbance. Pp. 287-296 of Air Pollution and Productivity of the Forest, D.D. Davis, ed. Washington, D.C.: Izaak Walton League of America.

Conkey, L. E. 1984b. Dendrochronology and forest productivity: red spruce wood density and ring width in Maine. Pp. 69-75 of Research in Forest Productivity, Use, and Pest Control, M. M. Harris and A. M. Spearing, eds. USDA Forest Service Gen. Tech. Rep. NE-90.

Conkey, L. E. 1985. Red spruce tree-ring widths and densities as indicators of past climate in Maine, USA. Unpublished manuscript.

Cook, E. R. 1985. The use and limitations of dendrochronology in studying effects of air pollution on forests. Paper presented at NATO Advance Research Workshop. Toronto, Ontario, Canada, May 13-17, 1985. Manuscript in press.

Cook, E. R., and G. C. Jacoby, Jr. 1977. Tree-ring-drought relationships in the Hudson Valley, New York. Science 198:399-401.

Curry, J. R., and T. W. Church. 1952. Observations on winter drying of conifers in the Adirondacks. J. Forestry 50:114-116.

Diaz, H. F., and J. Namias. 1983. Association between anomalies of temperature and precipitation in the western U.S. and western northern hemisphere and 700 mb height profiles. J. Climatol. Appl. Meteorol. 22:352-363.

Diaz, H. F., and R. G. Quayle. 1978. The 1976-77 winter in the contiguous United States in comparison with past records. Mon. Weather Rev. 106:1393-1421.

Environmental Protection Agency. 1984. The Acidic Deposition Phenomenon and Its Effects. Critical Assessment Review Papers, Vol. 2. Effects Sciences. Washington, D.C.: U.S. EPA.

Foster, J. R., and W. A. Reiners. 1983. Vegetation patterns in a virgin subalpine forest at Crawford Notch, White Mountains, New Hampshire. Bull. Torrey Bot. Club 110:141-153.

Friedland, A. J., A. H. Johnson, and T. G. Siccama. 1984a. Trace metal content of the forest floor in the Green Mountains of Vermont: spatial and temporal patterns. Water Air Soil Pollut. 21:161-170.

Friedland, A. J., A. H. Johnson, T. G. Siccama, and D. L. Mader. 1984b. Trace metal profiles in the forest floor of New England. Soil Sci. Soc. Am. J. 49:422-425.

Friedland, A. J., R. A. Gregory, L. Karenlampi, and A. H. Johnson. 1984c. Winter damage to foliage as a factor in red spruce decline. Can. J. For. Res. 14:963-965.

Groet, S. S. 1976. Regional and local variations in heavy metal concentrations of bryophytes in the northeastern United States. Oikos 27:445-456.

Hadfield, J. S. 1968. Evaluation of diseases of red spruce on the Chamberlin Hill Sale, Rochester Ranger District, Green Mountain National Forest. File Rep. A-68-8 5230. Amherst, Mass.: USDA-Forest Service Northeastern Area, State and Private Forestry Amherst FPC Field Office.

Hanson, D. W. 1980. Acid precipitation induced chemical changes in subalpine fir forest organic soil layers. M.S. thesis, The Graduate School, University of Maine, Orono.

Harries, H. 1966. Soils and vegetation in the alpine and subalpine belt of the Presidential Range. Ph. D. dissertation, Rutgers University, New Brunswick, N.J.

Holway, J. G., J. T. Scott, and S. Nicholson. 1969. Vegetation of the Whiteface Mountain region of the Adirondacks. Pp. 1-44 of Vegetation-Environment Relations at Whiteface Mt. in the Adirondacks, J. G. Holway and J. T. Scott, eds. Rep. No. 92, Atmospheric Sciences Research Center, State University of New York at Albany.

Hopkins, A. D. 1891. Forest and shade tree insects. II. W. Virginia Experiment Station Third Annual Rep. pp. 171-180.

Hopkins, A. D. 1901. Insect enemies of the spruce in the Northeast. U.S. Department of Agriculture, Division of Entomology Bull. No. 28, new series, pp. 15-29.

Houston, D. B. 1981. Stress triggered tree diseases, the diebacks and declines. Washington, D.C.: USDA Forest Service NE-INF-41-81.

Johnson, A. H., and T. G. Siccama. 1983. Acid deposition and forest decline. Environ. Sci. Technol. 17:294a-305a.

Johnson, A. H., T. G. Siccama, and A. J. Friedland. 1982. Spatial and temporal patterns of lead accumulation in the forest floor in the northeastern United States. J. Environ. Qual. 11:577-580.

Kelso, E. G. 1965. Memorandum 5220,2480, July 23, 1965. Amherst, Mass.:U.S. Forest Service Northern FPC Zone.

Likens, G. E., F. H. Bormann, R. S. Pierce, J. S. Eaton, and N. M. Johnson. 1977. Biogeochemistry of a Forested Ecosystem. New York: Springer-Verlag.

Lovett, G. M., W. R. Reiners, and R. K. Olson. 1982. Cloud droplet deposition in a subalpine balsam fir forest: hydrological and chemical inputs. Science 218:1303-1304.

Manion, P. D. 1981. Tree Disease Concepts. Englewood Cliffs, N.J.: Prentice Hall.

Marchand, P. J. 1984 Dendrochronology of a fir wave. Can. J. For. Res. 14:51-56.

McIntosh, R. P., and R. T. Hurley. 1964. The spruce-fir forests of the Catskill Mountains. Ecology 45:314-326.

McLaughlin, S. B., T. J. Blasing, L. H. Mann, and D. N. Duvick. 1983. Effects of acid rain and gaseous pollutants on forest productivity. J. Air Pollut. Control Assoc. 33:1042-1048.

Mollitor, A. W., and D. J. Raynal. 1983. Atmospheric deposition and ionic input in Adirondack forests. J. Air Pollut. Control Assoc. 33:1032-1035.

Mook, P. V., and H. G. Eno. 1956. Relation of heart rots to mortality of red spruce in the Green Mountain National Forest. Forest Res. Note 59. Upper Darby, Pa.: USDA Forest Service, Northeastern Forest Experiment Station.

Mosteller, F., and J. W. Tukey. 1977. Data Analysis and Regression: A Second Course in Statistics. Reading, Mass.: Addison-Wesley.

Mueller-Dombois, D. 1983. Tree-group death in North American and Hawaiian forests: a pathological problem or a new problem for vegetation ecology? Phytocoenologia 11:117-137.

Myer, W. H. 1929. Yields of second-growth spruce and fir in the Northeast. USDA Tech. Bull. No. 142. 52 pp.

Myers, O., and F. H. Bormann. 1963. Phenotypic variation in Abies balsamea in response to altitudinal and geographic gradients. Ecology 44:429-436.

Namias, J. 1970. Climatic anomaly over the United States during the 1960's. Science 170:741-743.

Oosting, H. J., and W. D. Billings. 1951. A comparison of virgin spruce fir forests in the northern and southern Appalachian system. Ecology 32:84-103.

Pielke, R. A. 1979. The distribution of spruce in west-central Virginia before lumbering. Report no. UVA-ENVSCI MESO-1979-1. University of Virginia, Charlottesville.

Pomerleau, R. 1962. Severe winter browning of red spruce in southeastern Quebec. Can. Dep. For. Bimon. Progr. Rep. 18:3.

Reiners, W. A., R. H. Marks, and P. M. Vitousek. 1975. Heavy metals in subalpine and alpine soils of New Hampshire. Oikos. 26:264-275.

Roman, J. R., and D. J. Raynal. 1980. Effects of acid precipitation on vegetation. Pp. 4-1 to 4-63 of

Actual and Potential Effects of Acid Precipitation on a Forest Ecosystem in the Adirondack Mountains, D. J. Raynal, A. L. Leaf, P. D. Manion, and C. J. K. Wang, eds. New York State ERDA Rep. 80-28, Albany, N.Y.

Scherbatskoy, T., and M. Bliss. 1984. Occurrence of acidic rain and cloud water in high elevation ecosystems in the Green Mts. of Vermont. Pp. 449-463 of Transactions of the APCA Specialty Conference: The Meteorology of Acidic Deposition (October 16-19, 1983 at Hartford, Conn.), P. J. Samson, ed. Pittsburgh, Penn.: Air Pollution Control Association.

Scott, J. T., and J. G. Holway. 1969. Comparison of topographic and vegetation gradients in forests of Whiteface Mountain, New York. Pp. 44-88 of Vegetation-Environment Relations at Whiteface Mountain in the Adirondacks, J. G. Holway and J. T. Scott, eds. Rep. No. 92, Atmospheric Sciences Research Center, State University of New York, Albany, N.Y.

Scott, J. T., T. G. Siccama, A. H. Johnson, and A. R. Breisch. 1984. Decline of Red Spruce in the Adirondacks, New York. Bull. Torrey Bot. Club 111:438-444.

Siccama, T. G. 1971. Presettlement and present forest vegetation in northern Vermont with special reference to Chittenden County. Am. Midland Naturalist 85:153-172.

Siccama, T. G. 1974. Vegetation, soil and climate on the Green Mountains of Vermont. Ecol. Monogr. 44:325-349.

Siccama, T. G., W. H. Smith, and D. L. Mader. 1980. Changes in lead, zinc, copper, dry weight and organic matter content of the forest floor of white pine stands in central Massachusetts over 16 years. Environ. Sci. Technol. 14:54-56.

Siccama, T. G., M. Bliss, and H. W. Vogelmann. 1982. Decline of red spruce in the Green Mountains of Vermont. Bull. Torrey Bot. Club 109:163-168.

Smith, W. H., and T. G. Siccama. 1981. The Hubbard Brook ecosystem study: biogeochemistry of lead in the northern hardwood forest. J. Environ. Qual. 10:323-333.

Tegethoff, A. C. 1964. High Elevation Spruce Mortality. Memorandum 5220, September 25, 1964. Amherst, Mass.: U.S. Forest Service Northern FPC Zone.

Wheeler, G. S. 1965. Memorandum 2400,5100 July 1, 1965. Laconia, N.H.: U.S. Forest Service Northern FPC Zone.

Whittaker, R. H. 1956. Vegetation of the Great Smoky Mountains, Ecol. Monogr. 26:1-80.

7

Streams and Lakes

James R. Kramer, Anders W. Andren, Richard A. Smith
Arthur H. Johnson, Richard B. Alexander, and *Gary Oehlert*

INTRODUCTION

Various attempts have been made to assess changes in the acidity of surface waters that may have occurred over the past 50 years. An excellent review of the methods used in the analyses as well as the basic conclusions from a large number of investigations may be found in the EPA's Critical Assessment Review Papers (Environmental Protection Agency 1984). For example, Henriksen (1979, 1980, 1982), Almer et al. (1978), Dickson (1980), Christophersen and Wright (1981), Thompson (1982), Galloway et al. (1983a), and Wright (1983) used basic geochemical concepts to derive general models that are designed to predict both the sensitivity and the degree of response of a water body to acid inputs. More complex mechanistic and time-dependent models that are intimately synchronized to hydrological, biological, and geomorphological factors have also been used (for example, Chen et al. 1982, Schnoor et al. 1982, Booty and Kramer 1984). These models have provided much useful information, but currently none can provide detailed, quantitative assessments of past or future trends in lake and stream acidification from acid deposition. (For an evaluation of the strengths and limitations of these models see Church (1984).)

Another approach has been to compare the acidic status of lakes, rivers, and ponds in North America using historical and current data on pH, alkalinity, and sulfate. Studies in the United States include, for example, those of Schofield (1976) for New York lakes, Davis et al. (1978) for lakes in Maine, Johnson (1979) for streams in New Jersey, Pfeiffer and Festa (1980) for lakes in New York, Hendrey et al. (1980) and Burns et al. (1981) for headwater streams in North Carolina and New Hampshire, Arnold et al. (1980) for Pennsylvania streams, Crisman et al. (1980) for lakes in northern Florida, Norton et al. (1981) for lakes in New Hampshire, Maine, and Vermont, Lewis (1982) for lakes in Colorado, Haines et al. (1983) for lakes in New England, and Eilers et al. (1985) for lakes in northern Wisconsin. Similar studies have also been published on lakes in Canada (Beamish and Harvey 1972, Beamish et al. 1975, Dillon et al. 1978, Watt et al. 1979, Thompson et al. 1980) and Scandinavia (Wright and Gjessing 1976, Rebsdrosf 1980). These studies analyzed water quality data using a variety of methods. Some sets include continuous data records and others only data at discrete points in time. Many of the analyses account for such complicating factors as changes in analytical methods and land use practices. The conclusions from a variety of studies are that at least some lakes exhibit a decrease in pH and alkalinity when historical and recent data are compared.

Similarly, a review by Turk (1983) compares historical and geographical trends in the acidity of precipitation with trends in North American surface waters. Based on the synthesis of a variety of data collected during discrete time periods (mainly the 1930s and late 1970s) and examined by Schofield (1976), Pfeiffer and Festa (1980), Davis et al. (1978), Norton et al. (1981), Burns et al. (1981), and Haines et al. (1983), he concludes that decreases in pH and alkalinity have occurred in lakes in New York and New England.

In addition, water quality data are available from the U.S. Geological Survey's Hydrologic Bench-Mark Network. These data, collected since the middle 1960s, provide a continuous record of high-quality data for pH, alkalinity, and sulfate as well as the major cations and anions for small watersheds in the United States. Smith and Alexander (1983) have studied these records and conclude that "In the northeastern quarter of the country, SO_2 emissions have decreased over the past 15 years and the trends in the cited chemical characteristics (alkalinity and sulfate) of Bench-Mark streams are consistent with a hypothesis of decreased acid deposition in that region. Throughout much of the remainder of the country, SO_2 emissions have increased and trends in stream sulfate, alkalinity and alkalinity/total cation ratios are consistent with a hypothesis of increased acid deposition."

There are, however, several acknowledged problems in the interpretation of "historical" lake water chemistry data. Perhaps the chief drawback is the lack of documentation of sampling and analytical procedures. A major problem encountered in our study has been the lack of sufficient documentation, before 1950, of the procedures for calculating pH and alkalinity. For example, whereas the current method for measuring pH and alkalinity employs standard electronic instruments, before about 1950 the method of choice was colorimetry using indicator dyes, which, when added in small quantities to solutions of unknown pH, cause color changes at specific pHs. Methyl orange, a commonly used indicator for measuring the alkalinities of lake waters in the 1930s, undergoes subtle color changes over a range of pH values, and the exact endpoint used by analysts is often not clearly specified in the historical records. In this chapter, we review some of these records and assess trends based on the most likely endpoints employed in the historical studies.

Furthermore, for reasons that are described in detail in Appendix D, titration of lake water solutions to a methyl orange endpoint requires a correction for "overtitration." Assumptions about the magnitude of the correction are especially important when analyzing low-alkalinity waters. In most previous studies, researchers have assumed that the correction factor is a constant independent of the original alkalinity. Values ranging from zero to greater than 100 microequivalents per liter (μeq/L) have been employed in such studies. A uniform method for analyzing these types of samples is highly desirable, and one goal of this chapter is to present a method for determining the historical corrections by calculations rather than by assumption. The method accounts for the fact that the relation between the correction factor and alkalinity is sensitive to the alkalinity of the test solution. In addition, the method provides a means for checking the internal consistency of the data set.

Additional factors that must be considered in interpreting current and historical data include (1) climatology, since the hydrology of lakes and rivers may vary considerably from year to year in response to fluctuations in precipitation and climatic factors affecting evaporation, (2) regional patterns of emissions of acid precursors, since natural waters downwind of sources may be expected to exhibit greater changes in water quality than those situated upwind, (3) changes in land use (although Drabløs et al. (1980) found that land use changes in Norway did not seem to be related to regional acidification patterns, local changes in alkalinity and pH regimes were evident), (4) the statistical validity of using data separated by 20 to 40 years to examine trends, and (5) representativeness of data sets for various regions. Inferences about trends obtained from data on a limited number of lakes within a region should reflect the distribution of values of alkalinity and pH (and other physical, chemical, and biological parameters) for the whole region.

In the current study, we have tried to overcome some of the difficulties mentioned above by examining temporal and spatial variations in several water quality parameters--pH, alkalinity, sulfate, and selected major anions and cations--that simultaneously give some information about the acidic status of lakes. The analysis is restricted to a few sets of data of high quality selected to provide both a determination of regional trends in lake and stream acidification and an illustration of a consistent methodologic approach.

In this chapter, we first examine the hypothesis that sulfate levels in lakes and wet precipitation in the northeastern United States and eastern Canada are correlated. Many concepts of lake acidification focus on the titration of a surface water body by sulfuric acid deposited from the atmosphere (e.g., Henriksen 1980). The result of this simple titration is to substitute sulfate ion for bicarbonate ion, thus lowering the alkalinity and the pH of the affected surface water. Wright (1983, 1984) has shown a linear correlation between averages of sulfate concentrations for selected lakes and average sulfate concentration in precipitation in eastern North America, and Thompson and Hutton (1985) show a similar relationship in eastern Canada.

To link sulfate ion concentration in atmospheric deposition to the sulfate ion concentration of surface waters, a direct relationship must exist between atmospheric and lake sulfate fluxes for a steady state assumption. A fundamental condition for this relationship to be valid is that there be no overriding internal sources or sinks of sulfate ions in the watersheds. If such sources and sinks do occur, then the method of averaging sulfate concentrations of large sets of lakes may mask the large variability in the data, particularly if averaging cancels factors equal in magnitude but opposite in sign. We examine data from a set of 626 lakes and assign to each a value of precipitation sulfate flux. We draw general conclusions about spatial associations between precipitation and lake sulfate fluxes from a statistical analysis of both the aggregated data and the data subdivided by regions.

Next we address sulfate-driven changes in stream-water chemistry, using data from the U.S. Geological Survey Bench-Mark Network. The network provides a continuous record of monthly data on major ions in stream water beginning in the middle 1960s. Data are from predominately undeveloped watersheds in 37 states. Continuous sampling and unchanged analytical methods for the period make the data particularly well suited for examining atmospheric influences on water quality as a function of spatial and temporal SO_2 emission patterns. The study of sulfate-driven changes in surface water chemistry is accomplished with the aid of a general paradigm that links acidic deposition to changes in the water quality in watersheds with acidic soil.

Finally, we compare pH and alkalinity data of lakes taken at discrete time intervals, i.e., from the 1920s and 1930s to the 1970s and 1980s. Three lake surveys were chosen for analysis; two in the northeastern United States (New York and New Hampshire) and one in the upper Midwest (Wisconsin). These lakes were selected because they provide, in our judgment, some of the most complete historical and recent data available, allowing us to test them for internal consistency according to the protocol of Kramer and Tessier (1982). We present this analysis to illustrate a chemically rigorous interpretation of historical pH and alkalinity measurements and to estimate the magnitude of changes in the acidic status of these lakes that may have occurred over the past 50 to 60 years.

In our approach we limited our data to those lakes for which historical data on pH, alkalinity, and free CO_2 acidity were available. The availability of simultaneous data for the three parameters allows us to perform a check for internal consistency of the data as an indicator of their reliability. Without this constraint a larger historical data base is available over wider geographic areas. Although our method allows us to screen the data for internal consistency, a larger (unscreened) data base may provide more expansive regional information and be amenable to more powerful statistical tests, although the data may be less reliable and hence the results of the tests less meaningful. Differences that may exist between the results obtained in this chapter and elsewhere must be viewed in light of the assumptions made and how the assumptions affect the selection and analysis of the data.

SULFATE FLUXES IN PRECIPITATION AND LAKES

In this section, we test the hypothesis that the fluxes of sulfate into lakes from atmospheric deposition generally determine the fluxes of sulfate outflow from lakes over a broad geographical area. Such a relationship must be demonstrated to exist before we can assume with any confidence that water chemistry and related data on fisheries records and lake sediment stratigraphy (described in the following chapters) are accurate indicators of atmospheric acid deposition.

Studies of the mass balance of sulfate in a few watersheds have been reported for which parameters such as atmospheric deposition, streamflow, and geology are sufficiently well determined to permit more detailed analyses (e.g., the Hubbard Brook watershed, Likens et al. 1985; and three lakes in the Adirondacks, Galloway et al. 1983b). The studies conclude that sulfate concentrations in these waters are controlled by atmospheric deposition of sulfur. In a different type of analysis, Wright (1983) found an approximately linear relationship between the mean concentration of sulfate in 15 groups of lakes in North American and mean excess (i.e., nonmarine) sulfate in wet precipitation.

In our analysis, we compare sulfate input fluxes in wet deposition with the estimated sulfate output fluxes from 626 lakes in New York, New England, Quebec, Labrador, and Newfoundland. Fluxes of sulfate from wet deposition were obtained by multiplying the concentration of sulfate in rain by the intensity of rainfall (meters/year). Lake output fluxes of sulfate were estimated as the product of sulfate concentration in lake water and the intensity of net precipitation (i.e., meters/year measured as rainfall minus evaporation). (See Appendix B for specific methods and references.) It was possible to perform a detailed statistical analysis to test the relation between sulfate inputs and outputs for linearity (or some other monotonic relationship) and regional variance on subdividing the area into smaller regions. Such an analysis suited the particular needs of this study, which addresses the question of broad regional patterns in eastern North America. However, we note that this analysis is subject to uncertainty because for each of these lakes we generally lack important information on dry deposition, transpiration, observed streamflow from the watershed, soil chemistry, and processes within the watershed that are sources of sulfur.

Results

Figure 7.1(a) is a plot of sulfate inputs to 626 lakes from wet-only deposition as a function of annual sulfate outputs from the lakes based on data for 1980-1982. It is apparent that there is a greater variability in lake sulfate outflow fluxes than in precipitation sulfate fluxes. The scatter is greatest for lakes in Massachusetts (Figure 7.1(b)) owing to some extreme outliers and is of similar, smaller magnitude for other regions. In some cases, the output flux of sulfate from lakes is greater than the flux from wet precipitation, suggesting that significant sources of sulfur other than from wet deposition exist in these watersheds. In other cases, the reverse is apparently true, perhaps suggesting that atmospheric sulfur may be retained in the watershed, or that sulfate is chemically reduced in the lake and deposited in the sediment.

The apparent lake sulfate output flux normalized to the wet-only precipitation sulfate input flux is analyzed by geographic region in Table 7.1. Deficiencies in lake sulfate outputs compared with inputs from wet-only deposition generally occur in remote regions such as Newfoundland and Labrador, whereas excesses occur in more populated areas such as New York, Connecticut, and Massachusetts. The origins of the deficiencies and excesses are not known on a lake-specific basis, but dry deposition, watershed processes involving sulfur cycling, and other sulfur sources of unknown origin are expected to be important contributors.

Analysis

In an effort to stabilize the variability in the sulfate output fluxes of lakes, the data were analyzed on the logarithmic (base e) scale (Figure 7.2). As a whole the data do not follow the $y = x$ line (i.e., the case in which sulfate outputs equal sulfate inputs). Linear regression yields an estimate of a line $\log y = a \log x + m_i'$ in which the intercept, m_i', is -0.306 (standard error, 0.041) and the slope, a, is 1.593 (standard error, 0.046). The slope of this line is significantly different from one, so the overall relationship on the natural scale, $y = m_i x^a$ in which m_i is +0.736 (antilogarithm (base e) of m_i') and a is 1.593, is significantly different from linear.

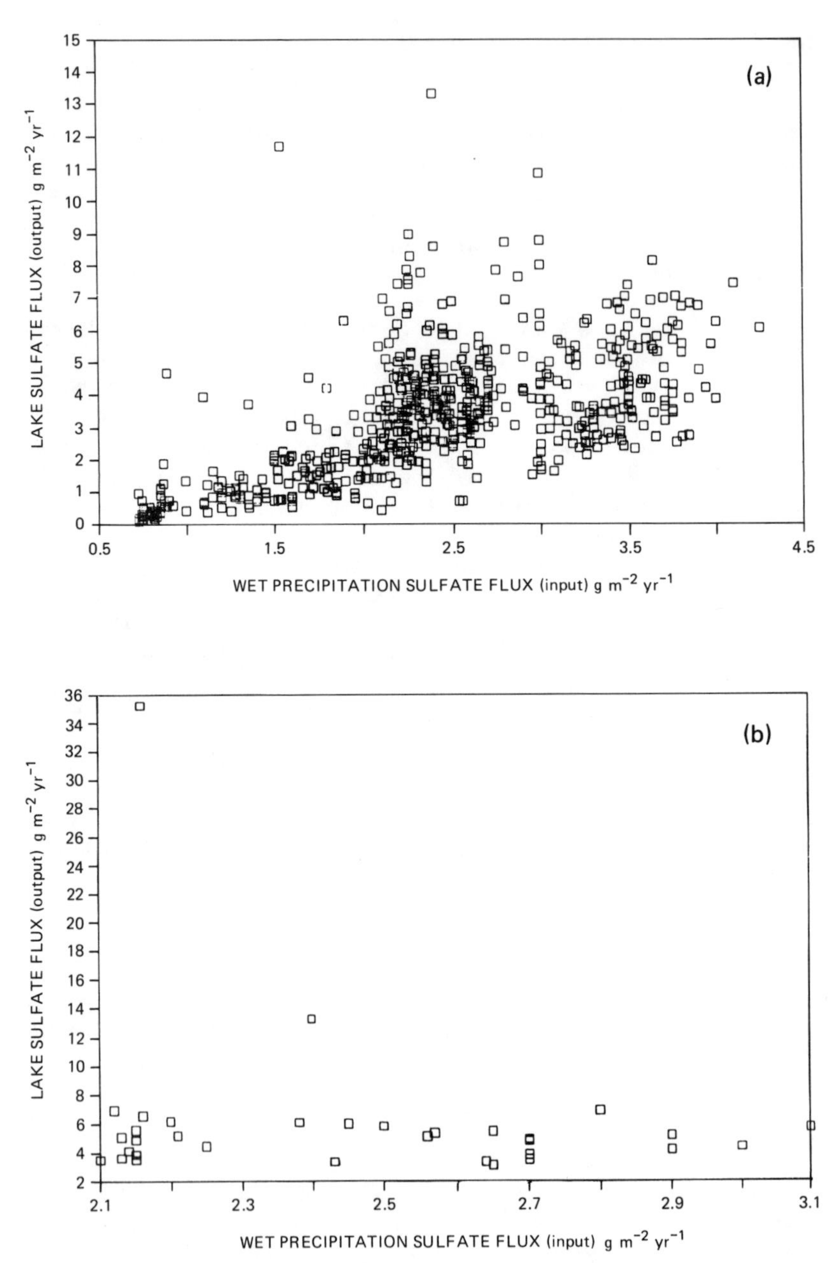

FIGURE 7.1 (a) Comparison of sulfate input fluxes to lakes from wet-only deposition with sulfate output fluxes from lakes in the northeastern United States and eastern Canada. (b) A similar plot for lakes in Massachusetts only.

TABLE 7.1 The Geographic Distribution of the Normalized Difference between Sulfate Inflow and Sulfate Outflow in Lakes of Northeastern North America as Determined by $x = \{ [SO_4] (lake) - [SO_4] (precip) \} / [SO_4] (precip)$[a]

	Excess Precipitation Sulfate Flux		Equal Sulfate Fluxes	Excess Lake Sulfate Flux						
Category[b]	1	2	3	4	5	6	7	8	9	Total Count (per state or province)
Range of x	$x<-0.5$	$-0.5 \leq x \leq -0.2$	$-0.2<x \leq 0.2$	$0.2<x \leq 0.5$	$0.5<x \leq 1.0$	$1.0<x \leq 1.5$	$1.5<x \leq 3.0$	$3.0<x \leq 5.0$	$x>5.0$	
State or Province[c,d]										
CT					${}_{4}6_{26}$	${}_{20}8_{35}$	${}_{31}9_{39}$			23
RI			${}_{1}1_{13}$	${}_{2}2_{25}$	${}_{3}4_{50}$	${}_{3}1_{13}$				8
MA				${}_{6}7_{21}$	${}_{7}9_{26}$	${}_{28}11_{32}$	${}_{17}5_{15}$	${}_{50}1_{3}$	${}_{50}1_{3}$	34
NY		${}_{5}4_{8}$	${}_{3}5_{10}$	${}_{10}12_{24}$	${}_{19}26_{51}$	${}_{10}4_{8}$				51
VT		${}_{2}2_{4}$	${}_{9}14_{29}$	${}_{12}14_{29}$	${}_{9}13_{27}$	${}_{5}2_{4}$	${}_{14}4_{8}$			49
NH		${}_{1}1_{2}$	${}_{7}11_{26}$	${}_{17}20_{47}$	${}_{7}10_{23}$	${}_{3}1_{2}$				43
ME		${}_{2}2_{6}$	${}_{6}10_{30}$	${}_{3}4_{12}$	${}_{9}12_{36}$	${}_{3}1_{3}$	${}_{14}4_{12}$			33
NF	${}_{4}2_{2}$	${}_{20}18_{19}$	${}_{25}40_{43}$	${}_{20}24_{26}$	${}_{4}6_{6}$	${}_{8}3_{3}$			${}_{50}1_{1}$	94
LB	${}_{85}40_{38}$	${}_{47}41_{39}$	${}_{10}16_{15}$	${}_{4}5_{5}$		${}_{3}1_{1}$	${}_{3}1_{1}$	${}_{50}1_{1}$		105
QU	${}_{11}5_{3}$	${}_{23}20_{11}$	${}_{40}64_{34}$	${}_{26}31_{17}$	${}_{38}52_{28}$	${}_{20}8_{4}$	${}_{21}6_{3}$			186
Total Count (per category)	47	88	161	119	138	40	29	2	2	626

[a] $[SO_4]$ (lake) is defined as the flux of the outflow of sulfate from a lake; $[SO_4]$ (precip) is the input flux of sulfate to a lake from wet-only precipitation.
[b] The lakes are grouped into 9 categories according to the value of x. A value of x close to zero (catergory 3) denotes approximately equal sulfate fluxes of lake output and wet-only deposition input. Lakes with values of x less than -0.2 (categories 1 and 2) retain sulfate in the watershed since the input of sulfate from wet-only precipitation exceeds the lake sulfate output. Lakes with values of x greater than 0.2 (catergories 4-9) have sources of sulfate inputs other than wet-only precipitation since the output of sulfate from the lake exceeds the input from wet-only precipitation.
[c] The states and provinces include: Connecticut (CT), Rhode Island (RI), Massachusetts (MA), New York (NY), Vermont (VT), New Hampshire (NH), Maine (ME), Newfoundland (NF), Labrador (LB), and Quebec (QU).
[d] Each row of numbers corresponds to the distribution by category of lakes within a given state or province. Each column of numbers corresponds to the geographical distribution, by state or province, of lakes in a given category. Subscripts to the right of the whole numbers correspond to the percent distribution of lakes by category within a given state or province. Subscripts to the left of the whole numbers correspond to the percent distribution by state or province of lakes in a given category. For example, 10 out of 43, or 23% of the lakes in New Hampshire are in category 5; 7% of all the lakes in category 5 are in New Hampshire.

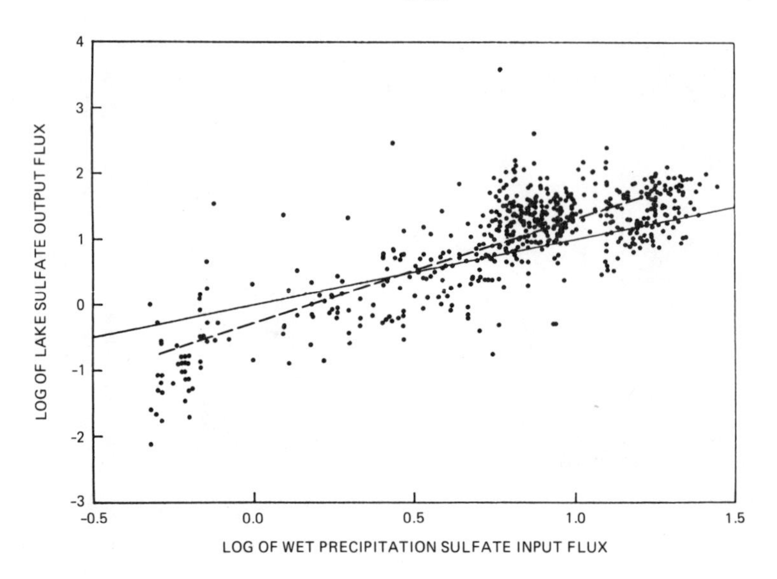

FIGURE 7.2 Comparison of the logarithm (base e) of sulfate fluxes to lakes from wet-only deposition with the logarithm (base e) of sulfate fluxes from lakes in the northeastern United States and eastern Canada. Solid line is the y = x line (i.e., condition in which sulfate flux from the lakes equals the sulfate flux to the lakes from wet-only deposition); dashed line is the line estimated from linear regression with slope of 1.593.

As noted earlier there is a tendency for lakes in some regions to lie above the line of equal fluxes (excess lake sulfate flux in Table 7.1) and other regions to lie below it (excess precipitation sulfate flux in Table 7.1). This tendency may result from regional differences in geology, dry deposition, or other factors causing the regional ratios of input to output fluxes to differ from the overall relationship. To account for regional differences we performed an analysis of covariance in which we allowed each region to have its own intercept, but fitted the same slope for all regions.

To analyze the data on a regional basis, we have chosen the following six regions: Connecticut, Massachusetts, and Rhode Island (CT-MA-RI); Maine, Vermont, and New Hampshire (ME-VT-NH); New York (NY); Newfoundland (NF); Labrador (LB); Quebec (QU). The linear regressions shown for each region in Figure 7.3 yield the following estimates:

Region	Intercept (m_i')	Standard Error
CT-MA-RI	0.691	0.0791
ME-VT-NH	0.257	0.0729
NY	0.280	0.1082
NF	-0.004	0.0626
LB	-0.515	0.0407
QU	0.098	0.0781
Slope (a)	1.089	0.0712

The improvement in fit in going from a single regression line to the six regional lines is highly significant, having a p value less than 10^{-6}. Furthermore, the common slope (a) is not significantly different from one. Thus, there is no evidence that the relationship on the natural scale ($y = m_i x^a$) is nonlinear once regional differences have been taken into account.

The multipliers m_i' are regionally specific. Thus lake sulfate output flux y is estimated to be a multiple of precipitation sulfate input flux x. (Note, that on the natural scale, our model will not in general estimate the arithmetic mean of the lake sulfate flux.) The multiple depends on such factors as the ratio of total deposition to wet-only deposition, the rate at which sulfur is demobilized in the watershed or sediments, the magnitude of sulfur sources in the watershed, and the magnitude of any bias in our estimate of wet-only sulfate flux. The values of m_i may be calculated from the antilogarithms of the intercepts (m_i') shown in Figure 7.3. The values are as follows: 2.00 (CT-MA-RI), 1.32 (NY), 1.29 (ME-VT-NH), 1.10 (QU), 1.00 (NF), and 0.60 (LB). These multiples are greatest in regions close to strong sulfur source areas and smallest in remote areas, suggesting that dry deposition plays a major role. Since the 0.60 value for Labrador indicates that output fluxes are smaller than input fluxes, watershed processes that retain sulfate may be important.

To obtain more detailed information of sulfur mass balance from these data, each location and perhaps each watershed needs to be considered separately. In some cases internal sources of sulfate may be important contributing factors to the sulfur mass balance (Wagner et al. 1982). For example, in Table 7.2, the geological setting is described for the locations of lakes in Massachusetts and Maine having large excesses of lake sulfate fluxes. Since strike zones, amphibolites,

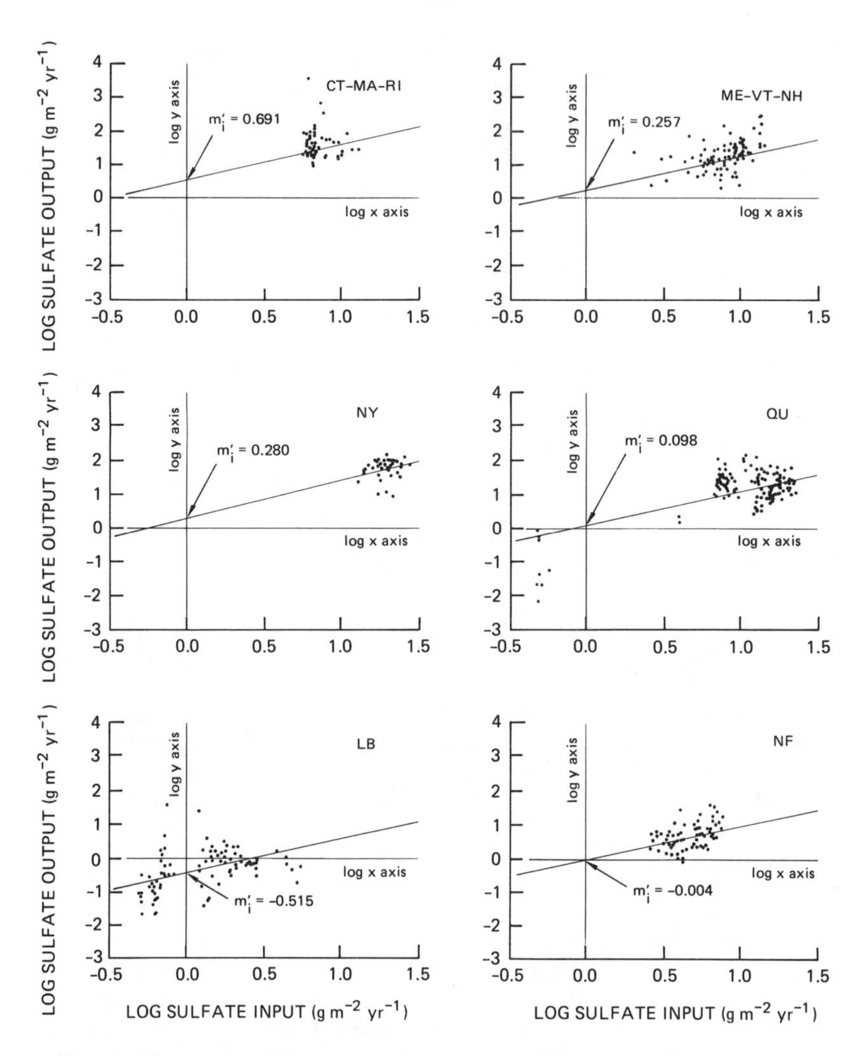

FIGURE 7.3 Regional relationships between inputs and outputs of sulfate for lakes in northeastern North America assuming that the only source of sulfate input is from wet deposition.

volcanics, and basalts are associated with sulfide minerals, it is quite possible that the excess sulfur in lakes with these associations are due to contributions from sulfide minerals. On the other hand, there are four lakes listed in Massachusetts (Big Bear Pond, Milham Street Pond, Round Pond, and Stearns Pond) for which

TABLE 7.2 Geological Characteristics of Watersheds in Massachusetts and Maine with Large Excess of Outflows of Lake Sulfate

Name	Location	Geological Setting
Massachuetts		
Wheeler Pond	42° 25′ 08″ 71° 31′ 17″	On a major strike fault, amphibolites and Andover granite.
Holts Pond	42° 09′ 46″ 71° 19′ 01″	On a major strike fault, amphibolites.
Bear Pond	41° 56′ 27″ 70° 59′ 38″	Molasse. Probable other anthropogenic source.
Little Farm Pond	42° 15′ 00″ 71° 22′ 41″	Mattapan volcanics.
Milham Street Pond	42° 19′ 17″ 71° 36′ 13″	Near major strike fault with diabase dikes, but probably other anthropogenic sources.
Round Pond	42° 35′ 53″ 70° 49′ 01″	Alkaline complex and fault. Sulfur excess probably anthropogenic.
Stearns Pond	42° 37′ 05″ 71° 04′ 04″	Fishbrook gneiss. Sulfur excess probably anthropogenic.
Maine		
Iron Bound Pond	45° 46′ 04″ 70° 05′ 23″	Devonian slate unit with fault.
Ferguson Pond	46° 51′ 58″ 68° 41′ 55″	Adjacent to recently found sulfide ore deposit.
Bluffer Pond	46° 22′ 30″ 69° 05′ 58″	Bluffer Pond volcanics.

SOURCE: G. M. Boone, Syracuse University, personal communication, 1985.

there are no obvious geological sources of sulfur, and these may be affected by anthropogenic contributions.

TRENDS IN THE CHEMISTRY OF HEADWATER STREAMS OVER THE PAST 15 TO 20 YEARS

In this section we consider trends in alkalinity, strong-acid anions, and nonprotolytic (base) cations* in

*Protolyte--any electrolyte that reacts with protons (H^+ ions) in the system of interest and especially in the pH range of interest. The proton condition is the ion balance condition for protolytes (e.g., Stumm and Morgan 1981, pp. 138-139).

small streams of the U.S. Geological Survey's Bench-Mark Network. The data are analyzed to assess trends and to determine whether the trends are consistent with the mechanisms thought to govern changes in surface water alkalinity driven by changes in sulfate deposition. In Chapter 1 these trends were related to other types of regional trends thought to be related to acid deposition. A general model linking acidic deposition to changes in surface water chemistry in watersheds with acid soils has been prominent for a number of years, (i.e., Seip 1980, Overrein et al. 1980, Cronan and Schofield 1979), and we use the version below (National Research Council 1984), which suggests that sulfate-driven changes in surface water chemistry occur as follows:

1. Sulfur deposition increases.
2. Sulfate concentrations in surface waters increase.
3. Concentrations of protolytic (H^+, Al^{3+}) and nonprotolytic (Ca^{2+}, Mg^{2+}, Na^+, K^+) cations in surface waters increase.
4. The increase in protolytic cations leads to decreases in surface water alkalinity.

This model is a consequence of the fact that charge balance must be maintained when the concentration of a mobile anion increases in a solution moving through an acid-dominated ion-exchange system. While the model does not take into account the complexities involved in the flow of SO_4^{2-} through the terrestrial portion of the watershed, it has been widely accepted on the strength of empirical support from studies of short-duration (e.g., a few years) time series (i.e., Galloway et al. 1983a) and in spatial comparisons of surface water chemistry in high versus low sulfate deposition areas (i.e., Mohn et al. 1980; Environmental Protection Agency 1984). The areas most prominently studied have been glaciated regions with thin, acidic soils, where clear-water lakes and streams having low alkalinity.

Base cation/acid anion--generally refers to a strong base cation/acid anion that for the system of interest can be considered to be totally dissociated, and hence its concentration is not a function of pH (e.g., base cations: Ca^{2+}, Mg^{2+}, Na^+, K^+; acid anions, Cl^-, NO_3^-, etc.).

In assessing trends in surface water alkalinity that span periods of decades, it is important to take into account the fact that internal changes can cause acidification of soils (particularly surface horizons) and perhaps acidification of surface waters. Major internal sources of H^+ in forest soils are dissociation of carbonic acid, cation accumulation in biomass, mineralization of organic sulfur and nitrogen compounds, nitrification, oxidation of reduced mineral compounds (e.g., pyrite), and humus formation (i.e., Driscoll and Likens 1982, Ulrich 1983). The major sinks for H^+ are reduction reactions, anion accumulation in biomass, mineral weathering, and destruction of organic acids. Any shifts or trends in the balance of processes governing the H^+ budgets of soils can change the net production of H^+ in the soil. In this regard Rosenquist et al. (1980) and Krug and Frink (1983) suggest correctly that the balance of H^+ production and consumption (particularly in the upper soil horizons) can be strongly acidifying during the development of maturing forests or after recovery of forests from major fires.

In acid forest soils not subjected to the input of strong-acid anions, most highly acidic solutions are the result of organic acids. This acidity is not easily transported through soils to surface water because the organic anions are retained or immobilized in the subsoil as a part of the regime of podsol development. Leachates are normally most acidic in the O and E horizons (the organic forest floor and uppermost mineral horizon, respectively). Acidification of the upper soil horizons does not necessarily translate into acidified surface waters because a mobile anion is needed to transport the protolytes. In cases where water from acidic surface horizons or from acid organic soils drains directly into streams, the stream water can be highly acidic because of organic acids. In theory, if water drains directly into streams through acidifying soil horizons, a trend of surface water acidification could be observed. However, to date, long-term changes in soil acidity owing to natural biogenic processes have not been shown in field studies to be paralleled by changes in surface water chemistry.

Decreases in surface water alkalinity that might result from the effect of runoff routed through naturally acidifying soils can be distinguished from sulfate-driven decreases in alkalinity in that long-term decreases in alkalinity resulting from acidifying soils should in most cases be accompanied by a decrease in the concentration of nonprotolytic cations. This follows because as the soil acidifies, cation exchange sites will be increasingly occupied by protolytic cations at the expense of non-protolytic cations, and the ratio of protolytic to non-protolytic cations in the soil solution is controlled by their ratio on exchange sites and by the anion concentration in the soil solution (Reuss 1985). Acidification of soils tends to increase the sulfate adsorption capacity, and reduce the dissociation of weak acids such as carbonic acid and organic acids, subsequently decreasing the soil-solution anion concentration and the effectiveness of cation leaching. The overall result of soil acidification from natural biogenic processes is that the soil solution becomes more acidic with lower concentrations of non-protolytic cations.

As a first approximation, alkalinity trends in surface waters can be due to changes in strong-acid anion flux plus changes in internal processes in accordance with Eq. (7.1):

$$\Delta Alk_{AA} + \Delta Alk_i = \Delta\Sigma B_i + \Delta\Sigma B_{AA} - (\Delta\Sigma A_{AA} + \Delta\Sigma A_i) \quad (7.1)$$

where ΔAlk_{AA} is the change in alkalinity due to changes in atmospheric strong-acid anion flux; ΔAlk_i is the change in alkalinity due to internal processes; $\Delta\Sigma B_i$ is the change in base cation concentrations due to internal processes; $\Delta\Sigma B_{AA}$ is the change in base cations due to changes in atmospheric acid anion flux; $\Delta\Sigma A_{AA}$ is the change in strong-acid anion concentrations due to changes in atmospheric flux; and $\Delta\Sigma A_i$ is the change in strong-acid anion concentrations due to changes in internal processes.

It should be noted that Eq. (7.1) ignores the possible secondary effects of acidic deposition on soil characteristics that affect alkalinity, and the addition of interactive terms may be appropriate. Equation (7.1) is, however, sufficient to demonstrate that several combinations of long-term changes in alkalinity, base cations, and strong-acid anions are possible. The balance of those processes will be governed by local soil and vegetation characteristics, disturbance history, and mobile anion deposition.

Having recognized that internal sources of anions and H^+ as well as atmospheric sources of mobile anions could affect trends in surface water alkalinity, and having suggested that there could be interactions between acid deposition and natural hydrogen-ion budgets, we point out that the deposition of strong acids or acidifying substances such as NH_4^+ to soils can promote soil acidification (Van Breeman et al. 1982, Ulrich 1983) by increasing proton input to the soil. The rate at which a soil will acidify under such influences is a function of the magnitude of the ionic input compared to the pool of exchangeable ions in the soil, and the ratio between acidic and basic cations in the solution added to the soil. In strongly acid forest soils, the annual ionic input is generally a very small fraction of the exchangeable pool, so the effect of atmospheric deposition in changing bulk soil chemistry would likely be slowly realized in most cases in the United States.

In summary, the pathways by which changes in surface water alkalinity may occur are through changes in the deposition of mobile anions and, under appropriate hydrologic conditions, changes in soil acidity, which can be natural or promoted by acidic deposition. Additional complexities are introduced by changes in processes that alter alkalinity within surface water bodies and by changes in weathering rates that may accompany changes in acid flux. Thus any trends in surface water alkalinity will be the result of numerous processes, some of which are interrelated. The balance of these processes is expected to be highly dependent on local conditions.

Methods

Since 1964 the U.S. Geological Survey has operated the hydrologic Bench Mark Network of 47 streamflow and water quality monitoring stations in small, predominantly undeveloped stream basins (Cobb and Biesecker 1971). The network includes stations in 37 states and was originally established to define baseline hydrologic conditions in a variety of climatic and geologic settings. Because of the application of consistent sampling and analytical methods (Skougstad et al. 1979) for a 15- to 20-year period at each site, records from the network are useful for investigating recent trends in water quality. Characteristics of the chemistry of the streams, and the geochemistry and land use histories of the basins are given in Appendix Tables C.2 and C.3, respectively.

Trends in anion and cation concentrations were estimated using the Seasonal Kendall test. This test is nonparametric and is intended for analysis of time trends in seasonally varying water quality data from fixed, regularly sampled monitoring sites such as those in the Bench-Mark Network (Hirsch et al. 1982, Smith et al. 1982). In addition to a test for trend, the statistical procedure includes an estimate of the median rate of change in concentration over the sampling period (called a trend slope in this report) and a method for adjusting the data to correct for effects of changing stream flow on trend in the water quality record. Trend is defined here simply as monotonic change with time occurring as either an abrupt or a gradual change in concentration. Quality-assured data sets for each stream typically contain anion (alkalinity, NO_3^-, SO_4^{2-}, Cl^-) and cation (pH, Ca^{2+}, Mg^{2+}, Na^+, K^+, NH_4^+) concentrations for 100 to 150 samples collected over 15 years, along with data on specific conductance and stream discharge, suspended sediment, etc. Data used in our analysis were screened using electroneutrality and conductivity balances (Σ anion equivalents = Σ cation equivalents $\pm$ 15 percent; and calculated specific conductance = measured specific conductance $\pm$ 15 percent) and by assessment of the quality-assurance measures used in sample collection, processing, and analysis.

Although our inability to account for aluminum and organic anions in some highly acidic streams (e.g., McDonalds Branch, New Jersey) may have introduced some bias into the analysis, we do not believe that this limitation is serious.

Sulfate output from the Bench Mark watersheds was estimated as the product of discharge-weighted mean sulfate concentrations and mean basin runoff for the 2-year period from 1980 and 1981. Sulfate deposition at each basin was estimated by linear interpolation of wet-only sulfate deposition data from the National Acid Deposition Program (NADP) for 1980 and 1981.

Results

Sulfate Input and Output

Critical to interpreting trends in stream sulfate is an understanding of watershed sulfur budgets and ecosystem sulfur cycling. Figure 7.4 indicates the major pathways of sulfur retention and release that need to be considered in determining how closely SO_4^{2-} inputs will be reflected in SO_4^{2-} yield in streams. Watersheds that

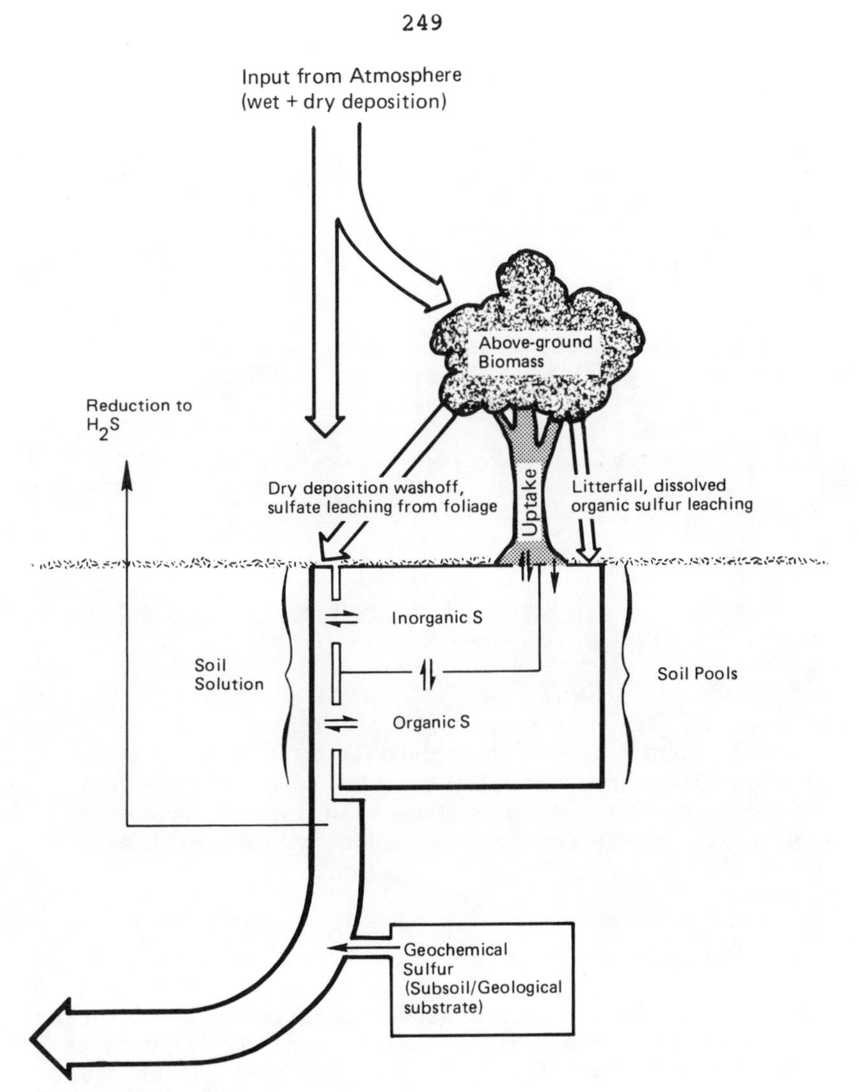

FIGURE 7.4 Flow diagram of the movement of sulfur in an undisturbed forested watershed.

are in a steady state with respect to sulfur would be the least ambiguous cases in which to judge the impact of changing rates of atmospheric sulfate deposition since changes in inputs are expected to be reflected in changes in output of similar magnitude and direction. Since

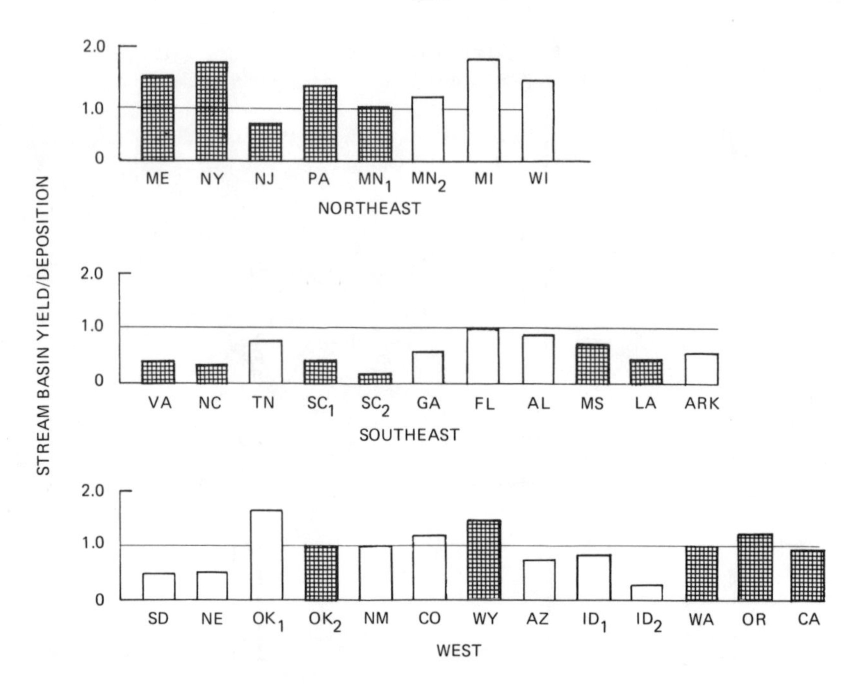

FIGURE 7.5 Ratio of stream basin sulfate yield to sulfate input from wet-only deposition at Bench-Mark stations that have ratios less than 2.0. Hatched bars correspond to streams with mean alkalinities less than 500 µeq/L. White bars correspond to streams with mean alkalinities greater than 500 µeq/L.

watershed-scale studies lack quantitative measurements of dry deposition, a precise understanding of sulfur cycling and budgets has not been achieved even in the most intensively studied sites (i.e., Likens et al. 1977, Richter et al. 1983). As a result, any explanation of the cause of trends in stream sulfate is subject to uncertainty since trends in internal sources and sinks as well as trends in atmospheric deposition may contribute to trends in surface water SO_4^{2-}.

For Bench-Mark watersheds in the northeastern United States, many variations of sulfur budgets are possible, with an overall tendency for output to exceed input owing to the effects of dry deposition and/or internal SO_4^{2-} sources (Figure 7.5).

Richter et al. (1983) showed that major amounts of SO_4^{2-} were retained in the B horizon of some of the

Paleudults at Walker Branch, Tennessee, attributable in part to SO_4^{2-} adsorption (Johnson and Henderson 1979). Thus, some soils (particularly Ultisols and Oxisols) will serve as effective sinks for atmospherically deposited SO_4^{2-} until the sulfate-sorption capacity becomes limited. Another means of sulfate retention is through incorporation into soil organic-S pools (Fitzgerald et al. 1982), and this mechanism could be quantitatively important in some instances.

Sulfate flux (as wet-only deposition) to and from all the U.S. Geological Survey's Bench-Mark watersheds for 1981 is shown in Appendix Table C.2. In trying to assess trends that might be related to changes in atmospheric sulfate flux, we have used watersheds whose ratios of yield to wet input were less than 2.0, judging that greater values indicated basins with important internal sources of SO_4^{2-} (Figure 7.5). We have done this without definitive information on what the input of sulfate from dry deposition might be, but, based on precipitation and throughfall studies in forests, it is reasonable to expect that inputs from dry deposition could be equal to the input from wet-only deposition in some cases (Morrison 1984). Similarities of sulfate budgets within some regions are notable. SO_4^{2-} output equals or is greater than wet input in watersheds of the Northeast, reflecting the influence of dry deposition, internal sources, or both. Sulfate appears to be strongly retained in watersheds of the Southeast. Soils in those watersheds are predominantly Ultisols (Appendix Table C.3), which typically have high SO_4^{2-} sorption capacities (Johnson 1980, Olson et al. 1982). Thus, sulfate-mass balance at the southeastern stations may be influenced by sorption sites in subsoil mineral horizons and/or by incorporation of sulfur into soil organic pools. Ratios of sulfate input to output at the western sites are variable, and they are difficult to account for in many cases given the information available on geological and soil characteristics.

Trends in Stream-Water Sulfate in Bench-Mark Streams

Figure 7.6 shows trends in stream sulfate concentrations over the periods of record for the watersheds that were judged not to have dominating internal SO_4^{2-} sources.

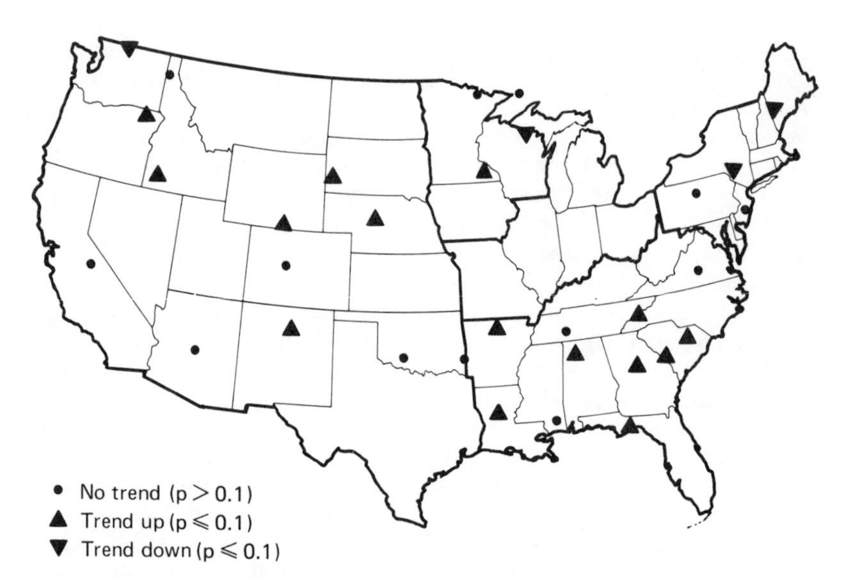

FIGURE 7.6 Trends in sulfate concentrations at Bench-Mark stations having a ratio of basin sulfate yield to basin sulfate deposition (wet-only) of less than 2.0. Regions B, C, D, and E are highlighted by darkened lines.

Sulfate concentrations tended to increase during the 15- to 20-year period at stations over a broad area of the continental United States extending from the Southeast to the mountain states and the Northwest. By contrast, stations in the northeastern quarter of the nation show either no trend or decreases in sulfate concentrations during the same period. This geographical pattern of trends occurs more or less independently of the significance criteria used in trend testing (Smith and Alexander 1983). Although the statistical significance of trends at many of the stations is high ($p < 0.01$), the magnitude of change has been small in most cases (Appendix Table C.2). Among stations showing a statistically significant trend in sulfate concentration (either a significant increase or a significant decrease), there is a median relative change in stream sulfate of 1.2 percent per year, or an overall change of about 20 percent over the period of record.

The trends in sulfate observed in Bench-Mark streams of the Northeast (Region B) and Southeast (Region C) regions are consistent with regional emission trends (Chapter 2, Figure 2.24) and with the contention that changing sulfate levels in small streams reflect changing emission and deposition patterns in watersheds that do not show evidence of dominating internal sulfate sources. Obviously, it would be preferable to have detailed sulfur cycling data for each watershed for further confirmation that sulfur output actually reflects atmospheric sulfur input. Insight into whether there is a cause-and-effect relationship between regional emissions and stream sulfate levels would be improved by site specific, process-level information regarding (1) the applicability of regional emission trends to the atmospheric sources contributing to deposition at each site and (2) the sources and sinks governing sulfate flux through each watershed.

Although considerable sulfate is retained in the Bench-Mark watersheds of the southeastern United States, there is a consistent pattern of increasing stream-water sulfate concentrations over the period of record. The trends in stream sulfate strongly reflect the regionwide trend in emissions, but we believe that more needs to be known about sulfur dynamics in the soils of those watersheds before we conclude that increased emissions have driven the increases in stream sulfate.

Trends in Alkalinity and Cations

The effect that increased sulfate flux through strongly acid soils may have in reducing surface water alkalinity is of particular practical concern. In acid soils charge balance requires that an increased cation flux must accompany an increased sulfate flux in the soil solution, and the balance between protolytic and nonprotolytic cations is determined by the ability of the soils to supply nonprotolytic cations through mineral weathering or cation exchange. In watersheds with thin, extremely acid soils, exchange sites are dominated by Al^{3+}, and H^+, and mineral weathering is usually limited. Thus, changes in SO_4^{2-} flux are expected to be accommodated mostly by changes in protolytic cation concentrations, thereby altering alkalinity in receiving water. In watersheds dominated by high base status soils and/or readily weatherable minerals that contain base cations, short-term changes in SO_4^{2-} flux are expected to be balanced mostly by changes in nonprotolytic cation flux. Thus, surface waters are expected to respond to increasing sulfate flux by both reduction of alkalinity and increases in non-

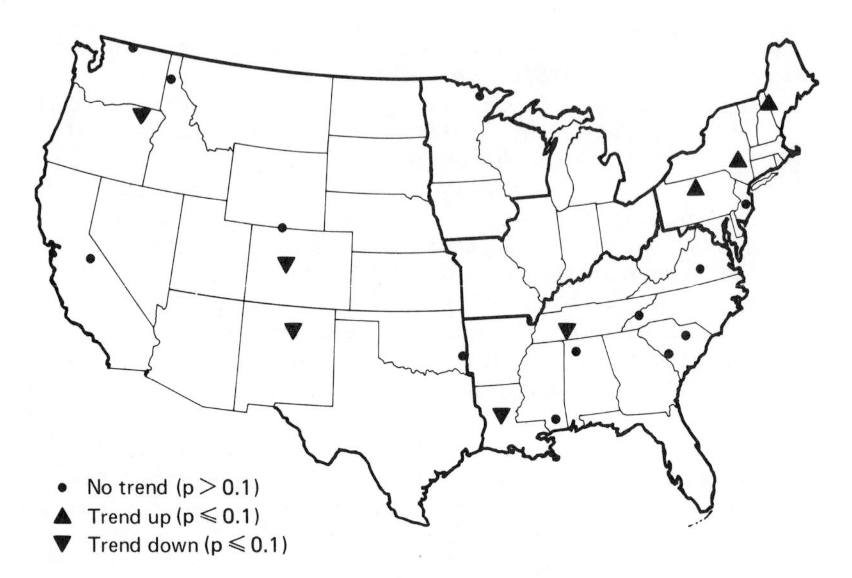

FIGURE 7.7 Trends in alkalinity at Bench-Mark stations having a mean alkalinity of less than 800 μeq/L. Regions B, C, D, and E are highlighted by darkened lines.

protolytic cation concentrations. In general, the magnitude of the alkalinity changes will therefore be less than the magnitude of the sulfate changes.

Figure 7.7 shows trends in alkalinity at stations with low to moderate mean alkalinity (<800 μeq/L). The regional pattern of trends is the approximate inverse of that of trends in sulfate (i.e., decreases in the South and West and increases in the Northeast) but station-by-station comparisons of statistically defined trends ($p < 0.1$) fail to show a consistent inverse relationship (see Appendix Table C.2). Following is a more detailed analysis of short-term changes in sulfate and alkalinity at these stations.

Nitrate is also a strong acid anion added from the atmosphere, and, although NO_3^- is apt to be strongly retained in the terrestrial ecosystem by plants and soil organisms in most cases, changes in NO_3^- concentrations in surface waters must be considered along with sulfate changes in interpreting alkalinity changes. For cases in which changes in strong-acid anion flux drive changes in stream chemistry, the relationship between changes in strong-acid anions, alkalinity, and cation concentrations given in Eq. (7.1) can be summarized in general terms with the equation

$$\Delta\Sigma A = \Delta\Sigma B - \Delta Alk, \quad (7.2)$$

where ΣA is the total strong-acid anion concentration (nitrate plus sulfate, µeq/L), ΣB is the total base (nonprotolytic) cation concentration (Ca^{2+} + Mg^{2+} + Na^{+} + K^{+}) (µeq/L), and Alk is the alkalinity (µeq/L). Equation (7.2) is essentially a statement of the charge balance that must prevail when a change in strong-acid anion flux is the variable affecting a change in stream chemistry.

Equation (7.2) can be differentiated with respect to the strong-acid anion concentration to give

$$\frac{d\Sigma B}{d\Sigma A} - \frac{dAlk}{d\Sigma A} = 1. \quad (7.3)$$

The derivatives $d\Sigma B/d\Sigma A$ and $dAlk/d\Sigma A$ are the rates of change of base cation concentration and alkalinity, respectively, per unit change in acid anion concentration. Their values will differ from one geochemical setting to another and provide a measure of the ability of a basin to accommodate short-term changes in sulfate flux by supplying base cations. Estimates of the derivatives for Bench-Mark stations were obtained from linear regressions of major cation concentrations and alkalinity on sulfate plus nitrate concentration. In contrast to the 15- to 20-year trends presented in Figures 7.6 and 7.7, these regressions tend to reflect short-term variations in stream chemistry. Values of the estimates are given in Table 7.3. The observed effect of strong-acid anion change on cations is due almost entirely to changes in SO_4^{2-}, since both concentrations and changes in nitrate are very small.

A plot of $d\Sigma B/d\Sigma A$ versus $d\,Alk/d\Sigma A$ is given in Figure 7.8 for streams with mean alkalinity of <500 µeq/L. For a majority of eastern stations and a number of lower-alkalinity western sites, $dAlk/d\Sigma A$ is negative and $d\Sigma B/d\Sigma A$ ranges from 0 to 1. Increases in acid anion concentrations are balanced, in part, by decreases in alkalinity and, in part, by increases in base cations. Alkalinity changes in the low-alkalinity streams (<500 µeq/L) balance about 30 percent of the change in SO_4^{2-}, and changes in base cations balance about 50 percent of the change in SO_4^{2-}. We have not conclusively determined why the residuals $(1 - (d\Sigma B/d\Sigma A - dAlk/d\Sigma A))$ are generally positive, but this may be due to the effect of ions that we did not use in the model (i.e., Cl^{-}, NH_4^{+}, Al^{3+}).

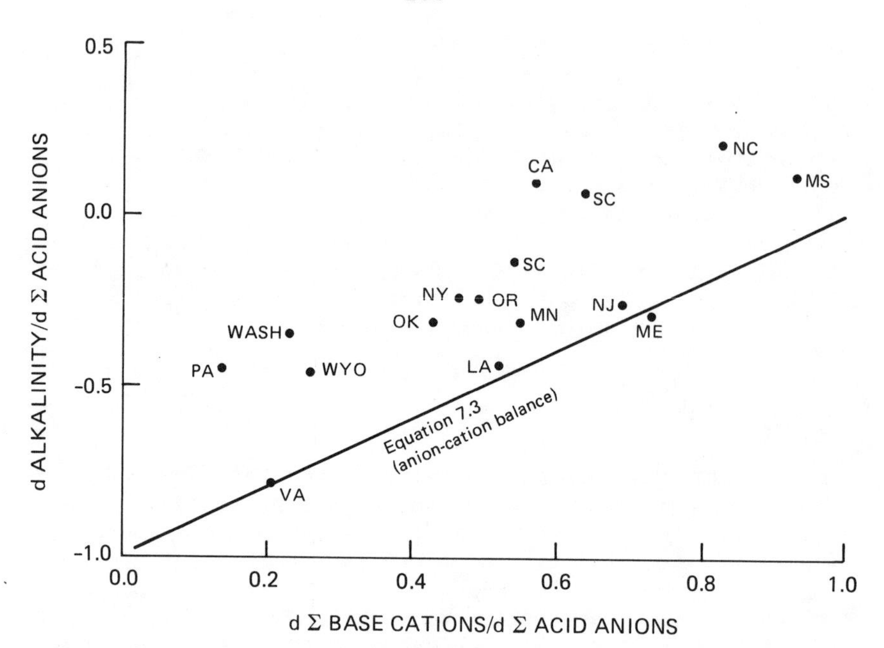

FIGURE 7.8 Plot of rate of change in base cation concentrations versus rate of change in alkalinity per unit change in acid anion concentration for stations with mean alkalinities less than 500 μeq/L.

Some stations in the group have small positive values of dAlk/dΣA that are not statistically distinguishable from zero (Table 7.3). At these stations, acid anion changes are essentially completely balanced by changes in base cations.

Figure 7.9 shows a larger set of stations, including for comparison sites with mean alkalinity values between 500 and 800 μeq/L. At higher-alkalinity sites, values of the derivatives are usually greater than 1.0, indicating that alkalinity and base cation changes cannot be entirely driven by changes in SO_4^{2-}. Mechanisms other than mobile anion flux must therefore be important at those sites. At low-alkalinity sites, the data are consistent with the hypothesis that changes in SO_4^{2-} flux drive changes in cation flux, resulting in changes in base cation concentrations and/or in alkalinity.

TABLE 7.3 Regression Estimates of Change in Base Cation Concentration and Alkalinity per Unit Changes in Acid Anion Concentration at Bench-Mark Stations of Moderate to Low Alkalinity

Station	$d\Sigma B/d\Sigma A$ (±1 Standard Deviation)	$d Alk/d\Sigma A$ (±1 Standard Deviation)	Anion-Cation Balance $1-(d\Sigma B/d\Sigma A - d Alk/d\Sigma A)$
Wild River at Gilead, Me.	0.73 (0.65, 0.81)	−0.29 (−0.37, −0.20)	−0.02
Esopus Creek at Shandaken, N.Y.	0.46 (0.39, 0.54)	−0.24 (−0.31, −0.18)	0.30
McDonalds Branch at Lebanon State For., N.J.	0.69 (0.66, 0.72)	−0.26 (−0.28, −0.24)	0.05
Young Womans Creek near Renovo, Pa.	0.14 (0.09, 0.19)	−0.45 (−0.52, −0.39)	0.41
Holiday Creek near Anersonville, Va.	0.21 (0.10, 0.33)	−0.78 (−0.86, −0.70)	0.01
Scape Ore Swamp near Bishopville, S.C.	0.54 (0.49, 0.59)	−0.13 (−0.19, −0.10)	0.33
Upper Three Runs near Ellenton, S.C.	0.64 (0.59, 0.69)	0.07 (−0.01, 0.15)	0.43
Sipsey Fork near Grayson, Ala.	2.50 (1.87, 3.22)	1.70 (1.08, 2.39)	0.20
Cypress Creek near Janice, Miss.	0.93 (0.86, 1.00)	0.12 (0.04, 0.19)	0.19
Cataloochee Creek near Cataloochee, N.C.	0.83 (0.76, 0.90)	0.21 (0.14, 0.28)	0.38
Kawishiwi River near Ely, Minn.	0.55 (0.47, 0.64)	−0.31 (−0.41, −0.22)	0.14
Encampment River near Encampment, Wyo.	0.26 (0.22, 0.31)	−0.46 (−0.52, −0.40)	0.28
Halfmoon Creek near Malta, Co.	0.53 (0.46, 0.60)	−0.16 (−0.24, −0.08)	0.31
Kiamichi River near Big Cedar, Okla.	0.43 (0.34, 0.52)	−0.31 (−0.38, −0.24)	0.26
Big Creek at Pollock, La.	0.52 (0.46, 0.59)	−0.44 (−0.49, −0.39)	0.04
Rio Mora near Terrero, N.M.	2.00 (1.92, 2.09)	1.15 (1.07, 1.24)	0.20
Merced River near Yosemite, Calif.	0.57 (0.40, 0.74)	0.10 (−0.02, 0.23)	0.53
Hayden Creek near Hayden Lake, Ida.	0.07 (−0.03, 0.17)	−0.06 (−0.20, 0.08)	0.87
Andrews Creek near Mazama, Wash.	0.23 (0.13, 0.33)	−0.35 (−.045, −0.25)	0.47
Minam River at Minam, Ore.	0.49 (0.38, 0.59)	−0.24 (−0.35, −0.14)	0.27

NOTE: Data from Buffalo River near Flat Woods, Tenn., may have been affected by land-use changes during the period of record and are not included.

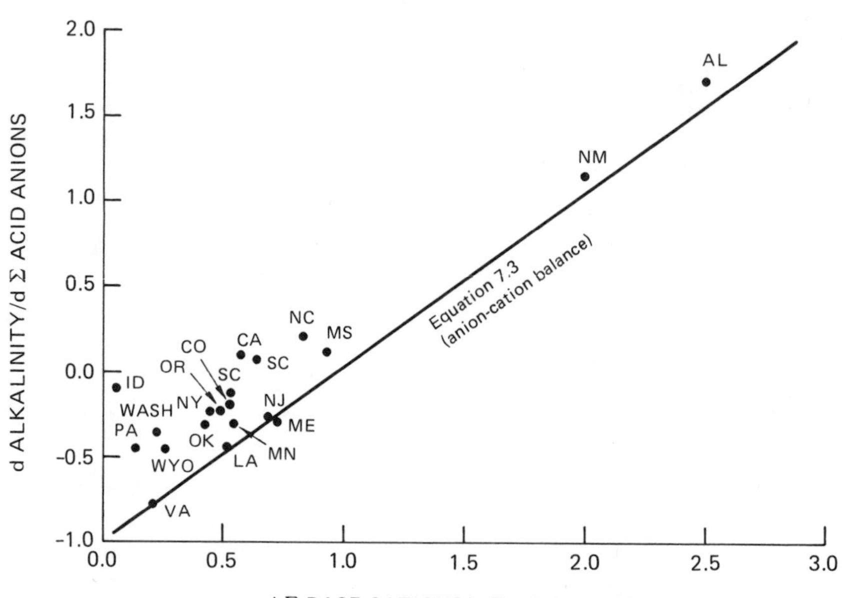

FIGURE 7.9 Plot of rate of change in base cation concentration versus rate of change in alkalinity per unit change in acid anion concentration for stations with mean alkalinities less than 800 μeq/L.

Interpretation of 15-Year Trends in Alkalinity

In addition to long-term trends in surface water chemistry, which might be caused by changes in atmospheric deposition of strong acids, processes in soils or in streams controlling hydrogen-ion budgets may have changed through time to produce changes in base cations, strong-acid anions, and alkalinity. Changes in mineralization rates of nitrogen and sulfur compounds, changes in weathering rates, and natural processes of soil acidification could result in such changes.

The trends in cations and alkalinity apparent in the Bench-Mark stream data afford an opportunity to evaluate the possible role of internal processes in recent (15-year) trends in alkalinity and strong-acid anions. Figure 7.10(a) shows trend slopes for alkalinity minus base cations plotted against trend slopes in sulfate plus nitrate for the 12 stations that display charge balance (Eq. (7.2)) errors of less than 1.5 μeq L^{-1} yr^{-1}.

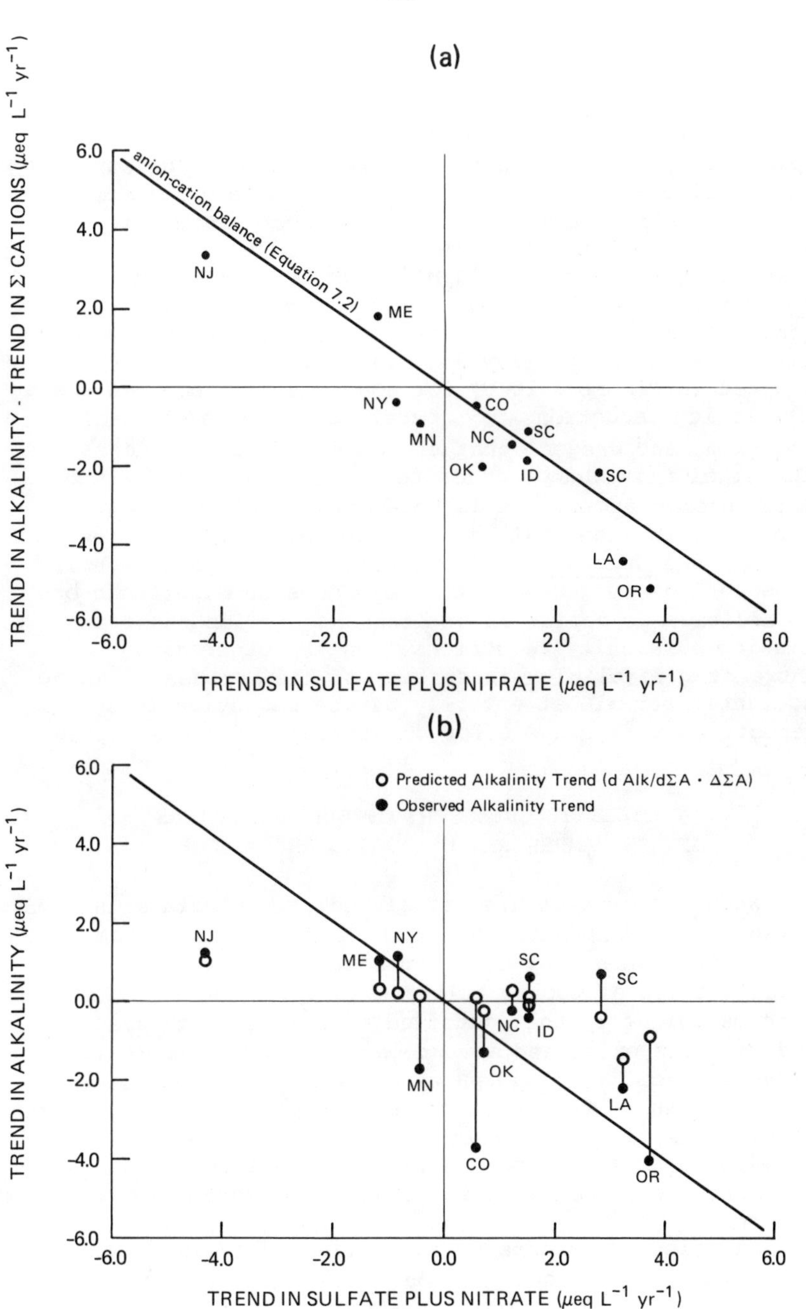

FIGURE 7.10 (a) Trend in alkalinity minus trend in base cations versus trends in sulfate plus nitrate. (b) Trend in alkalinity versus trend in sulfate plus nitrate.

The condition that charge balance is maintained within reasonable limits suggests that the trends are probably accurately modeled.

If trends in acid anion fluxes were the only cause of observed trends in stream chemistry, it would be expected that most alkalinity trend slopes would be considerably less than strong-acid anion trend slopes since some of the change in the latter is accompanied by changes in base cations. Figure 7.10(b) shows the alkalinity trend slope that would be predicted from the values of $(dAlk/d\Sigma A)\Delta\Sigma A$ (Table 7.3) for the 12 stations plotted against $\Delta\Sigma A$. Observed values of ΔAlk are also plotted in Figure 7.10(b) for comparison. In a few cases alkalinity reductions have occurred (in Minnesota, Colorado, and Oregon) that are considerably in excess of the alkalinity change predicted on the basis of strong-acid anion changes. It is noteworthy that the large decreases in alkalinity are mostly balanced by increases in SO_4^{2-} and <u>decreases</u> in base cations. The decrease in base cations suggests that some processes other than or in addition to mobile anion leaching are important in causing alkalinity reductions. On the other hand, increasing alkalinity in streams of the Northeast can be accounted for almost entirely by the reduction in sulfate deposition within the error constraints of the analysis.

PROTOLYTIC CHEMISTRY OF SURFACE WATERS OF NEW HAMPSHIRE, NEW YORK, AND WISCONSIN

Extensive sets of historical and recent data exist for lakes in New Hampshire, New York, and Wisconsin from which conclusions may be drawn regarding changes in the acidic status of poorly buffered lakes. These data are perhaps unique in that detailed records of both sampling and analysis exist and measurements of alkalinity, acidity (free CO_2 acidity), and pH were routinely made. Having measurements of the three parameters permits an assessment of the internal consistency--and hence the quality--of the data because the chemistry of the system is mathematically overdetermined if carbonate protolytes are assumed.

Analysis of the data is also illustrative of many of the factors that should be considered when attempting to assess time trends in data on aquatic protolytes.

The assessment procedure is as follows: (1) Define the underlying chemical principles. (2) Use the principles to adjust the data for bias owing to changes in analytical techniques. (3) Define a consistent data set by comparing independent estimates of alkalinity and pH. (4) Analyze the data for temporal changes.

Chemical Principles

Techniques for measuring alkalinity, pH, and acidity underwent significant changes between the 1930s and the 1980s. Early methods all relied on colorimetric techniques, which employ special chemicals, or indicators, that change color at specific values of pH. Recent methods entail electrometric techniques. Considerable uncertainties exist regarding the colorimetric data. First, they depend upon observations of color change that can vary from analyst to analyst. Also, we often do not know whether formulations of indicators have changed over time, thus altering the endpoints of titrations. Furthermore, the indicators themselves may be acids or bases, and even when they are added in small quantities they can alter the pH of the test solution and introduce systematic bias if unaccounted for. Such methodological biases may be large in historical data on poorly buffered waters. At the same time, recent data may also contain methodological biases or inaccuracies. For example, electrometric measurement of pH in dilute solutions is often unreliable (Blakar and Digernes 1984).

As will be shown, the pH at the true equivalence point in an alkalinity titration varies as a function of the original alkalinity of the test samples. Modern methods, such as the Gran titration technique (Gran 1952), take this into account, whereas the older titration technique, using the indicator methyl orange (MO), does not. That is, the endpoint pH in the MO technique is fixed, usually well below the true equivalence point. The MO indicator changes color according to the sequence yellow-orange-pink-red, spanning a pH range from about 4.6 to 3.1 (Clark 1928). The exact values of pH at which the color changes appear, as well as which color changes were used in particular circumstances, are not always evident (vide infra). Where corrections for this overtitration have been made in previous work, investigators have assumed values of the pH at the endpoint ranging from 4.04 to 4.5 (Kramer and Tessier 1982, Church 1984). Zimmerman and Harvey (1979) concluded that no corrections could be made because the true endpoint pH varies in an indeterminant manner. However, several other authors have assumed that the pH of the true equivalence point is invariant at pH = 5 for low-alkalinity waters, which is not precisely true (for a review, see Church 1984). The correction, simply made as the difference between the hydrogen-ion concentration at the indicator endpoint (about 10^{-4} N) and the assumed true endpoint (10^{-5} N) is a better approximation but not so good as can be achieved.

We offer an alternative--and chemically more rigorous--approach (Kramer and Tessier 1982). The calculations for this approach consider the correction of the MO and phenolphthalien endpoints for alkalinity and acidity and the correction of the colorimetrically determined pH to the actual pH that is due to the acid/base effect of the indicator. Furthermore, we employ the fact that a carbonate system is overdetermined if data on pH, alkalinity, and CO_2 acidity are available. Thus two independent values of alkalinity and three pH values are obtained. The following discussion outlines the approach, whereas Appendix D discusses the derivations and calculations in detail.

Alkalinity is the "excess base" relative to the ion-balance condition for the solubility of CO_2 in water and is zero when all the strong-acid concentrations equal all the strong base concentrations in the presence of CO_2:*

$$[Alk] = [HCO_3] + 2[CO_3] + [OH] - [H], \qquad (7.4)$$

whereas CO_2 acidity ([Acy]) is defined as the "excess" acid relative to the solution condition for $MHCO_3$:

$$[Acy] = [H_2CO_3] + [H] - [OH] - [CO_3], \qquad (7.5)$$

where square brackets [ ] indicate molar concentrations. Equations (7.4) and (7.5) are valid for a system of carbonate protolytes only. The possible influence of other protolytes is considered in Appendix D.

Addition of Eqs. (7.4) and (7.5) gives the total carbonate C_t:

---

*For simplicity, charges are not designated unless needed for meaning. $[H_2CO_3]$ is defined as $[CO_2]_{aq}$ + $[H_2CO_3]$.

$$C_t = [Alk] + [Acy] = [H_2CO_3] + [HCO_3] + [CO_3] \quad (7.6)$$

The pH for titration of [Alk] and [Acy] varies with C_t. The true endpoint values of pH for [Alk] and [Acy] can be determined by setting Eqs. (7.4) and (7.5) equal to zero and solving for [H]; the results for the endpoint concentrations of hydrogen, for $[H]_a$ and $[H]_b$ for [Alk] and [Acy], respectively, are approximately

$$[H]_a = (C_t K_1)^{1/2} - K_2/2 \quad (7.7a)$$

and

$$[H]_b = [K_1 (K_2 + K_w)/C_t]^{1/2} \quad (7.7b)$$

where K_1, K_2, and K_w are the first and second dissociation constants for carbonic acid and the dissociation constant for water, respectively.

Most fixed-endpoint titrations would not have been titrated to the correct value of [H] because of the dependence of the equivalence point [H] on total carbonate (Eq. (7.7a,b)), which would vary from sample to sample. Fixed endpoints can be adjusted to the correct endpoint value, however, if the actual titration endpoints pH_x[Alk] and pH_y [Acy] are known. The following equations relate the actual alkalinity and CO_2 acidity to the titrated fixed endpoint values $[Alk]_x$ and $[Acy]_y$. These relationships are derived knowing that [Alk] and [Acy] are each zero at their respective endpoints:

$$[Alk] = [Alk]_x + [HCO_3]_x + 2[CO_3]_x + [OH]_x - [H]_x \quad (7.8)$$

and

$$[Acy] = [Acy]_y + [H_2CO_3]_y + [H]_y - [OH]_y - [CO_3]_y, \quad (7.9)$$

where the subscripts x and y refer to the concentrations of each parameter at pH = x or pH = y. In practice, equations using C_t, K_1, K_2, K_w, and [H] are derived from Eqs. (7.8) and (7.9), and [Alk] and [Acy] are derived in an iterative fashion using these two equations (Appendix D, Eqs. (7e) and (7f)) and the definition of C_t.

MO was the indicator used for alkalinity titrations until about 1950 and sometimes even later. The pH of the

MO color change (and the phenolphthalein color change for [Acy]) must be known if one is to correct the historical alkalinity values for a comparison to values determined today. MO undergoes several subtle color changes from yellow to pink in the pH range of 4.5 to 3.1; this change is unaffected by ionic strength (Clark 1928, Kolthoff 1931).

Different pH values of the MO color change appear in the literature. Most titration procedures used before the mid-1940s followed the protocols stated in Standard Methods for Examination of Water and Wastewater (1926, 1930, 1933, 1936). The 1933 and 1936 volumes call for titration "until the faintest pink coloration appears; that is until the color of the solution is no longer pure yellow." These reference texts do not quote an exact value of the pH at the color change but do state that the pH is "about" 4.0. The 1960 edition of Standard Methods states more precisely: "The indicator changes to orange at pH 4.6 and pink at 4.0." Kolthoff and Stenger (1947) state "Thus with methyl orange . . . the color begins to change at a pH of about 4"; and later they note that the blank adjustment is 100 µeq/L which is equal to a pH of 4.0 for distilled water. They also note that the color changes from orange-yellow to orange at this point. Kramer and Tessier (1982) repeated the MO titration on a low-alkalinity sample; independent technicians following the Standard Methods (1933) procedure with the "faintest pink" directive found the pH for the color change to be 4.04.

Recently, several investigators have reported higher pH values for the MO color change. Eilers et al. (1985), using known solutions of $CaCO_3$, obtained a correction factor of 86 µeq/L using the "pink endpoint" reference; this correction is equivalent to a pH for this endpoint of 4.07 or less. In addition the authors reported the appearance of a "very faintest pink" color at a pH of about 4.3. G. E. Likens (personal communication, the New York Botanical Garden, 1985) reported the same color change at 4.25, and T. A. Haines (personal communication, Fish and Wildlife Service, Orono, Maine, 1985) reported a value for the "faintest pink" endpoint between 4.3 and 4.5 depending on the alkalinity of the original solution. Haines also noted a lower "definite orange" endpoint between about 4.0 to 4.27 again depending on the original alkalinity.

In an effort to determine which pH value of the MO endpoint was used in the historical surveys, we examined the relevant literature for documentation. We could find no records in the New York survey citing the value of pH used in the MO titration. In New Hampshire we could only find a reference to titration to "salmon pink" (R. E. Towne, New Hampshire Water Supply and Pollution Control Commission, personal communication, 1983). The best evidence is provided in the Wisconsin historical lakes survey. In a letter to J. Juday, a principal investigator of the survey from the University of Wisconsin, F. L. Taylor (1933), an analytical chemist on the survey team, describes his dissatisfaction with the "fainter endpoint" and discusses his rationale for changing the titration procedure from the "fainter endpoint" to an endpoint lower than a pH of 4.0:

> The titrations made in August for fixed CO_2 were considerably higher than those of previous years, or even of July, 1932, because they were the recorded titrations made to a deeper end-point of the indicator than that previously used. . . . I found that I could detect the color change at that point easier than at a fainter color, and hence could duplicate titrations easier. . .

Further, Taylor describes the pH range of the "fainter" endpoint:

> Titrations made previous to these were all made with this fainter end-point for which blanks probably ranged from 0.3 cc to 0.4 cc, i.e. from 1.5 to 2.0 ppm. In calculating earlier titrations by me I feel the safest blank to use is the lower (1.5 ppm).

Since Taylor did his blank study with boiled distilled water, the only adjustment should be for H^+. Thus the MO endpoints equivalent to 1.5 and 2.0 ppm (CO_2) would be 66 and 88 μeq/L, equal to values of the MO endpoint pH of 4.18 and 4.06 respectively.

Eilers et al. (1985) attempted to resolve the question of the MO endpoint by examining the relationship between specific conductance in an historical and a recent set of data. From regressions of alkalinity and specific conductance for Wisconsin lakes, they obtained a correction of 64 μeq/L for the MO endpoint, equal to a pH of 4.19 or less depending on total carbonate content (e.g., Eqs. (7g) and (7h), Appendix D).

At this time we have no explanation for the disparity in the values of pH reported for the MO color change. As noted above, some investigators have reported higher pH values (than 4.2) for the MO endpoint. However, we believe that the weight of the evidence suggests that a range of values between 4.0 and 4.2 is more probable. In the analyses of alkalinity changes presented in the following sections, we consider two possible endpoints for the MO adjustment: pH values of 4.04 and 4.19. The lower value is supported by the findings of Kramer and Tessier (1982), Kolthoff and Stenger (1947), Standard Methods (1960), and the recommended lower limit of the pH range noted by Taylor (1933); the higher value reflects the analysis by Eilers et al. (1985) and the upper limit of the pH range noted (and preferred) by Taylor (1933). The actual alkalinity correction, as noted before, also depends upon the total carbonate (C_t). Thus Eq. (7e) of Appendix D is used for the MO adjustment with values of $[H]_x$ of $10^{-4.04}$ and $10^{-4.19}$ µeq/L.

It is possible that the pH of color change may lie outside this range for specific cases owing to differences in human perception of the color change as well as other factors, such as analytical set-up, sample color interference, and the overall lack of sensitivity of the method. With respect to the last-named factor, it is important to realize that for the historical methods each 0.1 cm^3 of titrant is equivalent to 22 µeq/L alkalinity, which is about equal to the reported precision of the method. The lack of sensitivity of the early methods probably adds additional variance to the endpoint.

Phenolphtalein was the indicator used for base titration in order to determine "free" CO_2. The endpoint pH (pH_y = [AC]) for this titration is about 8.25, which results in a small adjustment (Eq. (7.9)) of less than 10 µeq/L.

The adjustment for MO-determined alkalinity can be quite large (up to about 91 µeq/L), whereas that for the free CO_2 acidity correction can be very small (about 2 µeq/L). A small variation in the MO endpoint of a few tenths of a pH unit can result in a variation in alkalinity of 10 to 20 µeq/L. Thus, one must have confidence that the MO titration was carried to the appropriate pH. A consistency check can be carried out whereby an independent estimate of [Alk] is made from the combination of [Acy] and pH measurements. The two values can be compared, and after establishing an acceptable margin of difference one can determine whether the values of [Alk], pH, and [Acy] obtained from the colorimetric pH, [MO], and [AC] are consistent or inconsistent.

Equation (13) of Appendix D gives the relationship for obtaining C_t from [Acy] and pH. C_t and [Acy] can be applied to Eq. (7.6) to obtain [Alk] by difference.

Colorimetric pH must also be corrected because the pH indicator dyes are weak acids and bases and alter the actual pH of a poorly buffered solution. The change for such a solution has been shown to be as large as 3 pH units (Blakar and Digernes 1984). For example, using one half of the indicator concentration as specified for old comparators, Blakar and Digernes show that an actual pH of 4.8 in a poorly buffered solution would be read as a pH of 6.0 if bromthymol blue, a commonly used indicator in historical surveys, were used. Kramer and Tessier (1982) showed that addition of bromthymol blue according to the comparator recipe to a sample with near zero alkalinity changes the pH from an actual value of about 5.2 to a value of nearly 7. At the same time, the same measurement carried out in a well-buffered solution will have a negligible effect on the pH. Therefore, to adjust for the indicator effect, one must know the colorimetric pH, the kind of indicator, details of concentrations and volumes, and the concentration of one other protolyte (usually [Alk]).

The actual pH can be determined in three ways by combinations of [Alk]/pH_c, [Acy]/pH_c and [Alk]/[Acy], where pH_c is the colorimetrically determined pH. The results of the three determinations are compared for equivalence as a consistency check, and the data may be accepted or rejected for an acceptable margin of difference. Equations (10) to (12) of Appendix D give the relationships for obtaining the pH from the three combinations of [Alk], [Acy] and pH_c.

To illustrate differences between the above method for correcting historical pH and using colorimetric pH values alone, we examined historical pH values of Wisconsin lakes. Some of the lakes were of high alkalinity and required little adjustment, whereas other lakes required a large correction to the colorimetric pH value. Figure 7.11 shows the magnitude of corrections of historical data, which ranged from +0.5 to -1.25 pH units.

It is also noteworthy that Birge et al. (1935) used MO alkalinity and phenolphthalein free CO_2 in a way similar to the use of [Alk]/[Acy] here to calculate a pH. In their use of the Kolthoff expression, they made the assumptions that $[HCO_3]$ = adjusted MO alkalinity and $[H_2CO_3]$ = free CO_2; they calculate pH, then, from $[H] = [H_2CO_3]K_1/[HCO_3]$.

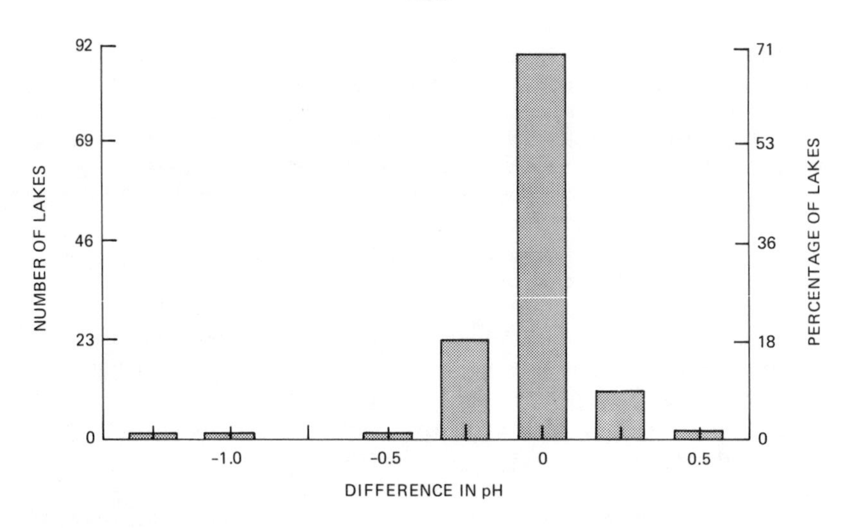

FIGURE 7.11 Correction of historical pH data for Wisconsin lakes, showing differences that can occur when pH is calculated by two methods of correction. pHs obtained from the method of applying a constant correction factor of 63 µeq/L are subtracted from pHs obtained from the method of Kramer and Tessier (1982).

In summary, (1) fundamental principles of ion balance are applied to a carbonate ion system in order to derive definitions of alkalinity and acidity; these concepts are also used to reevaluate data titrated to the uncorrected endpoint as well as to adjust colorimetric pH data. (2) If alkalinity, acidity, and pH data exist, two independent estimates of alkalinity and three estimates of pH can be made. If the estimates agree within an acceptable margin, the data may be considered internally consistent.

Data and Methods

For this report we analyzed historical data from New York (1929-1936), New Hampshire (1930s and 1940-1950), and Wisconsin* (1925-1932) and compared them with recent

---

*Records exist for 1925-1941. However, we included only data for 1925-1932 because of a change in titration to a lower MO endpoint that occurred in 1932. See earlier discussion.

TABLE 7.4 Analytical Aspects of Lakes Data from New York, New Hampshire and Wisconsin

Data Set	Type of Sample	Alkalinity	Acidity	pH	Comments
N.Y. 1930s	Surface Bottom	M.O.[a]	Phen.[b]	Col.[c]	One survey only at various locations in same lake — one data source. 248 lakes.
N.Y. 1970s and 1980s	Surface	Gran	—	Elec.[d]	Four data sources.
N.H. 1930, 1940-50	Profile	M.O.	Phen.[b]	Col.	Replicate depths common. One data source. 113 lakes.
N.H. 1970s	Profile	to pH 4.5	—	Elec.	One data source.
Wisconsin 1925-41	Surface Profile	M.O.	Phen.[e]	Col.	Multiple measurements on some lakes common. 161 lakes.
Wisconsin 1980	Surface	Gran	—	Elec.	One data source. Replicate measurements for many lakes.

[a] Methyl orange alkalinity.
[b] Phenolphthalein (titrant NaOH).
[c] Colorimetric.
[d] Electrometric.
[e] Phenolphthalein (titrant Na_2CO_3).

data. The historical data are described in detail for New York lakes in reports of the New York Department of Conservation (1930, 1931, 1932, 1933, 1934), for New Hampshire lakes in Hoover (1937, 1938), Warfel (1939), and Newell (1970, 1972, 1977), and for Wisconsin lakes by Juday et al. (1935). There are two recent surveys for New York lakes (Pfeiffer and Festa 1980, Colquhoun et al. 1984), one recent survey for New Hampshire lakes (Towne 1983), and one recent survey for Wisconsin lakes (Eilers et al. 1985). Table 7.4 summarizes some of the important aspects of the lake surveys.

As indicated above techniques for analysis of pH have changed over time. From the 1920s to the 1950s colorimetric comparators were used, and different indicators were used for different pH intervals. New York and New Hampshire lake surveys used the Hellige comparator, whereas pH was determined in Wisconsin by using the LaMotte kits. Specific indicator dyes available for use were generally named in New Hampshire and Wisconsin records; we assumed that the same indicators and ranges were employed for New York surveys. For example, methyl red ($pH < 6$) and bromthymol blue ($6 < pH < 7.6$) were available for use in the 1930s New Hampshire surveys, whereas chlorophenol red was substituted for methyl red in the 1940s New Hampshire surveys. Kramer and Tessier (1982) have published the available information on indicator pH range, indicator concentrations, and their acid dissociation constants. Additional details were compiled from laboratory notebooks and discussion with personnel who conducted the surveys in New Hampshire, New York, and Wisconsin.

Alkalinity and free CO_2 acidity were commonly measured following Standard Methods (1926, 1933) using MO and phenolphthalein, respectively, as indicators. Because there is ambiguity about the correct endpoint of MO, we incorporated two different values of pH, 4.04 and 4.19, in our analysis. The endpoint of 8.25 pH units was assumed for phenolphthalein.

For recent surveys in New York and Wisconsin the Gran Technique (Gran 1952) was used for analysis of alkalinity. The Gran function is a special case in the titration expression for alkalinity. When it is used correctly, there should be no requirement for endpoint adjustment. When $[H^+]$ predominates over all other terms in Eq. (7.4), the conventional acid-base balance becomes

$$F_1 = [H](V_s+v) = c_a(v - v_e) = b_0 + b_1 v; \qquad (7.10)$$

where V_s is sample volume, v and c_a are acid volume and concentration, v_e is the equivalence point volume, and b_0 and b_1 are linear coefficients obtained from regression of F_1 on v; $v_e = -b_0/b_1$ because $v_e = v$ when $F_1 = 0$. Only values of pH of about one unit less than pH_e should be used so that the assumptions of excess [H] are met. The recent alkalinity data for New York and Wisconsin were not adjusted since the Gran method was used.

Alkalinity was determined for recent New Hampshire data by titrating to a fixed endpoint pH of 4.5. Equation (7e) of Appendix D, calculated for the endpoint value of 4.5, was used to correct these data. Electrometrically determined pH in all of the recent surveys were accepted as given.

A specific consistency check can be applied when the alkalinity is less than about $-10\ \mu eq/L$. In this case,

Eq. (7.4) simplifies to: [Alk] = -[H]. This check can be applied to some of the recent New York and New Hampshire data.

Various calculations and consistency checks have been carried out on all of the historical data. These procedures were specified in the previous section and in Appendix D. They consist of correction of [MO], [AC], and pH_C to obtain [Alk], [Acy], and pH. Generally two independent estimates of alkalinity and three independent estimates of pH were obtained for the historical data.

In choosing the three data bases described here the primary consideration was the existence of data on pH, alkalinity, and free CO_2 acidity and detailed records of sampling and anaylsis that allowed us to check the internal consistency of the data and to evaluate their quality. Another important consideration, to which we have paid less attention, is representativeness, i.e., the degree to which the chosen lakes in a given region reflect the characteristics (morphology, water quality, history of land use in the watershed, etc.) of the complete set of lakes in the region. Two problems that we encountered were the lack of a consistent set of criteria to define representativeness and the lack of data for the complete set of lakes in the region required to characterize it accurately. Nevertheless, we note here a few general facts. The Wisconsin Department of Natural Resources has found that of the approximately 14,000 lakes in Wisconsin, 1,100 (or about 8 percent) are "acid sensitive." If we apply the definition to the 161 Wisconsin lakes chosen in this study, 43 (or 27 percent) of the lakes are acid sensitive. We thus believe that it is likely that the lakes chosen for study in this report do not underestimate the regional trend in acidity but may in fact overestimate it. In New York State, lakes have been classified by elevation and area groupings. Larger lakes tend to occur at lower elevations, whereas smaller and more acidic lakes tend to occur at higher elevations. The data set of New York lakes analyzed in this study includes only a limited number of high-elevation lakes; just one lake is above 800 m, and only 39 lakes are above 600 m in elevation. The data set overall includes about 10 percent of all the lakes within the Adirondack region, but it includes only about 4 percent of the lakes at the higher elevations. A different alkalinity trend may have been observed had we been able to assess more high-elevation lakes. Furthermore, the number of lakes increasing or decreasing in alkalinity could have been subdivided by lake-surface area because many of the larger lakes have high alkalinities. In this case, there might have been different trends for different area groupings.

Replication and Consistency

The laboratory precision of the MO titration method was reported as 20 μeq/L, and a recent study (Kramer and Tessier 1982) gives a similar value of 24 μeq/L. The colorimetric precision was reported to be 0.1 to 0.2 pH units. In the historical New Hampshire survey, investigators measured a depth profile for each lake and commonly duplicated the analysis at one depth, referred to here as a "depth replicate." In addition, in some cases entire depth profiles were repeated on the same date, or, in some instances, at a later date (up to 2 years). Averages of the duplicate profile data are referred to as "duplicate profile averages."

The procedure applied here is (a) to compare the variation in alkalinity for depth replicates with the laboratory precision, (b) to determine the variation in alkalinity of duplicate profile averages obtained on the same date to establish an acceptable difference for the two alkalinity estimates,* and (c) to determine the variation in duplicate profile averages obtained on different dates to establish a value reflecting natural short-time (2 years or less) variation in addition to variability introduced during analysis and sampling. In all the estimates the root-mean-square deviation was calculated.

Table 7.5 gives the results for the replication studies using New Hampshire data. The column headed MO refers to the average variations in alkalinity determined by MO titration. The variation in depth replicate measurements of 18 μeg/L is nearly the same as the laboratory precision of 20 to 24 μeq/L. Therefore it

---

*This difference estimate probably reflects a minimum value when applied to one method, whereas the consistency check considers the difference between two different methods of estimating alkalinity; thus there may be addition of errors for the multiple method technique that would result in a larger margin of error.

TABLE 7.5 Variation in Replicate Samples from Historic New Hampshire Lake Surveys

Sample	Sample Size	Alkalinity (μeq/L) by	
		MO	pHc/CO_2 Acidity
Depth replicate	10	18	
	10		91
Duplicate profile averages, same date	21	53	
	14		92
Duplicate profile averages, different date	21	67	
	10		70

NOTE: Results are shown for different groupings. They are the root mean square values for differences in duplicate sample results.

appears that the precision of field analyses was of the same magnitude as the laboratory analysis. The variation in duplicate profile averages of samples obtained on the same date was 53 μeq/L, more than two times the analytical precision and represents both analytical variation due to different analysts and spatial variations of alkalinity. Variation in duplicate profile averages of MO-determined alkalinity on different dates was 67 μeq/L. Compared with 53 μeq/L for samples obtained on the same date, this result suggests that historically the short-term (2 years or less) variation in alkalinity on average was about 41 μeq/L (root-mean-square deviation of 53 and 67 μeq/L), which is slightly less than the precision or the profile variability.

The column of Table 7.5 labeled "pH_C/CO_2 Acidity" shows the root-mean-square variations of alkalinity determined from colorimetric pH and free CO_2 measurements. The average differences for these estimates (91, 92, and 70 μeq/L) are larger than the corresponding differences for MO alkalinity, probably owing to the combination of free CO_2 and colorimetric pH errors in addition to the sampling and handling errors. It is also worth noting that the larger variations are due in part to three poor replicate measurements. If these data are ignored, the average replication for the second alkalinity estimate is quite similar to the values obtained by the MO technique.

In our analysis of the three sets of lakes in Wisconsin, New York, and New Hampshire, we consider the data to be consistent if the two estimates of alkalinity agree within 50 μeq/L. The tolerance of 50 μeq/L is the rounded equivalent of 53 μeq/L, the variation in alkalinity for duplicate profile averages of samples obtained on the same date. This value reflects, on average, the cumulative differences that are due to sampling and analysis errors and variability within the lake being surveyed.

A similar analysis of replicated data for Wisconsin indicates that the relative errors of the types presented in Table 7.5 are similar or less. Thus the same criterion of 50 μeq/L was used for Wisconsin data. There were no replication studies available for historical New York lakes data. The same criterion of 50 μeq/L was used in screening New York data, assuming that the same analytical techniques were employed as for the New Hampshire survey.

When the criteria for consistent data are applied to the historical data sets assuming a MO endpoint of 4.04, 68 percent of the New Hampshire data and 39 percent of the New York data are consistent. If a MO endpoint of 4.19 is assumed, 48 percent of the New Hampshire data, 52 percent of the New York data, and 90 percent of the Wisconsin data are consistent.

The method of consistency checking for alkalinity and pH is not without potential problems. First, the basic assumptions are that only carbonate protolytes are important and that the total carbonate remains unchanged during measurement of the colorimetric pH and free CO_2. Although the alkalinity will not change as a function of total CO_2, pH and acidity will change. The magnitude of possible effects on alkalinity that are due to other protolytes is discussed in Appendix D.

Inspection of Eqs. (7) and (10) to (12) of Appendix D shows that the relationship between alkalinity and pH_c/[Acy] is not linear. To assess the probable error in the second alkalinity estimation, various estimates of alkalinity were made using values of precision about given values of pH_c and [Acy]. Precisions for pH and free CO_2 were taken as ±0.2 and ±5 μeq/L, respectively. Figure 7.12 summarizes the results of these calculations. The average probable deviation for alkalinity was determined from the combinations of ranges of pH_c and free CO_2 as defined by the precisions. This potential error is compared with [Acy] for various colorimetric pH values; the actual alkalinity for the stated pH_c and [Acy] is shown in parentheses. The worst-case error of alkalinity (when both pH_c and [Acy] deviations give a maximum error) would be about five times that shown.

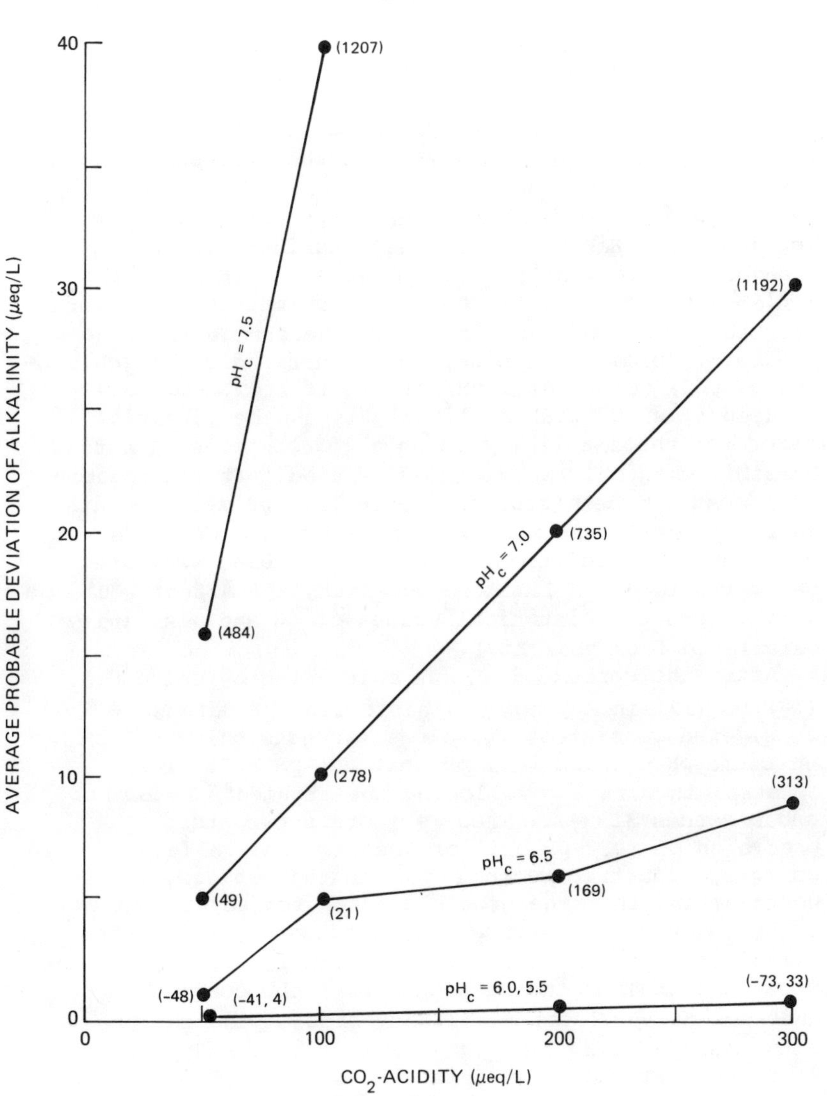

FIGURE 7.12 Average probable deviation to be expected for alkalinity calculated from free CO_2 acidity and colorimetric pH (pH_c). The deviation is obtained by varying pH_c by ±0.2 unit and CO_2 acidity by ±5 μeq/L about assumed values. Numbers in parentheses next to the data points are actual calculated alkalinities.

The plot shows that the error increases with increasing alkalinity; for the kinds of lakes discussed here, the probable error in the calculation would be about 20 μeq/L, equal to the analytical precision for the MO alkalinity titration. However, if the errors are cumulative, the probable error would be about 100 μeq/L; this may explain the few large variations for replicate New Hampshire samples described earlier.

Another potentially significant uncertainty in the analysis arises from the use of colorimetric indicators near the limits of their ranges. The change from one indicator (bromthymol blue) to another indicator (chlorophenol red) occurs at a pH_C of 6. If bromthymol blue is used at a pH of 6 or slightly less, the potential error for the alkalinity is enlarged compared with that possible when chlorophenol red is used. It is probable that most of the historic colorimetric pH measurements near a pH of 6 were made with bromthymol blue (Juday et al. 1935). In addition, if bromthymol blue were used below its lower pH limit on occasion, the effect would be to increase the historical values of pH and alkalinity calculated from pH_C/[Acy].

After the corrected pH was calculated by using Eqs. (10) to (12) in Appendix D, historical pH data were considered consistent if values agreed within ±0.2 pH unit for pH < 5 and ±0.5 pH unit for pH > 5. These constraints were set following the arguments by Blakar and Digernes (1984). These authors showed that a precision of ±0.2 pH unit or less is achievable by using colorimetric techniques. In this regard, it is worth noting that the probable error for adjustment of pH, using the same approach as discussed earlier for alkalinity, is less than 0.2 unit and is 0.5 unit in the worst case. Thus the data that pass the consistency check of 50 μeq/L for alkalinity should pass the pH consistency check.

Recent data for New Hampshire, New York, and Wisconsin could not be screened for consistency because of the lack of replicate data or redundant protolyte measurements. The New Hampshire alkalinity data were adjusted for a fixed pH endpoint of 4.5, as discussed earlier. Since the recent Wisconsin data also included major ions, the data were assessed for ion balance (Eilers et al. 1985). Baker and Harvey (1985) have compared the findings of two surveys of New York lakes (Pfeiffer and Festa 1980, Colquhoun et al. 1984). Noting that the pH measurements in the 1980 report were consistently low, they suggest that these measurements are in error, a finding also proposed by Schofield (1982). There are no means of confirming this hypothesis since there were no quality-assurance programs in any of the recent surveys. Rather than exclude the 1980 survey, we compared both sets of recent data with historical data. However, from this example it appears that in some cases recent electrometrically derived data may also exhibit variance in sample replication.

Trends in Alkalinity and pH of Lakes in New York, New Hampshire, and Wisconsin

To determine changes in the pH and alkalinity of lakes over the past 50 to 60 years, we corrected and screened the historical data as described earlier and compared them with comparable data for the 1970s and 1980s. The results are shown in Tables 7.6, 7.7, and 7.8 for lakes in New York, New Hampshire, and Wisconsin, respectively. In all cases negative values signify trends toward increasing acidification, and positive values signify trends toward increasing alkalization. To avoid excessive weighing of large outliers, the median and the 10th and 90th precentiles were calculated rather than the mean and the standard deviation.

In compiling the historical data, whenever possible we obtained the alkalinity averaged from the MO titration and from the colorimetric pH and free CO_2 acidity measurements. We determined the pH from the average of the three calculated values of pH.

Because of uncertainty in knowing the precise pH employed for the MO endpoint in the historical studies of New York and New Hampshire lakes, we have analyzed the data assuming two different pH values, 4.04 and 4.19. The lower value corresponds to the findings of Kolthoff and Stenger (1947) and Kramer and Tessier (1982). The value of 4.19 is supported by Taylor (1933) and Eilers et al. (1985). Since the latter two studies refer specifically to the Wisconsin data set, we assumed that the pH of 4.19 was the best estimate to use in the compilation of historical values of pH and alkalinity in Wisconsin lakes.

As is shown in Table 7.6 and Figures 7.13(a) and 7.13(b) for New York lakes, the choice of the MO endpoint may influence substantially the calculated changes in pH and alkalinity. Also, the results appear to be influenced

TABLE 7.6 Summary of Alkalinity and pH Changes for the Adirondack Region of New York

ALKALINITY				
pH (MO)	n	median (μeq/L)	10% (μeq/L)	90% (μeq/L)
4.04				
1984 report	97 (248)	1 (8)	−121 (−233)	166 (165)
1980 report	97 (248)	−28 (−14)	−141 (−244)	165 (140)
4.19				
1984 report	130 (248)	−44 (−69)	−143 (−377)	60 (61)
1980 report	130 (248)	−69 (−89)	−187 (−389)	41 (41)

pH				
pH (MO)	n	median (units)	10% (units)	90% (units)
4.04				
1984 report	97 (248)	−0.12 (0.10)	−1.04 (−1.05)	1.18 (1.32)
1980 report	97 (248)	−0.14 (0.02)	−1.60 (−1.21)	1.18 (1.11)
4.19				
1984 report	130 (248)	−0.63 (−0.55)	−1.66 (−1.66)	0.47 (0.53)
1980 report	130 (248)	−0.74 (−0.66)	−1.86 (−1.90)	0.04 (0.39)

NOTE: Values of the median, 10th, and 90th percentiles are given. Changes are defined as the recent minus the historical values. Changes are given for the consistent data, all the data (in parentheses), and relative to the 1980 and 1984 New York Department of Environmental Conservation reports (i.e., Pfeiffer and Festa 1980 and Colquhoun et al. 1984). The data are further compared for calculations assuming MO alkalinity endpoints of 4.04 and 4.19.

by the choice of the recent data set. When an endpoint pH of 4.04 is assumed and the historical data are compared with the 1984 data set, the median value for the change in alkalinity is +1 μeq/L, and the median value for the change in pH is -0.12 unit. Given the uncertainties in these analyses, the results are indistinguishable from zero. When the historical data are compared with the 1980 data set however, the calculated median change in alkalinity, -28 μeq/L, may be significant, although the median change in pH, -0.14 unit, is not. When the same data are calculated using 4.19 as the MO endpoint, there appear to be distinct trends toward decreasing alkalinity (-44 and -69 μeq/L for the 1984 and 1980 data, respectively) and decreasing pH (-0.63 and

TABLE 7.7 Summary of Alkalinity and pH Changes for Lakes in New Hampshire.

ALKALINITY				
pH (MO)	n[a]	median (μeq/L)	10% (μeq/L)	90% (μeq/L)
4.04	115 (170)	8 (8)	−47 (−59)	75 (80)
4.19	81 (170)	−7 (−21)	−73 (−157)	87 (90)
pH				
pH (MO)	n	median (units)	10% (units)	90% (units)
4.04	114 (169)	0.43 (0.44)	−0.29 (−0.23)	1.05 (1.05)
4.19	81 (170)	0.25 (0.20)	−0.84 (−0.46)	0.87 (0.90)

NOTE: Values of the median, 10th, and 90th percentiles are given for consistent data and all the data (in parentheses). Data are calculated for MO pH endpoints of 4.04 and 4.19. Differences are expressed as the recent value minus the historic value.
[a]In the case of New Hampshire, n refers to the number of samples, including duplicate samples for the same lake. The total number of lakes in the data set is 113.

-0.74 for the 1984 and 1980 data, respectively), regardless of the choice of recent data set.

To test whether liming activities may have biased our results, we examined the liming histories of New York lakes (data were provided through the courtesy of F. Vertucci, Cornell University, personal communication 1985). Our complete data set of 248 lakes included 16 limed lakes. Our consistent data set contained three limed lakes. These, however, had a negligible effect (about 1 μeq/L) on our analysis.

Changes in New Hampshire lakes do not appear to be so sensitive to the assumption of the MO endpoint (Table 7.7). If the 4.04 endpoint is assumed, the median value for the change in alkalinity over the 50 year time period in these lakes is +8 μeq/L, and the median for change in pH is +0.43 unit. If the higher endpoint is applied, median values of -7 μeq/L and +0.25 unit are predicted. Thus, although there are numerous cases of both increases and decreases in alkalinity and pH, no significant change in alkalinity is apparent. However, the pH on average may have increased.

The pH and alkalinity in the Wisconsin lakes apparently have increased significantly since the 1920s and 1930s. Table 7.8 and Figure 7.14 (derived from an MO endpoint pH

TABLE 7.8 Summary of Alkalinity and pH Changes for Lakes in Wisconsin.

ALKALINITY			
n	median (μeq/L)	10% (μeq/L)	90% (μeq/L)
145	38	−50	110

pH			
n	median (units)	10% (units)	90% (units)
145	0.51	0.01	1.12

NOTE: Values of the median, 10th, and 90th percentiles are given for consistent data. Data are calculated for a MO pH endpoint of 4.19 only. Differences are expressed as the recent value minus the historic value. Because data for 90% of the lakes passed the consistency check, calculations for all the data are not included.

of 4.19) indicate that the median of the change in alkalinity is +38 μeq/L and the median of change in pH is +0.51 unit. If an MO endpoint of 4.04 were applied, the calculated increases in alkalinity and pH would be even greater. However, an endpoint of 4.19 is the most likely value in this analysis, based on documentation provided by Taylor (1933) and the findings of Eilers et al. (1985).

Since a large percentage of the data was eliminated by our consistency checks in New York and New Hampshire, we tested whether use of the consistent data alone introduced any bias. Employing a nonparametric chi-square statistical test, we concluded that there is no evidence of such bias. The major effect of using consistent data only is demonstrated for alkalinity changes in New York lakes, assuming an MO endpoint value of 4.19 (see Figure 7.15). Calculations based on consistent data resulted in the elimination of most of the lakes with alkalinity changes greater in magnitude than 200 μeq/L. The median change in alkalinity increased from -69 μeq/L based on compilation of all the lakes to -44 μeq/L based on consistent data only. This difference reflects the fact that most of the lakes eliminated were those with high negative values for changes in alkalinity. Nevertheless, a few larger outliers remain, even for the consistent set of data.

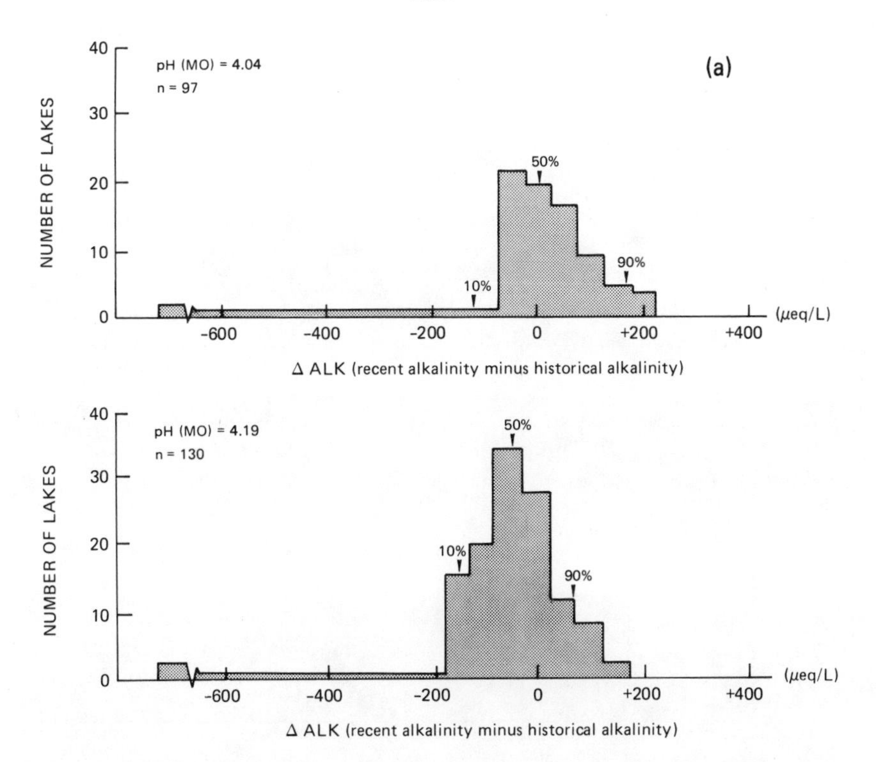

FIGURE 7.13 (a) Two scenarios for change in alkalinities of Adirondack lakes over five decades calculated from methyl orange endpoint pHs of 4.04 and 4.19. (b) Two scenarios for change in pHs of Adirondack lakes over five decades calculated from methyl orange endpoint pHs of 4.04 and 4.19.
NOTE: Analyses are based on comparison of historical data with data from Colquhoun et al. (1984).

Short-Term Variability

The maximum changes in alkalinity of lakes that can result from atmospheric inputs of acid are estimated to be approximately 100 μeq/L (Galloway 1984). Our analysis demonstrates that a number of lakes in the three regions of study have changes of this magnitude or higher. These lakes are listed in Table 7.9. We have not examined the factors responsible for these large changes, but such a study would be useful.

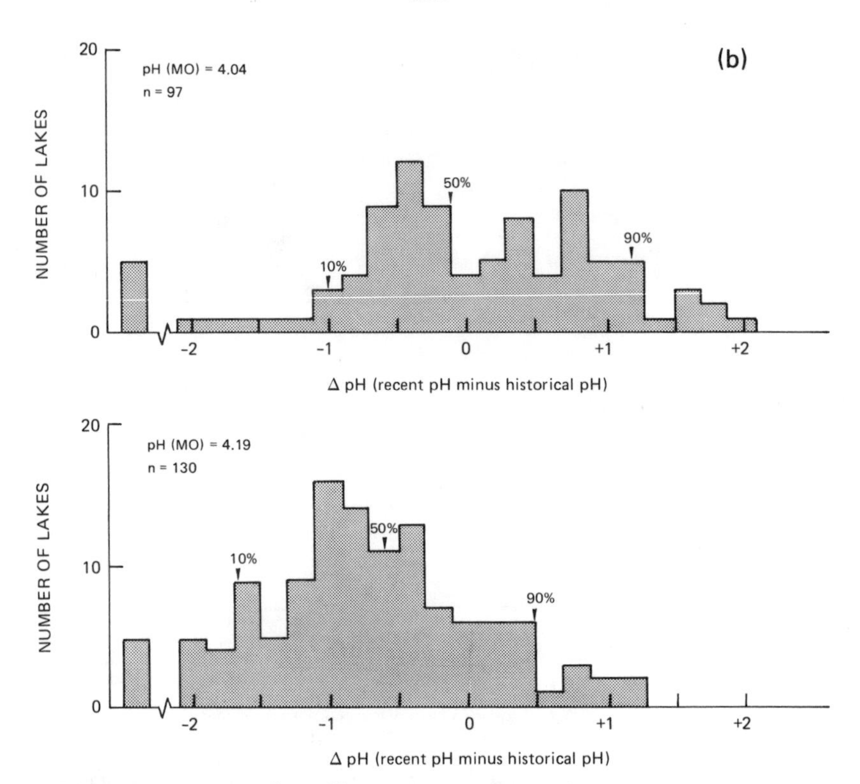

FIGURE 7.13 (continued).

Comparison of recent and historical data considers the possible changes in alkalinity of only two or three data points over about five decades. In assessing these results and the magnitude of changes, one must consider the magnitude of changes from short-term (seasonal) fluctuations and how such changes compare with the changes considered over five decades. This subject was addressed to some degree in the earlier discussion of the replication of historical data for New Hampshire. Average variation in the analyses of duplicate lake profiles collected at different times over short time periods (up to 2 years) was compared with variation in duplicate lake profiles collected on the same date. The conclusion was that the short-term variation was about 41 μeq/L on average compared to the variation in duplicate profile averages of about 50 μeq/L.

Intensively measured, low-alkalinity lakes may also provide an analysis of the magnitude of short-term

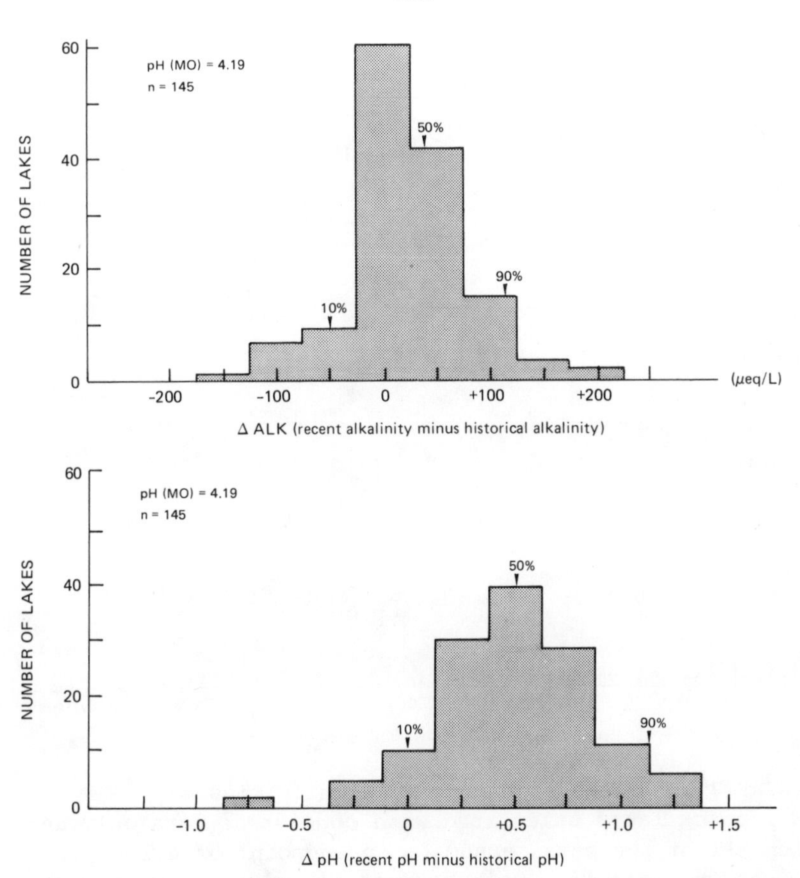

FIGURE 7.14 Changes in alkalinity and pH in Wisconsin lakes from the 1920s and 1930s to the 1980s, assuming a MO endpoint pH of 4.19.

changes. The Vermont Department of Water Resources and Environmental Engineering (1983) has been carrying out an intensive program of monitoring lakes since 1982. Lakes ranging in alkalinity from less than 0 to greater than 100 μeq/L were chosen for assessment. All data used were tested for consistency by calculating the departure from electroneutrality and comparing calculated and measured specific conductance; deviations of less than 15 percent were accepted. Monthly trends for alkalinity for each lake are shown in Figure 7.16 along with the means and standard deviations. The deviations in alkalinity are quite small for all lakes. The largest standard

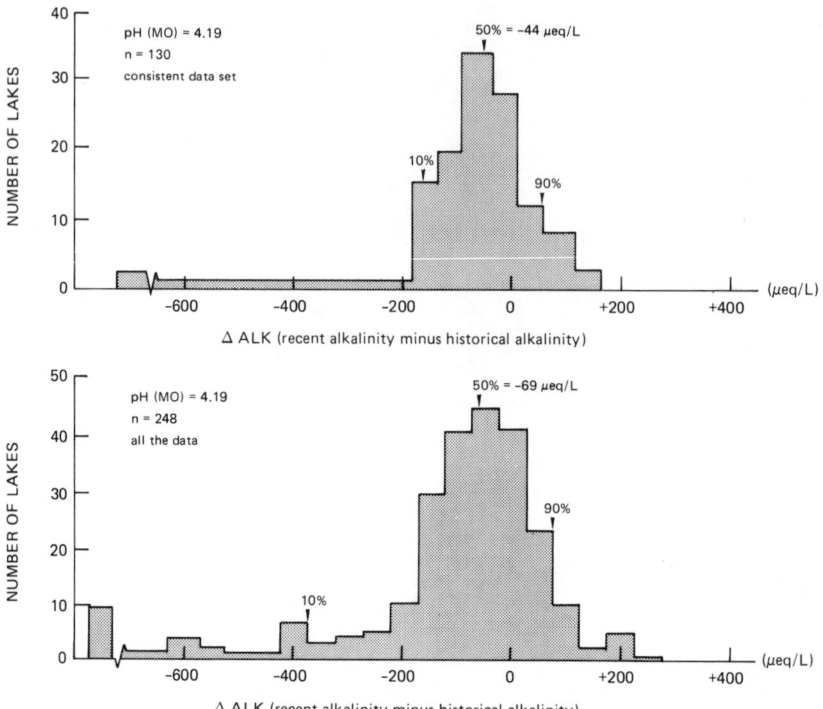

FIGURE 7.15 Plots of change in alkalinities of Adirondack lakes calculated with consistent data only and with all of the data, assuming an endpoint of 4.19. Recent data are from Colquhoun et al. (1984).

deviation of 52 μeq/L is about the same as the variation obtained in the historical duplicate profile analyses of New Hampshire lakes. Thus we may conclude that short-term variations in alkalinity may be quite small and are often of less than the tolerance permitted for consistent historical data. This conclusion would not deny the occurrence in individual lakes of large, episodic variations in alkalinity of perhaps 200 μeq/L or greater (Driscoll and Newton 1985).

Because changes in alkalinity caused by acid deposition may be of the same magnitude as the analytical error in our analysis of these lakes, the question arises whether any effects of acid deposition can be detected or rejected using the historical data from lakes in the 1920s and 1930s. However, the sample size and standard deviation

TABLE 7.9 Lakes That Have Experienced Large Changes in Alkalinities Since the 1930s

Lake	Increase (+) / Decrease (−)
New Hampshire:	
Opechee Bay, Laconia	−
Potanipo Pond, Brookline	−
Swain's Lake, Barrington	−
Beaver Lake, Derry	−
Canobie Lake, Windham	+
French Pond, Henniker	+
Great East Lake, Wakefield	+
Lovell Lake, Wakefield	+
Rocky Bound Pond, Croydon	+
Wash Pond, Hampstead	+
Wheelwright Pond, Lee	+
New York:	
Ballston Lake (7-1090)	−
Irving Lake (7-0732)	−
Pine Lake (7-0724)	−
Spruce Lake (7-0706A)	−
Elm Lake (5-0304)	+
Lilypad Pond (6-0025)	+
Silver Lake (3-03551)	+
Sterling Pond (6-0029)	+
Valentine Pond (5-0370)	+
Wisconsin:	
Dog Lake	−
Goodyear Lake	−
Virgin Lake	−
Big Crooked Lake	+
Booth Lake	+
Cranberry Lake	+
Diamond Lake	+
Frank Lake	+
Garth Lake	+
Katherine Lake	+
Little Arbor Vitae Lake	+
Little Muskie Lake	+
Little Portage Lake	+
Mill Lake	+
Silver Lake	+
Spirit Lake	+
Squash Lake	+
Sumach Lake	+
Yawley Lake	+

NOTE: The magnitudes of changes are greater than 100 μeq/L for New Hampshire and Wisconsin and 200 μeq/L for New York.

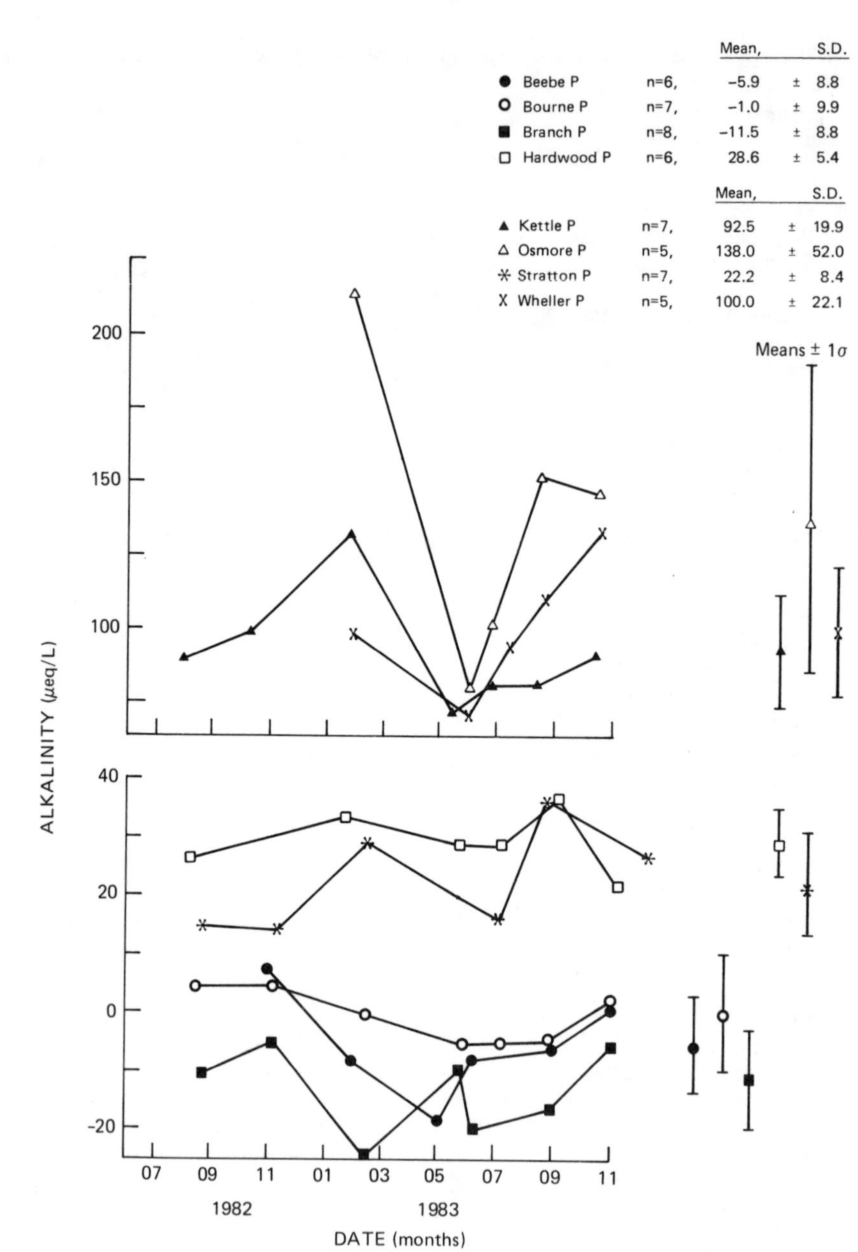

FIGURE 7.16 Monthly changes in alkalinity calculated from quality-assured data for lakes in Vermont.

of the alkalinity changes are such that with high probability (greater than 0.9), we would be able to detect a regional decrease of 30 to 50 μeq/L.

SUMMARY

Relationship Between Sulfate Input Flux to Lakes from Wet Deposition and Sulfate Output Flux from Lakes

In a large area of northeastern North America the mathematical relationship between fluxes of sulfate to lakes from wet deposition and fluxes of sulfate output from lakes is significantly linear when the data are examined on the basis of smaller contiguous regions (southern New England, New York, northern New England, Quebec, Newfoundland, and Labrador). Regional differences arise because of different multiplying factors in the regression analysis. One physical interpretation of this result is that the multiplying factor is proportional to the ratio of total atmospheric sulfur deposition (i.e., wet plus dry deposition) to wet-only sulfur deposition. Thus the aggregated data appear to exhibit nonlinearity because of the varying importance of dry deposition from region to region. In some cases internal sources of sulfur in the watershed may also contribute to the flux of sulfur into the lake. Some likely examples are cited.

Bench-Mark Streams

Trends in sulfate are discernible in Bench-Mark streams over a period extending from the mid-1960s to 1983. On a regional basis, those trends in stream sulfate are consistent with trends in SO_2 emissions.

Analysis of sulfate-mass balance in Bench-Mark watersheds of the southeastern United States, which are dominated by Ultisols, show a net retention of sulfate. Similar analyses for sites in the Northeast suggest that there is an excess of sulfate yield over wet-deposition sulfur input, but the contribution of sulfate to the excess from internal sources and from dry deposition cannot be determined with the current data.

In general in the northeastern United States (Region B) since the mid-1960s the Bench-Mark streams have experienced no change or decreases in sulfate concentrations and no change or increases in alkalinity. In the

Southeast (Region C), streams have shown increases in sulfate concentrations and have either shown no changes or have decreased in alkalinity.

In Bench-Mark streams with alkalinities less than 500 μeq/L, short-term changes in the flux of strong-acid anions (SO_4^{2-}, NO_3^-) are balanced by changes in alkalinity and nonprotolytic (base) cations. For these softwater Bench-Mark streams, short-term alkalinity changes balance from 0 to 78 percent of the sulfate change, averaging about 30 percent. This result supports the contention that changes in strong-acid anion fluxes affect surface water alkalinity in watersheds that have acid soils.

Alkalinity and nonprotolytic cation changes in low-alkalinity Bench-Mark streams over the 15- to 20-year period of record are consistent with changes caused by a changing flux of strong-acid anions from the atmosphere, but there is also evidence suggesting that internal watershed processes or in-stream processes govern trends in nonprotolytic cations and alkalinity. At many eastern low-alkalinity stations changes in strong-acid anions are sufficient to account for alkalinity and nonprotolytic cation changes, but at some western stations that receive low rates of SO_4^{2-} deposition substantial reductions in alkalinity have occurred that have a large component not accounted for by changes in strong-acid anions. In these cases the alkalinity reductions are accompanied by reductions in nonprotolytic cation concentrations and may result primarily from changes in internal-terrestrial or in-stream processes.

Lakes in Wisconsin, New York, and New Hampshire

Data from historical surveys of lake water parameters in Wisconsin, New York, and New Hampshire are of varying quality, but sufficient data and documentation are available to test for internal consistency. Applying the criteria of Kramer and Tessier (1982), we obtained consistent sets of pH and alkalinity data for approximately 300 lakes for which recent data also exist. Comparing historical with recent data, it is possible to estimate pH and alkalinity changes over the past half-century.

This method is not without problems. Use of the MO indicator in the historical studies called for titration to the "faintest pink" endpoint. Estimates in the literature of the pH of this endpoint range between about 4.0 to higher than 4.3. The color of the endpoint in this range has been variously described as "faint pink," "pink," and "salmon pink." The Wisconsin survey provides the best documentation of the pH of the MO endpoint, citing a range of pH values from 4.06 to 4.18 for the "fainter" pink color and recommending that a value of 4.18 be used in calculating titrations.

Our analysis shows that changes in alkalinity and pH in New York lakes are sensitive to the assumption made about the pH of the MO endpoint. Assuming an endpoint pH of 4.2 (or greater), New York lakes, on average, appear to have increased substantially in acidic status over the past 50 years, as reflected in reductions in both alkalinity and pH. Alternatively, if a value of the pH endpoint close to 4.0 is assumed and if the historical data are compared with the 1984 data set, then there appears to have been little change, on average, in the acidic status of New York lakes over the past 50 years. However, if the historical data are compared with the 1980 New York data set, there may have been an overall decline in alkalinity. New Hampshire lakes do not appear to be so sensitive to the choice of the endpoint pH of MO. If an endpoint pH close to 4.2 is assumed, these lakes, on average, show little or no decrease in alkalinity and show a slight increase in pH. With an endpoint near 4.0 there is again little or no change, on average, in alkalinity, but an increase in pH of about 0.4 pH unit. Wisconsin lakes, on average, appear to show a significant increase in alkalinity (38 μeq/L) and pH (+0.5 unit) if an MO endpoint of 4.2 is assumed. The increases are even larger if an endpoint close to 4.0 is assigned.

Currently, the question regarding the correct endpoint pH in historical alkalinity titrations is unresolved. In the judgment of the authors of this chapter the endpoint lies in the range from 4.19 to 4.04, but evidence is not sufficient to specify a most likely value. Tests done with indicator dyes which have been preserved for the past 50 years may be valuable in resolving this question.

We have not attempted to indicate the mechanisms for change in acidic status of these lakes. Any discussion of mechanisms must consider each lake separately and include all the hydrological and biogeochemical factors. For most lakes, data are not available currently to permit this kind of analysis.

One possible avenue of future research is to conduct a detailed study of the outliers in the three data sets. If the maximum change in lake alkalinity from deposition of atmospheric sulfur is about 100 μeq/L, then study of the larger changes subjected to the most careful scrutiny for quality may offer a rational explanation for other acidification mechanisms.

REFERENCES

Almer, B., W. Dickson, C. Eckstrom, and E. Hornstrom. 1978. Sulfur pollution and the aquatic ecosystems. Pp. 273-311 of Sulfur in the Environment, Part II. Ecological Impacts. J. O. Nriagu, ed. New York: J. Wiley and Sons.

Arnold, D. E., R. W. Light, and V. J. Dymond. 1980. Probable effects of acid precipitation on Pennsylvania waters. U.S. Environmental Protection Agency. EPA-600/3-80-012. 21 pp.

Baker, J., and T. Harvey. 1985. Critique of acid lakes and fish population status in the Adirondack region of New York State. Draft final report for NAPAP Project E3-25, U.S. Environmental Protection Agency.

Beamish, R. J., and H. H. Harvey. 1972. Acidification of the La Cloche Mountain Lakes, Ontario and resulting fish mortalities. J. Fish. Res. Board Can. 29:1131-1143.

Beamish, R. J., W. L. Lockhart, J. C. van Loon, and H. H. Harvey. 1975. Long-term acidification of a lake and resulting effects of fishes. Ambio 4:98-102.

Blakar, I. A., and I. Digernes. 1984. Evaluation of acidification based on former colorimetric determination of pH: The effect of indicators on pH in poorly buffered water. Verh. Int. Ver. Limnol. 22:679-685.

Booty, W. G., and J. R. Kramer. 1984. Sensitivity and analysis of a watershed acidification model, Phil. Trans. R. Soc. London Ser. B 305:441-449.

Burns, D. A., J. N. Galloway, and G. R. Hendry. 1981. Acidification of surface waters in two areas of the eastern United States. Water Air Soil Pollut. 16: 277-285.

Chen, C. W., J. D. Dean, S. A. Gherini, and R. A. Goldstein. 1982. Acid rain model: hydrologic module. J. Env. Eng. Div. Am. Soc. Civ. Eng. 108:455-472.

Christophersen, N., and R. F. Wright. 1981. Sulfate budget and a model for sulfate concentrations in stream water at Birkenes, a small forested catchment in southernmost Norway. Wat. Resour. Res. 17:377-389.

Church, M. F. 1984. Predictive modeling of acidic deposition on surface waters. Pp. 4-113--4-128 of The Acidic Deposition Phenomenon and Its Effects. Critical Assessment Review Papers. Vol. II. Effects Sciences. A. P. Altshuller and R. A. Linthurst, eds. EPA-600/8-83-016 BF. U.S. Environmental Protection Agency.

Clark, W. M. 1928. The Determination of Hydrogen Ions. Baltimore, Maryland: The Williams and Wilkins Company.

Cobb, E. D., and J. E. Biesecker. 1971. The national hydrologic Bench-Mark network. Circular 460-D, U.S. Geological Survey. 38 pp.

Colquhoun, J., W. Kreter, and M. Pfeiffer. 1984. Acidity states of lakes and streams in New York State. New York Department of Environmental Conservation, Raybrook, New York. 49 pp + viii app.

Crisman, T. L., R. L. Schulze, P. L. Brezonik, and S. A. Bloom. 1980. Acid precipitation: the biotic response in Florida lakes. Pp. 296-297 of Ecological Impact of Acid Precipitation. D. Drabløs, and A. Tollan, eds. Proceedings of an international conference, Sandefjord, Norway. Sur Nedbørs Virkning Pa Skog Og Fisk (SNSF) Project, Oslo.

Critical Assessment Review Papers (CARP). 1984. The acidic deposition phenomenon and its effects. A. P. Altschuller and R. A. Linthurst, eds. Report EPA-600/8-83-016AF. U.S. Environmental Protection Agency.

Cronan, C. S., and C. L. Schofield. 1979. Aluminum leaching response to acid precipitation: effects on high-elevation watersheds in the Northeast. Science 204(20):304-306.

Davis, R. B., M. O. Smith, J. H. Bailey, and S. A. Norton. 1978. Acidification of Maine (U.S.A.) lakes by acidic precipitation. Verh. Int. Ver. Limnol. 10:532-537.

Dickson, W. 1980. Properties of acidified waters. Pp. 75-83 of Proceedings of an International Conference. Ecological Impact of Acid Precipitation, D. Drabløs and A. Tollan, eds. Sur Nedbørs Virkning Pa Skog Og Fisk (SNSF) Project, Oslo, Norway.

Dillon, P. J. 1983. Chemical alteration of surface water by acidic deposition in Canada. Pp. 275-286 of Ecological Effects of Acid Deposition. Report P, 1636. National Swedish Environmental Protection Board.

Dillon, P. J., D. S. Jeffries, W. Snyder, R. Reid, N. D. Yan, D. Evans, J. Moss, and W. A. Scheider. 1978. Acid precipitation in south-central Ontario: recent observations. J. Fish. Res. Board Can. 35:809-815.

Drabløs, D., D. I. Sevaldrud, and J. A. Timberlid. 1980. Historical land-use changes related to fish status development in different areas in southern Norway. Pp. 367-369 of Proceedings of an International Conference. Ecological Impact of Acid Precipitation, D. Drabløs and A. Tollan, eds. Sur Nedbørs Virkning Pa Skog Og Fisk (SNSF) Project, Oslo, Norway.

Driscoll, C. T., and G. E. Likens. 1982. Hydrogen ion budget of an aggrading forested ecosystem. Tellus 34:283-292.

Driscoll, C. T., and R. M. Newton. 1985. Chemical characteristics of Adirondack lakes. Environ. Sci. Technol. 19:1018-1024.

Eilers, J. M., G. E. Glass, and A. K. Pollack. 1985. Water quality changes in northern Wisconsin lakes, 1930-1980. RL-D Draft 539. U.S. Environmental Protection Agency.

Environmental Protection Agency. 1984. The Acidic Deposition Phenomenon and Its Effects. In Critical Assessment Review Papers, Vol. II. Effects Sciences. A. P. Altshuller and R. A. Linthurst, eds. EPA-600/8-83-016 BF.

Fitzgerald, J. W., T. C. Strickland, and W. T. Swank. 1982. Metabolic fate of inorganic sulfate in soil samples from undisturbed and managed forest ecosystems. Soil Biol. Biochem. 14:529-536.

Galloway, J. N. 1984. Long-term acidification. In The Acidic Deposition Phenomenon and Its Effects. Critical Assessment Review Papers. Vol. II. Effects Sciences. A. P. Altshuller and R. A. Linthurst, eds. EPA-600/8-83-016 BF.

Galloway, J. M., S. A. Norton and M. R. Church. 1983a. Freshwater acidification from atmospheric deposition of sulfuric acid: A conceptual model. Environ. Sci. Technol. 17:541A-545A.

Galloway, J. N., C. L. Schofield, N. E. Peters, G. R. Hendrey, and E. R. Altwicker. 1983b. The effect of atmospheric sulfur on the composition of three Adirondack lakes. Can. J. Fish. Aquat. Sci. 40:799-806.

Gran, G. 1952. Determination of equivalence point in potentiometric titrations. Analyst 77:661-671.

Haines, T. A., J. J. Akielagzek, and P. J. Rago. 1983. A regional survey of chemistry of headwater lakes and streams in New England: vulnerability to acidification. Air Pollution and Acid Rain Series. U.S. Fish and Wildlife Service Report.

Hendrey, G. R., J. N. Galloway, S. A. Norton, C. L. Schofield, P. W. Schaffer, and D. A. Burns. 1980. Geological and hydrochemical sensitivity of the eastern United States to acid precipitation. EPA-60013-80-024. U.S. Environmental Protection Agency. 100 pp.

Henriksen, A. 1979. A simple approach for identifying and measuring acidification of freshwater. Nature 278:542-545.

Henriksen, A. 1980. Acidification of freshwaters--a large scale titration. Pp. 68-74 of Ecological Impact of Acid Precipitation. D. Drabløs and A. Tollan, eds. Sur Nedbørs Virkning Pa Skog Og Fisk (SNSF) Project, Oslo.

Henriksen, A. 1982. Changes in Base Cation Concentrations Due to Freshwater Acidification. Acid Rain Research Report 1/1982. Norwegian Institute for Water Research (NIVA), Oslo.

Hirsch, R. M., J. R. Slack, and R. A. Smith. 1982. Techniques of trend analysis for monthly water quality data. Water Resour. Res. 18:107-121.

Hoover, E. E. 1937. Biological Survey of Androscoggin Saco and Coastal Watersheds. New Hampshire Fish and Game Commission, Concord.

Hoover, E. E. 1938. Biological Survey of the Merrimack Watershed. New Hampshire Fish and Game Commission, Concord.

Johnson, A. H. 1979. Evidence of acidification of headwater streams in the New Jersey pinelands. Science 206:834-836.

Johnson, D. W. 1980. Site susceptibility to leaching by H_2SO_4 in acid rainfall. Pp. 525-536 of Effects of Acid Precipitation on Terrestrial Ecosystems. T. C. Hutchinson and M. Havas, eds. New York: Plenum Press.

Johnson, D. W., and G. S. Henderson. 1979. Sulfate absorption and sulfur fractions in a highly weathered soil under mixed deciduous forest. Soil Sci. 128:34-40.

Juday, J., E. A. Birge, and V. W. Meloche. 1935. The carbon dioxide and hydrogen content of the lake waters of northeastern Wisconsin. Trans. Wis. Acad. Sci. Arts Letters 29:1-82.

Kolthoff, I. M. 1931. Pp. 30-31, 52-53 of The Colorimetric and Potentiometric Determination of pH. New York: John Wiley and Sons.

Kolthoff, I. M., and V. A. Stenger. 1947. Titration methods. Pp. 61-62 of Volumetric Analysis, Vol. II. New York: John Wiley and Sons.

Kramer, J. R., and A. Tessier. 1982. Acidification of aquatic systems: a critique of chemical approaches. Environ. Sci. Technol. 16:606A-615A.

Krug, E. C., and C. R. Frink. 1983. Acid rain on acid soil: a new perspective. Science 221:520-525.

Lewis, W. M. 1982. Changes in pH and buffering capacity of lakes in the Colorado Rockies. Limnol. Oceanogr. 27:161-171.

Likens, G. E., F. H. Bormann, R. S. Pierce, and N. M. Johnson. 1977. Biogeochemistry of a Forested Ecosystem. New York: Springer-Verlag. 146 pp.

Likens, G. E., F. H. Bormann, R. S. Pierce, and J. S. Eaton. 1985. The Hubbard Brook Valley. Chapter II of An Ecosystem Approach to Aquatic Ecology: Mirror Lake and Its Environment, G. E. Likens, ed. New York: Springer-Verlag.

Mohn, E., E. Joranger, S. Kalvenes, B. Sollie, and R. Wright. 1980. Regional surveys of the chemistry of small Norwegian lakes: a statistical analysis of the data from 1974-1978. Pp. 240-241 of Ecological Impact of Acid Precipitation, D. Drabløs and A. Tollens, eds. Sur Nedbørs Virkning Pa Skog Og Fisk (SNSF) project.

Morrison, I. K. 1984. Acid rain: a review of literature on acid deposition effects in forest ecosystems. For. Abstr. 4S:483-506.

National Research Council. 1984. Acid Deposition: Processes of Lake Acidification. Washington, D.C.: National Academy Press.

Newell, A. E. 1970. Biological survey of the lakes and ponds in Cheshire, Hillsborough, and Rockingham Counties. New Hampshire Fish and Game Commission, Concord.

Newell, A. E. 1972. Biological survey of the lakes and ponds in Coos Grafton and Carrol Counties. New Hampshire Fish and Game Commission, Concord.

Newell, A. E. 1977. Biological survey of the lakes and ponds in Sullivan, Merrimack, Belknap and Strafford Counties. New Hampshire Fish and Game Commission, Concord.

New York Department of Conservation. 1930. Biological Survey of the St. Lawrence Watershed, Vol. V. Albany, New York.

New York Department of Conservation. 1931. Biological Survey of the Oswegatchie and Black River Systems, Vol. VI. Albany, New York.

New York Department of Conservation. 1932. Biological Survey of the Upper Hudson Watershed, Vol VII. Albany, New York.

New York Department of Conservation. 1933. Biological Survey of the Raquette Watershed, Vol VIII. Albany, New York.

New York Department of Conservation. 1934. Biological Survey of the Mohawk-Hudson Watershed, Vol. IX. Albany, New York.

Norton, S. A., R. B. Davis, and D. F. Braekke. 1981. Responses of Northern New England Lakes to Atmospheric Inputs of Acids and Heavy Metals. Completion Report: Project A-048-ME. U.S. Department of the Interior, Office of Water Research and Technology. 90 pp.

Oliver, B. G., E. M. Thurman, and R. L. Malcolm. 1983. The contribution of humic substances to the acidity of colored natural waters. Geochim. Cosmochim. Acta 47:2031-2035.

Olson, R. J., D. W. Johnson, and D. S. Shriner. 1982. Regional assessment of potential sensitivity of soils in the eastern U.S. to acid precipitation. Oak Ridge National Laboratory, Environmental Sciences Division, Oak Ridge, Tenn. Publication Number 1899. 50 pp.

Overrein, L. M., H. M. Seip, and A. Tollan. 1980. Acid Precipitation-Effects on Forests and Fish. Final report. Sur Nedbørs Virkning Pa Skog Og Fisk (SNSF) project, 1972-1980. Oslo. 175 pp.

Pfeiffer, M. H. and P. J. Festa. 1980. Acidity status of lakes in the Adirondack region of New York in relationship to fish sources. New York Department of Environmental Conservation, Albany, N.Y. FW-P168(10/80). 36 pp.

Rebsdorf, A. 1980. Acidification of Danish softwater lakes. In Proceedings of the International Conference on Ecological Impact of Acid Precipitation, D. Drabløs and A. Tollan, eds. Sur Nedbørs Virkning Pa Skog Og Fisk (SNSF) project, 1432 As, Norway. 383 pp.

Reuss, J. O. 1985. Implications of the Ca-Al exchange system for the effect of acid precipitation on soils. J. Environ. Qual. 14:26-31.

Richter, D. D., D. W. Johnson, and D. E. Todd. 1983. Atmospheric sulfur deposition, neutralization and ion leaching in two deciduous forest ecosystems. J. Environ. Qual. 12:263-269.

Rosenquist, I. Th., P. Jorgensen, and H. Rueslatten. 1980. The importance of natural H^+ production for acidity in soil and water. Proceedings of an International Conference. Pp. 240-241 of Ecological Impact of Acid Precipitation, D. Drabløs and A. Tollan, eds. Sur Nedbørs Virkning Pa Skog Og Fisk (SNSF) project, Oslo.

Schnorr, J. L., G. R. Carmichael, and F. A. Van Schepen. 1982. An integrated approach to acid rainfall assessments. Chapter 13 of Vol. II. Energy and Environmental Chemistry: Acid Rain, L. H. Keith, ed. Ann Arbor, Michigan: Ann Arbor Science.

Schofield, C. L. 1976. Lake acidification in the Adirondack mountains of New York: causes and consequences. In Proceedings of the First International Symposium on Acid Precipitation and the Forest Ecosystem, L. S. Dochinger and T. A. Seliga, eds. U.S. Department of Agriculture Forest Service Technical Report NE-2.

Schofield, C. L. 1982. Historical fisheries changes as related to surface water pH changes in the United States. Pp. 57-68 of Acid Rain/Fisheries, R. E. Johnson (ed.). Bethesda, Md.: American Fisheries Society.

Seip, H. M. 1980. Acidification of Freshwater--Sources and Mechanisms. Pp. 358-366 of Ecological Impacts of Acid Precipitation, D. Drabløs and A. Tollan, eds. Proceedings of an international congress in Sandefjord, Norway. Sur Nedbørs Virkning Pa Skog Og Fisk (SNSF) Project, Oslo.

Skougstad, M. W., M. J. Fishman, L. C. Friedman, D. E. Erdmann, and S. Duncan, Eds. 1979. Methods for Determination of Inorganic Substances in Water and Fluvial Sediments Techniques of Water-Resources Investigations of the U.S. Geological Survey. Book 5, Chapter A1. Washington, D.C.: U.S. Government Printing Office.

Smith, R. A., and R. B. Alexander. 1983. Evidence for acid precipitation induced trends in stream chemistry at Hydrologic Bench-Mark stations. Circular 910. U.S. Geological Survey. 12 pp.

Smith, R. A., R. M. Hirsch, and J. R. Slack. 1982. A study of trends in total phosphorus measurements at NASQAN stations. USGS Water-Supply Paper 2190.

Standard Methods for the Examination of Water and Sewage. 1926. 4th ed. American Public Health Association, New York.

Standard Methods for the Examination of Water and Sewage. 1930. 6th ed. American Public Health Association, New York.

Standard Methods for the Examination of Water and Sewage. 1933. 7th ed. American Public Health Association, New York.

Standard Methods for the Examination of Water and Sewage. 1936. 8th ed. American Public Health Association, New York.

Standard Methods for the Examination of Water and Sewage. 1960. 11th ed. American Public Health Assocation, New York.

Stumm, W., and J. J. Morgan. 1981. Aquatic Chemistry. New York: John Wiley and Sons.

Taylor, F. L. 1933. Letter of February 23, 1933, from F. Lowell Taylor to J. Juday, Department of Chemistry, University of Wisconsin. Archives of the University of Wisconsin, Madison.

Thompson, M. E. 1982. The cation denudation rate as a quantitative index of sensitivity of eastern Canadian rivers to acidic atmospheric precipitation. Water Air Soil Pollut. 18:215-226.

Thompson, M. E., and M. B. Hutton. 1982. Sulfate in lakes of eastern Canada: calculated atmospheric loads compared with measured wet deposition. National Water Research Institute, Environment Canada, Burlington, Ontario.

Thompson, M. E., and M. B. Hutton. 1985. Sulfate in lakes of Eastern Canada: calculated yields compared with measured wet and dry deposition. Water Air Soil Pollut. 24:77-83.

Thompson, M. E., F. C. Elder, A. R. Davis, and S. Whitlow. 1980. Evidence of acidification of rivers of Eastern Canada. Pp. 244-245 of Ecological Impact of Acid Precipitation, D. Drabløs and A. Tollan, eds. Proceedings of an international conference. Sur Nedbørs Virkning Pa Skog Og Fisk (SNSF) project, Oslo.

Towne, R. E. 1983. Past and present pH and alkalinity of New Hampshire lakes and ponds, New Hampshire Water Supply Pollution Control Commission, Concord. Unpublished manuscript. 20 pp.

Turk, J. T. 1983. An evaluation of trends in the acidity of precipitation and the related acidification of surface water in North America. Water Supply Paper 2249. U.S. Geological Survey. 18 pp.

Ulrich, B. 1981. Okosystemare Hypothese uber die Ursachen des Tanneusterbens. Forstwiss. Centralbl. 100:228-238.

Ulrich, B. 1983. Stabilitat von waldokosystemen unter dem einfluss des "Sauren regens." Allg. Forstz. 26/27:670-677.

Van Breeman, N., P. A. Burrough, F. J. Velthorst, H. F. van Dobben, T. deWitt, T. B. Ridder, and H. F. R. Reijuders. 1982. Soil acidification ammonium sulfate in forest canopy throughfall. Nature 299:548-550.

Vermont Department of Water Resources and Environmental Engineering. 1983. Vermont acid precipitation program--long-term lake monitoring. Vermont Department of Water Resources and Environmental Engineering, Montpelier.

Wagner, D. P., D. S. Fanning, J. E. Foss, M. S. Patterson, and P. A. Snow. 1982. Morphological and mineralogical features related to sulfide oxidation under natural and disturbed land surfaces in Maryland. Chapter 7 of Acid Sulfate Weathering, J. A. Kittrick, D. S. Fanning, and L. R. Hossner, eds. SSSA Special publication 10. Soil Society of America, Madison, Wisc.

Warfel, H. E. 1939. Biological survey of the Connecticut watershed. New Hampshire Fish and Game Commission, Concord.

Watt, W. D., D. Scott, and S. Ray. 1979. Acidification and other chemical changes in Halifax County lakes after 21 years. Limnol. Oceanogr. 24:1154-1161.

Wright, R. F. 1983. Predicting Acidification of North American Lakes. Norwegian Institute for Water Research (NIVA) report 4/1983. Oslo, Norway. 165 pp.

Wright, R. F. 1984. Regional pattern of sulfate concentrations in lakes in Eastern North America. Paper submitted to Committee on Monitoring and Assessment of Trends in Eastern North America, National Research Council. May 1984.

Wright, R. F., and E. T. Gjessing. 1976. Acid precipitation: changes in the chemical composition of lakes. Ambio 5:219-223.

Zimmerman, A. P., and H. H. Harvey. 1979. Sensitivity to acidification of waters of Ontario and neighboring states. Final report to Ontario Hydro. University of Toronto, Ontario. 118 pp.

8

Fish Population Trends in Response to Surface Water Acidification

Terry A. Haines

INTRODUCTION

The most widely reported consequence of acid deposition is the reduction or elimination of fish populations in response to surface water acidification (Overrein et al. 1980, Haines 1981, Altshuller and Linthurst 1984, Dillon et al. 1984). To demonstrate conclusively that fish population trends in a body of water are related to atmospheric acid deposition, temporal associations between surface water chemistry and fish populations are required: It must be shown that the water body formerly supported a viable fish population, either self-sustaining or hatchery-maintained; that one or more fish species formerly present have been reduced or eliminated; that the water body is more acidic now than when fish were present and that the increased acid level was not caused by local factors; and that other adverse factors are either absent or unimportant. The number of available data sets meeting these criteria is very small, and most have not been published in the peer-reviewed literature.

A larger number of data sets exist that demonstrate spatial associations between surface water chemistry and fish populations. In some cases, laboratory or field experiments serve to confirm the spatial associations between fish species with water chemistry parameters linked to acidification (e.g., pH and aluminum). In other cases temporal data on fish populations are available, but there are no temporal data on water quality; or the converse may be true. Other than temporal association studies with data on both fish populations and water quality, the strongest cases for demonstrating adverse effects of acidification on fish population trends are those that combine spatial data with one or more of the above types of evidence (Magnuson et al. 1984). These studies are also very scarce. In this chapter I review available North American data sets that purport to relate fish population status or trends to acid deposition.

LABORATORY INVESTIGATIONS

Laboratory experiments have confirmed that chemical parameters associated with acidification, such as pH, dissolved organic carbon, levels of aluminum, and calcium, affect the survival of fish. These laboratory studies have been reviewed in Altshuller and Linthurst (1984) and Haines (1981).

Exposing fish to various pH levels reveals intraspecific and interspecific differences in acid tolerance (Tables 8.1 and 8.2). The environment, however, is much more complex than this, and acid tolerance is also affected by other factors. Varying concentrations of toxic trace metals, such as aluminum and cadmium, can act synergistically with acid, while divalent cations, especially calcium, and organic ligands can act antagonistically with acid and trace metals. In general, reduced pH and increased trace metal concentrations are toxic to fish, and this toxicity is reduced by increased levels of dissolved organic carbon and divalent cations. Toxicity varies in a complex manner with changing proportions of the toxic and mediating factors, fish life history stage, and fish species. Given the complexity of the interactions it is generally not possible to predict the response of a fish species population to acidification in the absence of considerable experimental evidence. Such evidence is currently available for only a few fish species, such as brook trout (*Salvelinus fontinalis*) and white sucker (*Catostomus commersoni*), although data are now being collected on additional species.

FIELD EXPERIMENTS

A few researchers have studied the effects of acidification on fish using manipulation experiments. Hall et al. (1980) acidified a small stream in New Hampshire, but the stream did not support a permanent fish population. A few adult brook trout were present in the acidified reach during part of the year, but they did not exhibit morphological symptoms of acid stress.

TABLE 8.1 Median Survival Time (hours) for Fish Exposed to Acutely Toxic pH Levels

Species		Size or Age	pH Range 2.0-2.5	2.6-2.8	3.0-3.1	3.2-3.3	3.4-3.5	3.6-3.7	3.8-3.9	4.0-4.1	4.2-4.3	4.4-4.5	References[b]
Brook trout	*	10-60 g	< 1			7							a
		fngl[a]			2-3	3-6	6-18	10-38	14-51	20-270			b
		2 g	1	2	3-6	12-14	45						c
		90 g	1	4	9	18	61-66	334					c
		60-130 g	< 1		1		5-9						d
		50 g					25	66-70					e
		50-90 g					10-32						f
Rainbow trout	*	1 g						8	23	37			g
	*	130 g		1-4	4	8	18						g
	*	200-300 g	2	2	5								h
		2-5 g			1	2	3	6	17	83	117	133	i
	*	2-5 g			< 1	1	2	6	27	133			i
	*	4.5-15 cm			1		2	3	8	22	70		j
		4.5-15 cm			2		3	7	18	55			j
Brown trout	*	1-5 g					25	40		120			k
	*	6 g	1-2		3-7					2-4			l
	*	60-80 g	3	4	9								h
Arctic char	*	100-170 g	3	3	4								h
White sucker		7 mo			1	2	5	10	30-200	350	1000		m
Roach		7-13 cm			< 1	1		3	12				j

NOTE: Asterisks denote experiments using low-alkalinity water.
[a] fngl: Fingerling, age 0+ (less than 1 yr), weight usually < 50 g.
[b] Key to references:

- a Daye and Garside 1976
- b Johnson 1975
- c Robinson et al. 1976
- d Packer and Dunson 1972
- e Swarts et al. 1978
- f Falk and Dunson 1977
- g Kwain 1975
- h Edwards and Hjeldnes 1977
- i McDonald et al. 1980
- j Lloyd and Jordan 1964
- k Brown 1981 with 0.1 millimols of calcium
- l Edwards and Gjedrem 1979
- m Beamish 1972

SOURCE: Altshuller and Linthurst (1984)

TABLE 8.2 Percent Survival of Fish Exposed to Chronically Toxic pH Levels.

Species		Size or Age	Length of Exposure (days)	pH Range									References[a]
				3.2	3.6	4.2-4.4	4.5-4.6	4.8-5.0	5.2-5.6	5.9-6.2	6.5-6.8	7.0-7.5	
Brook trout		100-300 g	5			60-90						100	a
	*	10-60 g	7	0	85	100		100	100		100		b
	*	5 g	11			100		100					c
		50 g	65			0-36							d
		150-360 g	150				0	75			75	100	e
Rainbow trout	*	200-300 g	100					93	96	97			f
Brown trout	*	60-80 g	100					94	98	95			f
Arctic char	*	100-170 g	100					90	100	100			f
Fathead Minnow		1 yr	400				80		75	85	75	85	g
Flagfish	*	Mature Adult	20				36	86	79	100	93		h

NOTE: Asterisks denote experiments using low-alkalinity water.

[a]Key to references:

a Dively et al. 1977
b Daye and Garside 1975
c Baker 1981
d Swarts et al. 1978
e Menendez 1976
f Edwards and Hjeldnes 1977
g Mount 1973
h Craig and Baski 1977

SOURCE: Altshuller and Linthurst (1984).

Mills (1984) acidified Lake 223 of the Experimental Lakes Area, Ontario, by direct addition of sulfuric acid. The lake pH was reduced from 6.49 (1976) to 5.17 (1983). Lake trout (*Salvelinus namaycush*), white sucker (*Catostomus commersoni*), fathead minnow (*Pimephales promelas*), and slimy sculpin (*Cottus cognatus*) were abundant at the start of the experiment, and pearl dace (*Semotilus margarita*) were rare. Fathead minnow and slimy sculpin abundance declined in 1979 (pH 5.64), lake trout declined in 1980 (pH 5.59), and white sucker declined in 1981 (pH 5.16). Pearl dace first increased rapidly in 1980, following the decline of fathead minnow, then declined in 1982 (pH 5.11). No fish species successfully reproduced in 1982. Adults of some species survived but were in generally poor condition because of the loss of forage species. It is difficult, however, to extrapolate these results to areas that receive acidic deposition. Inasmuch as the acid was added directly to the lake, the dissolved organic carbon and trace metal (especially aluminum) concentrations in the lake may not reflect those in lakes in which the watershed also receives acid inputs.

SPATIAL ASSOCIATIONS

South Central Pennsylvania

Personnel from the Institute for Research on Land and Water Resources, Pennsylvania State University, have conducted a number of studies in the Laurel Hill area of south central Pennsylvania. In one, researchers conducted a spatial association study in 61 streams on Laurel Hill (W. Sharpe, Pennsylvania State University, personal communication, 1985; Figure 8.1). For 12 of the streams on-site surveys identified significant cultural disturbances in the watershed that were expected to affect water quality, such as agricultural, highway, and industrial discharge, and these streams were therefore excluded from further consideration. The remaining streams were located in watersheds with only minor land use activities. These streams were surveyed for water chemistry and fish population (by electrofishing a 100-m section) and were divided into three groups: those with reproducing fish populations (young-of-the-year fish of nonhatchery origin present), those with remnant fish populations (only 2-year classes), and those with no fish (Table 8.3).

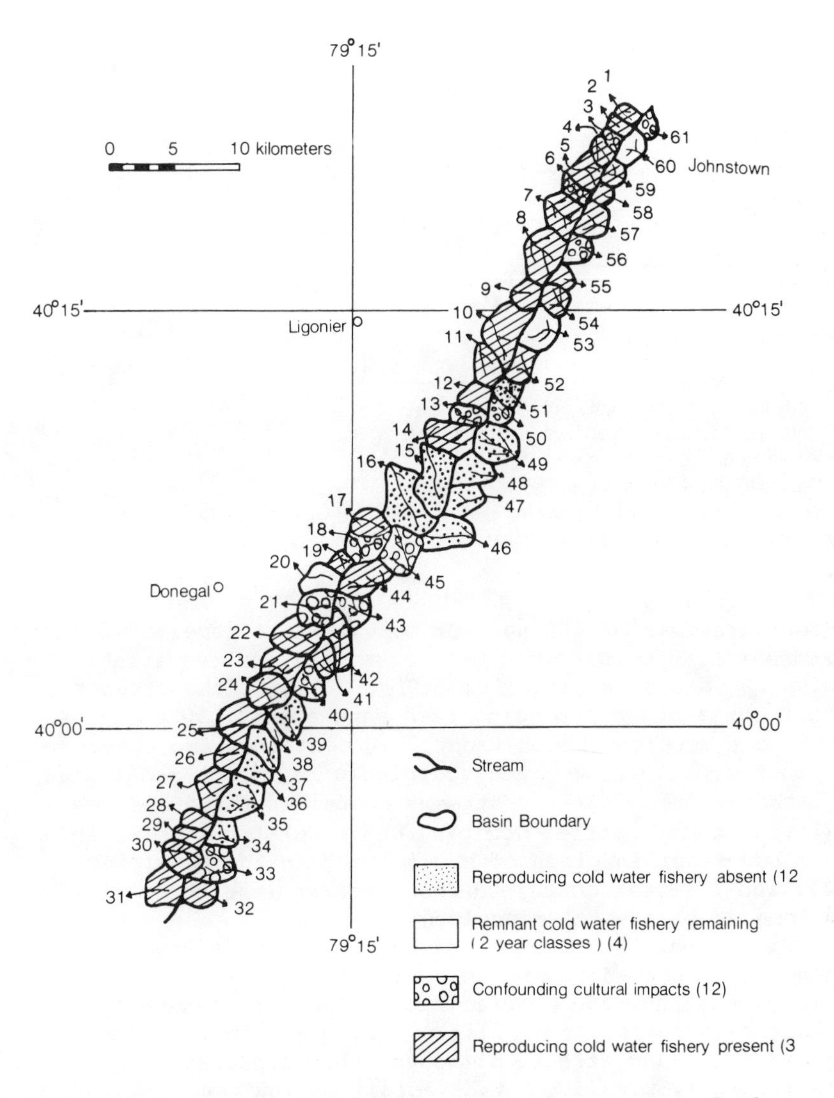

FIGURE 8.1 Location and fish population status of 61 streams on Laurel Hill, south central Pennsylvania. Source: W. Sharpe, Pennsylvania State University.

The 33 streams that contained reproducing fish populations all were of pH 6.0 or higher, and aluminum concentrations were less than 100 µg/L. Of the 12 streams lacking reproducing fish populations, only one (Beaver Dam Run, pH 6.22) had a pH above 6.0. Eight of the 12 had pH values less than 5.0 and aluminum

TABLE 8.3 Fish Population Status and Water Chemistry of 49 Streams on Laurel Hill, South Central Pennsylvania

Fish Population Status	Number of Streams	Mean pH (range)	Mean Alkalinity[a] (range)	Mean Aluminum[b] (range)
Reproducing[c]	33	6.73 (6.0-7.2)	139 (27-340)	21 (5-97)
Remnant[d]	4	6.36 (5.7-7.1)	88 (6-262)	49 (18-100)
Not reproducing[e]	12	4.97 (4.3-6.2)	7 (0-43)	454 (15-1,000)

[a] Units are microequivalents per liter.
[b] Units are micrograms per liter total filterable.
[c] Young-of-the-year fish present.
[d] Only two year-classes present.
[e] No young-of-the-year fish present.
SOURCE: W. Sharpe, Pennsylvania State University.

concentrations of 400 µg/L or more. The 4 streams with remnant fish populations were chemically quite variable, with pH values ranging from 5.7 to 7.1. Of the streams that contained reproducing fish populations, 10 contained only one species (brook trout), 18 contained two species (brook trout plus mottled sculpin in 17, brook trout plus blacknose dace in 1), 4 streams contained three species (brook trout, mottled sculpin, and brown trout in 2, and rainbow trout in place of brown trout in the remaining 2), and 1 stream contained four species (the three salmonids plus mottled sculpin).

Fish population status in these streams is thus generally related to stream chemistry. Although there are exceptions, the more acidic streams are generally those with sparse or no fish population. This does not prove cause and effect, although other explanations involving factors other than acidification have been ruled out.

Vermont

Langdon (1983, 1984) conducted a spatial association study of fish populations and water chemistry in 29 Vermont lakes. The lakes selected for the fisheries survey were chosen from a larger set of lakes (Clarkson 1982) and represented the lowest alkalinity lakes in the survey. Fish were collected with experimental (variable mesh size) gill nets, baited minnow traps, and seines. Two of the lakes were fishless, and 27 contained 1 to 10 fish species.

For this report I conducted a stepwise multiple-regression analysis for physical and chemical factors that could potentially affect the number of fish species in these lakes. The variables considered were pH, specific conductance, color, sum of divalent cations (calcium and magnesium), surface area, maximum depth, and elevation. The only significant ($p \leq 0.05$) variables were pH ($p = 0.0001$), sum of divalent cations ($p = 0.026$), and surface area ($p = 0.028$). The regression equation is

$$\log(\text{number of species} + 1) = -1.33 + 0.41\,(\text{pH}) - 0.004\,(\text{sum of divalent cations}) + 0.0004\,(\text{surface area});$$

$$r^2 = 0.62,\ p = 0.0001,\ n\ (\text{number of lakes}) = 29.$$

I also compared fish abundance, expressed as catch per day (a standard unit of effort), with the same physical and chemical variables. Similar to the above analysis the three statistically significant variables of fish abundance were pH ($p = 0.048$), sum of divalent cations ($p = 0.006$), and surface area ($p = 0.070$). The regression equation is

$$\log(\text{catch per unit effort} + 1) = -1.17 + 0.57\,(\text{pH}) - 0.023\,(\text{sum of divalent cations}) + 0.001\,(\text{surface area});$$

$$r^2 = 0.51,\ p = 0.001,\ n = 29.$$

Generally, smaller lakes have lower pH and contain fewer numbers of fish species and individuals. Surprisingly, the relationship with divalent cations was reversed; larger lakes were lower in divalent cations than expected. Attempts to partition the variability between pH and surface area were unsuccessful. Apparently these two variables are highly correlated.

The lakes surveyed were generally not affected by cultural disturbances. Most contained no permanent structures in the watershed, although five contained seasonal dwellings and one contained year-round dwellings. Many of the lakes, however, were stocked with salmonid fish, but only one lake contained stocked salmonids alone. All the lakes were accessible to fishing and, fishing pressure was not believed to be excessive for any lake.

The previous presence of fish populations in the two lakes that are now fishless is equivocal. One lake is known to have contained fish previously but may have been affected by beaver activity that blocked fish access to a spawning stream. There are no reliable data to confirm that the other lake ever contained fish, but there are anecdotal records that suggest that it did (Langdon 1983).

Maine

Haines (1985) conducted a spatial association survey of water chemistry and fish population in 22 lakes in Maine. The lakes were all low in color and contained no human habitation or other recent land disturbance in the watershed. None had been stocked, reclaimed, or otherwise manipulated, and they were sufficiently remote from vehicle access as to make casual introduction of fish species or intensive fishing pressure unlikely. A survey team visted each lake at least three times. The team collected water samples in spring, summer, and fall to analyze for pH, alkalinity, specific conductance, color, and all major ions. Fish populations were surveyed once, in summer, using a standard collecting protocol. Two experimental gill nets (6 ft deep containing five 25-ft long panels of 0.5-, 0.75-, 1-, 1.25-, and 1.5-inch mesh square measure) were set from late afternoon until midmorning the following day. One net was set in shallow water and one in the deepest area. In addition, six standard wire mesh (0.25 inch square measure) minnow traps were set in various locations in the littoral area for the same time period. The team counted and identified the species of all fish collected.

A stepwise multiple-regression analysis was conducted following the same procedure used with the Vermont spatial association data. For number of fish species the significant variables were sum of divalent cations ($p = 0.045$), and surface area ($p = 0.042$). The next most important variable was pH ($p = 0.21$). The regression equation (including pH) is

$$\log(\text{number of fish species} + 1) = -0.659 + 0.003\ (\text{sum of divalent cations}) + 0.008\ (\text{surface area}) - 0.124\ (\text{pH});$$

$$r^2 = 0.66,\ p = 0.0002,\ n = 22.$$

Fish abundance was also analyzed as for Vermont. The only significant variable was sum of divalent cations ($p = 0.066$). The next most important variable was pH ($p = 0.23$). The regression equation (including pH) is

$$\log(\text{catch per unit effort} + 1) = -1.27 + 0.011\ (\text{sum of divalent cations}) + 0.353\ (\text{pH});$$
$$r^2 = 0.53,\ p = 0.0007,\ n = 22.$$

The number of significant variables for the Maine data was less than for the Vermont data, but the same variables were identified as most important in both cases. In the Maine lakes the relationship of divalent cations was direct, as expected, rather than inverse. In general, small lakes were more acidic, lower in divalent cations, and contained fewer numbers of species and individuals. Attempts to partition the variability among pH, divalent cations, and area were unsuccessful, apparently because of the high degree of correlation among these variables.

Wisconsin

Several authors have reported results of fish population and water chemistry surveys in northern Wisconsin lakes. Rahel and Magnuson (1980) reported no relationship between lake pH and number of fish species present in lakes that were similar in size and habitat. Conversely, the same authors (Rahel and Magnuson 1983) reported that cyprinids and darters were generally absent from Wisconsin lakes of pH < 6.2. However, oligotrophic lakes with pH near 7.0 also lacked acid-sensitive fish species because of species interactions and biogeographical factors.

Wiener et al. (1984) surveyed fish populations in two groups of lakes in northern Wisconsin. All lakes were seepage lakes, but one group was acidic with pH levels near 5.0 and the other consisted of lakes with near-neutral pH levels. The acidic group contained significantly fewer fish species than the near-neutral group, and the absent species consisted primarily of cyprinids and darters. However, it is doubtful that these lakes have been acidified by atmospheric deposition. (See Chapter 7.) These data confirm that fish species distribution is affected by water chemistry factors related to acidification but do not demonstrate trends related to atmospheric deposition.

TEMPORAL ASSOCIATIONS

Ontario

Lakes located in the LaCloche Mountain region of Ontario constitute one of the best North American data sets documenting the adverse effects of acid deposition on fish populations, although the major source of acid is certainly the sulfur dioxide emissions from metal smelters located 65 km to the northeast in Sudbury, Ontario (Beamish 1976). Acidification has occurred with sufficient rapidity that declines and in some cases extinctions of fish populations have been documented since the late 1960s (Beamish and Harvey 1972, Beamish 1974a,b, Beamish et al. 1975, Harvey 1975, Beamish 1976, Beamish and Van Loon 1977, Harvey and Lee 1982). The acidic deposition is accompanied by increased deposition of certain heavy metals (e.g., copper, nickel), which may also be toxic to fish (Beamish 1976, Beamish and Van Loon 1977). Of the 212 lakes in the LaCloche Mountains, 68 have been surveyed for fish populations (Harvey 1975). Of these, 38 lakes are known or suspected to have had reductions in the number of fish species, and 54 fish populations are known to have been lost (Harvey and Lee 1982). The number of fish species per lake was significantly ($p = 0.005$) correlated with lake pH, even after correction for differences in lake area. Explanations for declines in fish populations other than changes in water chemistry related to acid deposition have been ruled out (Altshuller and Linthurst 1984).

Adirondack Mountains, New York

The first reports of acid-deposition-induced acidification of lakes and concomitant reductions in fish populations in North America were for the Adirondack Mountain region (Schofield 1965, 1973, 1976a,b). Several recent survey reports (Pfieffer and Festa 1980; Colquhoun et al. 1984) summarized recent data concerning water chemistry and fish populations in lakes within the Adirondack ecological zone. Schofield (1976b) surveyed 40 lakes that had been initially surveyed in the 1930s. In the 1930s three lakes had pH < 5.0 and no fish, and one lake with pH between 6.0 and 6.5 also had no fish. In 1975, 19 of these 40 lakes had pH < 5.0 and had no fish. Two additional lakes with pH between 5.0 and 5.5

TABLE 8.4 Summary of Fish Distribution by pH Class for 289 Waters with Concurrent Fish Surveys and Water Chemistry (1975-1982)

Fish Population Status	pH Classification		
	< 5.0	5.0 to 6.0	> 6.0
All water samples	33	93	163
Waters without fish	13	5	6
Waters with only nontrout species	5	11	35
Waters with trout/salmonid species	15	77	122

SOURCE: Colquhoun et al. (1984).

also lacked fish. Thus 17 lakes apparently lost fish populations during the interval between the 1930s and 1975. Pfeiffer and Festa (1980) concluded that at least 180 lakes in the Adirondack zone had lost fish populations as a result of acidification. However, the data supporting this conclusion were not published. Colquhoun et al. (1984) presented a summary of pH and fish population status data for 289 Adirondack lakes (Table 8.4).

Baker and Harvey (1985) recently completed an exhaustive analysis of all available water chemistry and fish population status data for Adirondack lakes. To evaluate changes in fish population status over time they developed semiquantitative indicators of population status that incorporate the uncertainties associated with varying sampling techniques. These indicators were then used to test hypotheses by means of ordered classification statistics. Because fish sampling procedures have varied significantly over time, quantitative indicators such as catch per unit effort and population presence/absence cannot be used. Indicators were developed to express stocking data, the quantity of the survey data, the status of the population of 14 common fish species, and the fish community's status as a whole.

Each indicator consisted of an ordered classification variable that assigned a numerical scale to qualitative ratings. For example, the quality of fish survey data was assigned a classification variable ranging from 0 to 8, as follows:

Rating	Number of Years Surveyed	Number of Years between First and Last Survey
0	0	-
1	1	-
2	2	<10
3	2	>10
4	3	<10
5	3	>10
6	4	>10
7	$\geq$5	10-20
8	$\geq$5	>20

The classifications were made by a single experienced biologist to ensure uniformity and were done blind, i.e., without knowledge of the lake identity. Fish population data were classified with no knowledge of lake chemistry or acidity status. Ordered classifications are used in cases in which different levels of some factor can be recognized but not quantified with a standard unit of measurement on a continuous scale (Kleinbaum and Kupper 1978). Standard statistical tests are available for the classified data.

The data assessment involved three phases: (1) evaluating fish population status, (2) evaluating possible explanations for observed population declines or losses, and (3) analyzing statistical associations between decline or loss of fish populations and chemical factors related to acidity.

Of the 604 lakes that had sufficient data for rating fish community status, 49 (8.1 percent) were rated with high probability as having populations reduced or eliminated as the result of acidification. An additional 64 (10.6 percent) lakes were so rated with marginal probability. A number of lakes did not have previous survey data to demonstrate changes in fish communities over time but currently have either no fish or only sparse populations of acid-tolerant species. The estimated number of Adirondack lakes with fish communities adversely affected by acidification varies from 50 to 200-300, depending on the degree of certainty accepted by the evaluator. The numbers of populations of the 14 species of fish that were investigated and that were apparently lost as a result of acidification are summarized in Table 8.5.

TABLE 8.5 Numbers of Populations Classified by Ratings of Population Status and Likely Adverse Effects of Acidification

Species	No Evidence for Loss of Population	Apparently Unrelated to Acidification	Apparently as a Result of Acidification	
			Marginal Evidence	Adequate Evidence
Brook trout	409	86	114[a]	98
Lake trout	68	25	10	8
Rainbow trout	54	34		2
Lake whitefish	28	24		1
Smallmouth bass	92	42		2
Largemouth bass	49	12		1
Chain pickerel	29	12		0
White sucker	336	66		10
Brown bullhead	420	74	13	13
Pumpkinseed sunfish	240	104		7
Yellow perch	185	80		2
Golden shiner	213	83	18	12
Creek chub	147	100		7
Lake chub	6	12		0

[a]Includes 71 lakes considered typical brook trout habitat, with only recent survey data available and no fish caught.
SOURCE: Baker and Harvey (1985).

Fish community status ratings were significantly correlated with lake pH (Figure 8.2). The lake pH levels were not significantly different ($p > 0.05$) for community status ratings 0, 1, and 2 and for community status ratings 3, 4, and 5, but the two groups were significantly ($p \leq 0.01$) different from each other. Although this correlation does not prove cause and effect, it lends support to the hypothesis that declines or losses in fish populations in these lakes are the result of acidification. An investigation of the relationship between lake pH and fish community status (Figure 8.3) indicated that lakes with pH 5.6 or greater had a high probability of having a healthy fish community, whereas lakes with pH below 5.6 had a decreasing probability of having a healthy fish community.

Eight New York lakes have been identified for which there are several sources of information about trends in chemistry and biota (Table 8.6). Measurements of water chemistry and observations of fish population status have been made for these lakes, which have been cored for analysis of sediment diatoms and trace elements. (See

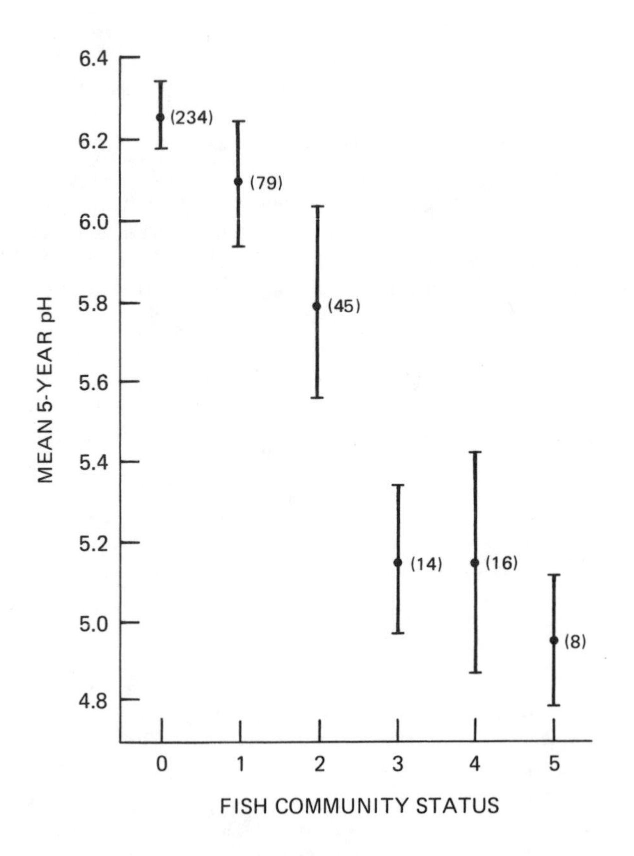

FIGURE 8.2 (Opposite) Mean pH (and 95 percent confidence interval) for lakes with fish community status ranked 0 to 5. Mean pH based on average value over 5-year period centered on year of last fish survey. Number of lakes in parentheses. Source: Baker and Harvey (1985).

Community Status	Key Community Characteristics
0	Community appears "healthy." Any changes in species composition through time appear random and are apparently due to normal fluctuations. For lakes with only recent survey data available, species diversity is high and/or characteristic for the Adirondacks. Species considered acid sensitive are present.
1	One or two species have apparently declined in abundance and/or disappeared from the lake. These species are not, however, considered particularly acid sensitive (relative to other species in the lake), and it is _unlikely_ that population declines or losses resulted from acidification.
2	One or two species have apparently declined in abundance and/or disappeared from the lake. The species affected _may_ be acid sensitive (relative to other species in the lake); thus the community decline _may_ be a result of acidification. Neither the evidence for loss of populations nor the indications of the potential influence of acidification are particularly strong, however. A number of populations have maintained constant or increased levels of abundance over time.
3	Several species have disappeared from the lake. These species are expected to be acid sensitive relative to other species remaining in the lake, suggesting the possibility that community decline may be a result of acidification.
4	All species or the majority of species have disappeared. Any species left (e.g., brown bullhead, yellow perch) are expected to be acid tolerant. No questions regarding sampling techniques, although the absence of species is confirmed by only one sample.
5	All species have disappeared. No fish caught in two or more samples (in 2 or more years). No questions regarding sampling techniques.

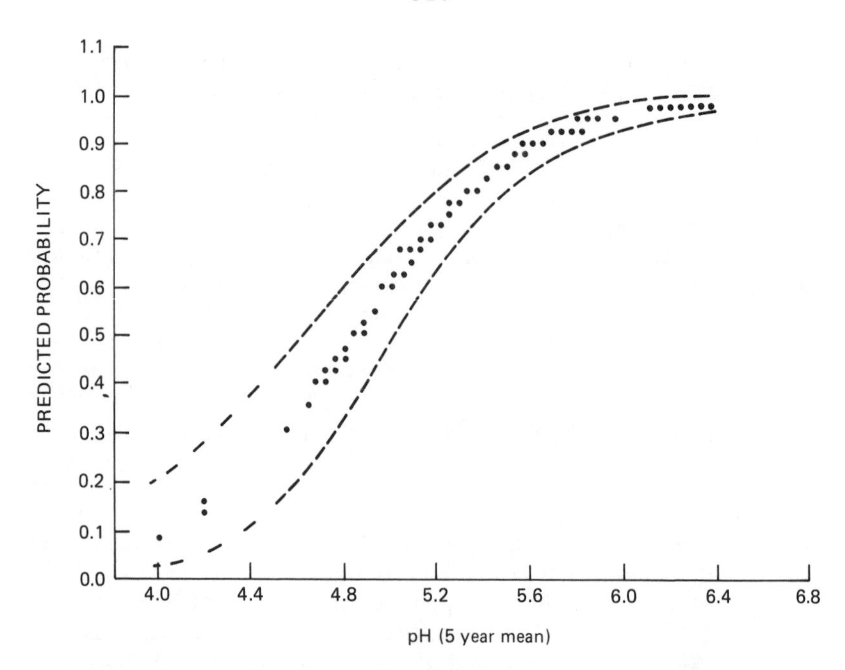

FIGURE 8.3 Predicted probability (and 95 percent confidence interval (dashed lines)) that status = 1 (fish community "healthy") as a function of mean 5-year pH. Status = 1 for lakes with fish community status rated 0 or 1. Source: Baker and Harvey (1985).

Chapter 9.) Lakes Sagamore and Panther exhibit little or no evidence of acidification. Fish populations reflect only local management activities and are otherwise unchanged. Woodhull Lake chemistry is ambiguous, but the fish population may be experiencing acid stress as demonstrated by the disappearance of smallmouth bass and a recent decline in lake trout abundance. However, brook trout also disappeared, which cannot be attributed to acidification, lake whitefish are still present, and declines in lake trout abundance may reflect changes in hatchery introductions. Honnedaga, Big Moose, Woods, Upper Wallface, and Deep lakes, however, also show concomitant declines in pH and fish populations. Disappearance of fish generally lags behind pH decline by a few years. This may be variation in the dating of pH decline or may result from survival of adult fish for several years after reproduction ceases.

TABLE 8.6 Comparisons of Water-Chemistry and Fish-Population Trends with Trends Predicted from Analysis of Sediment Diatoms[a]

		Diatom Stratigraphy		
Lake[a]	Present pH	Present/Background[b] pH	Onset of pH Change	Fish Population Histories
Woodhull	5.6	6.0/6.1	No obvious trend	6 species in 1931; 8 species collected plus 3 reported in 1954, 8 species in 1974-1981; brook trout last collected 1954 and smallmouth bass last collected 1974, but lake trout and lake whitefish still present.
Sagamore	5.6	6.3/6.1	No obvious trend	9 species in 1933, 8 species in 1976; no changes in any major species.
Panther	6.2	6.1/6.4	Slight steady decline	Fish population eradicated in 1951 and restocked; no changes in fish survival.
Honnedaga	4.8	5.2/6.1	Unclear	Lake trout became rare and emaciated in early 1950s, last collected in 1954; brook trout declined in mid-1960s and were rare by the mid-1970s.
Woods	4.7	4.8/5.2	Not dated	Lake trout reported present previously but never collected; formerly known for large brook trout, brook trout abundant in 1961, scarce since 1966, and all of hatchery origin; golden shiner last collected in 1966.
Big Moose	4.9	4.9/5.8	~1950	10 species in 1948, 5-6 species since 1962; smallmouth bass and lake whitefish last collected in 1948; lake trout last collected in 1966.
Deep	4.7	4.8/5.0	1940-1950	Brook trout present, moderately abundant in 1954, absent in 1964; no survival of stocked fish since 1964.
Upper Wallface	5.0	4.7/5.1	1945-1955	Brook trout moderately abundant in 1963-1968, absent in 1975.

[a] Data are also available for a ninth lake, Little Echo, but because it is a colored bog-pond, it is not considered here.

[b] Background pH values determined from pre-1850 estimates.

SOURCES: Diatom data are from Chapter 9 of this report; water-chemistry and fish-population data are from Baker and Harvey (1985) and D. Webster, Cornell University, personal communication.

Although Woods, Deep, and Upper Wallface lakes experienced relatively small decreases in pH (0.4, 0.2, 0.4 units, respectively), the sensitivity of fish to acidity increases markedly over the range of pH from about 5.2 to 4.8. This contention is supported by the observation that many lakes in New York and New England with pHs of 5.0 to 5.2 currently support fish, while lakes with lower pHs generally do not. A contributing factor may be the increase in dissolved aluminum, a metal toxic to fish, at pH values lower than about 5 owing to the decrease in dissolved organic chemical species that can effectively bind (and thus immobilize) aluminum at higher pHs (Davis et al. 1985).

South Central Pennsylvania

An intensive study of precipitation chemistry, stream chemistry, and fish populations was conducted during February 1981, on four streams in the Laurel Hill, Pennsylvania, area, three of which were included in the survey discussed in the section of this chapter on spatial associations (Sharpe et al. 1984). Three of the streams (McGinnis Run, Linn Run, and Card Machine Run) were acidic, with pH declining that month to <5.0 (Figure 8.4), and contained only a few stocked trout. One stream (Wildcat Run) was well buffered (minimum pH 5.6) and contained self-sustaining populations of mottled sculpin, brook trout, and rainbow trout. Historical fisheries data document the presence of brook trout in three of the streams (Wildcat, McGinnis, and Card Machine) before 1930 and the survival of stocked fish in Linn Run in 1932. Fish kills at fish-rearing facilities on two of the acidic streams (Card Machine Run and Linn Run) were recorded in the 1960s and of adult stocked fish in all three streams in the 1970s. Two of the acidic streams (Linn Run and McGinnis Run) contain a spruce bog in the headwaters, but the bog covers less than 1 percent of the area of either stream basin, and comparing main-stream (bog influenced) with tributary stream (not bog influenced) chemistry indicated no significant differences attributable to organic acids.

An in situ fish toxicity bioassay was conducted on one acidic stream (McGinnis Run) and on the nonacidic stream (Wildcat Run) (Sharpe et al. 1983). Trout fry survived only 4 to 9 days in the acidic stream but lived for the duration of the experiment (36 days) in the nonacidic

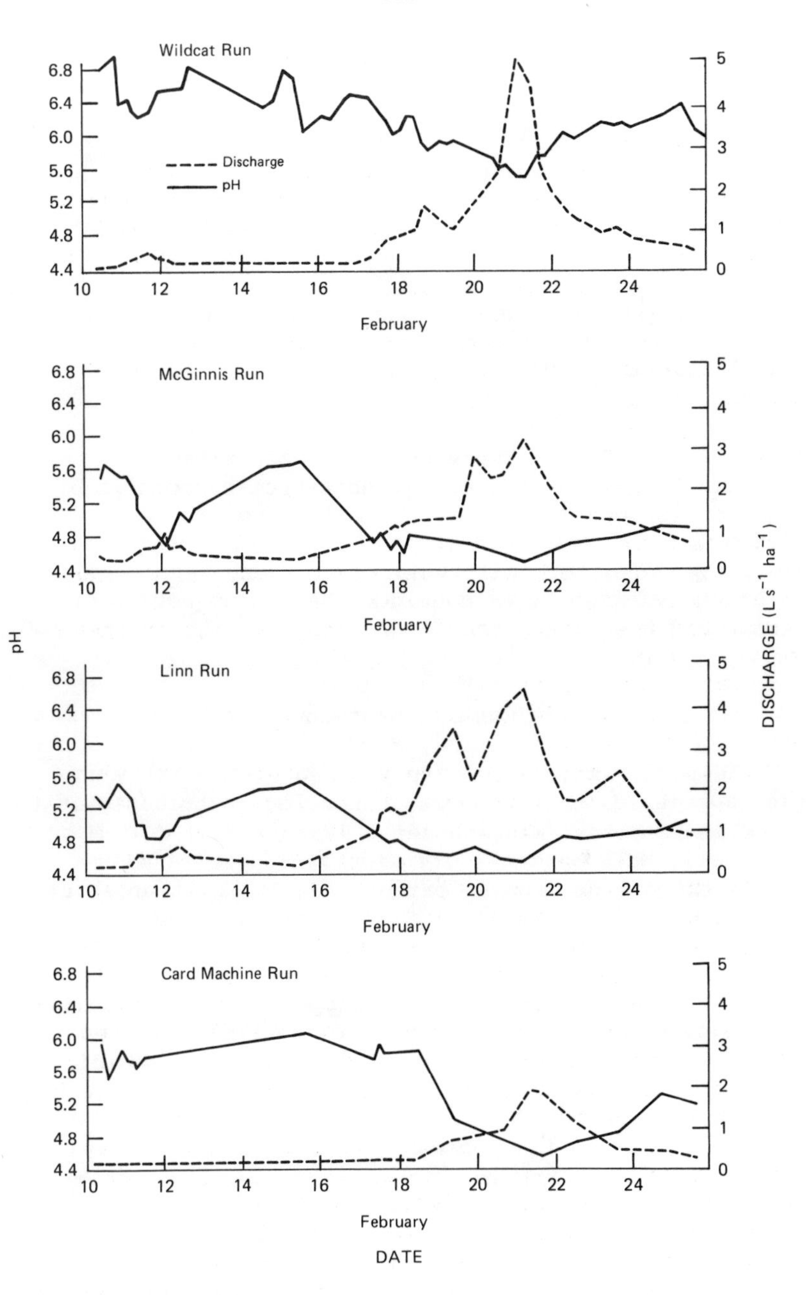

FIGURE 8.4 Changes in pH and discharge (in liters per second per hectare) over time for four streams on Laurel Hill, south central Pennsylvania. Source: Sharpe et al. (1984).

TABLE 8.7 Results of 36-Day In Situ Bioassay for Brook, Brown, and Rainbow Trout Fry in Two Streams on Laurel Hill, South Central Pennsylvania

	Survival Time (Days)	
Species	Low pH[a]	High pH[b]
Rainbow fry	4	36
Brown fry	9	36
Brook fry	9	36

[a]Low-pH stream: pH, 4.8-5.9; aluminum concentration, 0.18-1.1 mg/L.
[b]High-pH stream: pH, 6.1-7.0; aluminum concentration, 0.01-0.14 mg/L.
SOURCE: Sharpe et al. 1983.

stream (Table 8.7). These results confirm that the streams that are now without fish cannot support viable fish populations because of the acidic conditions. Inasmuch as these streams formerly did support viable populations (either self-sustained or long-term survival of hatchery fish) it is concluded here that acidification has caused the disappearance of fish from these streams.

North Central Massachusetts

A temporal association study of water chemistry and fish species distribution was conducted in tributaries to the Millers River, Massachusetts (Figure 8.5) (D. Halliwell, Massachusetts Division of Fisheries and Wildlife, personal communication, 1985). Fifteen streams that had been surveyed in 1953 and 1967 were surveyed again in 1983 (Table 8.8). Fish surveys were conducted by using a fish toxicant (rotenone) in 1953 and by electrofishing in 1967 and 1983. Species were identified but individuals were not enumerated in 1967. Stream pH was measured colorimetrically in 1953 and 1967 and electrometrically in 1983.

Streams with pH $\geq$ 6.0 in 1983 generally had relatively stable fish communities and pH over time (Table 8.8). One stream, Riceville Brook, lost three species over the study interval, but the other streams lost no more than two species. Of the streams that had pH < 6.0 in 1983, most lost three or more species, although one (Osgood) was unchanged and one (Mormon Hollow) lost only two species. These streams generally had substantial pH reductions over time ($\geq$1 unit), with most of the decline occurring between 1953 and

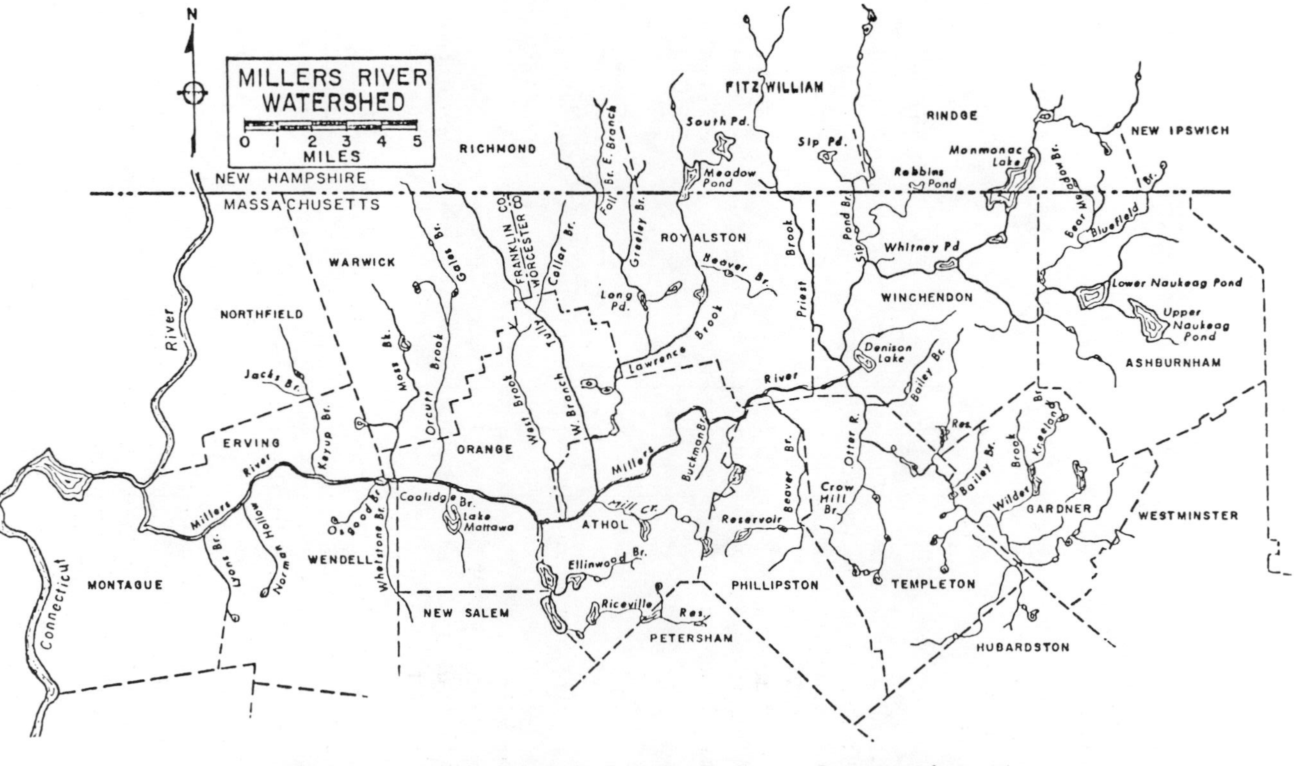

FIGURE 8.5 Map of Millers River watershed, north central Massachusetts.

TABLE 8.8 Comparisons over Time of Number of Fish per Acre by Species and Summer pH in 15 Tributaries of Millers River, North Central Massachusetts[a]

	Brook Trout			White Sucker			Blacknose Dace			Brown Trout			Brown Bullhead		
Stream	1953	67	83	53	67	83	53	67	83	53	67	83	53	67	83
Moss	39	+		9	+	78	70		744	16	+	56		+	
Orcutt				37	+	22	9	+		13		11		+	11
Ellinwood	13		37	38		11				5			18		26
Jacks	236	+	1,067			167	71		667						
Riceville	11		72	98		211	339		306						
Lyons	73	+	382	15			127	+		58	+	127	4		+
Osgood	111	+	+	56			289			78	+	340		+	+
Whetstone	257	+	100	56			215			172	+	70			+
U. Keyup	420	+	820	24	+		216	+		4	+				
West	62	+	113	49	+	88	14	+	1,775	32	+		14	+	13
Boyce	676		1,242	148		463	1,190			52					68
Mormon Hollow	103	+	71	12			77			72	+	14			7
Lawrence	6		3[b]	69		76	19			21		7[b]	7		35
Wilder				31									138		
Templeton	30												5		
Number of population[c]	13	8	11	13	4	8	12	4	4	11	7	7	6	4	9

Stream	Chain Pickerel			American Eel			Longnose Dace			Fallfish			Pumpkinseed		
	1953	67	83	53	67	83	53	67	83	53	67	83	53	67	83
Moss	66			2		11		+	100	41		556	5		
Orcutt	10	+	44	2	+	33	51		456	138	+	44	16	+	
Ellinwood	98		5							30		16	8		
Jacks															
Riceville	11												45		17
Lyons							4	+							
Osgood						40	144								+
Whetstone				2		10	88			17					
U. Keyup				16	+			+							
West	165			3				+		41	+	275			
Boyce			3				86			48					
Mormon Hollow				5	+	+	15								
Lawrence	231		79				3			52					
Wilder	23												8		
Templeton	10														
Number of population	8	1	4	6	3	5	7	4	2	7	2	4	5	1	2

TABLE 8.8 (continued)

Stream	Tesselated Darter			Common Shiner			Largemouth Bass			Number of Species			Summer pH		
	1953	67	83	53	67	83	53	67	83	53	67	83	53	67	83
Moss										8	4	6	7.1	5.5	6.3
Orcutt	3		33	9	+		3	+	22	11	9	9	6.9	6.5	6.3
Ellinwood							3		11	8	–	6	6.7	6.0	6.3
Jacks										2	1	3	–	6.0	6.2
Riceville									17	5	–	5	6.9	5.5	6.2
Lyons										6	4	3	6.9	6.5	6.1
Osgood										5	3	5	–	5.0	5.8
Whetstone										7	2	4	6.5	5.5	5.7
U. Keyup										5	6	1	6.5	6.0	5.7
West	84	+	88		+	125	3			10	9	7	6.9	6.0	5.7
Boyce	38			29						8	–	4	–	–	5.6
Mormon Hollow										6	3	4	6.9	5.5	5.5
Lawrence	2			8						10	–	5	6.7	5.5	5.4
Wilder										4	0	0	6.3	4.5	4.7
Templeton										3	0	0	–	3.5	4.5
Number of population	4	1	2	3	2	1	3	1	3						

[a] A plus sign indicates species present, but not enumerated; a blank space indicates the species was not found during sampling; a minus sign indicates the stream was not sampled.
[b] Hatchery fish.
[c] Number of population is the number of streams in a given year supporting a given species of fish.
SOURCE: D. Halliwell, Massachusetts Division of Fisheries and Wildlife, 1985.

1967. Inasmuch as pH was measured colorimetrically both times, the difference seems unlikely to be a result of methodological differences. (See Chapter 7.)

Two streams, Wilder and Templeton, had pH < 5.0 in 1983 and had lost all fish species by 1967. These streams were of different character than the others surveyed, being of lower gradient (i.e., drop in elevation per stream mile), and higher color (organic acids), and containing primarily warm-water fish species. Nevertheless, pH apparently has declined and fish communities have been lost in these streams. There is no apparent source of acid other than atmospheric deposition.

Nova Scotia

Watt et al. (1983) compared temporal trends in angler harvest of Atlantic salmon with current water chemistry in 22 rivers in Nova Scotia. Angler harvest data were available from 1936 to the present and were nearly continuous. Earlier data exist, but the methodology used is not comparable with that used subsequent to 1936. Previous (1954-1955) water chemistry data of high quality (multiple pH values measured electrometrically) were available for five rivers. Twelve of the rivers now have pH > 5.0 (Table 8.9). Three of these have previous pH data available; two (St. Marys and Musquodobit) have not changed significantly, and one (Medway) declined 0.8 unit. Ten of the twelve rivers had no significant change in salmon harvest, one (Gold) increased significantly, and one (Musquodobit) decreased significantly although its pH did not change appreciably.

Ten of the rivers now have pH < 5.0. Two of these (Clyde and Tusket) have previous pH data and both declined in pH, but statistical tests of the decline could not be performed because of inadequate data. (There is only a single historical value in one case, and there was a change in sample collection site in the other.) Nine of the ten rivers with low pH exhibit a significant decline in angler harvest of salmon, which generally began in the 1950s; the other's decline (Lisomb) was not significant. The rivers included in this comparison were carefully selected to eliminate any that were affected by other pollution sources, fish stocking, or change in physical barriers to fish migration. It is presumed that fish from rivers with geographical proximity have similar migratory patterns and are subjected to similar levels

TABLE 8.9 Water Chemistry and Atlantic Salmon Angler Harvest Data for 22 Nova Scotia Rivers

	St. Marys	Gold	Gaspereau	Moser	Port Dufferin	Ecum Secum	Quoddy	Ship Harbor	Kirby	Petit	Musquodoboit	Medway	East	Ingram	Tangier	Middle	Clyde	Lawrencetown	Tusket	Liscomb	Nine Mile	Isaacs Harbor
pH																						
Mean pH 1954-55	6.3										7.0	6.1					5.0[a]		5.2			
Mean pH 1979-1983	6.0	5.3	6.3	5.3	5.3	5.7	5.4	5.5	5.2	5.5	6.7	5.3	4.7	4.9	4.9	4.9	4.6	4.7	4.7	4.9	4.7	4.9
Difference	−0.3										−0.3	−0.8					−0.4		−0.5			
Significance level	ns										ns	b					c		c			
Fish																						
Slope of regression	+6.9	+2.2	+0.4	+1.0	+0.6	+0.4	0	−0.4	−0.3	−1.6	−1.7	−5.2	−1.3	−4.0	−2.3	−1.4	−1.3	−2.0	−2.3[d]	−1.3	−0.8	−0.6
% variance explained	6.2	1.4	6.0	2.1	2.3	1.4	0	2.0	2.2	7.2	2.0	7.4	69.0	38.0	29.1	59.6	31.2	45.4	23.4	6.3	38.6	17.1
Significance level	ns	e	ns	ns	ns	ns	ns	ns	ns	ns	f	ns	b	b	g	b	b	b	f	ns	b	h

NOTE: ns means not significant.
[a] Single value.
[b] $p < 0.001$.
[c] Not tested because of various data inadequacies.
[d] Excluding recently stocked fish.
[e] $p < 0.05$.
[f] $p < 0.01$.
[g] $p < 0.0001$.
[h] $p < 0.02$.
SOURCE: Watt et al. (1983).

and sources of mortality away from the home rivers. Other studies indicate a significant mortality of early feeding fry of Atlantic salmon in a Nova Scotia river with pH < 5.0 (Lacroix et al. 1983).

SUMMARY

The hypothesis that increased acidity of surface waters caused by long-range atmospheric transport has reduced or eliminated fish populations in northeastern North America was evaluated by examining fishery survey data. The number of statistically valid data sets located was remarkably low. The strongest evidence in support of the hypothesis consists of temporal association data from Adirondack Mountain lakes and Nova Scotia rivers. Both data sets clearly demonstrate declines in acid-sensitive fish species populations over the past 20 to 40 years. Limited water chemistry data indicate that the lakes and streams in question are currently more acidic than they were previously, and fish population status is clearly correlated with present pH. Waters that are now acidic (pH < 5.0) support few or no fish populations. A temporal association study in Massachusetts streams provides similar results, although the number of streams is small and the data are of lesser quality. The LaCloche Mountain area data set is of high quality but does not reflect effects from long-range transport.

The remaining data sets consist largely of spatial associations of surface water chemistry and fish populations, supported in some cases by temporal association data or field experiments. Data from Pennsylvania streams, Vermont lakes, and Maine lakes demonstrate that fish population status is related to present water chemistry. Generally, waters with summer pH less than 5.0 to 5.5 support few or no fish populations. Limited data suggest that at least some of these waters formerly supported fish. Limited field experimental data demonstrate that fish will not now survive in these acidic waters and that addition of acid to surface waters will eliminate fish populations. Although these data demonstrate a relationship between surface water acidity and fish population status, they cannot be used to determine the source of the acidity.

Critics have suggested that other factors could reduce or eliminate fish populations. These factors include chemical pesticides use, change in fish hatchery production, change in angler pressure, and increased beaver activity. Direct effects of human activities (obstruction of fish migrations by dams, degradation of water quality by agricultural and urban runoff, municipal sewage, and industrial wastes, for example) greatly reduced fish populations in accessible waters in the colonial and post-Civil War periods.

However, remote lakes were not directly affected by these factors, and while lumbering and subsequent burning of the watersheds undoubtedly affected fish populations in less accessible lakes, these factors generally were most important in the late 1800s and early 1900s. Concern for declining fish resources resulted in developing artificial propagation of fish in hatcheries and led to the founding of the American Fisheries Society to advance knowledge concerning fish resources (Kendall 1924, Thompson 1970). Early attempts to supplement fish populations by introducing hatchery-reared fish generally failed and fell into disfavor, however. For example, Smallwood (1918) documents the failure of hatchery introduction in Lake Clear, New York. Lake Clear now has a pH > 7.0 and contains at least five species of fish (Colquhoun et al. 1984).

Beaver activity can either enhance or degrade fish populations, depending on the particular circumstances. Beaver were reintroduced into the Adirondack Mountains of New York beginning in 1905 and by the 1920s had reached population densities high enough to raise concern about beaver damage to various resources (Johnson 1927).

Chemical pesticides have been used in remote areas of northeastern North America for control of spruce budworm and blackfly populations, with detrimental effects on fish populations (Burdick et al. 1964, Anderson and Everhart 1966, Elson 1967, Kerswill and Edwards 1967, Locke and Havey 1972). Organochlorine compounds are generally no longer used for these purposes, and most affected fish populations have recovered (Dean et al. 1979). Analysis of brook trout from a series of remote lakes of varying pH in northern New England failed to detect significant organochlorine residues in any fish (Haines 1983).

It is impossible to rule out factors other than acidification as important in the decline or loss of fish populations except by intensive, case-by-case investigations. Such investigations have seldom been made, and in fact the quantity of data on which to base estimations of trends in fish populations is exceedingly sparse. However, a few such data sets do exist, and in the cases cited here the involvement of factors other than acidification in the decline of fish populations appears to be negligible.

REFERENCES

Altshuller, A., and R. Linthurst, eds. 1984. The Acidic Deposition Phenomenon and Its Effects: Critical Assessment Review Papers. Vol. II. Effects Sciences. EPA-600/8-83-016BF. U.S. Environmental Protection Agency.

Anderson, R., and W. Everhart. 1966. Concentrations of DDT in landlocked salmon (Salmo salar) at Sebago Lake, Maine Trans. Am. Fish. Soc. 95:160-164.

Baker, J. 1981. Aluminum toxicity to fish as related to acid precipitation and Adirondack surface water quality. Ph.D. dissertation. Cornell University, Ithaca, N.Y.

Baker, J., and H. Harvey. 1985. Critique of acid lakes and fish population status in the Adirondack region of New York State. Draft final report for NAPAP Project E3-25. U.S. Environmental Protection Agency.

Beamish, R. 1972. Lethal pH for the white sucker, Catostomus commersoni (Lacepede). Trans. Am. Fish. Soc. 101:355-358.

Beamish, R. 1974a. Growth and survival of white suckers (Catostomus commersoni) in an acidified lake. J. Fish. Res. Bd. Can. 31:49-54.

Beamish, R. 1974b. Loss of fish populations from unexploited remote lakes in Ontario, Canada as a consequence of atmospheric fallout of acid. Water Res. 8:85-95.

Beamish, R. 1976. Acidification of lakes in Canada by acid precipitation and the resulting effects on fishes. Water Air Soil Pollut. 6:501-514.

Beamish, R., and H. Harvey. 1972. Acidification of the La Cloche Mountain lakes, Ontario, and resulting fish mortalities. J. Fish. Res. Bd. Can. 29:1131-1143.

Beamish, R., and J. Van Loon. 1977. Precipitation loading of acid and heavy metals to a small acid lake near Sudbury, Ontario. J. Fish. Res. Bd. Can. 34:649-658.

Beamish, R., W. Lockhart, J. Van Loon, and H. Harvey. 1975. Long-term acidification of a lake and resulting effects on fishes. Ambio 4:98-102.

Brown, D. 1981. The effects of various cations on the survival of brown trout, Salmo trutta at low pHs. J. Fish. Biol. 18:31-40.
Burdick, G., E. Harris, H. Dean, T. Walker, J. Skea, and D. Colby. 1964. The accumulation of DDT in lake trout and the effect on reproduction. Trans. Am. Fish. Soc. 93:127-136.
Clarkson, B. 1982. Vermont acid precipitation program winter lake surveys 1980-1982. Vermont Department of of Water Resources and Environmental Engineering, Montpelier.
Colquhoun, J., W. Kretzer, and M. Pfeiffer. 1984. Acidity status update of lakes and streams in New York State. New York State Department of Environmental Conservation Report WM (P-83(6/84)). Albany.
Craig, G., and W. Baksi. 1977. The effects of depressed pH on flagfish reproduction, growth and survival. Water Res. 11:621-626.
Davis, R., D. Anderson, and F. Berge. 1985. Loss of organic matter, a fundamental process in lake acidification: paleolimnological evidence. Nature 316:436-438.
Daye, P., and E. Garside, 1975. Lethal levels of pH for brook trout, Salvelinus fontinalis (Mitchill). Can. J. Zool. 53:639-641.
Daye, P., and E. Garside. 1976. Histopathologic changes in surficial tissues of brook trout, Salvelinus fontinalis (Mitchill), exposed to acute and chronic levels of pH. Can. J. Zool. 54:2140-2155.
Dean, H., J. Skea, J. Colquhoun, and H. Simonin. 1979. Reproduction of lake trout in Lake George. N.Y. Fish Game J. 26:188-191.
Dillon, P., N. Yan, and H. Harvey. 1984. Acidic deposition: effects on aquatic ecosystems. CRC Critical Reviews in Environmental Control 13(3):167-194.
Dively, J., J. Mudge, W. Neff, and A. Anthony. 1977. Blood Po_2, Pco_2 and pH changes in brook trout (Salvelinus fontinalis) exposed to sublethal levels of acidity. Comp. Biochem. Physiol. 57A:347-351.
Edwards, D., and T. Gjedrem. 1979. Genetic variation in survival of brown trout eggs, fry, and fingerlings in acidic water. Research Report 16, Acid Precipitation--Effects on Forest and Fish Project. Sur Nedbørs Virkning Pa Skog Og Fisk (SNSF), Aas, Norway.

Edwards, D., and S. Hjeldnes. 1977. Growth and survival of salmonids in water of different pH. Research Report 10, Acid Precipitation--Effects on Forest and Fish Project. Sur Nedbørs Virkning Pa Skog Og Fisk (SNSF), Aas, Norway.

Elson, P. 1967. Effects on wild young salmon of spraying DDT over New Brunswick forests. J. Fish. Res. Bd. Can. 24:731-767.

Falk, D., and W. Dunson. 1977. The effects of season and acute sub-lethal exposure on survival times of brook trout at low pH. Water Res. 11:13-15.

Haines, T. 1981. Acidic precipitation and its consequences for aquatic ecosystems: a review. Trans. Am. Fish. Soc. 110:669-707.

Haines, T. 1983. Organochlorine residues in brook trout from remote lakes in the northeastern United States. Water Air Soil Pollut. 20:47-54.

Haines, T. 1985. U.S. Fish and Wildlife Service, Orono, Maine, Unpublished data.

Hall, R., G. Likens, S. Fiance, and G. Hendrey. 1980. Experimental acidification of a stream in the Hubbard Brook Experimental Forest, New Hampshire. Ecology 61:976-989.

Harvey, H. 1975. Fish populations in a large group of acid-stressed lakes. Verh. Int. Ver. Limnol. 19:2406-2417.

Harvey, H., and C. Lee. 1982. Historical fisheries changes related to surface water pH changes in Canada. Pp. 45-55 of Acid Rain/Fisheries, R. Johnson, ed. Bethesda, Md.: American Fisheries Society.

Johnson, C. 1927. The beaver in the Adirondacks: its economics and natural history. Roosevelt Wild Life Bull. 4:501-641.

Johnson, D. 1975. Spawning behavior and strain tolerance of brook trout (*Salvelinus fontinalis*) in acidified water. M.S. thesis. Cornell University, Ithaca, N.Y.

Kendall, W. 1924. The status of fish culture in our inland public waters, and the role of investigation in the maintenance of fish resources. Roosevelt Wild Life Bull. 2(3):205-351.

Kerswill, C., and H. Edwards. 1967. Fish losses after forest sprayings with insecticides in New Brunswick, 1952-62, as shown by caged specimens and other observations. J. Fish. Res. Bd. Can. 24:709-729.

Kleinbaum, D., and L. Kupper. 1978. Applied Regression Analysis and Other Multivariable Methods. Boston, Mass.: Duxbury Press.

Kwain, W. 1975. Effects of temperature on development and survival of rainbow trout, Salmo gairdneri, in acid waters. J. Fish. Res. Bd. Can. 32:493-497.

Lacroix, G., D. Gordon, and D. Johnson. 1983. Water chemistry and ecology of fish populations in streams of the Westfield drainage in Nova Scotia. Pp. 1-10 of 1983 Workshop on Acid Rain, R. Peterson and H. Hord, eds. Can. Tech. Rep. Fish. Aquat. Sci. 1213.

Langdon, R. 1983. Fisheries status in relation to acidity in selected Vermont lakes. Montpelier, Vt.: Vermont Agency of Environmental Conservation Report.

Langdon, R. 1984. Fishery status in relation to acidity in selected Vermont lakes, 1983. Montpelier, Vt.: Report of the Department of of Water Resources and Environmental Engineering, Agency of Environmental Conservation.

Lloyd, R., and D. Jordon. 1964. Some factors affecting the resistance of rainbow trout (Salmo gairdneri Richardson) to acid waters. Int. J. Air Water Pollut. 8:393-403.

Locke, D., and K. Havey. 1972. Effects of DDT upon salmon from Schoodic Lake, Maine. Trans. Am. Fish. Soc. 101:638-643.

Magnuson, J., J. Baker, and F. Rahel. 1984. A critical assessment of effects of acidification on fisheries in North America. Phil. Trans. R. Soc. London Ser. B 305:501-516.

McDonald, D., H. Hobe, and C. Wood. 1980. The influence of calcium on the physiological responses of the rainbow trout, Salmo gairdneri, to low environmental pH. J. Exp. Biol. 88:109-131.

Menendez, R. 1976. Chronic effects of reduced pH on brook trout (Salvelinus fontinalis). J. Fish. Res. Bd. Can. 33:118-123.

Mills, J. 1984. Fish population responses to experimental acidification of a small Ontario lake. Pp. 117-131 in Early Biotic Responses to Advancing Lake Acidification. G. Hendrey, ed. Boston, Mass.: Butterworth.

Mount, D. 1973. Chronic effect of low pH on fathead minnow survival, growth and reproduction. Water. Res. 7:989-993.

Overrein, L., H. Seip, and A. Tollan. 1980. Acid precipitation--Effects on forest and fish. Research Report 19, Acid Precipitation--Effects on Forest and Fish Project. Sur Nedbørs Virkning Pa Skog Og Fisk (SNSF), Aas, Norway.

Packer, R., and W. Dunson. 1972. Anoxia and sodium loss associated with the death of brook trout at low pH. Comp. Biochem. Physiol. 41A:17-26.

Pfeiffer, M., and P. Festa. 1980. Acidity status of lakes in the Adirondack region of New York in relation to fish resources. New York State Department of Environmental Conservation Report FW-P168 (10/80). Albany, N.Y.

Rahel, F., and J. Magnuson. 1980. Fish in naturally acidic lakes of northern Wisconsin, U.S.A. Pp. 334-335 of Ecological Effects of Acid Precipitation, D. Drabløs and A. Tollan, eds. Acid Precipitation--Effects on Forest and Fish Project. Sur Nedbørs Virkning Pa Skog Og Fisk (SNSF), Aas, Norway.

Rahel, R., and J. Magnuson. 1983. Low pH and the absence of fish species in naturally acidic Wisconsin lakes: inferences for cultural acidification. Can. J. Fish. Aquat. Sci. 40:3-9.

Robinson, G., W. Dunson, J. Wright, and G. Mamolito. 1976. Differences in low pH tolerance among strains of brook trout (Salvelinus fontinalis). J. Fish. Biol. 8:5-17.

Schofield, C. 1965. Water quality in relation to survival of brook trout, Salvelinus fontinalis (Mitchill). Trans. Am. Fish. Soc. 94:227-235.

Schofield, C. 1973. The ecological significance of air-pollution-induced changes in water quality of dilute-lake districts in the northeast. Trans. N.E. Fish Wildl. Conf. Pp. 98-111.

Schofield, C. 1976a. Lake acidification in the Adirondack Mountains of New York: causes and consequences (abstract). P. 477 in Proceedings of the First International Conference on Acid Precipitation and the Forest Ecosystem, L. Dochinger and T. Seliga, eds. General Technical Report NE-23. U.S. Department of Agriculture Forest Service.

Schofield, C. 1976b. Acid precipitation: effects on fish. Ambio 5:228-230.

Sharpe, W., W. Kimmel, E. Young, and D. DeWalle. 1983. In-situ bioassays of fish mortality in two Pennsylvania streams acidified by atmospheric deposition. Northeast Environ. Sci. 2:171-178.

Sharpe, W., D. DeWalle, R. Leibfried, R. DiNicola, W. Kimmel, and L. Sherwin. 1984. Causes of acidification of four streams on Laurel Hill in southwestern Pennsylvania. J. Environ. Qual. 13:619-631.

Smallwood, W. 1918. An examination of the policy of restocking the inland waters with fish. Am. Naturalist 52:322-352.

Swarts, F., W. Dunson, and J. Wright. 1978. Genetic and environmental factors involved in increased resistance of brook trout to sulfuric acid solution and mine acid polluted waters. Trans. Am. Fish. Soc. 107:651-677.

Thompson, P. 1970. The first fifty years--the exciting ones. Pp. 1-2 of A Century of Fisheries in North America. N. Benson, ed. Special Publication 7. Washington, D.C.: American Fisheries Society.

Watt, W., C. Scott, and W. White. 1983. Evidence of acidification of some Nova Scotian rivers and its impact on Atlantic salmon, *Salmo salar*. Can. J. Fish. Aquat. Sci. 40:462-473.

Wiener, J., P. Rago, and J. Eilers. 1984. Species composition of fish communities in northern Wisconsin lakes: relation to pH. Pp. 133-145 of Early Biotic Responses to Advancing Lake Acidification, G. Hendrey, ed. Vol. 6 of Acid Precipitation Series. Boston, Mass.: Butterworth.

9

Paleolimnological Evidence for Trends in Atmospheric Deposition of Acids and Metals

Donald F. Charles and *Stephen A. Norton*

INTRODUCTION

Paleolimnological analyses of lake sediments have traditionally been used to reconstruct many aspects of the evolution of lake/watershed ecosystems, including terrestrial and aquatic vegetational succession (Davis et al. 1975), fire history (Patterson 1977), trophic status (Davis and Norton 1978, Stockner and Benson 1967), lake acidification (Battarbee 1984), and even the occurrence of blight or disease (Bradstreet and Davis 1975). Long-term changes in meteorology, morphology of the lake basins, soil characteristics, land use, and surface water chemistry can be partially determined from the sediment record.

The sediments of a lake contain information on the lake's past: its biota, water chemistry, watershed characteristics, and material deposited directly from the atmosphere (Frey 1969, Pennington 1981). The information is provided by the organic and inorganic substances, in dissolved or particulate form, that entered or were formed within a lake and were deposited in its sediments. The primary materials are controlled by watershed geology, climate, and biological processes. Once deposited at the bottom of lakes, sediments can be affected by secondary

*The introduction and the section on comparison of diatom and chemical data were jointly authored by Donald F. Charles and Stephen A. Norton. D. Charles authored the section on diatoms and chrysophytes. S. Norton prepared the sections on chemical stratigraphy of lake sediments and peat bogs.

processes, such as transport (horizontal and vertical) and a variety of chemical and biological activities.

Sediments can be sampled by taking cores; the core samples represent the time history of deposition of sediments, with the older sediments lying at greater depth. The times when particular intervals of sediment were deposited can be determined either from analysis of radioactive decay products (lead-210, cesium-137) or by using changes in sediment characteristics that correspond to well-dated local events, such as pollen and charcoal as indicators of logging and forest fires.

Nearly all lakes in the Northern Hemisphere, except those in karst topography, were formed by Pleistocene glaciation. Thus, in most lakes in the North Temperate Zone, the stratigraphic record represents the period from formation of the lake to the present (spanning 10,000 to 15,000 years or less). The period of concern in terms of recent anthropogenic acidification is about 50 to 200 years. In lakes with typical sedimentation rates this smaller interval is usually represented within the top one-half meter of sediment, with the time resolution possible in sediment studies depending on the sedimentation rate, extent of mixing, subsampling interval, and type and quality of the dating of the sediment. Resolution may range from 1 year or season for annually laminated (varved) sediment to more than 20 years for lakes with slow sedimentation rates.

Several components of sediments provide information on factors related to lake acidification. Concentrations of polycyclic aromatic hydrocarbons (PAHs), soot particles, lead, sulfur, vanadium, sulfur isotope ratios, and magnetic particles can be interpreted to indicate trends in atmospheric deposition of substances derived from combustion of fossil fuels. The pH of lake water during the past can be inferred by analyzing assemblages of diatoms and chrysophytes. Remains of chydorids (littoral crustaceans) and chironomids (midge larvae) may also provide insight into changes in aquatic biota related to acidification.

Acidification of lakes and their watersheds can be inferred from changes in concentrations of common and trace metals, such as calcium (Ca), magnesium (Mg), sodium (Na), potassium (K), zinc (Zn), lead (Pb), aluminum (Al), manganese (Mn), and iron (Fe). Various disturbances of watersheds can be indicated by the common and trace metals, pollen, charcoal, and changes in sedimentation rate based on lead-210 dating. Indications of watershed disturbance should be corroborated if at all possible by thorough investigation of historical records and other studies, such as tree ring analysis.

Of the above characteristics of sediment, the most information available for acid-sensitive lakes is derived from data on sediment diatom assemblages and sediment chemistry, particularly trace metals. Thus, in this chapter we emphasize these data.

DIATOM AND CHRYSOPHYTE SEDIMENT ASSEMBLAGES

Analyzing and interpreting sediment diatom and chrysophyte assemblages is the best paleolimnological technique available for reconstructing past lake-water pH. Investigators are using this approach increasingly to assess changes that may have been caused by atmospheric deposition of strong acids, because patterns of change in lake-water pH can be used to help determine trends in acid deposition.

Diatoms and Chrysophytes as Indicators of Lake Chemistry

Diatoms make up a large group of single-celled freshwater and marine algae (division Bacillariophyta). They have siliceous cell walls and are formed of two halves or valves. Chrysophytes (Chrysophyceae) are primarily freshwater plankton. In this review the term chrysophyte refers to only one family, the Mallomonadaceae, also known as the scaled chrysophytes. Its members have flagella and an external cell covering of overlapping siliceous scales and bristles. The scales are used for paleoecological reconstructions.

The distributions of diatom taxa are closely related to water chemistry (Cleve 1891, Kolbe 1932, Hustedt 1939, Jørgensen 1948, Cholnoky 1968, Patrick and Reimer 1966, 1975, Patrick 1977). For this reason, diatoms are commonly used as indicators of pH, nutrient status, salinity, and other water quality characteristics (e.g., Lowe 1974). Stratigraphic analysis of fossil diatom assemblages can be used to investigate changes in lakes resulting, for example, from shifts in climate, development of watershed soils and vegetation, local human disturbance of watersheds, and acid deposition (e.g., Battarbee 1979, 1984, Pennington 1981, Fritz and Carlson 1982, Brugam 1983, 1984, Del Prete 1972).

Diatom assemblages in sediment are good indicators of past lake pH because (1) diatoms are common in nearly all freshwater habitats, (2) distributions of diatom taxa are strongly correlated with lake-water pH (Hustedt 1939, 1927-1966; Meriläinen 1967; Battarbee 1979, 1984; Gasse and Tekaia 1983; Gasse et al. 1983; Huttunen and Meriläinen 1983; Davis and Anderson 1985; Charles 1985a; Anderson et al. in press), (3) diatom remains are preserved well in sediment and can be identified to the lowest taxonomic level, (4) their remains are usually abundant in sediment (10^4 to 10^8 valves/cm^3 of sediment) so that rigorous statistical analyses are possible, and (5) many taxa are usually represented in sediment assemblages (20 to 100 taxa per count of 500 valves is typical) so that inferences are based on the ecological characteristics of many taxa.

Some disadvantages in using diatoms as pH indicators are that (1) diatom identification requires considerable taxonomic expertise, (2) occasionally diatoms are not well preserved because of dissolution (e.g., in some peaty and some calcareous sediments), (3) sometimes the number of taxa is low (e.g., in some bog lakes), (4) calibration data sets (the current relationship between water chemistry and surface sediment diatom assemblages) are not always available for the lake region studied, and (5) good ecological data are not always available for all dominant taxa. Other problems associated with interpretation of diatom data are discussed at the end of this section.

In general, the use of chrysophyte scales for pH reconstructions involves the same advantages and disadvantages as for diatoms (Smol 1985a,b; Smol et al. 1984a), except that the number of chrysophyte taxa in a sediment assemblage is in the range of one-tenth the number of taxa of diatoms and most chrysophyte taxa are euplanktonic (normally suspended in the water). The latter characteristic provides an advantage over diatoms in the study of acidic lakes because euplanktonic diatoms are usually rare or nonexistent in lakes with a pH below about 5.5 to 5.8 (Battarbee 1984, Charles 1985a). In these cases, chrysophytes may be more sensitive indicators of water chemistry changes than diatoms because they live in direct contact with the open water, whereas most diatoms grow in the shallower water of the littoral zone, which may be chemically different from the open water.

Techniques for Determining pH Trends

Several techniques based on diatom assemblages have been used to assess trends in acidification and to derive equations for inferring lake-water pH. These are described briefly below. Further descriptions and discussion of details and uncertainties associated with the reconstruction of lake-water pH are covered in depth by Gasse and Tekaia (1983), Battarbee (1984), Davis and Anderson (1985), Charles (1985a), and Smol et al. (1985).

The simplest and most straightforward approach is to count sediment-core diatom and chrysophyte assemblages and prepare depth profiles of percentages of the dominant taxa. Changes in the profiles are then interpreted in light of the ecological data available on the taxa. At the present time, this is the only technique used to analyze chrysophyte scale data.

Hustedt (1939) made one of the first significant steps toward establishing a more quantitative approach for using diatoms as pH indicators. He recognized the strong relationship between diatom distributions and lake-water pH and defined the following pH occurrence categories:

Acidobiontic--optimum distribution at pH below 5.5
Acidophilic--widest distribution at pH less than 7
Circumneutral/indifferent--distributed equally above and below pH 7
Alkaliphilic--widest distribution at pH greater than 7
Alkalibiontic--occurs only at pH greater than 7

Assignments of diatom taxa to these categories can be based on literature references and on the distribution of taxa within waters of particular geographic regions. Changes in the percentages of diatom valves in each pH category in a sediment core can be used to estimate trends in lake-water pH. This method makes use of data on most of the taxa within a core, not just the dominant species.

Nygaard (1956) took the next major step with the development of a set of indices. These indices are based on ratios of the percentages of diatom valves in Hustedt's pH categories. First, acid units and alkaline units are calculated.

acid units = 5 (% acidobiontic) + (% acidophilic),
alkaline units = 5 (% alkalibiontic) + (% alkaliphilic)

The percentages of valves in the two extreme pH categories, acidobiontic and alkalibiontic, are arbitrarily weighted by a factor of 5 because diatoms in these categories are presumably stronger indicators of pH.

The formulas for Nygaard's indices are

$$\alpha = \frac{\text{acid units}}{\text{alkaline units}},$$

$$\omega = \frac{\text{acid units}}{\text{number of acid taxa}},$$

$$\varepsilon = \frac{\text{alkaline units}}{\text{number of alkaline taxa}}.$$

Of these, index α is the best predictor of current pH in acidic lakes (Meriläinen 1967, Davis and Anderson 1985, Charles 1985a; Figure 9.1) and is used most frequently.

Renberg and Hellberg (1982) derived a new index (Index B), also based on pH categories:

$$\text{Index B} = \frac{(\%\ \text{indifferent}) + 5\ (\%\ \text{acidophilic}) + 40\ (\%\ \text{acidobiontic})}{(\%\ \text{indifferent}) + 3.5\ (\%\ \text{alkaliphilic}) + 108\ (\%\ \text{alkalibiontic})}.$$

Index B has advantages over index α, including the use of more information and less reliance on alkaline taxa, which are typically rare or absent in acid lakes. New indices incorporating Hustedt's (1939) pH categories have been developed by Watanabe and Yasuda (1982) and Brakke (1984).

Meriläinen (1967) refined quantitative techniques even further by developing an approach to predict lake-water pH from the index values. The relationship between the $\log_{10}$ of index values and measurements of lake-water pH for several lakes is determined by using regression analysis (e.g., Figure 9.1). Predictive equations are then derived directly from the slope and intercept of the regression equations.

Predictive equations can also be developed from multiple linear-regression analysis of measured lake-water pH with the percentages of diatoms in each pH category (e.g., Davis and Anderson 1984, Charles 1985a, Figure 9.2). Other approaches for inferring pH involve the use of multiple regression of selected taxa and

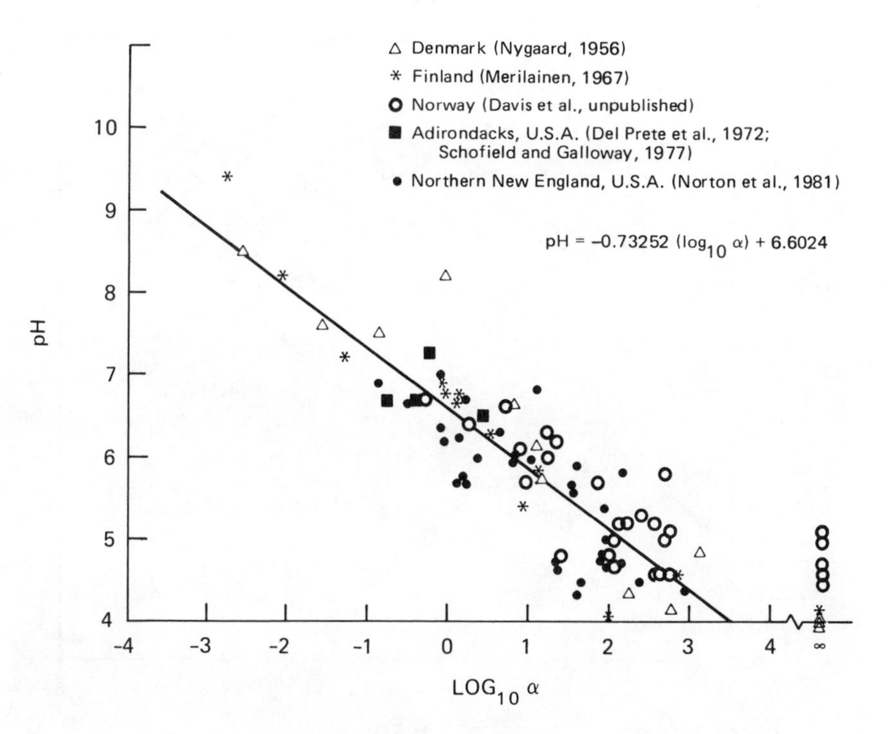

FIGURE 9.1 Logarithm of Nygaard's alpha index for surface sediment diatom assemblages versus lake surface water pH (Norton et al. 1981). Data for Norway are from R. B. Davis, F. Berge, and D. Anderson, University of Maine, Orono, unpublished data.

multiple regression of principal components of taxa data sets (Davis and Anderson 1984, Gasse and Tekaia 1983). The standard error for inferred pH ranges between ±0.25 and ±0.5 pH units (Battarbee 1984, Davis and Anderson 1984, Charles 1985a).

Usually, trends in pH curves are not analyzed statistically. Instead, subjective interpretations are made that account for the nature of the diatom assemblages, the error associated with the predictive techniques, evidence of sediment mixing, and other factors. This is not a problem if pH changes are great, but relatively small changes, for example, within the standard error of the predictive equation, must be interpreted cautiously, especially if the changes do not show a consistent trend over several sediment intervals. Esterby and El-Shaarawi

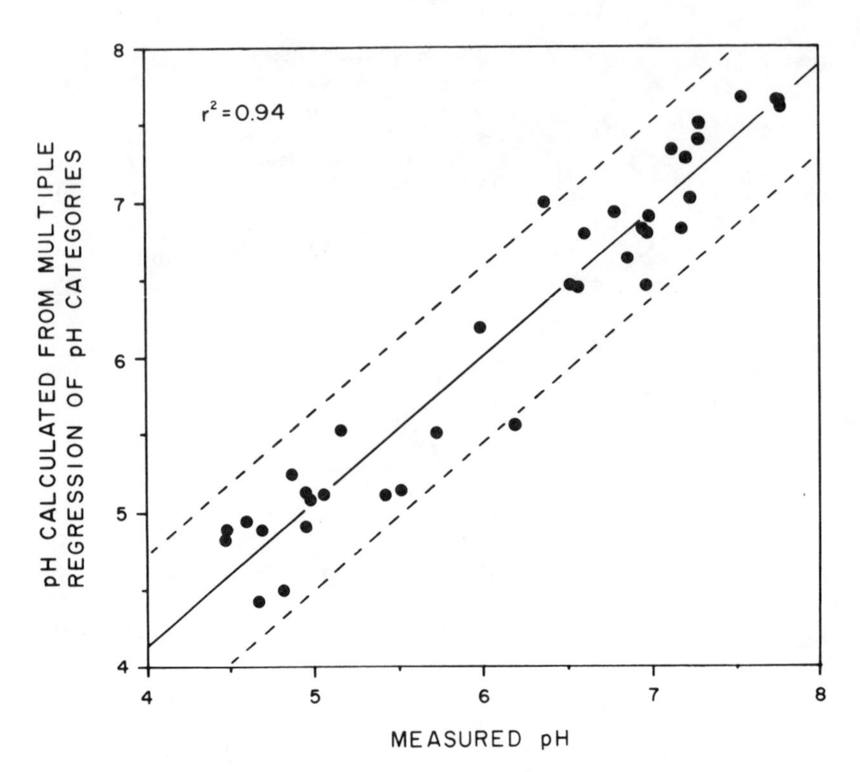

FIGURE 9.2 Predicted lake pH calculated using the equation derived from multiple linear regression of pH categories (assignments based primarily on the distribution of taxa in Adirondack lakes) versus measured surface pH for 37 Adirondack lakes. The dashed lines represent the 95 percent confidence intervals for an individual prediction of pH from diatom data (Charles 1985a).

(1981a,b) have developed a point-of-change technique to determine the point of maximum rate of change in a profile of dominant diatom taxa and whether the change is statistically significant. This technique has been used to evaluate taxa profiles from at least one lake (Delorme et al. 1984). The technique has also been applied to pH profiles (G. W. Oehlert, University of Minnesota, personal communication.) New statistical approaches should be developed that account for different sources of uncertainty (e.g., Oehlert 1984).

Diatom and chrysophyte data can be used to address questions such as: Has a lake become more acidic or alkaline? How great were the changes? When did they occur? What were the causes? The extent to which these questions can be answered depends on (1) the quality of sediment cores, (2) the preservation and diversity of sediment diatoms, (3) the quality of diatom slide preparation and counting methods, (4) the quality of taxonomic identifications, (5) the precision and accuracy of the pH inference techniques and the applicability of the equations to a study region or lakes, (6) the accuracy and precision of dating and other information on sediment characteristics, and (7) the availability of historical watershed and atmospheric deposition data.

Evaluation of Lake Acidification Causes

Diatom and chrysophyte data can be used not only to infer the past pH trend of a lake but in many cases to suggest the causes of the changes.

There are three major potential causes of acidification of relatively undisturbed, acid-sensitive lakes in eastern North America: (1) long-term natural acidification, (2) watershed disturbances, such as logging and fires, and ensuing responses of vegetation and soils, and (3) atmospheric deposition of strong acids from distant sources. Other factors may affect lake pH, but not on a regional scale. These include nearby emission sources, discharge of factory effluent, cultural development (roads and houses, for example), land clearance for agriculture, afforestation (more common in Europe), acid mine drainage, mixing with seawater, liming, paludification, drainage of wetlands, and water-level changes such as those resulting from small man-made dams, beaver activity, or changes in climate. The last factor is probably important only when a significant proportion of a watershed is wetland and net sulfur reduction-oxidation and cation exchange processes are affected. Because these other factors have not, with few exceptions, affected the lakes evaluated in this report they are not considered further.

The three primary potential causes of lake acidification are addressed below. Diatom studies of long-term acidification in both Europe and North America are also briefly reviewed. Following this is a summary of patterns of diatom changes to be expected in response to each major acidification cause.

Long-Term Natural Acidification

Long-term trends in the postglacial development of Temperate Zone European and North American lakes have been studied using sediment diatom, pollen, and chemical analyses. These studies indicate that acidic to weakly alkaline lakes in areas having bedrock that is relatively resistant to weathering have undergone a gradual long-term acidification process. This process has been recognized at least since the studies of Lundquist (1924) and has been observed in many geographic areas. In Europe these areas include Sweden (Digerfeldt 1972, 1975, 1977, Renberg 1976, 1978, Renberg and Hellberg 1982, Salomaa and Alhonen 1983, Tolonen 1972), Finland (Alhonen 1967, Tyrni 1972, Tolonen 1967, 1980, Tolonen et al. 1985), Denmark (Nygaard 1956, Foged 1969), northwestern England (Evans 1970, Haworth 1969, Round 1957, 1961, Pennington 1984), Scotland (Alhonen 1968, Pennington et al. 1972), Wales (Crabtree 1969, Evans and Walker 1979, Walker 1978), Czechoslovakia (Rehakova 1983), and Greenland (Foged 1972). Fewer regions in North America have been investigated, but diatom data indicate that long-term acidification has occurred in Mirror Lake, New Hampshire (Sherman 1976); Cone Pond, New Hampshire (Ford 1984); Bethany Bog, Connecticut (Patrick 1954); Berry Pond, Massachusetts (Rochester 1978); Heart Lake, Upper Wallface Pond, and Lake Arnold in the Adirondack Mountains, New York (Reed 1982; Whitehead et al. in press; Figure 9.3); Crystal Lake, Wisconsin (Conger 1939); Lake Mary, Wisconsin (J. C. Kingston, University of Minnesota at Duluth, personal communication); Vestaberg Bog, Michigan (Colingsworth et al. 1967); Red Rock Lake, Colorado (Norton and Herrmann 1980); and on Ellesmere Island, above the arctic circle (Smol 1983; J. Smol, Queens University at Kingston, personal communication).

We can make some generalizations based on these studies. The magnitude and the rate of long-term acidification vary among lakes and within lakes. Diatom-inferred pH, when it has been calculated, indicates declines from 0.5 pH unit or less to about 2.5 pH units; for example, from pH 7.5 to pH 5.0 for Upper Wallface Pond (Whitehead et al. in press; Figure 9.3). Rates of change are gradual; declines of 1 pH unit take hundreds to thousands of years. In general, lakes with the highest current pH have acidified the least, and data for lakes with pH currently above about 7.5 indicate little

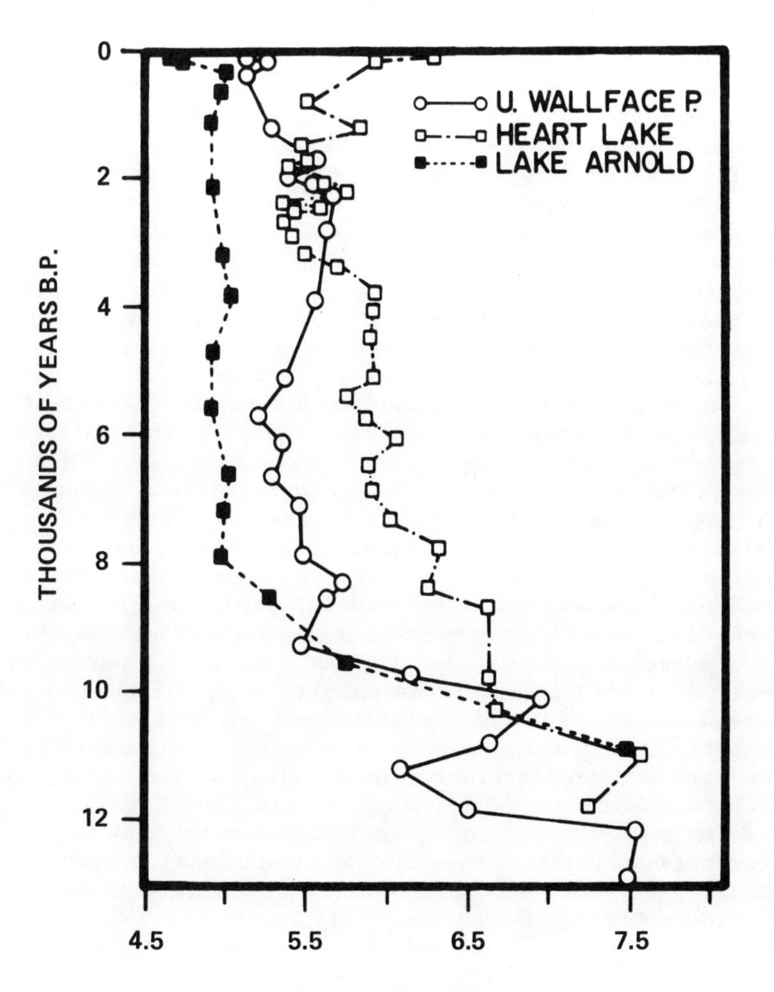

FIGURE 9.3 Weighted average diatom-inferred pHs for Heart Lake, Upper Wallface Pond, and Lake Arnold, Adirondack Mountains High Peaks Region. Averages were determined from index α, index B, and multiple-regression equations. Each pH value was weighted by the standard error of each predictive equation. B. P. signifies before present. Whitehead et al. in press.

or no long-term acidification (e.g., Linsley Pond (Patrick 1943), Pickerel Lake (Haworth 1972), Sunfish Lake (Sreenivasa and Duthie 1973), Kirchner Marsh (Brugam 1980), South King Pond (Ford 1984)).

The most rapid acidification generally occurred during the early postglacial period. Following deglaciation, the lakes were commonly alkaline (pH 7.5 or above), the major reason for the high pH being the presence of unweathered till and outwash deposits with large supplies of easily leachable base cations. Alkalinity declined after easily leached bases were removed and as soil and vegetation developed. The roles of these processes in soil and lake development have been shown by Crocker and Major (1955), Livingstone et al. (1958), Andersen (1966), Berglund and Malmer (1971), and Jacobson and Birks (1980). The extent of the decline in pH and alkalinity depended largely on the characteristics of the soil, bedrock, climate, and vegetation. Acidification was greater in regions where geologic materials are more resistant to weathering, where soils and glacial deposits are thinner, where rainfall is greater, and where vegetation results in greater output of organic acids. The early period of most rapid acidification was usually followed by long periods of relative pH stability or very gradual decline (usually 1.0 pH unit or less). Some fluctuations in pH occurred and are attributed to changes in such factors as climate, watershed vegetation, and lake-level changes. For European lakes especially, other events such as seawater intrusion (connection with saltwater environments) and land use changes caused by early inhabitants also affected lake-water pH.

Watershed Disturbance

Logging, fire, blowdown, and similar disturbances can potentially change lake pH and alkalinity, although there is evidence that these changes are often relatively minor and short lived (Rosenqvist 1978, Gorham et al. 1979, Wright 1981, Nilsson et al. 1982, Martin et al. 1984). A pattern of pH change that might be expected is an initial, rapid increase in pH as the flux of cations released from a watershed increases. This would be followed by gradual decline to below predisturbance pH as cation flux decreases because of landscape stabilization and increased uptake by rapidly regrowing vegetation (Nilsson et al. 1982, Gorham et al. 1979). Alternatively, the loss of a forest canopy could change the type and amount of dry deposition, thereby altering the depositional flux of acidifying compounds. Factors affecting the response of a lake to watershed disturbance include (1) the nature, extent, and severity of the disturbance; (2) past frequency of similar disturbances; (3) the influence of drainage pattern and density, geology, soil type and texture, steepness of watershed slopes, and hydrology; and (4) taxonomic composition and rate of vegetation regrowth. If initial major changes are short lived (a few years at most), they might not be detected by analysis of sediments. This could occur if the appropriate sediment interval representing the disturbance time were not analyzed or if altered diatom and chrysophyte assemblage composition were not discernible because of sediment mixing or some other factor.

Atmospheric Deposition of Strong Acids

Precipitation containing excess strong acids is falling on large areas of eastern North America (Chapter 5) and western Europe. Available evidence suggests that the quantities of strong acids now being deposited are significantly greater than they were before the year 1800 (Chapter 2) and that this precipitation can cause acidification of lakes. Some authors suggest, for example, that the concentration of sulfate in many Adirondack lakes may be five times higher than before 1800 (Galloway et al. 1983, Holdren et al. 1984).

Diatom-based pH reconstructions have been used to assess the effect of acid deposition on lakes in Europe and North America. For Europe, recent declines in diatom-inferred lake-water pH related to acid deposition have been reported for Sweden (Almer et al. 1974, Renberg and Hellberg 1982), Norway (Berge 1976, 1979, 1983, Davis and Berge 1980, Davis et al. 1983, Davis and Anderson 1985), Finland (Tolonen and Jaakkola 1983, Tolonen et al. 1985), Scotland (Flower and Battarbee 1983, 1985), The Netherlands (van Dam et al. 1981, van Dam and Kooyman-van Blokland 1978), and the Federal Republic of Germany (Arzet et al. 1985). Most of these studies, and some on North American lakes, were reviewed by Battarbee (1984). He concludes that

> Despite some ambiguous situations associated with land use changes the weight of evidence

favours acid precipitation as the main cause of recent acidification in the lakes so far studied. Its effect has yet to be disproved at any site and the temporal and spatial patterns of acidification within NW Europe, limited though the data are, are consistent with such an hypothesis. In all cases the onset of acidification postdates about 1800, after the development of coal as a major power source during the industrial revolution, and the later acidification of Swedish and Finnish lakes may be associated with a postwar increase in emissions from oil combustion (Ottar, 1977) as well as a change in the pattern of emissions.

Evaluating Causes of Recent Acidification

Long-Term Acidification Declines in lake pH caused by natural, long-term acidification processes should be continual throughout the period studied. Based on studies of long-term lake histories the overall decline should be small, no more than about 0.1 pH unit in a 20-year period. There should be no major changes in taxa composition or at least no major shifts in the percentage of diatom valves assigned to each pH category.

Watershed Disturbance If a pH decline were caused only by watershed disturbance and processes associated with vegetation recovery, the period of pH changes should be logically related to the time of the disturbance; the larger and more rapid the pH changes, the closer they should have been to the time of disturbance. A pattern that might be expected is a sudden change in pH followed by a gradual return to previous conditions, although there are few data to substantiate this contention. In the case of fires there may be an increase in pH followed by a decrease (e.g., Dickman and Fortescue 1984).

In addition to changing the pH, watershed disturbance may affect the nutrient input to lakes. This effect, in and of itself, can cause changes in the composition of diatom assemblages.

Atmospheric Deposition of Strong Acids Lake acidification can be attributed to acid deposition if (1) the lake is in a region receiving deposition of excess strong acids and there is evidence of industrial pollution in the upper levels of the sediment; (2) inferred pH declines occurring after an increase in acid deposition could reasonably be expected to have taken place, based on knowledge of current water chemistry and watershed characteristics; (3) the pH decline is more rapid than could be accounted for by natural acidification processes; and (4) changes are not correlated with, and cannot be attributed to, watershed changes or other local factors.

When these criteria are used to evaluate causes of acidification, several factors must be emphasized. First, more than one of the above processes may be affecting a lake simultaneously. Second, the history of each lake-watershed system and nearby emission sources should be thoroughly investigated so that all potential causes of acidification can be evaluated. Third, interpretations based on diatom and chrysophyte data should be consistent with interpretations of other sediment data, historical water chemistry and fish population data, and current models of lake acidification processes. Fourth, assessments of regional acidification trends should be based on studies of several lakes within a region.

ASSESSMENT OF RECENT LAKE ACIDIFICATION TRENDS IN EASTERN NORTH AMERICA

To evaluate recent acidification trends in eastern North America, 27 diatom data sets meeting minimum criteria were assembled and reviewed. Locations of these lakes are shown in Figure 9.4. In addition, four lakes in Rocky Mountain National Park, Colorado, were selected as controls. Characteristics of these lakes and their diatom assemblages are summarized in Appendix Tables E.1 to E.5. The steps in analyzing the data were as follows: (1) assemble all relevant data sets, (2) evaluate pH and acidification trends by type of lake and by geographic region, (3) evaluate potential causes of the trends, and (4) draw conclusions concerning trends in acid deposition.

Data Selection

The criteria for selecting lakes and diatom assemblage data were as follows:

FIGURE 9.4 Locations of 27 lakes in eastern North America for which data on diatoms in sediment are available. The number next to each lake corresponds to the order in which it appears in Appendix Table E.1 and in subsequent tables. The depicted regions have been defined in Chapter 1.

1. Lake-water alkalinity less than 200 microequivalents per liter (μeq/L). Acid neutralizing capacity is low enough so that lake pH and diatom assemblage composition could have been altered as a result of significant change in watershed characteristics or atmospheric deposition of acids.

2. Minimal or, if not minimal, well-known watershed disturbance or influence from local emission sources, industrial effluents, or other factors.

3. Sufficient quality and quantity of diatom data, with taxonomic identification to lowest possible level, an adequate number of time intervals analyzed, and access to primary data including diatom counts.

4. Extent to which diatom data have been related to lake-water pH and alkalinity (e.g., profiles of percent of dominant diatom taxa only versus inferred pH calculations).

5. Quality of sediment dating and the availability of data for other sediment characteristics related to acid deposition and lake acidification.

Chrysophyte data were available only for Delano Lake, Jake Lake, and Found Lake in Algonquin Park (Smol 1980) and for Big Moose Lake (Smol 1985a), Deep Lake, Lake Arnold, and Little Echo Pond (J. P. Smol, Queens University at Kingston, personal communication) in the Adirondack Mountains. Major changes in diatom stratigraphy were usually accompanied by changes in chrysophyte stratigraphy and they both indicated the same pH trend.

Many of the data represented in Appendix E (Tables E.4 and E.5) are as yet unpublished. This is because investigators have collected the data only recently, and their analyses are currently incomplete.

Most of the data listed in Appendix E are for lakes in the Adirondack Park, New York, or in New England. In general the number of lakes is too few for making regional assessments, and the lakes are not broadly representative of each region, even of the poorly buffered lakes. All the Canadian lakes reported, except B and CS, have a pH above 6.0; a large percentage of eastern Canadian lakes, however, have pH values less than 6.0. The nine New England lakes are among the most acidic in the region and are among the 20 lowest-pH lakes of the 94 lakes studied by Norton et al. (1981), all of which had a pH less than 7.0.

With a few exceptions, the surface sediment diatom-inferred pH values agree within about 0.1 to 0.4 unit of current lake-water measurements (Appendix E, Table E.5).

Trends in pH

On a regional basis, only diatom and chrysophyte assemblages in the Adirondack lakes indicate significant recent acidification. The three lakes with recent measured pH above 5.5 acidified only slightly (Panther

Lake; Del Prete and Galloway 1983, Davis et al. in preparation) or not at all (Seventh Lake, Del Prete and Schofield 1981; Sagamore Lake, Del Prete and Galloway 1983). The pH of Big Moose Lake (Figure 9.5) and Honnedaga Lake declined about 1 pH from before 1800 to the present. The decrease in the pH of Big Moose Lake is associated with recent increases in acidobiontic taxa, including *Fragilaria acidobiontica* Charles and *Stauroneis gracillima* Hustedt, and decreases in the indifferent, euplanktonic taxon *Cyclotella stelligera* Cleve and Grunow (Figure 9.6). The increase in chrysophyte taxa indicating acidic conditions, such as *Mallomonas hamata* Asmund and *M. hindonii* Nicholls (Smol et al. 1984a,b; Figure 9.7), also indicates a lowering of pH. Diatoms and chrysophytes in Deep Lake, Lake Arnold, and Upper Wallface Pond indicate a recent drop in pH from about 5.0 to around 4.7. Shifts in taxonomic composition (e.g., Charles 1985b) suggest that increased concentrations of aluminum or other metals may have accompanied the pH decline. Trends in pH for three other lakes are more difficult to interpret. Diatoms and chrysophytes in Little Echo Pond, a bog pond, suggest a relatively small pH change. Little change in average pH over time would be expected in a highly colored acidic bog that is buffered by organic acids. However, one taxon, *Asterionella ralfsii* var. *americana* Körner (acidophilic to acidobiontic), dramatically increases in abundance (to >90 percent of the diatom assemblage) in the top few centimeters of the core. This may represent a natural fluctuation or may be an indication of acidification. Interpretation based on changes in the percentage of only one taxon are generally much less reliable than those based on change in the percentages of several taxa.

There is no obvious pH trend for Woodhull Lake. However, there is an unexplained difference of 0.7 to 0.8 pH unit between diatom-inferred pH and available measurements of recent lake-water pH. Del Prete and Galloway's (1983) analysis of a Woods Lake core indicated a gradual, though somewhat irregular, pH decrease continuing to recent times. The investigation by Davis et al. (in preparation) of a different Woods Lake core infers an increase in pH in the period 1875-1910 followed by a pH decrease from 1920 to the present. The surface sediment diatom flora suggest more acidic conditions than do the pre-1800 flora.

Changes in sediment diatom assemblage composition in New England lakes indicate recent acidification (Norton

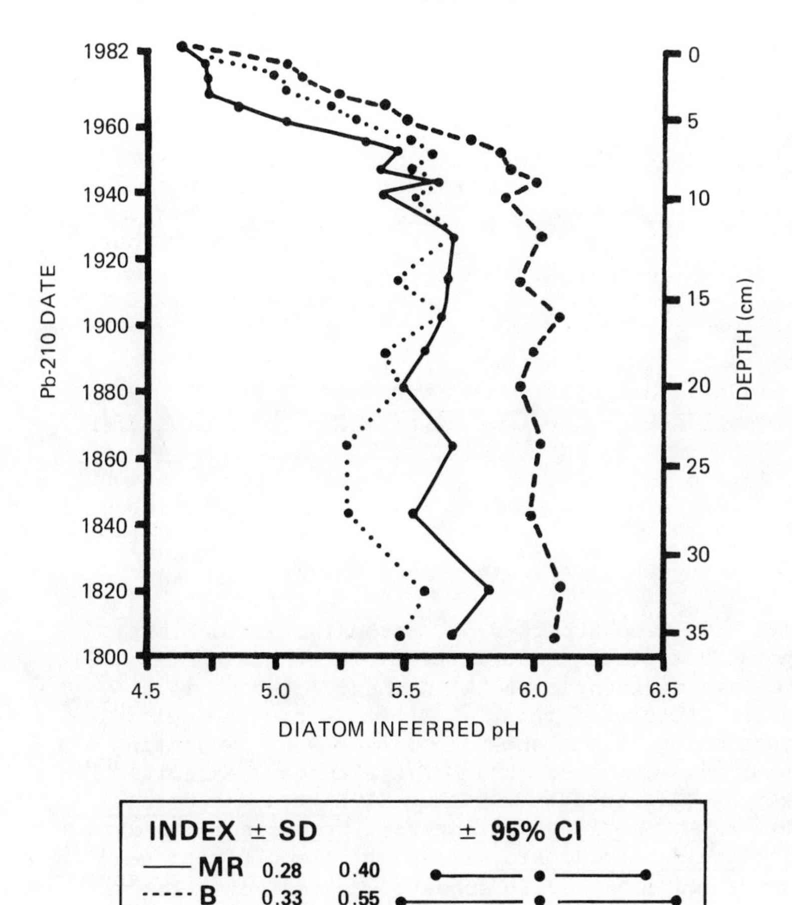

FIGURE 9.5 Profiles of inferred pH from Big Moose Lake (core 2) based on a multiple-regression equation (MR, solid line), index B (dashed line), and index α (dotted line). Standard deviations are 0.28, 0.33, and 0.33, and the 95 percent confidence intervals are 0.40, 0.55, and 0.56 for the MR, index B, and index α equations, respectively. Equations used for pH inference, and strength of relationship between measured and predicted pH: MR equation, pH = 8.14 - 0.041 ACB - 0.034 ACF - 0.0098 IND - 0.0034 ALK, r^2 = 0.94; index B, pH = 7.04 - 0.73 $\log_{10}$ index B, r^2 = 0.91; index α, pH = 6.81 - 0.70 $\log_{10}$ index α, r^2 = 0.91. ACB, acidobiontic; ACF, acidophilic; IND, indifferent, circumneutral; ALK, alkaliphilic. Adapted from Charles (1984).

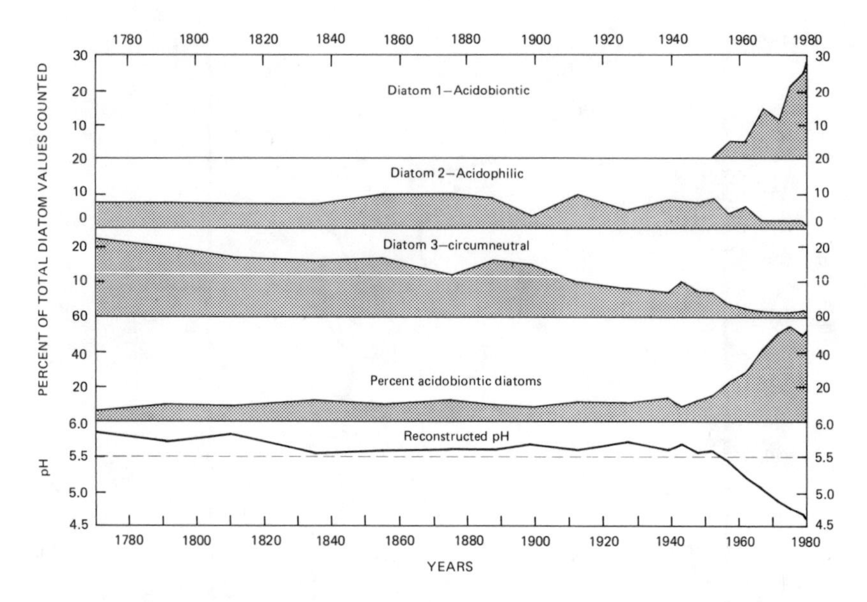

FIGURE 9.6 Reconstructed pH, percent of acidobiontic diatoms, and percent abundance of three diatom taxa, Big Moose Lake sediment core 2. Reconstructed pH is a weighted average of index α, index B, and the multiple-regression equations shown in Figure 9.2. Weighting was based on the standard error of the three predictive equations (Figure 9.5). Diatom 1, *Stauroneis gracillima*, diatom 2, *Anomoeoneis serians* var. *bracysira*, and diatom 3, *Cyclotella stelligera*, represent the dominant taxa in the core and occurred in subsets of 38 study lakes with mean pH values of 4.8, 5.9, and 6.8, respectively (Charles 1984). Percentage of euplanktonic taxa within the core followed the pattern of *C. stelligera* abundance very closely. Years are based on lead-210 dating (Figure 9.5).

et al. 1985); however, most of the pH-trend data indicate no change or a slight decrease (e.g., in Speck Pond; Figure 9.8) in pH from pre-1800 to the present. At Ledge Pond, a total decrease of 0.6 to 0.7 unit, starting slowly during the 1800s but accelerating after 1960, has been inferred (Davis et al. 1983, Norton et al. 1985). There are fluctuations in some of the profiles that appear related to watershed events (Davis et al. 1983).

The Algonquin Park lakes and Batchawana Lake have apparently undergone little change in pH in the last 200

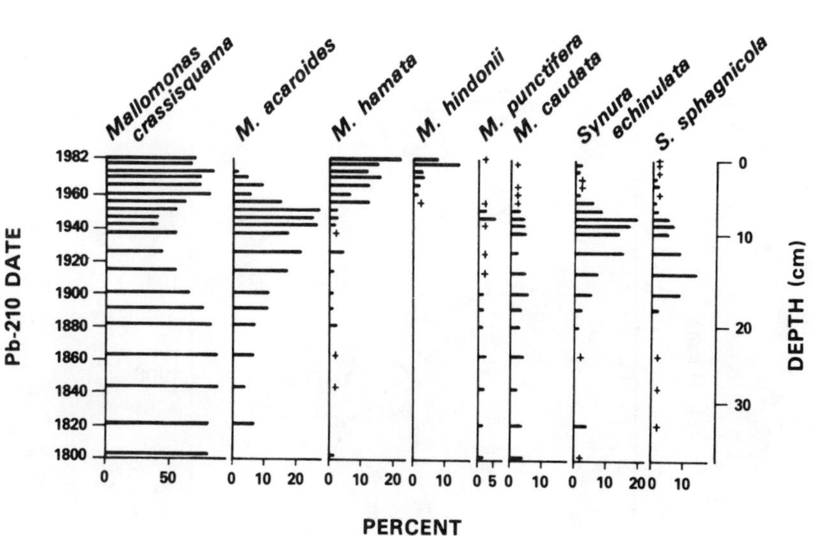

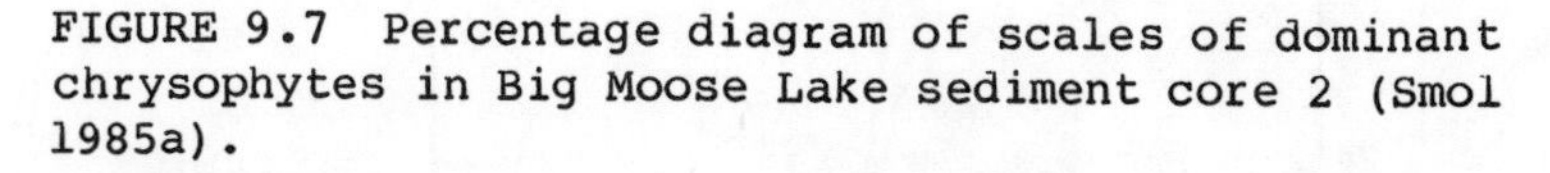

FIGURE 9.7 Percentage diagram of scales of dominant chrysophytes in Big Moose Lake sediment core 2 (Smol 1985a).

years. This is not unexpected given that their current alkalinities are generally greater than about 50 μeq/L. It is difficult to distinguish no change in pH from a slight change by using diatom data when alkalinities are this amount or above.

Lakes B and CS have apparently become more acidic in the past 30 to 50 years. Much of the cause could be local sources of sulfur emission near Wawa, Ontario, that increased output during the 1940s and 1950s (Dickman et al. 1983, Dickman and Fortescue 1984, Fortescue et al. 1981, 1984, Somers and Harvey 1984). Therefore, the pH trends cannot be used to help interpret changes in long-range transport of strong-acid precursors. However, the data are useful and are included here because they provide further evidence that sediment diatom assemblages can indicate changes caused by increased deposition of strong acids.

The Rocky Mountain lakes are considered to be controls. They are all low-alkalinity lakes (three with less than 60 μeq/L) sensitive to increased input of strong acids, but they do not currently receive precipitation with an annual average pH lower than about 5.0 (Gibson et al. 1984). Changes in the diatom stratigraphy are relatively

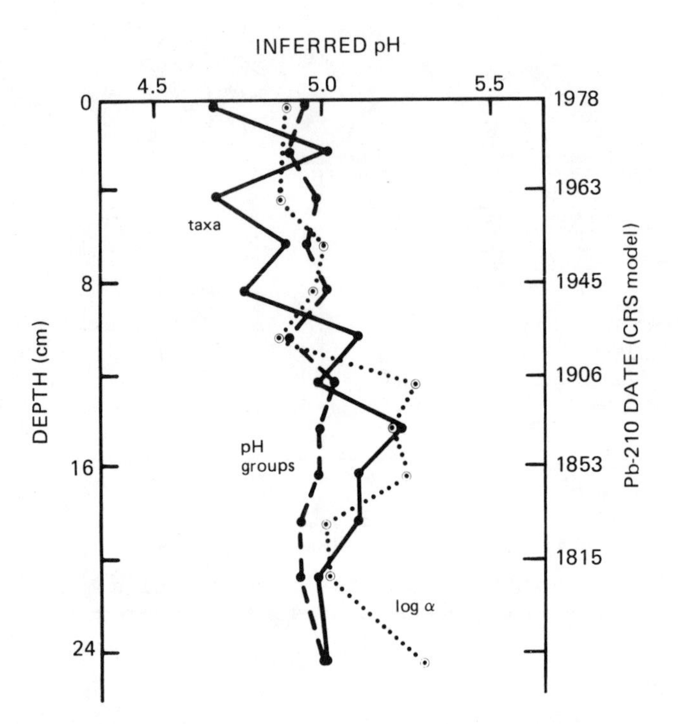

FIGURE 9.8 Comparison of downcore pH for Speck Pond, Maine, calculated from multiple-regression analysis of taxa and pH groups and from $\log_{10}$ index α (from Battarbee 1984). Lead-210 dates have been added to published figure; data from R. B. Davis, University of Maine, Orono, personal communication. CRS refers to Constant Rate of Supply model.

minor, occur gradually, and indicate no sign of acidification. In Husted Lake there is even a slight increase in the percentage of alkaliphils toward the sediment surface.

The 27 eastern North American lakes for which historic pH has been inferred can be divided into two groups, based on their background (pre-1800) pH. Thirteen had a pre-1800 diatom-inferred pH of 5.3 or less, and fourteen had a pH at that time of 5.7 or greater. The two groups represent the dominant buffering systems in the studied regions: bicarbonate (HCO_3^-) in the higher-pH lakes, organic acids and aluminum species in the lower-pH lakes. The differences in diatom assemblages are generally greater between the two groups than among individual lakes within any one group, and these specific differences are similar to those observed in surface sediment assemblages and discussed by Norton et al. (1981), Battarbee (1984), Davis and Anderson (1985), and Charles (1985a). One of the greatest differences in composition is that, with one exception, euplanktonic diatom taxa are present in pre-1800 assemblages from all the lakes that had a diatom-inferred pH above 5.9 and they are present only rarely in assemblages of lakes that had a pH less than 5.4. The exception is _Asterionella ralfsii_ var. _americana_, which may be one of the few euplanktonic taxa in lakes with pH < 5.5 in eastern North America (e.g., Sweets 1983). Recent studies of surface sediment diatom assemblages of lakes in the Killarney region of Ontario show that euplanktonic _Cyclotella_ species are common in lakes with a pH less than 5.5. The role of morphometric factors may be an important reason for the difference in occurrence of euplanktonic diatoms in these lakes compared with others in the northeastern United States (M. Taylor and H. Duthie, University of Waterloo, personal communication).

The important conclusions from the distribution of pre-1800 inferred pH values are that (1) the lakes considered in this study were about equally divided overall (but not by geographic region) on the basis of the pre-1800 buffering system and (2) the number of lakes with a pre-1800 diatom-inferred pH value less than 5.5 suggests that these types of lake were relatively common in the Adirondack Mountains and New England before the Industrial Revolution.

The diatom-inferred pH profiles examined for this study can be classified into four patterns, although several showed fluctuations that were probably caused by natural variation in diatom composition, watershed events, or inconsistencies introduced by pH reconstruction methodologies. The patterns are the following: (1) background pH greater than about 6.0 and no overall trend in pH, (2) relatively rapid decline in pH from the range of 5.7-6.0 to less than 5.0 (some pH declines start in pH ranges above 6.0), (3) pH about 4.8 to 5.2 throughout entire profile, and (4) decline from the range of pH 4.8-5.3 to as low as about 4.5.

The biggest changes in pH and sediment core diatom composition occur when diatom-inferred pH decreases from a pH above approximately 5.5 to a pH below 5.5 (pattern 2). A smaller pH drop was observed for lakes with a pre-1800 pH of less than 5.5 (pattern 4). The same pattern of decline in diatom-inferred pH for currently clearwater lakes from about pH 5.0 to pH 4.5 or above has been observed in Norway (Holmvatn, Davis et al. 1983; Hovvatn, Davis and Berge 1980; Nedre Maalmesvatn, Davis and Anderson 1985), Finland (Kairajarvi and Valkeislampi, Tolonen et al. 1985), and Scotland (Loch Enoch, Battarbee 1984). Inferred pH declines in lakes with a low background pH must be interpreted with caution. Even a slight decline in pH could be associated with a significant decrease in acid-neutralizing capacity and vice versa (Chapter 7). This is because the change in alkalinity per pH unit is greater in the pH range 4.5 to 5.0 than the range 5.5 to 6.0 because of the logarithmic nature of the pH scale.

Changes in Water Chemistry Associated with pH Trends

Three important questions arise with respect to the above discussion:

1. What was the chemistry of lakes that apparently had a pre-1800 pH less than 5.5?
2. In particular, what anions were primarily associated with the low pH?
3. What specific chemical changes could have occurred in these lakes in response to increased deposition of strong acids, and how would those affect hydrogen-ion concentration and alkalinity?

In regions with low-alkalinity waters that receive precipitation that is not acidic or is only slightly so, (e.g., northeastern Minnesota and Wisconsin, western Ontario, and Labrador; Wright 1983), clearwater lakes with pH less than 5.5 are rare. However, in these regions and in regions now receiving excess acid deposition there are lakes that have pH less than 5.5 and are moderately to highly colored (platinum-cobalt units > 25) (Lillie and Mason 1983, Gorham et al. in press). These lakes are typically associated with peatland vegetation and are only slightly influenced by groundwater input. Plants such as _Sphagnum_ can remove cations from surrounding water (Clymo 1963), and decay products of plants yield organic acids, which may or may not be colored (e.g., Vitt and Bayley 1984). These organic acids can account for a significant portion of total anions in lake water and cause pH values to be less than 5.0 (Oliver et al. 1983). It seems probable, therefore, that organic acids were an important cause of acidic conditions in many lakes in acid-sensitive areas before 1800 (Patrick et al. 1981).

Also, if pH was about 5, aluminum concentrations may have been moderate at that time, although much of the aluminum would have been complexed with organic compounds (Driscoll et al. 1984).

As low-pH (5.0 to 5.5) lakes became more acidic, changes in aluminum and organic acid concentrations were probably the most important ones affecting pH and alkalinity. The output of dissolved aluminum from these lakes' watersheds should have increased with increases in inputs of strong acids (Almer et al. 1978, Johnson et al. 1981, Driscoll et al. 1984), except perhaps for seepage lakes in outwash deposits. This is consistent with the fact that most of the lakes examined for this report that have a current pH less than 5.5 have concentrations of total aluminum greater than 100 to 200 μg/L; the value in some lakes is as high as 600 to 700 μg/L (Appendix Table E.3).

The increase in percentages of diatom taxa such as *Stauroneis gracillima* and *Fragilaria acidobiontica* toward the top of some Adirondack lake sediment cores suggests that recent increases in aluminum concentration may have occurred. In the Adirondacks these taxa are abundant only in lakes with pH less than 5.5 and total aluminum concentrations greater than 200 μeq/L (Charles 1985a,b). Also, computer-modeling efforts based on knowledge of current lake-water chemistry and diatom-inferred pH trends suggest that the concentration of aluminum in Big Moose Lake increased as pH decreased (Charles 1984). Increased aluminum loading could lead to reduced acid-neutralizing capacity and reduced organic acids and water color. A relatively minor decline of pH in the range 4.5 to 5.5 can cause a major change in aluminum speciation with a substantial increase in total aluminum charge (Driscoll et al. 1984), resulting in reduced acid-neutralizing capacity. In this way, aluminum acts as a buffer (Johannessen 1980), minimizing pH change as a result of increased acid loading. Increased concentration of anions is balanced more by increased positive aluminum charge than by hydrogen ion. Decreasing pH and increasing aluminum concentration could cause a decrease in organic acids and water color. This is supported by theoretical analysis, laboratory studies, and observations that water color has decreased over time in lakes that have apparently acidified recently (Schofield 1976, Almer et al. 1978, Dickson 1980). Also, studies of limed lakes indicate that as alkalinity increases, color increases and concentration of aluminum decreases (Hultberg and Andersson 1982, Yan 1983, Wright 1984).

Analysis of the data discussed here suggests that when significant lake acidification has occurred in acid sensitive regions of eastern North America, it has followed two major patterns: (1) a decrease in pH of currently low-color lakes from about 5.7-6.4 to 5.0 or less and changes in alkalinity owing initially to loss of bicarbonate alkalinity and later to an increase in concentration of positively charged aluminum species; and (2) a decrease in pH from approximately 4.8-5.2 to 4.5 or above, with almost all the associated reduction in acid-neutralizing capacity caused by increases in aluminum and decreases in organic acid concentrations. In reality, patterns of pH decrease probably represent a continuum, but the above cases are likely typical of most responses.

Evaluation of Trends in Acid Deposition

Trends in diatom-inferred pH can be used to evaluate hypotheses concerning trends in acid deposition for eastern North America. For example, one can test the null hypothesis that atmospheric deposition of strong acids has not increased sufficiently in the past 100 years to cause significant pH and alkalinity changes in lakes with low buffering capacity. If the hypothesis is true, no evidence should exist for acidification of any type of lake in any region that cannot be accounted for by natural acidification, response to watershed changes, or another local factor. In order to reject this null hypothesis:

1. There should be evidence of decreasing pH (acidification) in poorly buffered lakes that cannot be accounted for by natural causes or response to watershed changes.
2. The above decreases should have been relatively rapid and have occurred sometime after the earliest date that independent evidence suggests acid deposition could have increased.

3. Decreases in lake-water pH for lakes with similar pre-1800 diatom-inferred pH should have been proportional to acid loading. For most regions in eastern North America acid deposition should have increased the most in areas that currently receive the largest loading.

The quantity and the quality of diatom and chrysophyte data now available are insufficient to permit definite conclusions to be drawn regarding all aspects of the hypothesis for any geographic area. However, there is strong evidence for certain aspects. First, diatom profiles for some Adirondack and New England lakes indicate recent acidification, with the greatest pH decreases in the Adirondack lakes. Big Moose Lake represents the best case of an acidified lake in which increased acid deposition is the only reasonable cause to which recent acidification can be attributed. Diatom related changes in other lakes, such as Upper Wallface Pond and Deep Lake, are greater than would generally be expected from natural variation and are also probably due primarily to increased acid deposition, although watershed disturbance may play a minor role.

There are additional data for Adirondack lakes indicating recent acidification, but either the cores were not dated or the role of watershed disturbances is unclear. These data obviously cannot be used to reject the null hypothesis, but they are consistent with the alternative hypothesis.

For lakes for which adequate data are available (see Appendix E), decreases in pH were relatively rapid and occurred since about 1900, when acid deposition in the Northeast probably began increasing (Chapter 2). Changes in the Adirondacks appear to have been greater than in New England. The Adirondacks receive the greater amount of acid deposition.

Thus, testing of the above hypothesis using available diatom data for the Adirondacks and New England indicates that deposition of strong acids has increased sufficiently in the past 100 years to cause a decline in lake-water pH. Data are insufficient to permit evaluation of trends in other regions, such as eastern Canada, now receiving acid precipitation.

The maximum calculated pH decline for all lakes evaluated was about 1 pH unit (pH range 5 to 6) since 1900, which is similar to observations for acidified European lakes (Battarbee 1984), and the onset of rapid acidification began after 1940. However, studies of the sediments of more lakes will be necessary before the magnitude and timing of recent acidification in regions can be estimated with any certainty. This research is now under way.

Ongoing Studies

Several lake acidification studies are now under way that include analysis of diatom and chrysophyte assemblages as a means of reconstructing lake-water pH. The largest is the Paleoecological Investigation of Recent Lake Acidification (PIRLA) project funded by the Electric Power Research Institute. Five to fifteen lakes in each of four geographic regions (Adirondack Mountains (New York), New England, northern Minnesota-Wisconsin-Michigan, and northern Florida) have been cored and the sediments are being analyzed for diatoms, chrysophytes, and other characteristics. Patterns of pH change in lakes in these regions will be related to patterns of acid deposition to determine extent and important causes of lake acidification. A similar type of study of Rocky Mountain lakes has been completed (Baron et al. 1984, D. R. Beeson, U.S. Park Service, Fort Collins, Colorado, personal communication).

Ronald Davis and Dennis Anderson of the University of Maine and other investigators are completing analysis of diatom assemblages from sediment cores of three Adirondack lakes (Panther, Sagamore, and Woods) and nine New England lakes (the first nine New England lakes listed in Appendix Table E.1). Studies of surface sediments and cores are also being conducted on lakes of the Pocono Plateau in Pennsylvania (John Sherman, Academy of Natural Sciences of Philadelphia, personal communication). The relationships between surface sediment diatom assemblages and limnological characteristics in eastern Canada are being studied at the University of Waterloo (at least 40 lakes; Hamish Duthie and Mark Taylor, personal communication) and at the Freshwater Institute in Winnipeg, Manitoba (about 25 lakes; Geoff Davidson, personal communication). Equations for predicting water chemistry characteristics will be developed and used in future pH reconstruction studies.

Cores from lakes in the Matamek region of Quebec are being investigated (H. Duthie, University of Waterloo, personal communication), and investigations of pH histories of lakes in the Wawa region of Ontario are continuing (Michael Dickman, Brock University, personal communication). The Long Range Transport of Atmospheric Pollutants program, coordinated by F. C. Elder at the Canada Centre for Inland Waters, Burlington, Ontario, is continuing in Nova Scotia and in the Algoma calibrated watershed. A study of diatom-pH-alkalinity relationships and sediment stratigraphy of Sierra Nevada lakes began in the fall of 1985 (R. Holmes, University of California, Santa Barbara, personal communication). Chrysophyte assemblages in surface sediments of many lakes in eastern Canada are being studied also (J. P. Smol, Queens University at Kingston, personal communication). This will increase the ecological data base for chrysophytes so they can be better used in stratigraphic studies to interpret past lake changes.

UNCERTAINTIES IN THE INTERPRETATION OF SEDIMENT DIATOM AND CHRYSOPHYTE ASSEMBLAGE DATA AND RECOMMENDATIONS FOR FURTHER RESEARCH

Diatom and Chrysophyte Taxonomy

One of the most important sources of error in diatom and chrysophyte analyses is incorrect identification of taxa. The need to resolve taxonomic problems and for better communication and more consistency in taxonomic designations among investigators is great. Comparing data among investigators is often difficult because different synonyms are used and there is inconsistency in how taxonomic levels are split or lumped. One of the major potential uses of diatom and chrysophyte analyses is to compare data among many types of lake/watershed systems in many parts of the world. More consistency in taxonomy and data presentation would aid greatly in these comparisons.

Recommendation: Further taxonomic studies should be made of difficult groups and unidentified taxa, and more floras of diatoms from acidic and circumneutral lakes should be published. In addition, communication among investigators working with acid-water floras should be improved with the goal of better agreement on taxonomic categories and data presentation.

Diatom and Chrysophyte Physiology

There are some major questions concerning distribution of diatoms and chrysophytes in acidic lakes. Answers to these questions would aid in better understanding the relationships between changes in assemblages and acidification of lakes.

Recommendation: Culture and bioassay studies (e.g., Gensemer and Kilham 1984) should be performed to address at least the following questions:

1. Why do euplanktonic diatoms appear, in many regions, to be generally limited to lakes with a pH > 5.8?
2. What are the physiological mechanisms underlying the strong relationship between diatom and chrysophyte distributions and factors related to lake-water pH?
3. What are the roles of aluminum, silicon, and organic materials in determining the occurrence of taxa?
4. What interactions occur between factors related to pH and those, such as phosphorus and nitrogen concentrations, related to trophic state?

Diatom and Chrysophyte Ecology

Interpretation of diatom data is limited by a lack of complete knowledge of diatom autecology, including information on habitats, seasonality, and relationships to water chemistry characteristics. More information on how diatoms and chrysophytes interact among themselves and with other organisms would help to explain seasonal and long-term successional patterns. Such information would also contribute to our understanding of how important factors other than those related to pH are in causing variation in assemblage composition.

Recommendations: 1. Conduct more field and laboratory studies on the spatial and temporal distributions of diatoms and chrysophytes in different habitats within and among a variety of geographical regions. In particular these should be designed to establish (a) the distribution of individual taxa along specific environmental gradients and (b) the relationship between euplanktonic and periphytic diatoms and how their relative abundance is affected by changing conditions within a lake ecosystem, such as changes in alkalinity, trophic state, color, dissolved organic carbon, transparency, aluminum, trace metals, water level, substrate, and other biota.

2. Study diatom and chrysophyte changes in lakes that are experimentally acidified or limed.

Sediment Processes

Transport of diatoms and chrysophytes to lakes (e.g., Battarbee and Flower 1984) and within lakes, such as through differential transport of taxa (Sweets 1983) and sediment focusing, and changing patterns of processes within sediments, such as dissolution and mixing, affect the degree to which sediment assemblages represent those living in a lake. The extent to which these factors influence historical reconstructions is difficult to determine.

Recommendations: 1. Perform more field work to determine how diatoms and chrysophytes are transported to lake sediments and which taxa or types of diatoms tend to be overestimated or underestimated within the sediments. Statistical analyses may be needed to evaluate the sensitivity of reconstruction techniques to this factor.

2. Perform more field and laboratory work to learn the mechanisms of diatom and chrysophyte dissolution and to evaluate its influence on reconstructions. Although dissolution usually has a negligible effect on assemblages (mostly only those in peaty or calcareous sediments) it can be important, especially when it influences the percentage of dominant taxa.

3. Investigate further the influence of sediment mixing on reconstructions. This can be and has been done by studying mixing processes, such as physical transport and bioturbation, within lakes and in the laboratory; use of mathematical models incorporating dating and other data; comparison of stratigraphy of varved versus non-varved sediments; comparison of long-term plankton and sediment-trap assemblages with sediment core assemblages; and the addition of large quantities of foreign marker particles to a lake followed by subsequent study to learn how they are mixed within sediments.

Inference Techniques

The precision and accuracy of current inference techniques, while generally good for most lakes, could be improved to give better predictions for a wider variety of types of assemblages. Also, quantitative inference techniques are currently limited to determination of pH. Changes in other water chemistry parameters, such as alkalinity, color, aluminum concentration, and habitat characteristics, currently cannot be quantitatively reconstructed, although some qualitative interpretations can be made.

Recommendations: 1. The size and quality of data sets of diatoms, chrysophytes, and lake characteristics should be improved. More lakes in more regions should be studied, and as many lake characteristics as are relevant to diatom and chrysophyte ecology should be measured frequently and over long periods of time. This work could be coordinated with other limnological studies. All data collected should be as compatible as possible with other calibrational data sets.

2. Inference techniques should be improved by trying new approaches. The most promising is to develop transfer functions that are multiple-regression equations using percents of selected taxa. The taxa chosen should correlate better statistically than those now used with lake characteristics to be inferred. To be widely applicable these equations need to include many taxa (more than 50), which is not possible today because of the limited size of calibration data sets.

3. Inference techniques should be developed for lake characteristics other than pH, such as alkalinity, color, aluminum content, and total organic carbon (e.g., Davis et al. 1985).

Statistical Analysis

It is often difficult to determine if diatom-inferred pH changes within a core are statistically significant because the numbers have not been subjected to statistical analysis.

Recommendation: Full use should be made of existing statistical procedures, and new techniques should be developed to evaluate and express the following:

1. The precision error (standard error, standard deviation, 95 percent confidence interval) associated with individual inferred pH values. To date, error analysis has been based solely on the regression analysis used to calculate predictive equations from the calibration data set. They do not account for characteristics of assemblages in sediment cores being analyzed. Error

estimate procedures should be developed that account for (a) variation in the calibration data set, including the pH measurements, and (b) characteristics of core assemblages, including diversity, the percentage of valves and diatom taxa that are used in the pH inference calculations, and the probability that taxa have been placed in the correct pH category.

2. The existence and significance of trends and points of change within individual pH profiles. This type of analysis can be accomplished by using a variety of existing techniques, including regression and change-point analysis (e.g., Esterby and El-Shaarawi 1981a,b).

3. Variability among the pH values inferred by using a variety of indices. It is often best to present several inferred pH curves derived by using a variety of techniques, but this can sometimes be confusing. The alternative is to use only one or two that are considered to be the best, but this involves subjective judgments and the temptation to use the curve that best fits an investigator's interpretation. Objective criteria for selecting and presenting data are necessary.

4. The similarity of patterns expressed by more than one diatom-inferred pH profile from either the same or several cores. The data can be analyzed statistically by using regression analysis or Monte Carlo simulation.

Acidification Processes and Selection of Study Sites

In some cases, diatom and chrysophyte data sets may not yield much information about lake acidification or atmospheric deposition because (1) the study lakes are well buffered and have changed little in alkalinity or diatom composition even though atmospheric loading of acids may have increased substantially; (2) there was a major watershed disturbance, such as fire, logging, and cultural development, that obscures the relationship between acid deposition and the watershed changes as causes of acidification; and (3) the lake sediments are very mixed (physical mixing or bioturbation) or there is some problem with the core or dating of the sediment.

Recommendations: 1. Choose study lakes that have a current alkalinity between about -10 and 50 µeq/L. These lakes would probably have had low enough buffering that their alkalinity could have changed enough under the influence of acidic precipitation to cause a change in diatom assemblages. Changes in pH in lakes with lesser or greater alkalinities would be less likely to be observable because of greater buffering.

2. If possible, choose lake/watershed systems that have not been disturbed significantly within the time period of interest. If this is not possible, choose systems with the least amount of disturbance and for which a good historical record of the changes exists.

3. Investigate several sites within a region. Choose sensitive systems with different watershed, morphologic, and hydrologic characteristics.

4. Investigate many sediment characteristics within cores. The additional data aid greatly in evaluating the nature of lake changes and assessing causes and mechanisms of acidity change.

5. Select study systems and evaluate results in the context of current lake-acidification models. Interpretations of diatom and chrysophyte data with respect to changes in acid deposition (or other factors influencing lake-water acidity) should account for limnological processes (e.g., sediment buffering, sulfate reduction in the hypolimnion and sediments, and change in organic acids) and watershed biogeochemistry changes (e.g., aluminum leaching, sulfate adsorption on soils, and cation leaching). Diatom inference data can be used to evaluate acidification hypotheses and models (e.g., the amount by which pH or alkalinity should have changed over time given certain model assumptions). After minimal lake selection criteria are met, base the final selection of study lakes on the hypothesis to be tested. Ideally, paleoecological, limnological, and watershed studies should be integrated.

ACKNOWLEDGMENTS

Donald Charles thanks Dennis Anderson, Klaus Arzet, David Beeson, Ronald Davis, L. Denis Delorme, Anthony Del Prete, Hamish Duthie, Jesse Ford, John Kingston, John Smol, and Kimmo Tolonen for providing access to their unpublished data. They also provided interpretation of their results and many useful comments on the manuscript. Jill Baron, Charles Driscoll, Jesse Ford and Stephen Norton contributed water chemistry and other information on many of lakes mentioned in this section.

Some of the unpublished Adirondack region data were obtained as part of research efforts supported by the Electric Power Research Institute, Palo Alto, California.

THE CHEMICAL STRATIGRAPHY OF LAKE SEDIMENTS

The unpolluted atmosphere consists largely of the gases nitrogen (78 percent), oxygen (20 percent), argon (1 percent), water, and carbon dioxide with variable but trace amounts of other gases, such as ammonia, carbon monoxide, mercury, methane, and nitrous oxide. In addition, natural processes inject particulate organic and inorganic material into the atmosphere, matter that ultimately sediments out. Scientists have long known (e.g., Smith 1872) that human activities have resulted in alteration of the chemistry of the atmosphere with respect to both gaseous and particulate material. Nonurban regional pollution of the atmosphere was clearly identified as occurring as early as the 1900s in Scandinavia (e.g., Oden 1968), in the 1950s in Great Britain (Gorham 1958) and the United States (e.g., Junge and Werby 1958), the 1960s in Europe (Oden 1968), and in the 1970s in Canada (Beamish and Harvey 1972). Gaseous pollutants (sulfur dioxide, sulfur trioxide, nitrous oxide, and nitrogen dioxide) were the earliest recognized pollutants. Significant pollution by airborne metals of regional atmospheres in the United States was suggested by the pioneering work of Lazrus et al. (1970), who measured lead, zinc, copper, iron, manganese, and nickel in precipitation over a period of six months. Whereas the collection and analytical techniques of this study may be challenged in terms of implied concentration and deposition rate of pollutants, the geographic trends are clear. The northeastern United States was receiving significantly more trace metals through precipitation than was the Southeast, Midwest, or West.

Lakes receive a flux of gaseous, liquid, and particulate material directly from the atmosphere as both dry and wet deposition. Additional terrestrial fluxes of these materials derived from the atmosphere, altered in quantity and quality by terrestrial processes, may also reach the lake. Physical, biological, and chemical processes in the lake sequester a fraction of the flux of these pollutants in lake sediments.

In the absence of long-term, regional studies of atmospheric fluxes and deposition of pollutants the chemistry of lake sediments may serve as one of the few surrogate indicators of atmospheric pollution. However, the faithfulness of reproduction of the atmospheric deposition signal by sediments is highly variable, depending on the constituent being deposited and the ecosystem characteristics. Important variables include water chemistry, focusing of sediment, resuspension of sediment, the flux of material from the drainage basin, and organic productivity within the lake. After initial deposition, sediment may be resuspended and resedimented (not necessarily in the same place) or partially mixed vertically with older (or younger) sediment by inorganic or organic processes; it may be chemically altered, too, by changing overlying water conditions (e.g., acidification and eutrophication), deoxygenation within the sediments, sulfate reduction within the sediment (producing hydrogen sulfide and precipitation of otherwise mobile metals), and so on. Thus, the chemistry of a particular vertical increment of sediment is determined by the following:

1. The chemistry of the material deposited from the atmosphere as well as the chemistry of the material eroded from the drainage basin, some of which may be resuspended from other parts of the lake and deposited at the site. Redeposited sediment places older material with possibly different characteristics in a younger interval.
2. Net sedimentation of material derived from within the lake, including organisms (diatom valves, invertebrate remains) and resuspended sediment.
3. Additions to, losses from, or relocation of inorganic and organic matter within the sediment after deposition. These changes are generally greatest at or near the sediment/water interface for oligotrophic lakes and decrease abruptly to insignificance at a depth of between 5 and 10 cm. (See Carignan and Tessier 1985, Holdren et al. 1984.) Thus, some of the sediment within an interval will not have the same age as the interval. The presence of laminated sediments in some lakes indicates that at least in situ mechanical redistribution of deep water sediment after one year is essentially nonexistent for a few lakes.

A lake-drainage basin in a long-term steady state typically should have sediment chemistry (except for water content) that is relatively constant below a depth of no more than about 10 cm. However, changes in focusing of sediment (Dearing 1983) may result in long-term changes in sediment chemistry (Hakanson 1977). Also, lakes with greatly increased deposition rates caused by accelerated erosion and deposition with no accompanying change in sediment chemistry have been identified. Thus, invariant chemistry is not a foolproof indication of steady state (Hanson et al. 1982). Changes in the above factors affecting a single chemical component result in reciprocal percentage composition changes for all other constituents in the sediment. The concentration of one constituent is generally not an independent variable. However, deposition rates for the other constituents would be unchanged by an independent change in one component, unless there were a chemical link between the two.

Mechanical disturbance of the Earth's surface by activities such as deforestation, tillage, and construction results in accelerated erosion of detritus from drainage basins and increased lacustrine sedimentation rates. The chemistry of the deposited material may (Davis and Norton 1978) or may not (Hansen et al. 1982) change. Coincident with or later in time than the disturbances is the increased injection of particles into the atmosphere, followed by increased atmospheric deposition. Although the composition of the particulate matter may differ from place to place and time to time, the material may be indistinguishable from normal sediment.

If one can establish chronology within the sediment stratigraphy, one can evaluate the deposition rate (absolute net flux) of material at a site at the lake bottom. The distinction between deposition rate (i.e., grams per square centimeter per year) and sedimentation rate (i.e., centimeters of sediment per year) needs to be emphasized. The latter is controlled largely by water content and organic content, which are, in turn, a function of a number of variables, especially lake pH. Lakes with lower pH (oligotrophic to mesotrophic) tend to have profundal sediments with higher organic content (Norton et al. 1981) and may have high sedimentation rates but low deposition rates.

Modern sediment deposition rates of metals from atmospheric deposition can be evaluated approximately if one knows (1) the average background deposition rates for a particular metal; (2) the increases in the gross (bulk) deposition rate, caused by increased terrestrial erosion and other factors of all elements combined, and (3) the modern deposition rate for the particular constituent of interest. Only five to eight years ago did researchers begin to assess deposition rates accurately. This was made possible because of advances in the accurate dating of lake sediments.

Any particular interval of nonlaminated sediment, when collected and analyzed, is a mixture of older, contemporaneous, and younger material. Laminated sediment may contain older and contemporaneous, but not younger, particulate sediment. The age of the interval is generally thought of as the age of deposition, not necessarily as the age of the material. (Chemical diagenetic processes deny even this assumption.) The absolute age of sediment intervals can be established using varves (e.g., Tolonen and Jaakkola 1983), radionuclides associated with atmospheric nuclear weapons testing (especially cesium-137) (Pennington et al. 1973), and the naturally occurring radionuclide lead-210 (e.g., Appleby and Oldfield 1978). Dating with cesium-137 is imprecise in lakes suitable for studying the effects of acid rain (Longmore et al. 1981, Davis et al. 1984). It is useful in dating sediments in lakes with higher pH, higher sedimentation rates, and more abundant clay minerals, which fix the cesium-137 in the sediment. Lead-210 is the method of choice, but the method is not without uncertainty. Lead-210 chronology is, however, commonly in agreement with other dating techniques of known validity, such as varves.

Errors in collection (especially in coring techniques, such as gravity coring (Baxter et al. 1981) and small-diameter push coring (Hongve and Erlandsen 1979)), subsampling of the sediment, dating of the sediment, chemical analysis of the sediment, and drainage-basin and lake-basin processing of atmospheric pollutants make it difficult to reconstruct precisely historic atmospheric deposition rates from net deposition rates. However, in lakes with little or no disturbances other than changing atmospheric chemistry, trends can be established. In addition, if mixing of sediments is minimal the chronology of trends can be established. Furthermore, proportional changes in the atmospheric deposition rate of certain immobile constituents, such as lead and vanadium, can be estimated.

Smelting, burning fossil fuels, and other human activities inject a variety of gaseous, liquid, and particulate materials into the atmosphere that differ in composition and physical characteristics from soil dust. These materials include trace metals, such as lead, zinc, vanadium, selenium, and arsenic (Taylor et al. 1982, Galloway et al. 1982), fly ash (Griffin and Goldberg 1981), soot (Renberg and Wik 1984), magnetic spherules (Oldfield et al. 1981), organic combustion products such as polycyclic aromatic hydrocarbons (or more simply,

PAHs) (Heit et al. 1981), and excess sulfate, nitrate, and metals with different isotopic ratios from those that occur in soils (Nriagu and Harvey 1978). If the rates of atmospheric deposition of these substances are significantly different from natural (background) deposition rates, trends in deposition in sediments may be observed.

The types of data gathered for sediment chemistry are variable but include the following:

1. Concentration of a particular constituent per unit mass of wet (least common), dry, or ignited (most common) sediment. The last-named involves total digestion of the sediment (see Buckley and Cranston 1971).
2. Concentration of a particular constituent in a particular fraction of the total sediment, such as the strong-acid-soluble fraction and the ammonium acetate-exchangeable fraction. These derived values are defined operationally, and methodological problems (see Campbell and Tessier 1984) rule out rigorously interpreting data within a sediment core and comparing sediment cores from different lakes (see Reuther et al. 1981).

The most common way to represent chemical data is as concentration per unit volume or mass of sediment. This parameter, however, is not an independent variable; for any particular component it is inversely related to the concentrations of all other constituents. Even trends in concentration of some anthropogenically linked constituent, such as lead derived from the atmosphere, could be diluted or enhanced in the sedimentary record by increases or decreases in the bulk sedimentation rate. In the absence of accurate age dating of sediment, concentration trends are the optimal representation of chemical data from cores.

The data are commonly shown as departures from average background values. Alternatively, various types of enrichment factors (see Galloway and Likens 1979) may be used. A constituent of interest is normalized to background concentrations, or a constituent is normalized to an element whose concentrates are relatively unaffected by anthropogenic activity. This approach is valid if the normalizing parameter is constant.

If sediments can be dated we can calculate an absolute net flux for a constituent at a given locality at the lake bottom. For a lake receiving a constant supply of detritus it is possible to determine deposition rates for atmospherically derived substances as a bulk rate that either includes autochthonous and allochthonous sources or isolates only the atmospheric component. It is clear from many studies in which multiple cores from the same lake have been studied that only fortuitously would a deposition rate at the lake bottom precisely mimic the atmospheric deposition rate.

Dillon and Evans (1982) and Evans et al. (1983) have developed the concept of a whole-lake burden for constituents that are atmospherically derived, not delivered to the lake from the terrestrial part of the drainage basin, and not exported from the lake. Using numerous widely scattered cores, they demonstrated that the areal burden for lead did not vary appreciably from lake to lake in a small region. The logical extension of this approach is to find the representative sediment record that is characteristic of the total anthropogenic burden for the region and use it to reconstruct directly the history of atmospheric deposition. This can be done only for chemically immobile elements, such as lead, and in sedimentary environments with little mixing.

Secondary effects of acid deposition may affect the sedimentary record. Effects include adsorption of chemicals by sediment, such as sulfate adsorption analogous to sulfate adsorption in soils (Johnson and Cole 1977), desorption or increased mineral weathering caused by lowered pH (Schindler et al. 1980, Hongve 1978, Oliver and Kelso 1983, Kahl and Norton 1983, Norton 1983), reduced organic decay (Andersson et al. 1978), and altered diagenesis, for example, increased sulfate reduction followed by sulfide precipitation within either the sediment or the water column (Schindler et al. 1980). Initial deposition of constituents clearly can be modified by postsedimentation processes. As previously discussed, these effects probably become ineffective below a sediment depth of 5 cm or less.

THE SEDIMENT RECORD

As a result of human activity numerous elements are injected into the atmosphere. If the return flux of these elements approaches that of natural processes, detection of the perturbation becomes possible. For abundant elements, such as iron, manganese, calcium, and aluminum, natural sediment fluxes in lakes are probably in excess of atmospheric fluxes. However, processes related to various human activities cause atmospheric fluxes of certain acid precursors and trace metals that are in excess of natural fluxes.

The relationship between local point sources of pollution and enrichment of these various substances in sediments of adjacent lakes is well established. The most classic example of a point source is the smelter at Sudbury, Ontario, that emits several pollutants associated with the processing of nickel and copper ore. Nriagu (1983), Nriagu and Wong (1983), and Nriagu et al. (1982) found that in dated cores from lakes proximal to the smelter the variations in concentrations (and also deposition rates) for arsenic, selenium, nickel, copper, zinc, and lead paralleled the history of ore production at Sudbury. Figure 9.9 demonstrates the sharp rise in the concentrations of arsenic in the recent sediments in a number of these lakes. The source of pollutants could be identified unambiguously as the Sudbury smelter. Similarly, Crecelius (1975) demonstrated a correspondance between arsenic in sediment and the chronology of operation of a copper smelter that emitted arsenic-rich dust 35 km upwind from the lake.

Other studies of lake sediment chemistry have indicated that lakes may receive excess metal inputs from multiple pathways. For example, Tolonen and Meriläinen (1983) suggest that both atmospheric inputs and direct inputs from municipal sewage have contributed to increases in the accumulation rates of chromium, copper, and zinc in some Finnish lakes. Skei and Paus (1979) cite evidence suggesting that a saltwater fjord (Ranafjord) receives excess lead, zinc, and copper from atmospheric inputs and direct surface inputs, both related to nearby mining activities.

Early studies performed before the importance of long-range transport of atmospheric pollutants was fully recognized presented interpretations of sediment records based on atmospheric and direct inputs from local sources. For example, Edgington and Robbins (1976), using lead-210-dated cores from southern Lake Michigan, attempted to model the sediment record (for lead) in light of local atmospheric emissions as well as inputs from rivers. They concluded that fluxes of lead from anthropogenic sources overwhelmed fluxes from natural sources by nearly an order of magnitude. Downstream, in Lake Erie, Nriagu et al. (1979) concluded that the major portion of the lead, zinc, copper, and cadmium in the modern sediments was derived from direct surface inputs to the lake, not from the atmosphere. These reports and many others like

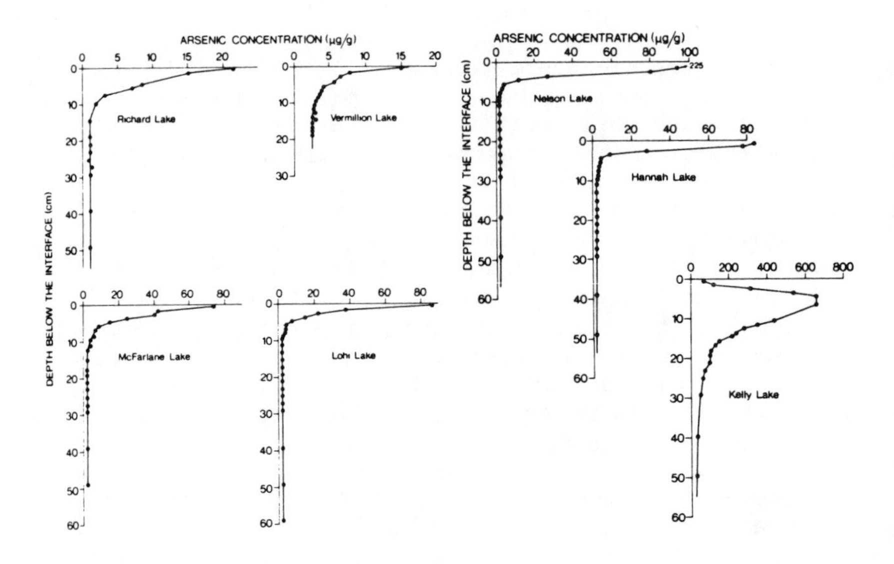

FIGURE 9.9 Arsenic concentrations in sediments of lakes near Sudbury, Ontario (from Nriagu 1983).

them were not initiated to study atmospheric fluxes of metals. However, there is a striking uniformity in the finding of increased concentrations of metals in modern sediments reached by a variety of researchers using a variety of techniques on a wide variety of lakes. These conclusions are consistent with data from a restricted class of lakes that over the past 10 years has been studied specifically to address the question of atmospheric deposition of acids and metals and associated biological effects.

These lakes are oligotrophic, low-alkalinity lakes with a history of relatively stable land use and with watersheds that have been continuously forested and undeveloped. Many of the lakes are at a relatively high elevation in the northeastern part of the United States. Researchers study this type of lake assuming that changes in sediment chemistry could be assigned unambiguously to changes in atmospheric chemistry.

One of the earliest studies of heavy metals in sediment cores from remote lakes was made by Iskandar and Keeney (1974) in Wisconsin. The metal concentrations in sediment were based on the technique of nitric acid extraction of the sediment, not total digestion. Concentrations of extractable lead and zinc increased in

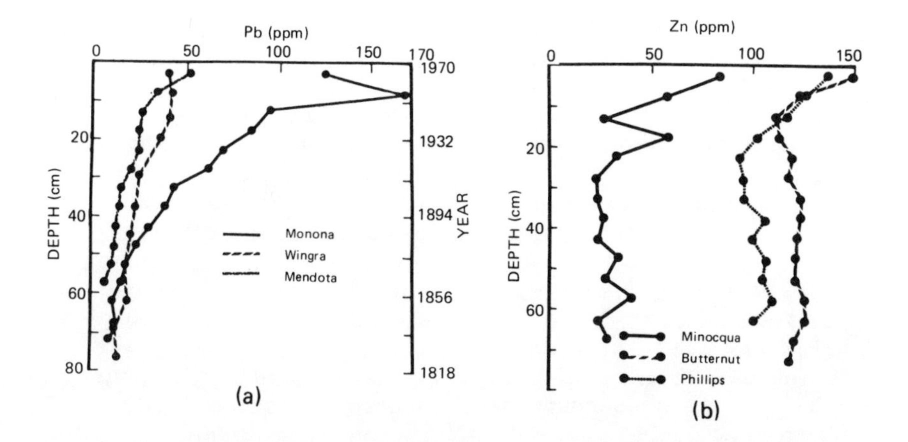

FIGURE 9.10 Vertical distributions of (a) lead and (b) zinc in selected Wisconsin lake sediments (from Iskandar and Keeney 1974).

younger sediments in all lakes (Figure 9.10) regardless of watershed activities, including some pristine lakes in northern Wisconsin. Iskandar and Keeney postulated that lead was delivered through atmospheric processes but did not attempt to explain the elevated zinc concentrations. Norton et al. (1978), using sediment chemistry of pollen-dated cores from New England lakes ranging from culturally eutrophic to pristine and oligotrophic, suggested that the ubiquitous rise in the concentration of zinc in lake sediments in the 1800s could best be explained as having been caused by atmospheric deposition.

Galloway and Likens (1979) reported the detailed sediment chemistry of a single core from Woodhull Lake in the Adirondack Mountains of New York. The coring technique, the analytical methods used, and poor dating prevented quantification of sediment deposition rates. Nonetheless, increasing concentrations of many elements associated with fossil fuel burning, such as silver, gold, cadmium, chromium, copper, lead, antimony, vanadium, and zinc, led them to conclude that the increases were caused by increased atmospheric deposition of these metals. Heit et al. (1981) reported on two lake sediment cores from the same region. Although both cesium-137 and lead-210 dating were done, information from the latter method was not utilized. The cesium-137 chronology suggested that PAHs and fossil-fuel-related elements started increasing about 1948, in agreement with the estimate of Galloway and Likens, also based on cesium-137. This dating methodology is now believed to be inaccurate (Davis et al. 1984), giving ages of sediment intervals that are too young and consequently deposition rates that are too high. This was the first serious attempt at evaluating atmospheric deposition rates of trace metals from analysis of lake sediments.

At about the same time, in Quebec, Ouellet and Jones (1983), also utilizing cesium-137 for dating, came to conclusions similar to those reached by Galloway and Likens as well as by Heit et al. with respect to the timing of the increases of heavy metal loading (Figure 9.11). This consensus "demonstrated" a regional and apparently synchronous air pollution phenomenon.

Perhaps the best way to date sediments is to use varves or varves in conjunction with other dating techniques. Johnston et al. (1982), using varves, pollen, cesium-137, and lead-210, concluded that for three lakes in Maine accelerated sedimentation of lead and zinc started well before 1950, perhaps as early as the late 1880s. Tolonen and Jaakkola (1983) have nearly perfected these varve-dating techniques, studying varved sediments in Finland and calculating deposition rates for metals with great precision. I. Renberg (Umea University, Sweden, personal communication) has been able to evaluate sediment strata year by year using this approach. Unfortunately, lakes containing varved sediment are rare in the United States, and few have been well studied.

Norton et al. (1982) and Kahl et al. (1984), studying a variety of lakes in northern New England as well as in the Adirondack Mountains (Norton 1984), have utilized lead-210 chronology for sediment dating and have concluded that the increase in heavy metal loading, as indicated by increasing concentrations of lead, zinc, and copper, across the New York-New England region, started in the period 1850 to 1900; Evans et al. (1983) and Evans and Dillon (1982) reached the same conclusion for Ontario. Starting about 1930 to 1940, the concentration of vanadium increased concurrently with the rise of consumption of oil, much of which is vanadium rich. Deposition rates for elements that are particularly enriched (e.g., lead), increase as well (Figures 9.12, 9.13). The concurrence of the increase in deposition rates of lead, zinc, and vanadium in a variety of lakes in the northeastern United States and eastern Canada

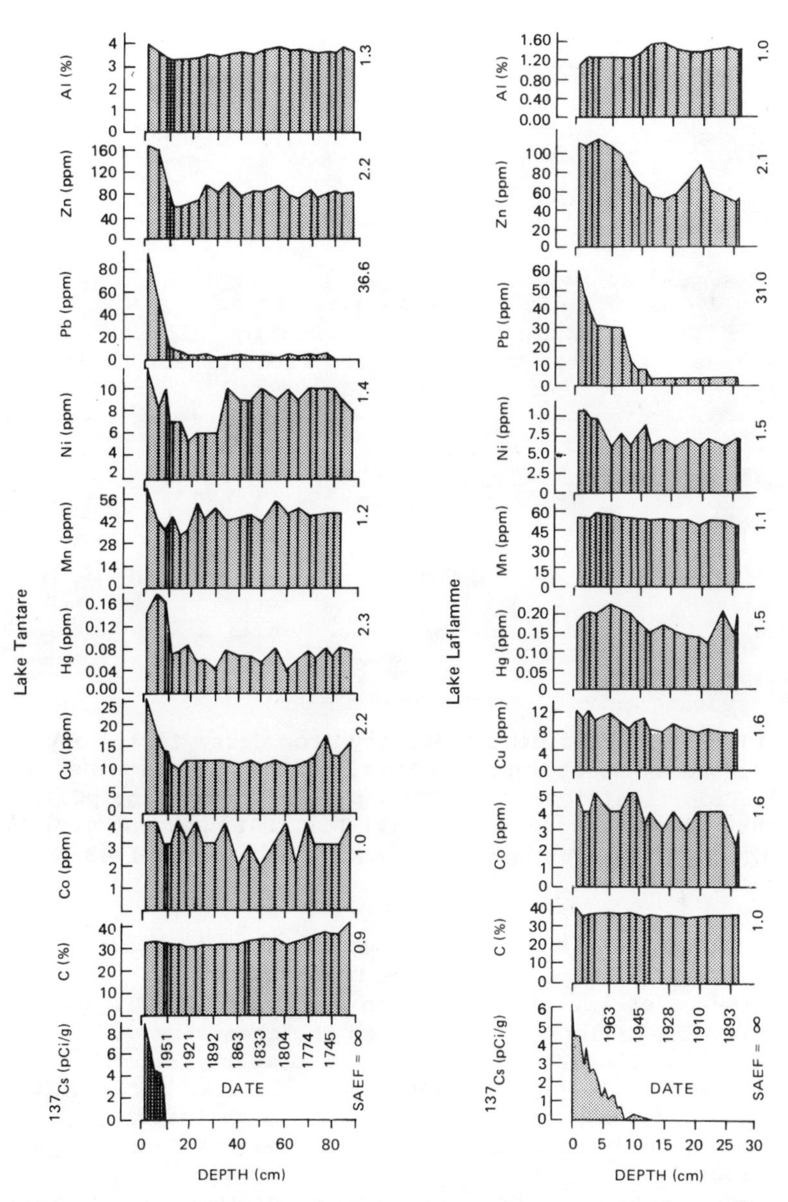

FIGURE 9.11 Sediment chemistry from lakes Tantare and Laflamme, Quebec, Canada (from Ouellet and Jones 1983). SAEF, sedimentary anthropogenic enrichment factor; pCi/g, picocuries per gram; ppm, parts per million. The value of SAEF is obtained by dividing the measured concentration by the background concentration. Thus, because the background value of picocuries per gram of cesium-137 is negligibly small, the SAEF value was assigned a value of infinity (∞).

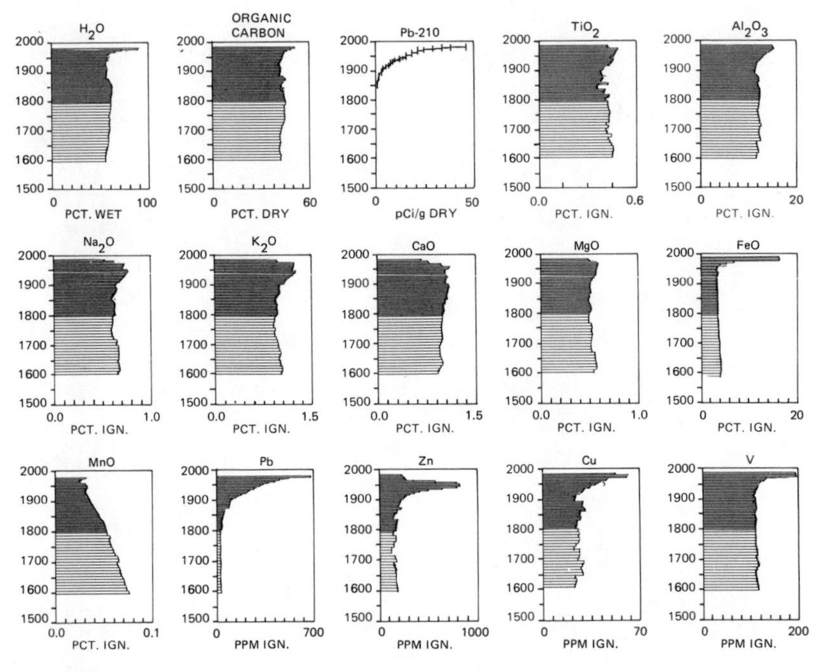

FIGURE 9.12 Sediment chemistry from Jerseyfield Lake, Adirondack Mountains, New York. PPM IGN., parts per million, ignition; PCT. IGN., percent, ignition; pCi/g DRY, picocuries per gram, dry; PCT. DRY, percent dry; PCT. WET, percent wet. (From Kahl and Norton 1983.)

suggests that atmospheric deposition, among other factors, is the source of the metals.

Evans et al. (1983), Dillon and Evans (1982), and Evans and Dillon (1982) studied the chemistry of sediments from 10 lakes. They used multiple cores, dated them with lead-210, and assessed total lake burdens for zinc, cadmium, and lead. (Coring and analytical techniques differed slightly from those of Norton et al. (1982).) There was no relationship between watershed size and metal burden. Although concentrations of trace metals in sediment varied appreciably with water depth, where the cores were taken (Figure 9.14) in the various lakes the whole-lake burdens of these metals varied only

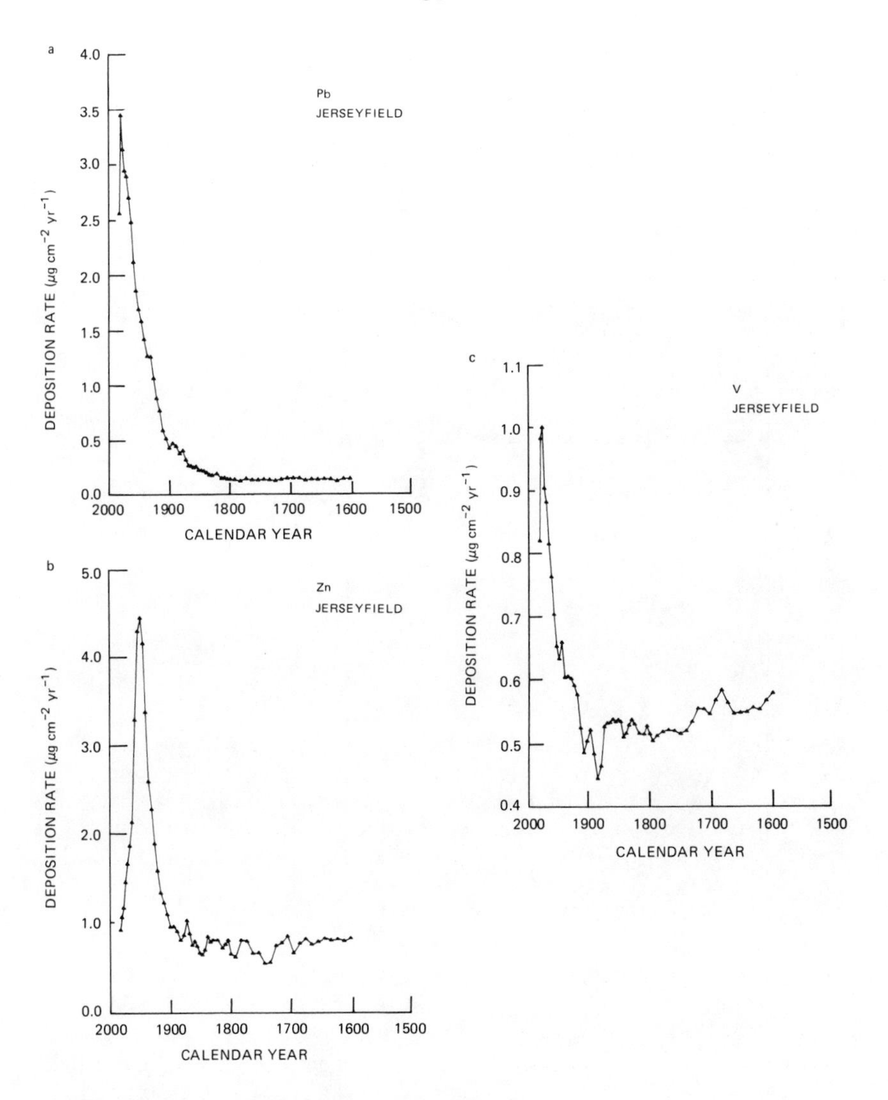

FIGURE 9.13 Calculated deposition rates for lead, zinc, and vanadium over time from Jerseyfield Lake, Adirondack Mountains, New York (from Norton in press).

over a range of about 100 percent, suggesting significant atmospheric input and retention of the atmospherically derived metals. Of these metals lead is retained the most efficiently, zinc the least efficiently. Evans and Dillon (1982) reconstructed the approximate atmospheric loading for lead, based on the multiple-coring approach,

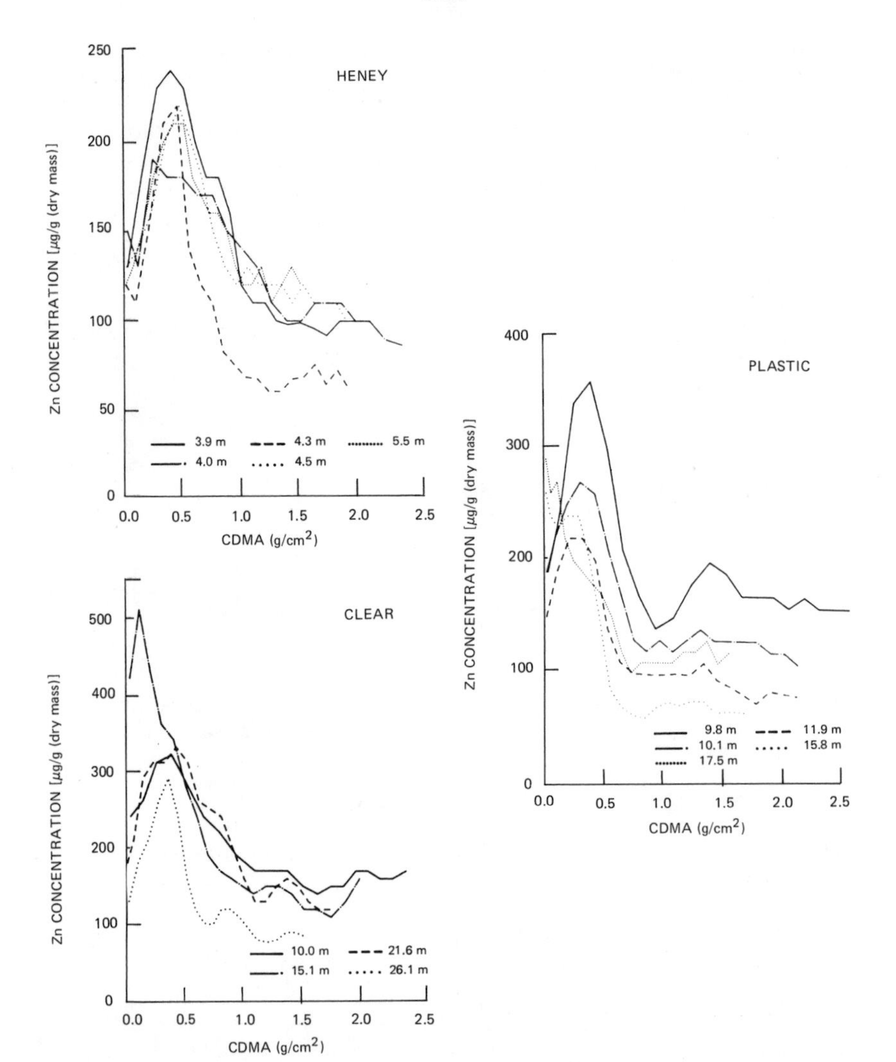

FIGURE 9.14 Zinc concentrations versus CDMA (cumulative dry mass (of sediment) per unit area) (from Evans et al. 1983). CDMA, an alternative way to express depth profile, is employed to take into account sediment compaction. Thus g/cm^2 corresponds to sediment depth; e.g., the lower the value of CDMA, the more recent the sediment.

using whole-lake lead burdens through time. The chronology of the increasing concentrations and deposition rates is similar to that reported for remote lakes in New England and New York. For example, Tables 9.1 to 9.3 list data for copper, lead, and zinc, respectively, showing enrichment factors as well as increased accumulation rates for 10 lakes in the Adirondack Mountains of New York. In spite of widely differing limnological and watershed characteristics, the sediment records give comparable conclusions.

This selection of representative data from the remote lakes of eastern North America demonstrates the influence of an improving analytical technology on the evolution in our understanding of the record of changing atmospheric deposition of heavy metals as revealed in lake sediments.

Neither the concentration nor the deposition rates of the heavy metals can be directly related to the acidity of precipitation, however. Elevated concentrations of certain trace metals in sediment only indicate the existence of polluted air masses. The acidity of precipitation in the eastern part of North America is due predominantly to the presence of sulfuric acid in precipitation (National Atmospheric Deposition Program 1984) with geographically and seasonally variable contributions of nitric acid. The proportional contribution of nitric acid to the acidity of precipitation has probably increased in recent decades (see Chapter 2), while regional sulfur emissions in the northeastern United States have probably not changed substantially in the past 50 years (see Chapter 2). Several studies (e.g., Mitchell et al. 1981, 1985) have attempted to relate the distribution of sulfur in sediments to atmospheric deposition of sulfate. Sulfur may reach the sediment through a variety of pathways, including sedimentation of allochthonous and autochthonous organic matter, adsorption of sulfate on sedimenting detritus, sulfate reduction with precipitation and sedimentation of sulfides from the water column (Mayer et al. 1982, Schindler et al. 1980), and sulfate reduction within the sediment and precipitation of sulfides at depth within the sediment (Norton et al. 1978, Carignan and Tessier 1985, Holdren et al. 1984). Thus the sulfur content of a particular interval of sediment, and even the speciation of sulfur in that interval, may bear no direct relationship to atmospheric depositional values. Sulfur profiles in sediment vary widely from lake to lake (Figure 9.15), although the general history of sulfur deposition is

TABLE 9.1 Copper Concentration in Sediment Cores from 10 Lakes in the Adirondack Mountains, New York

Lake	Cu Concentration (ppm)			Ratio of Maximum to	Cu Deposition Rate ($\mu g\ cm^{-2}\ yr^{-1}$)			Ratio of Maximum to
	Background[a]	Maximum	Year of Max.	Background	Background[a]	Maximum	Year of Max.	Background
Brooktrout	21	73	1979	3.5	0.10	0.26	1976	2.6
Deep	42	94	1979	2.2	0.13	0.29	1979	2.2
Fourth	22	57	1976	2.6	0.20	0.54	1976	2.7
Jerseyfield	24	61	1974	2.5	0.12	0.37	1970	3.1
Merriam	20	52	1980	2.6	0.07	0.42	1980	6.0
Panther	21	41	1980	2.0	0.11	0.48	1981	4.4
Sagamore	20	46	1983	2.3	0.23	0.69	1983	3.0
(BS) Silver	25	48	1962	1.9	0.09	0.41	1981	4.6
T	38	86	1982	2.3	0.17	0.48	1982	2.8
Upper Wallface	15	56	1979	3.7	0.08	0.26	1982	3.3

[a] Pre-1850 average.
SOURCE: S. A. Norton, University of Maine, unpublished data, submitted in report supported by Environmental Protection Agency.

TABLE 9.2 Lead Concentration in Sediment Cores from 10 Lakes in the Adirondack Mountains, New York

Lake	Pb Concentration (ppm)			Ratio of Maximum to Background	Pb Deposition Rate ($\mu g\ cm^{-2}\ yr^{-1}$)			Ratio of Maximum to Background
	Background[a]	Maximum	Year of Max.		Background[a]	Maximum	Year of Max.	
Brooktrout	27	558	1982	21	0.14	2.06	1968	15
Deep	25	640	1973	26	0.06	2.01	1973	34
Fourth	15	270	1976	18	0.10	2.55	1976	26
Jerseyfield	26	675	1979	26	0.12	3.42	1979	29
Merriam	20	459	1980	23	0.07	3.70	1980	53
Panther	40	498	1967	12	0.25	5.5	1981	22
Sagamore	65	335	1979	5	0.8	4.82	1983	6
(BS) Silver	26	307	1978	12	0.12	2.50	1980	21
T	37	299	1982	8	0.16	2.54	1974	16
Upper Wallface	24	759	1979	32	0.09	3.36	1979	37

[a] Pre-1850 average.

SOURCE: S. A. Norton, University of Maine, unpublished data, submitted in report supported by Environmental Protection Agency.

TABLE 9.3 Zinc Concentration in Sediment Cores from 10 Lakes in the Adirondack Mountains, New York

Lake	Zn Concentration (ppm)			Ratio of Maximum to Background	Zn Deposition Rate ($\mu g\ cm^{-2}\ yr^{-1}$)			Ratio of Maximum to Background
	Background[a]	Maximum	Year of Max.		Background[a]	Maximum	Year of Max.	
Brooktrout	145	783	1963	5	0.7	3.20	1958	5
Deep	120	791	1967	7	0.35	2.49	1967	7
Fourth	95	629	1976	7	0.8	5.93	1976	7
Jerseyfield	160	816	1950	5	0.8	4.47	1950	6
Merriam	100	696	1948	7	0.35	3.07	1960	9
Panther	150	668	1962	4	0.9	6.84	1983	8
Sagamore	250	746	1958	3	2.8	10.6	1983	4
(BS) Silver	240	978	1958	4	0.8	4.75	1962	6
T	105	1035	1960	10	0.5	9.65	1960	19
Upper Wallface	130	804	1951	6	0.5	3.70	1973	7

[a] Pre-1850 average.

SOURCE: S. A. Norton, University of Maine, unpublished data, submitted in report supported by Environmental Protection Agency.

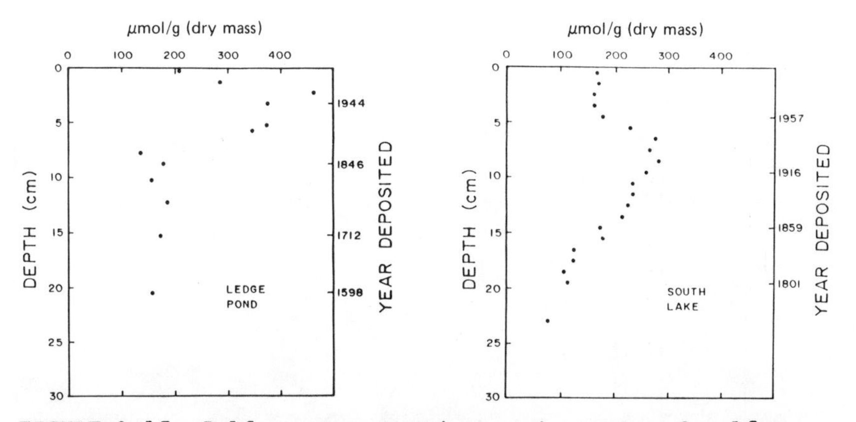

FIGURE 9.15 Sulfur concentrations (micromols of sulfur per gram of dry mass) in sediments of Ledge Pond, Maine, and South Lake, New York (from Mitchell et al. 1985).

similar. Concentrations of total sulfur in sediment do not appear to reflect the atmospheric deposition history as inferred from sulfur emission estimates (see Chapter 2).

Studies of the isotopic composition of sulfur in cores of sediment have been attempted to ascertain the history of atmospheric deposition of anthropogenic sulfur (Nriagu and Harvey 1978). This has been done largely to corroborate the chronology of deposition of other constituents, such as lead, and also to shed light on terrestrial and aquatic processing of sulfur. The studies are in their infancy and do not yet permit a clear understanding of processing. Shiramata et al. (1980) have studied lead-isotope ratios to discriminate between anthropogenic atmospheric lead and lead derived from the drainage basin. These studies are also not yet sufficiently developed to be useful in studying trends in deposition.

A number of solid products are formed and injected into the atmosphere as a result of the combustion of fossil fuels. Griffin and Goldberg (1981) demonstrated a variable trend in the concentration of charcoal particles in the sediment of southern Lake Michigan. An increase in concentration began about 1900, peaked around 1960, and has declined slightly since then, possibly due to increased retention of fly ash at the sources of combustion (Figure 9.16). They interpret this trend as a reflection of emission rates for particulate matter in

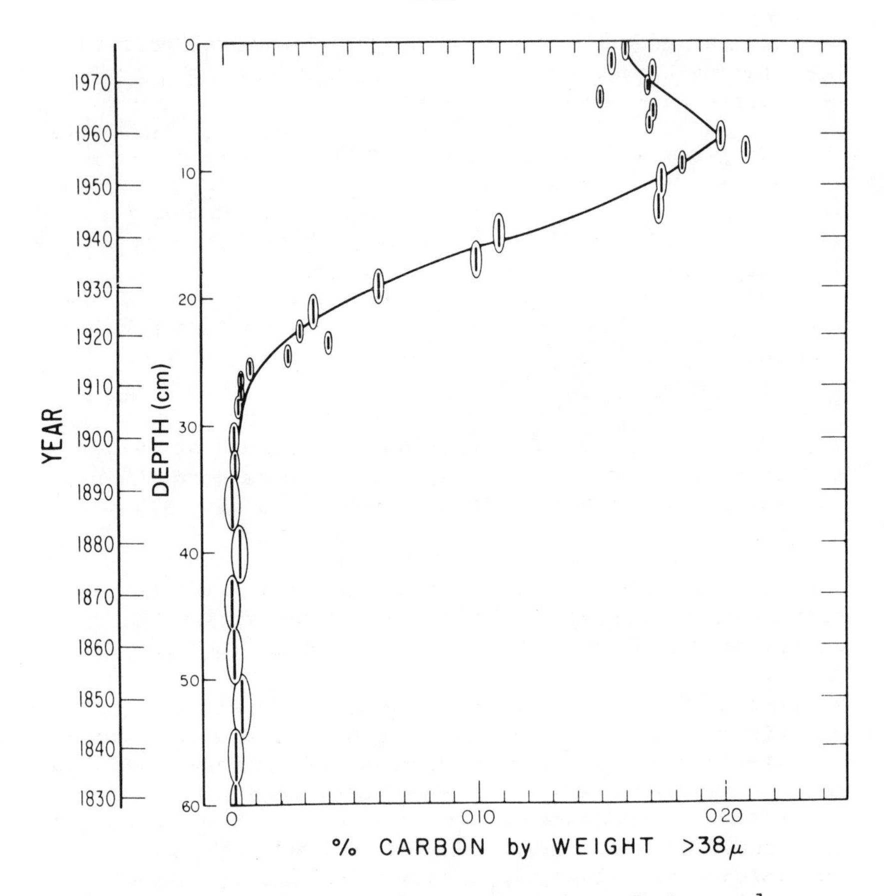

FIGURE 9.16 The record of concentration of charcoal particles in a sediment core from southern Lake Michigan. Particle sizes are greater than 38 microns (from Griffin and Goldberg 1981).

that region. They also note that particles from coal combustion have been dominant since 1900. However, particles derived from combustion of oil first appear around 1950 and have increased since then. Renberg and Wik (1984) have found similar trends and used their results to develop a chronology, using soot particles, in varved lake sediments in Sweden. The soot chronology is so ubiquitous in Swedish lakes that it can be used as an independent chronological tool. Concurrent with the increase in fossil-fuel-related charcoal is the concentration of magnetic spherules produced by high-temperature combustion.

PAHs in sediments may be produced by natural processes or enhanced by bioaccumulation of trace amounts produced by other natural processes, such as diagenesis of organic matter and natural fires (Hase and Hites 1976). However, it is clear from analysis of sediments predating the industrial revolution that concentrations of natural PAHs in sediment are very low, and certain PAHs are known to be produced only by fossil fuel consumption. Numerous investigators have detected high levels of anthropogenically produced PAHs in surface sediments and cores from polluted lakes, but the sources of the PAHs have not been determined unequivocally (e.g., Lakes Lucerne, Zurich, and Greifensee in Switzerland; Wakeham et al. (1980)).

Heit et al. (1981) and Tan and Heit (1981) reported biogenic and anthropogenic PAHs in sediment cores from two relatively pristine lakes in the Adirondack Mountains of New York. Although the chronologies of these cores are not precisely known, the authors demonstrated dramatic downcore changes in PAH concentrations. Correlations with atmospherically derived regional indicators of atmospheric pollution, such as lead and zinc concentrations, are positive, so the trends in the PAH concentrations are likely to represent trends in regional emissions.

Several other studies report comparable results from a variety of lakes in the Northeast. R. Hites (Indiana University, personal communication) is now evaluating PAH data from well-dated cores in New York and New England for which ancillary data, on heavy metals, for example, are available (Figure 9.17). There seems to be little doubt about the general time relationships between PAH abundance and other indicators of polluted precipitation, particularly the onset of pollution. M. Heit (Department of Energy, Environmental Measurements Laboratory, New York, personal communication) has suggested that the decrease in PAH deposition rate over the past 20 to 30 years (see Figure 9.17) may be related to a reduction in low-temperature burning of coal, specifically from home heating.

Although profiles of metals and other pollutants do not reflect directly the acidity of precipitation, sediment chemistry is influenced by lake-water pH. This influence has been studied experimentally and empirically.

Acidification of laboratory microcosms (e.g., Hongve 1978, Norton 1983, Kahl and Norton 1983, Norton and Wright 1984) and experimental whole-lake acidification

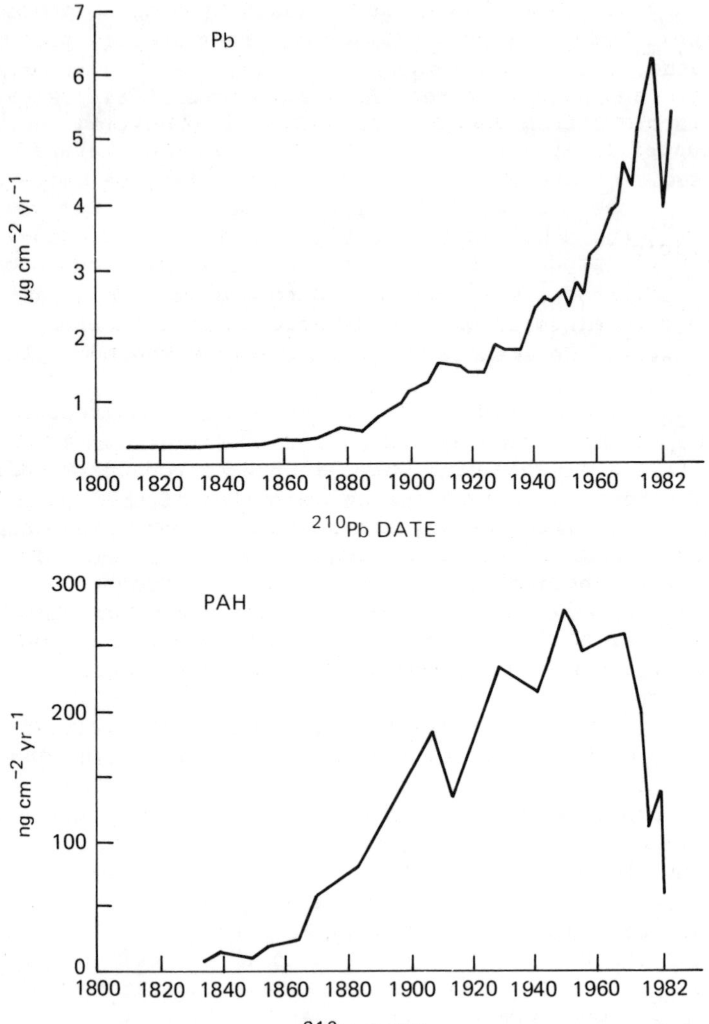

FIGURE 9.17 PAH and lead deposition rates for a core from Big Moose Lake, New York (lead data from Norton, unpublished; PAH data, courtesy of R. Hites, Indiana University, Bloomington, unpublished). Unit of mass given for PAH deposition is nanogram (ng) equal to one-billionth of a gram.

(Schindler 1980) suggest that several elements, at least zinc, manganese, and calcium, are lost preferentially from sediment as pH is lowered. Kahl and Norton (1983) were able to deplete substantially the tops of sediment cores of zinc, calcium, and manganese during experimental acidification of microcosms. Other elements, such as aluminum and iron, were released in substantial amounts, but the reservoir of these elements in the sediment is sufficiently large so as not to be substantially altered by changes in the pH of the overlying water.

Dickson (1980), Norton et al. (1980), and Evans et al. (1983) arrived independently at the conclusion that zinc concentrations were lower in recent sediments in lakes with low pH (<5.5). Concentrations of calcium and manganese also seem to be lower in recently deposited sediment in acidified lake sediment cores (see Figure 9.12). These declines start in sediment as old as 100 years. It is not clear whether the observed reductions occurred as a result of postdepositional processes causing accelerated rates of release of metals from the sediment, a synchronous process causing leaching of metals from sedimenting particles, or predepositional conditions causing reduced levels of metals to deposit on sedimenting particles. Carignan and Nriagu (1985) have observed active loss of manganese from sediments, in situ, where the overlying water had been acidified by acidic precipitation. If the loss of metal is in situ it could affect sediment deposited as much as 10 to 20 years older. If the loss of metal is predepositional and not caused by post-depositional diagenesis of sediment, trends in zinc, calcium, and manganese sediment deposition rates, if they are independent, may be useful in tracking the process of lake acidification.

Peaks in concentration and apparent deposition rates of zinc below the surface sediment in acidic lakes may be related partially to sulfate reduction in the sediment and precipitation of zinc sulfide below the sediment/water interface, i.e., in older sediment. This effect has been observed by Carignan and Tessier (1985) in Clear Lake, near Sudbury, Ontario, but the general occurrence of this phenomenon has not been demonstrated; it cannot explain calcium and manganese depletion in sediment nor reduction of the concentration of zinc to values lower than background values. This has been observed in several lakes with pH below 5.

In summary, multiple lines of chemical evidence derived from lake sediment indicate a history of changing atmospheric chemistry over a time frame of about 100 years. Sediment constituents related to coal burning and smelting, which include soot spheres, magnetic spherules, PAHs, charcoal, metals, and excess sulfur, increase in concentration dramatically through the first half of the twentieth century. Pollutants associated with combustion of fuel oil (glassy spherules and excess vanadium) and gasoline (lead) increase over the past 40 to 50 years.

The synchronous rise of these constituents in sediments in both remote and urban lake sites in the northeastern United States and eastern Canada implicates atmospheric pollution as the likely source. Evidence for the actual acidity of precipitation cannot be inferred from the chemistry of sediments but decreases in the pH of aquatic systems are suggested by recent declines (relative to immobile elements) in the deposition rates of elements mobilized by increasing lake-water acidity. The elements include zinc, manganese, and calcium.

THE CHEMICAL STRATIGRAPHY OF PEAT BOG DEPOSITION

Ombrotrophic bogs, by definition, receive all their nutrients from the atmosphere. *Sphagnum*, a dominant genus in these bogs, has a very high capacity for cation exchange and adsorbs and incorporates considerable quantities of metals, releasing hydrogen ions (acidity) in the process. Consequently, many researchers have utilized the chemistry of surface peats or peat land plants to establish a regional pattern of atmospheric deposition of pollutants. For example, Pakarinen and Tolonen (1976) and Steinnes (1977) demonstrated regional trends in heavy metal concentrations in lichens and mosses for Finland and Norway, respectively. They concluded that spatial concentration gradients were related to emission gradients. Data from Groet (1976) indicate high concentrations of metals in mosses in the northeastern United States; regional emissions of metals are believed to be responsible for the high concentrations of metals. Neiboer et al. (1972) give evidence for a strong relationship between concentrations of heavy metals in lichens and metal emissions from the nickel smelter at Sudbury, Ontario. Thus the spatial chemical pattern displayed by modern lichens and mosses may be used to demonstrate short- and long-range transport of metal pollutants.

The precise relationships between depositional flux of various metals and their retention by the contemporary moss or lichen surface are not known. Atmospheric concentrations, precipitation amount and type, and plant species community composition are probably all important factors in the relationships. Analysis of data in the literature for Scandinavia (Pakarinen 1981a) suggests that over the decades of the 1960s and 1970s, metal concentrations increased in mosses/lichens, implying that atmospheric deposition rates increased, too. Pakarinen (1981a) repeated measurements of contemporary *Sphagnum* in Finland from 1975-1976 to 1979-1980 with the same analytical protocol and found significantly declining concentrations of lead, zinc, and iron; copper was unchanged. These declines occurred in a time of reduced emissions and declining use of lead in gasoline in northern Europe.

This evidence suggests that trends in atmospheric deposition of metals may be monitored using mosses or lichens. However, total retention of metals after initial sorption by the living surface has not been demonstrated and for most elements is probably unlikely. Pakarinen (1981b) has demonstrated that the order of decreasing mobility of elements in a recently formed *Cladonia arbuscula* profile is potassium > nitrogen > phosphorus > magnesium > zinc > copper > manganese > calcium > lead > iron. Damman (1978), in a classic study, demonstrated that in a *Sphagnum*-dominated hummock profile the order of decreasing mobility is potassium > sodium > lead > phosphorus > nitrogen > magnesium. It is clear that the chemistry of cores of accumulating ombrotrophic bogs do not accurately portray atmospheric deposition chemistry nor the chemistry of the initially formed organic matter. These conclusions were anticipated by the experiments of Clymo (1967), who showed that aerobic decay of at least *Sphagnum* proceeds at considerable and measureable rates, releasing nutrients associated with the biomass.

Mosses (especially *Sphagnum*) do not obtain all their nutrients directly from the atmosphere. Much of the nutrient pool is recycled, and because of decay the chemistry of these mosses is not fixed after death. Therefore, reconstructions of trends in atmospheric deposition are difficult to establish from profiles of chemistry. The best opportunities for reconstructions exist where the living surface is just at or only slightly above the water table (i.e., a bog hollow).

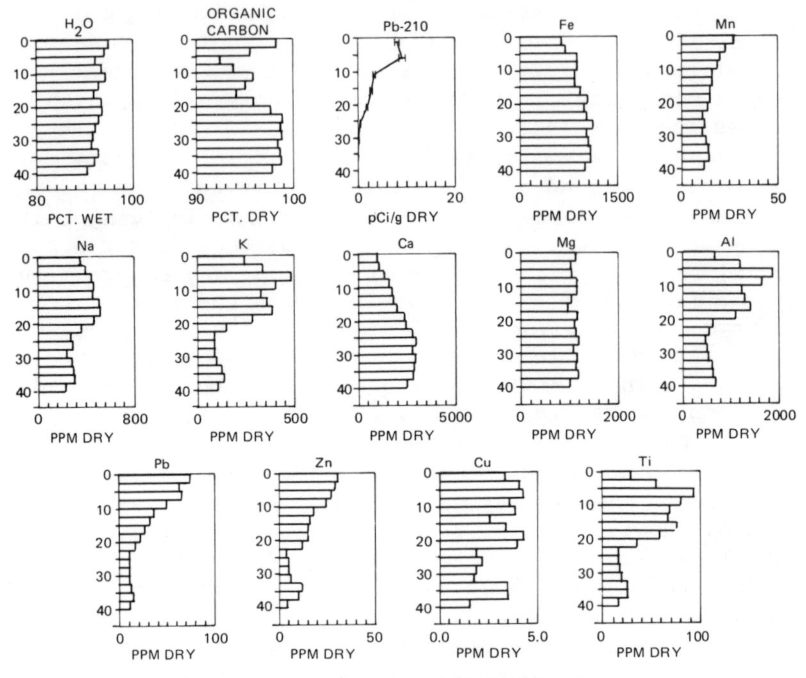

FIGURE 9.18 Chemistry of a hollow peat core from Big Heath, Mount Desert Island, Maine. Water table at about 5 cm. PPM DRY, parts per million, dry; pCi/g DRY, picocuries per gram, dry; PCT. DRY, percent dry; PCT. WET, percent wet. (From Norton 1983.)

There, the older moss is below the water table where decay and chemical alteration are retarded. Damman (1978) showed that a hollow core in Norway yielded a lead profile that resembled those derived from lake sediments (see, e.g., Norton and Hess 1980). In a similar study in Maine Norton (1983) (Figure 9.18) found nearly identical results. However, comparison of hummock and hollow peat profiles suggests that multiple processes are affecting their chemistries (see Figures 9.18 and 9.19). These processes probably include the following as the most important: upward and downward biological pumping, involving, for example, cesium-137, potassium, and manganese; downward or upward migration of mobile elements during precipitation events or capillary

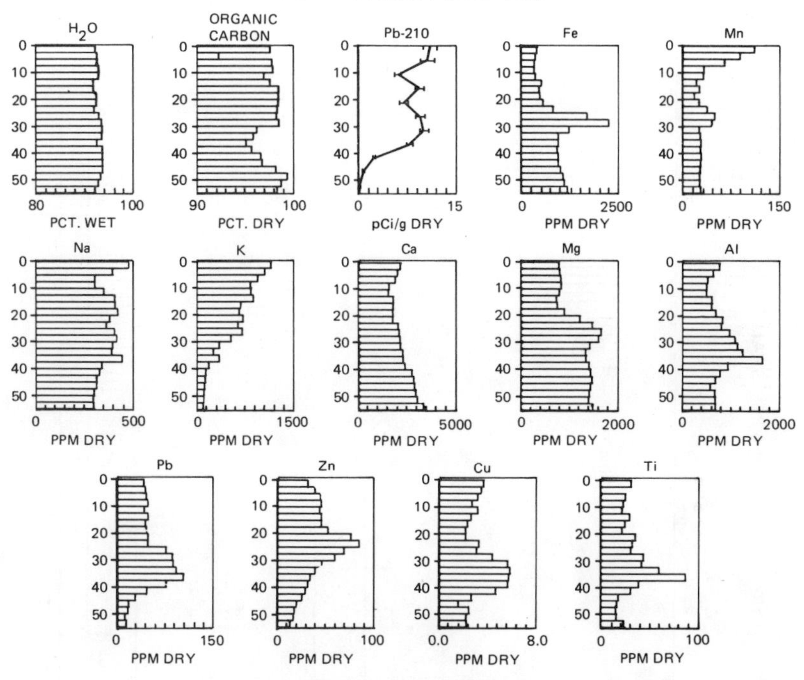

FIGURE 9.19 Chemistry of a hummock peat core from Big Heath, Mount Desert Island, Maine. Water table at about 30 cm. See caption, Figure 9.18 for definitions. (From S. A. Norton, University of Maine, Orono, unpublished data.)

wicking; concentration of immobile elements, such as iron, manganese, and titanium by biological decay and selective removal of mobile elements; and redox reactions related to the water table affecting, for example, zinc, iron, and possibly manganese. The profiles shown in Figures 9.18 and 9.19 are similar to those given by Damman (1978) and probably represent end members for chemical profile development. Intermediate water table conditions probably produce intermediate-type profiles, and the presence of certain vascular plants drastically alters the mobility and distribution of some metals. For example, high concentrations of surface manganese are associated with shrubs. In their absence, manganese in surface peats is low in concentration.

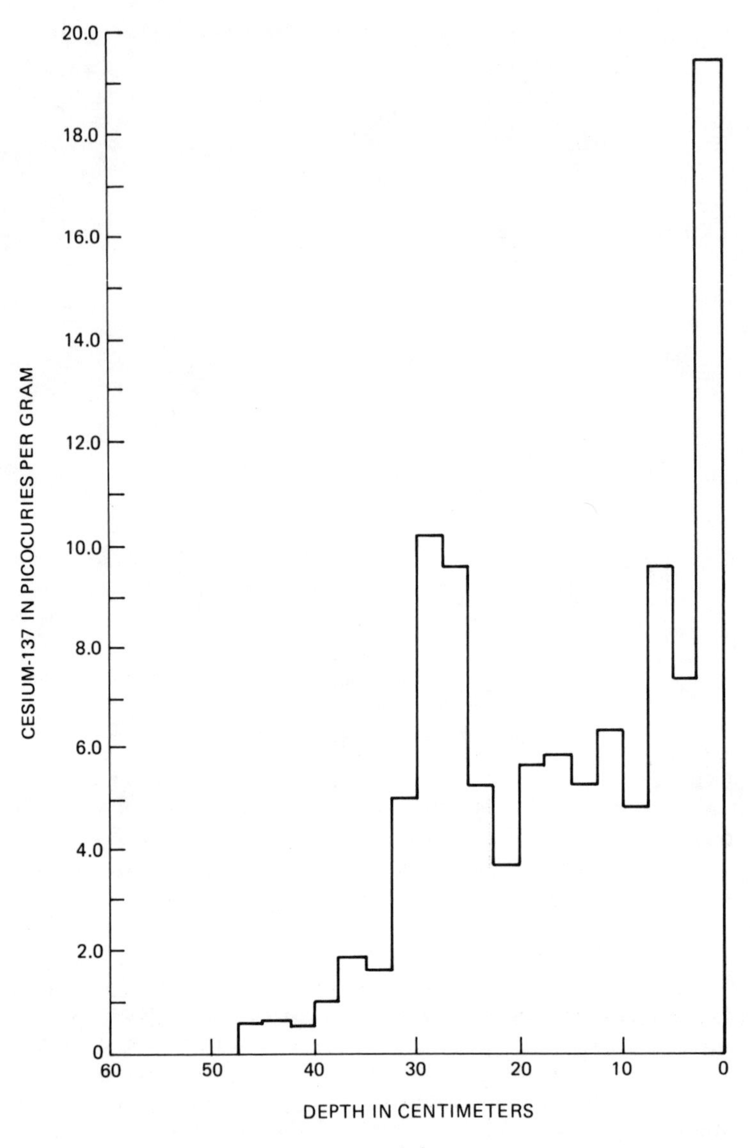

FIGURE 9.20 (a) Cesium-137 (Cs-137) activity in a hummock peat profile, Bull Hummock, in Bull Pasture Plain, New Brunswick (August 29, 1981) (from Olson 1983); (b) annual deposition of strontium-90 (Sr-90) in New York City. Cs-137 deposition = 1.7 x Sr-90 deposition. (From Larsen 1984.)

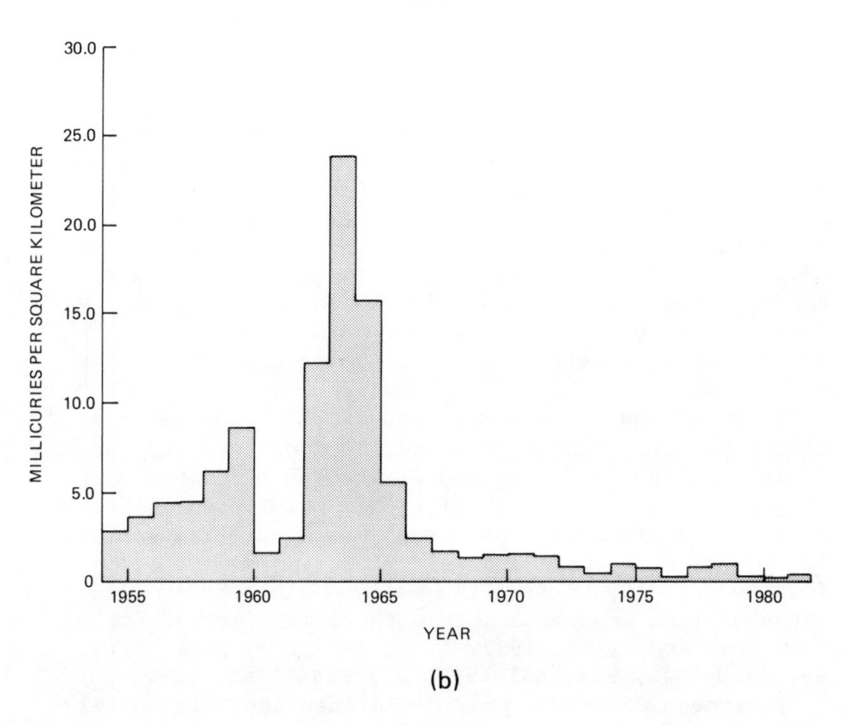

FIGURE 9.20 (continued).

The development of a reliable dating tool for peat is an additional problem in utilizing the chemistry of peat profiles for establishing chronological trends in atmospheric deposition of pollutants. Dating methods include pollen stratigraphy; chronostratigraphic markers, such as charcoal, cesium-137, and lead-210; increment dating; and magnetic stratigraphy. Pollen stratigraphy and the use of charcoal are typically imprecise for dating purposes because atmospheric particulate material can penetrate through the living surface into older peat. Continuous age-depth relationships cannot be established; only one or several time lines are established in the stratigraphic section, and extrapolation and interpolation of ages are necessary. (See Livett et al. 1979.) This shortcoming also arises with the use of cesium-137, a product of thermonuclear weapons testing, as a stratigraphic marker. It gives one (the 1963 maximum) or at best two ages. (See Clymo 1978.) Also, it has been well established that cesium as well as potassium is strongly recycled by or into living moss, and pumped downward into older peat by vascular plants (see Oldfield et al. 1979). The resulting cesium-137 profiles may bear little resemblance to the known deposition history of cesium-137 (Figure 9.20). During the past 10 years, several workers have used lead-210, a naturally occurring radionuclide deposited from the atmosphere, to establish the chronology of cores and thereby interpret trends in atmospheric deposition of certain constituents. Lead-210 dating assumes no penetration of this element through the living surface during initial deposition and no vertical displacement of it from the stratum in which it is deposited.

Views on the conservative and immobile nature of lead differ widely. One of the first uses of lead-210 for the dating of peat in the United States was by Hemond (1980). Although the stratigraphic resolution of the peat profile from Thoreau's Bog, Massachusetts, was poor for purposes of trend analysis (four increments of peat for the past 100 years), the derived atmospheric loading rates of lead are comparable with values derived from lake sediments (1971-1977, 43 mg m^{-2} yr^{-1}; 1948-1971, 44; 1920-1948, 33; 1881-1920, 11; 1844-1881, 12).

Increment dating of peat (Pakarinen and Tolonen 1977) is based on counting annual nodes in moss species. Comparison between increment dating and lead-210 dating suggests that in hummocks, lead-210 may be mobile with downward dislocation of lead. Lead-210 analysis gives ages that are older than increment ages and are thus too old if the increment dating method is correct (Figure 9.21). This effect would artificially elevate deposition rates based on lead-210 in recent years. Increment dating cannot be utilized in hollows, where biological decay apparently destroys the evidence. On the other hand, pollen dating and lead-210 chronology appear to agree for hollow profiles (Tolonen et al. in press, Norton in press). At best, it appears that dating of peat profiles is somewhat imprecise. Consequently, the chronology of trends in peat chemistry, and thus inferred changes in atmospheric deposition of metals as well as net deposition rates, are imprecise.

Madsen (1981) showed increasing levels of mercury over the past 200 years in lead-210-dated cores from two Danish bogs and suggested that this trend in mercury concentration relates to increased emissions from industrial or agricultural activity. Pakarinen et al. (1980) used increment dating to establish net loading

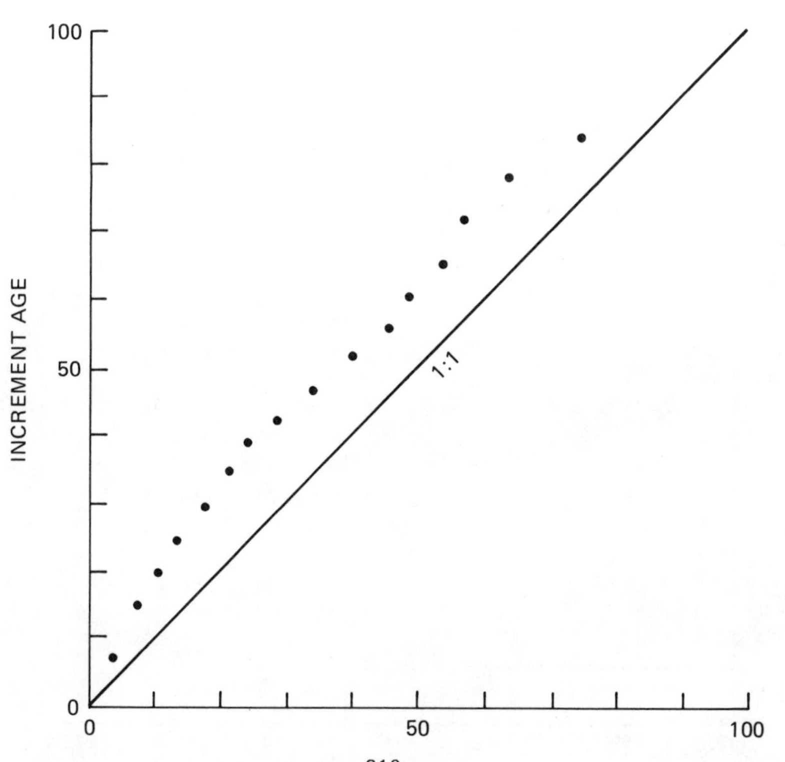

FIGURE 9.21 Increment age versus lead-210 age for a hummock peat core from Great Heath, Cherryfield, Maine (from Norton 1984).

rates of a variety of metals including lead, zinc, and copper, over the past 10 to 20 years. The values are similar to values of measured atmospheric deposition. However, strong redistribution of zinc and iron appears to occur. S. Norton (unpublished data) has dated, by lead-210, and chemically analyzed 10 pairs of cores from adjacent hummock-hollow associations from ombrotrophic bogs in Maine. Only cores from hollows have heavy-metal stratigraphy and deposition rates, especially for lead and zinc, that resemble those of lake sediments in the eastern United States (Figure 9.22); the resemblance is crude at best.

Oldfield et al. (1978) have developed techniques for measuring the magnetic particle content of peat profiles

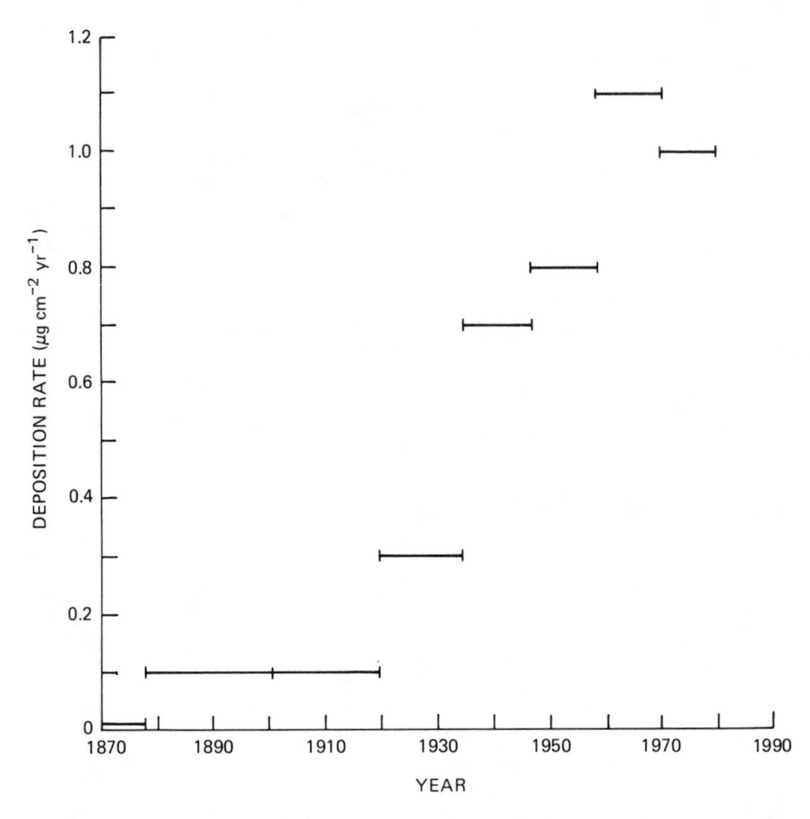

FIGURE 9.22 Deposition rate of lead for a peat core from a hollow in Crystal Bog, Maine (from Norton 1983).

and found a stratigraphy mimicking reconstructed histories of emissions for the local regions near the bogs (Figure 9.23). The particles are generated as part of the fly ash in coal-based emissions. However, Oldfield et al. believe that the transport distance is relatively small. The metals lead and zinc, however, are known to be transported in large proportions for hundreds to a few thousands of kilometers. Thus, cores may give information about both short- and long-range transport of atmospheric pollutants.

In summary, modern peats record spatial variation in recent atmospheric deposition of chemical and particulate pollutants; such peats typically have higher concentrations of pollutants than preindustrial peat. However,

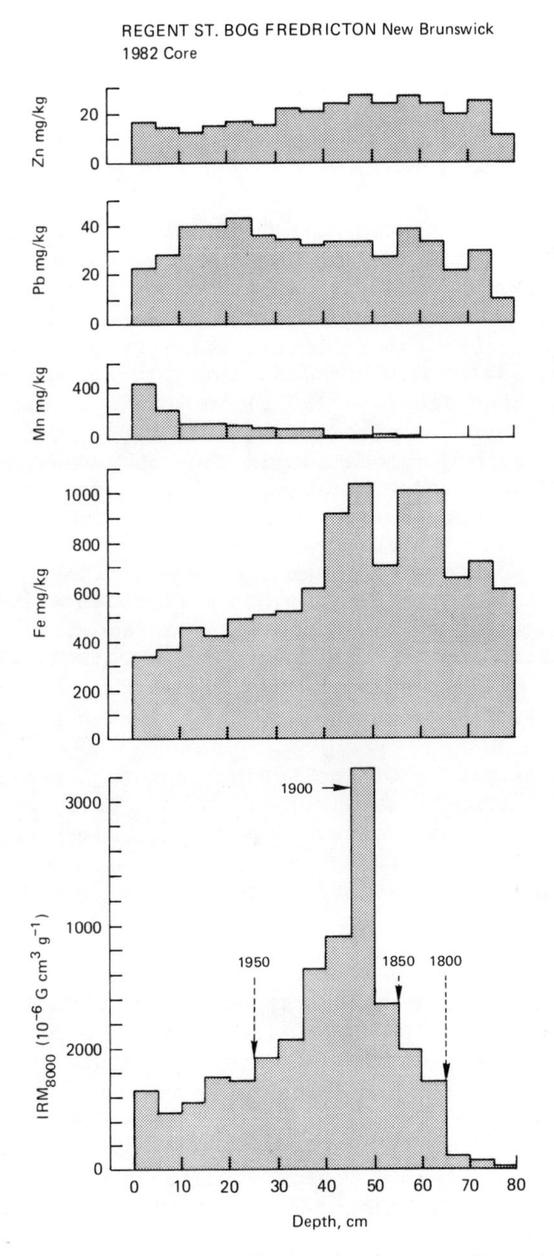

FIGURE 9.23 Magnetic and chemical stratigraphy from a hummock peat core (1982) from Regent Street Bog, Fredricton, New Brunswick; dating by increments. IRM is isothermal remanent magnetism in units of Gauss, centimeters cubed per gram. (From F. Oldfield, University of Liverpool, U. K., and K. Tolonen, University of Oulu, Oulu, Finland, unpublished data.)

many physical, chemical, and biological processes make precise reconstruction of the deposition history of metals and other pollutants difficult.

DIRECT COMPARISON OF DIATOM AND CHEMICAL DATA

We have screened approximately 60 lakes in the United States for sediment data relating to pH reconstructions based on diatoms and about 40 lakes for sediment metals. Both chemical and biological data as well as lead-210 dating are available for 18 lakes, permitting direct comparison of diatom and chemical interpretations (Table 9.4). Tolonen and Jaakkola (1983) and Davis et al. (1983) also present comparisons of this type. Four of the lakes are at high elevation in the Rocky Mountain National Park in an area not currently receiving high levels of acid precipitation (Gibson et al. 1984). The remaining lakes in the sample are in the New England-New York region. Most of the lakes are low-alkalinity oligotrophic lakes and thus are not a representative sample of the lakes in the Northeast. In addition, the sample is biased toward lakes sensitive to atmospheric deposition of acids, and most were selected on the basis of being little disturbed and are thus at high altitude and remote.

Diatom stratigraphy of the Rocky Mountain lakes shows no evidence of acidification (D. R. Beeson, U.S. Park Service, Fort Collins, Colorado, personal communication, 1984; Appendix Table E.1). The sediment chemistry of these lakes (Baron et al. 1984; Figure 9.24) indicates an influence of mining activities in the region as shown by

TABLE 9.4 Lakes for Which Data on Diatoms, Sediment Chemistry, and Dating Are Available for Analysis

Adirondack Park, N.Y.	N. New England	Rocky Mountain National Park, Colo.
Big Moose Lake	Branch Pond, Vt.	Emerald Lake
Deep Lake	E. Chairback Pond, Me.	Lake Haiyaha
Panther Lake	Klondike Pond, Me.	Lake Husted
Sagamore Lake	Ledge Pond, Me.	Lake Louise
Woods Lake	Mountain Pond, Me.	
	Lake Solitude, N.H.	
	Speck Pond, Me.	
	Tumbledown Pond, Me.	
	Unnamed Pond, Me.	

NOTE: For more detailed data on these lakes, see Appendix E, Tables E.1-E.5.

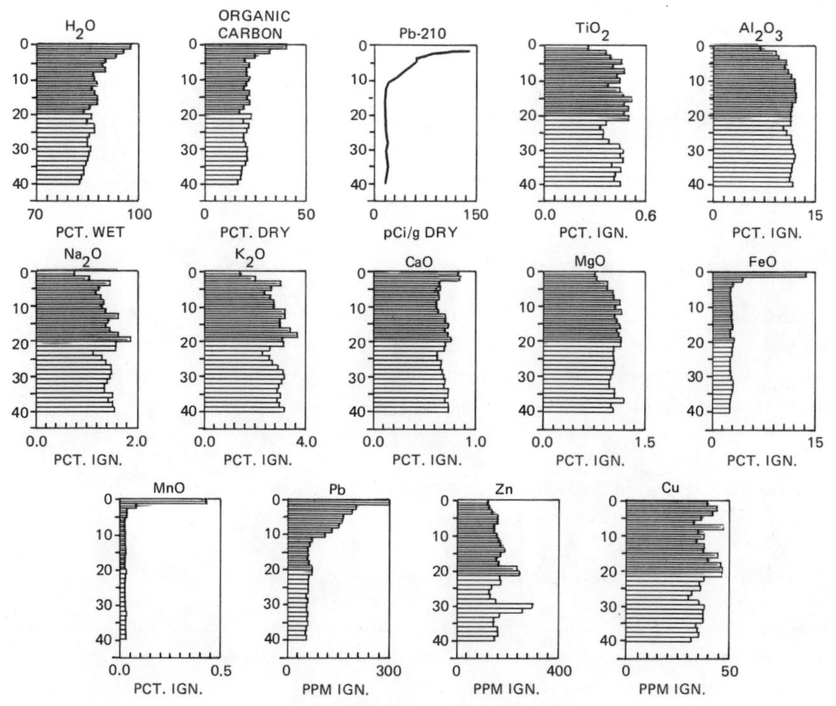

FIGURE 9.24 Sediment chemistry of Lake Husted, Rocky Mountain National Park. The recent increase in lead is indicative of mining activities in the region. See caption, Figure 9.12, for definitions. (From Baron et al. 1984.) Diatom-inferred pH for this sediment core has been stable for at least the past 150 years, fluctuating around 6.6 to 6.8 pH units (see Appendix Table E.5; Beeson 1984).

elevated lead concentrations but no history of fossil-fuel-related atmospheric inputs. This is indicated by the lack of elevated concetrations of certain diagnostic elements such as zinc. Surface water chemistry of lakes from this region indicates little or no acidification from excess atmospheric sulfate (see Appendix Table E.3; Wright 1983, Gibson et al. 1984).

All study lakes in the northeastern United States have evidence of increased atmospheric deposition of metals, such as lead and copper, that are associated with combustion of fossil fuels and mining activities, but not

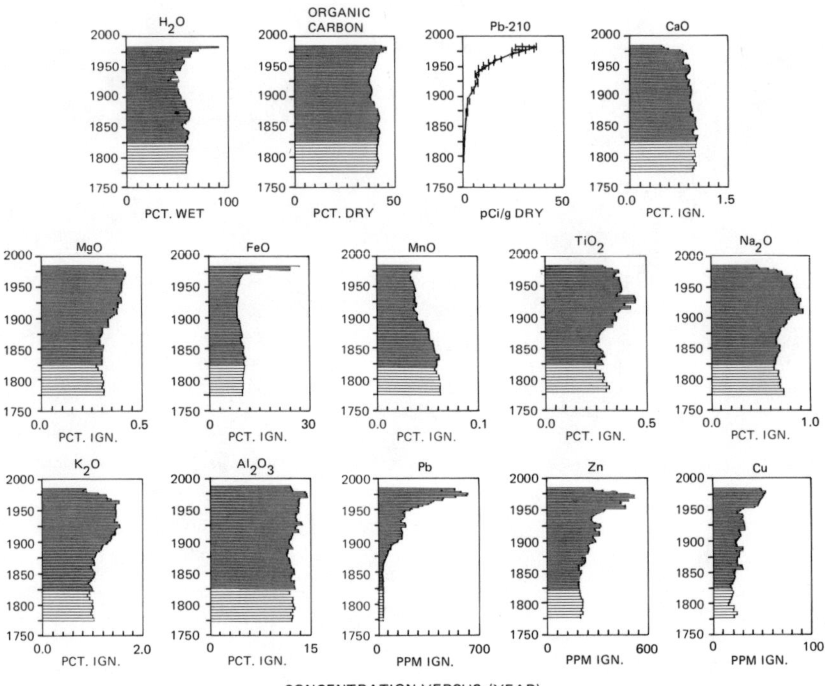

FIGURE 9.25 Concentrations versus year for various chemical species in Big Moose Lake (core 2). See caption, Figure 12, for definitions. Unpublished data from D. Charles (Indiana University) and S. Norton (University of Maine).

all show evidence of recent acidification. Three case studies are presented here with comparative data. Big Moose Lake, New York, contains biological and chemical evidence that is indicative of acidification. Sediment from Speck Pond, Maine, has seemingly contradictory evidence. Panther Lake, New York, although receiving deposition of strong acids, shows little evidence of acidification.

For Big Moose Lake, analyses of data on diatoms, chrysophytes, and sediment chemistry all indicate recent acidification (Appendix Table E.5; Figures 9.5 to 9.7, and 9.25). Diatom-inferred pH declines beginning around 1950, and falls rapidly in the 1960s. Changes in the

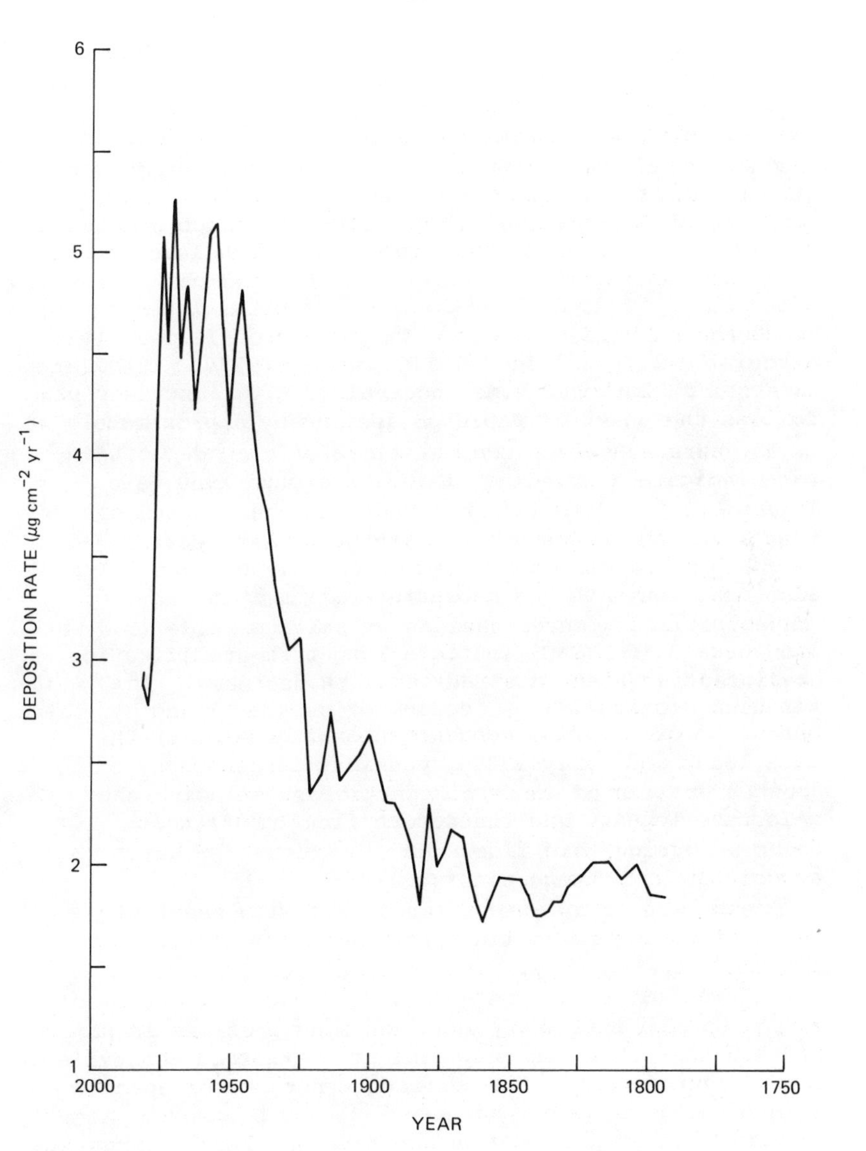

FIGURE 9.26 Deposition rate of zinc versus year in Big Moose Lake (core 2). Unpublished data from S. Norton, University of Maine, Orono.

abundance of chrysophyte taxa (Figure 9.7) indicate the same pattern of pH change. The diatom-inferred pH also suggests that pH had been declining slowly since at least the early 1800s, possibly resulting from long-term gradual acidification from natural processes. Sediment chemistry data (Figures 9.25 and 9.26) are consistent with the diatom-inferred pH profiles, although changes do not generally occur at the same time. Both lead and zinc concentrations begin to increase in the sediment intervals dated late 1800s, as they do for most other lakes in the northeastern United States. The zinc-profile peak is associated with a lead-210 date of 1965-1970. The diatom-inferred pH for that time interval is 5.0. The zinc peak follows the onset of rapid pH decline by approximately 10 to 15 years. However, calculations of zinc deposition rate indicate a leveling of influx around 1950 (see Figure 9.26). This is consistent with the hypothesis that lower pH is associated with processes resulting in reduced influx of zinc to the sediment. Diagenesis may also contribute to the subsurface maximum in zinc concentration. Concentrations of calcium oxide (CaO) and manganese oxide (MnO) indicate long-term acidification consistent with an inferred early pH decrease. The titanium dioxide (TiO_2), sodium oxide (Na_2O) and potassium oxide (K_2O) concentrations increase in the late 1800s and early 1900s, possibly in response to greater erosion of inorganic particles following the selective logging and the forest fire (1903) that occurred during this time. No significant pH shift is associated with these events.

Speck Pond is an example of a lake with seemingly conflicting interpretations for microfossil evidence and chemical stratigraphy. The chemical stratigraphy of Speck Pond (S. A. Norton and R. B. Davis, University of Maine, unpublished data; lead and zinc profiles in Davis and Anderson 1984) is very similar to that of Jerseyfield Lake (Figure 9.12). The pH reconstruction for Speck Pond (Figure 9.8) suggests that the lake has been only slightly acidified if at all; the pH has remained in the range 5.2 to 4.8 for at least the past 100 years. Whereas the chemical profile of lead in Speck Pond is typical for nearly all lakes in the northeastern United States, the zinc profile has been interpreted by S. Norton (unpublished) to be caused by acidification of some parts of the ecosystem. Neither in situ leaching nor leaching of sedimenting particles (Kahl and Norton 1983, Schindler et al. 1980) by an acidifying water column is reflected in the diatom data. Leaching by acid deposition of zinc and other metals bound to detritus before entry to the lake and subsequent sedimentation of detrital particulate matter is a plausible explanation for the impoverished concentration of sedimentous zinc. This explanation assumes that little readsorption of dissolved zinc on sedimenting particles occurs in the lake.

The present chemistry of Speck Pond is dominated by hydrogen ion, aluminum, calcium, and sulfate (Appendix Tables E.2 and E.3). No sources of sulfate have been identified in the watershed. Thus, before the input of excess sulfate, an anion must have been present to balance the hydrogen ions and some unknown amount of calcium and aluminum. Bicarbonate alkalinity would be virtually zero at pH = 5, but Lazerte and Dillon (1984) and others have postulated the importance of organic acids in determining lake acidity. One hypothesis is that organic acidity is replaced by sulfate-balanced acidity with a corresponding decline in dissolved organic carbon and increased water clarity (Davis et al. 1985), and an increased level of aluminum uncomplexed by organic anions. This untested model would explain the observations of relatively invariant diatom-based pH with a concurrent decline of sediment deposition rates for calcium, manganese, and zinc (relative to TiO_2 and Al_2O_3) caused by leaching of particulates on the land before deposition; increased water clarity; and hydrogen-, calcium-, aluminum-, and sulfate-rich waters. Liming experiments in Sweden (H. Hultberg, Swedish Water and Air Pollution Research Institute, Gothenburg, personal communication) and in Norway (Wright 1984) have shown that when the pH of lakes is raised the reverse of some of these processes occurs (those involving diatoms, water clarity, dissolved organic carbon, and uncomplexed aluminum).

For Panther Pond, diatom-inferred pH (Appendix Table E.5; Del Prete and Galloway 1983) is nearly constant (around 6.5) throughout most of the core.* Concentrations

---

*Analysis of diatom assemblages suggests a decline in pH of about 0.3 to 0.4 unit near the top of the sediment core based on the reduction in percent of euplanktonic diatoms within the core, a decrease in _Cyclotella stelligera_ (indifferent), and an increase in _Melosira distans_ var. _distans_ (acidophil). However, reexamination

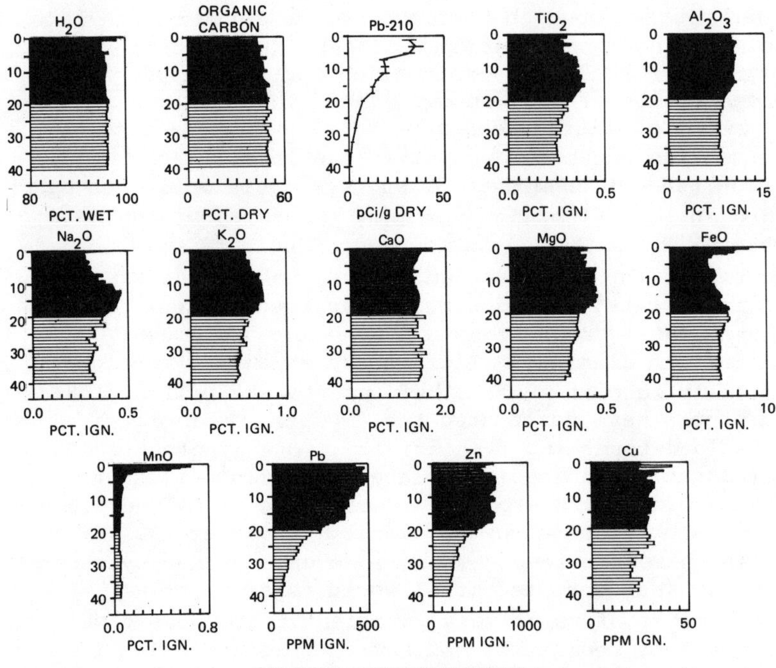

FIGURE 9.27 Concentration versus depth for various chemical species in Panther Lake. See caption, Figure 9.12, for definitions. (From Norton 1984.)

of lead and zinc (Figure 9.27) decrease toward the sediment surface. Based on these declines plus the shape of the lead-210 curve, it appears that the atmospheric inputs of lead-210, lead and zinc are diluted by increased deposition of major elements such as silicon dioxide. There is no decline of zinc, calcium, and manganese relative to other metals that are not mobilized by significantly lower pH. Deposition rates (not shown) of

---

of the diatom assemblages at the top of the Panther Lake core, in light of recent taxonomic revisions, indicates a significant percent of the Melosirae should be reclassified as indifferent rather than acidophilic, and the diatom-inferred pH decrease may not be as great as first determined (D. Charles, unpublished data).

such metals as titanium, manganese, aluminum, sodium, and potassium increase dramatically from the late 1800s to the early 1900s, when logging occurred in the watershed. Because Panther Lake currently has an alkalinity about 150 μeq/L (it was probably greater earlier), it is relatively well buffered against changes that might result from changing atmospheric deposition or watershed disturbance. The sediment chemistry data are consistent with the relatively constant diatom-inferred pH. Although the chemistry data indicate an influx of material associated with fossil fuel combustion and watershed disturbance, there is no strong evidence for increased lake-water acidification.

CONCLUSIONS

Diatoms and Chrysophytes

1. Diatom and chrysophyte data for 31 lakes were evaluated to assess regional trends in lake acidification. The lakes in each region (Adirondack Mountains, eleven; New England, ten; eastern Canada, six; and the Rocky Mountains, four) were generally not representative of the range of chemistries displayed by lakes in those regions.

2. There is evidence that certain poorly buffered lakes have become more acidic recently. The lakes for which evidence of acidification is strongest are in the Adirondack Mountains, New York. Of the eleven lakes for which there are data, six of the seven lakes with a current pH at or below 5.0 show evidence of recent acidification, while one is a naturally acidic bog lake. None of the four lakes with current pH above about 5.0 shows any obvious evidence of substantial pH decline. Where dating is available, evidence indicates that the most rapid pH changes (decrease of 0.5 to 1.0 pH unit) occurred between 1930 and 1970. For some lakes there have been watershed disturbances that make sediment data more difficult to evaluate.

3. The diatom data for New England lakes indicate either a slight or no decrease in pH. Diatom data for lakes in eastern Canada indicate no change in pH (four lakes) or a significant decrease, which is probably caused primarily by local smelting activities (two lakes). There is no evidence to indicate declines in pH in the Rocky Mountain lakes.

4. Before 1800, several lakes in the Adirondack Mountains and New England had pH values less than 5.5. These lakes now have a pH of only 0.1 to 0.4 pH unit lower and total aluminum concentrations greater than 100 μg/L. Because of the potential importance of organic acid and aluminum buffering, a small decline in pH could be associated with significant decreases in alkalinity.

5. Analysis of the Adirondack lake/watershed data sets indicates that the recent rapid declines in lake-water pH are most likely related to increased acid deposition. However, watershed disturbances may also play a role.

6. Analysis of the three lakes in the Adirondacks (Big Moose Lake, Deep Lake, and Upper Wallface Pond) for which data are adequate for the timing of acid-deposition-related events to be assessed suggests that the onset of decline began in the 1930s to 1950s.

7. Analysis of sediment diatom and chrysophyte assemblages is the best paleolimnological technique currently available for reconstructing past lake-water pH. However, there are important uncertainties in the technique, and further research is necessary to improve procedures and interpretation of the results.

Geochemistry

1. Lead deposition from the atmosphere started increasing sometime in the middle to late 1800s, as shown by increasing concentrations in sediments and increasing sediment-deposition rates of lead, independent of most other elements.

2. The history of zinc sedimentation roughly parallels that of lead. However, in recent sediment in acidic (pH < 5.5) lakes, concentrations and apparent deposition rates of zinc and other easily mobilized metals decrease toward the surface of the sediment.

3. Copper and cadmium increase less than lead or zinc. Other metals commonly not measured, such as chromium, arsenic, and antimony, also increase in this same period.

4. The concentration and the deposition rate of vanadium increase around 1940.

5. PAHs and variously shaped carbonaceous particules related to fossil fuel consumption are found in lake sediments. Their concentrations and deposition rates are interpretable in terms of estimated emissions resulting from various processes.

6. Trace metals, PAHs, and magnetic spherules are present in ombrotrophic peat profiles. Their concentrations are elevated in peat dating from current times to more than 100 years old. Trends in deposition can be estimated only for cores from hollows in the bogs and then only crudely. The abundance of spherules appears to mimic most closely the known emission histories in the local (<100-km) region.

Interpretations of lake acidity and atmospheric deposition changes based on stratigraphic analysis of both diatom assemblages and sediment chemistry are generally consistent.

REFERENCES

Alhonen, P. 1967. Paleolimnological investigations of three inland lakes in southwestern Finland. Acta Bot. Fenn. 76:1-59.

Alhonen, P. 1968. On the late-glacial and early post-glacial diatom succession in Loch of Park, Aberdeenshire, Scotland. Mem. Soc. Fauna Flora Fenn. 44:13-20.

Almer, B., W. Dickson, C. Ekström, E. Hörnström, and U. Miller. 1974. Effects of acidification on Swedish lakes. Ambio 3:30-36.

Almer, B., W. Dickson, C. Ekström, and E. Hörnström. 1978. Sulfur pollution and the aquatic ecosystem. Pp. 271-311 of Sulfur in the Environment, Part II, Ecological Impacts, J. O. Nriagu, ed. New York: John Wiley and Sons.

Andersen, S. T. 1966. Interglacial vegetational succession and lake development in Denmark. Paleobotanist 15:117-127.

Anderson, D. S., R. B. Davis, and F. Berge. In press. Relationships between diatom assemblages in lake surface-sediments and limnological characteristics in southern Norway. Hydrobiologia.

Andersson, G., S. Fleischer, and W. Graneli. 1978. Influence of acidification on decomposition processes in lake sediment. Verh. Int. Ver. Limnol. 20:802-807.

Appleby, P. G., and F. Oldfield. 1978. The calculation of lead-210 dates assuming a constant rate of supply of unsupported ^{210}Pb to the sediment. Catena 5:1-8.

Arzet, K., D. Krause-Dellin, and C. Steinberg. 1985. Acidification of selected lakes in the Federal

Republic of Germany as reflected by subfossil diatoms, Cladoceran remains, and sediment chemistry. In Diatoms and Lake Acidity, J. P. Smol, R. W. Battarbee, R. B. Davis, and J. Meriläinen, eds. The Hague, The Netherlands: W. Junk.

Baron, J. 1983. Comparative water chemistry of four lakes in Rocky Mountain National Park. Water Res. Bull. 19:897-902.

Baron, J., D. R. Beeson, S. A. Zary, P. M. Walthall, W. L. Lindsay, and D. M. Swift. 1984. Long-term research into the effects of acidic deposition in Rocky Mountain National Park: Summary Report, 1980-1984. U.S. Park Service. Tech. Rep. No. 84-ROMO-2.

Battarbee, R. W. 1979. Diatoms in lake sediments. Pp. 177-225 of Paleohydrological Changes in the Temperate Zone in the Last 15000 Years, Subproject B, Lake and Mire Environments, B. E. Berglund, ed. Vol. II, Lund University, Sweden.

Battarbee, R. W. 1984. Diatom analysis and the acidification of lakes. Phil. Trans. R. Soc. London 305:451-477.

Batterbee, R. W., and R. J. Flower. 1984. The inwash of catchment diatoms as a source of error in the sediment-based reconstruction of pH in an acid lake. Limnol. Oceanogr. 29:1325-1329.

Baxter, M. S., J. G. Farmer, I. G. McKinley, D. S. Swan, and W. Jack. 1981. Evidence of the unsuitability of gravity coring for collecting sediment in pollution and sedimentation rate studies: Environ. Sci. Technol. 15:843-846.

Beamish, R. J., and H. H. Harvey. 1972. Acidification of the La Cloche Mountain lakes, Ontario, and resulting fish mortalities. J. Fish. Res. Board Can. 29:1131-1143.

Beeson, D. R. 1984. Historical fluctuations in lake pH as calculated by diatom remains in sediments from Rocky Mountain National Park. Bull. Ecol. Soc. Am. 65:135.

Berg, C. O. 1966. Middle Atlantic States. Pp. 191-237 of Limnology in North America, D. G. Frey, ed. Madison, Wisc.: University of Wisconsin Press.

Berge, F. 1976. Kiselalger og pH i noen elver og innsjøer i Agder og Telemark. En sammenlikning mellom arene 1949 og 1975. [Diatoms and pH in some rivers and lakes in Agder and Telemark (Norway): a comparison between the years 1949 and 1975.] Sur Nedbørs Virkning Pa Skog Og Fisk. IR18/76. Norwegian Forest Research Institute, Aas, Norway.

Berge, F. 1979. Kiselalger og pH i noen innsjoer i Agder og Hordaland. Sur Nedbørs Virkning Pa Skog Og Fisk. [Diatoms and pH in some lakes in the Agder and Hordaland Counties, Norway.] IR42/79. Norwegian Forest Research Institute, Aas, Norway.

Berge, F. 1983. Diatoms as indicators of temporal pH trends in some lakes and rivers in southern Norway. Nova Hedwigia 73:249-265.

Berglund, B., and N. Malmer. 1971. Soil conditions and late-glacial stratigraphy. Geol. Foeren. Stockholm Foerh. 93:575-586.

Bradstreet, T. E., and R. B. Davis. 1975. Mid-postglacial environments in New England with emphasis on Maine. Arct. Anthropol. 12:7-22.

Brakke, D. F. 1984. Acidification history and prediction using biological indices. Verh. Int. Ver. Limnol. 22:1391-1395.

Brugam, R. B. 1980. Postglacial diatom stratigraphy of Kirchner Marsh, Minnesota. Quat. Res. (N.Y.) 13:133-146.

Brugam, R. B. 1983. The relationship between fossil diatom assemblages and limnological conditions. Hydrobiologia 98:223-235.

Brugam, R. B. 1984. Holocene paleolimnology. Pp. 208-221 of Late Quaternary Environments of the United States, Vol. 2: The Holocene, H. E. Wright, Jr., ed. Minneapolis, Minn.: University of Minnesota Press.

Buckley, D. E., and R. E. Cranston. 1971. Atomic absorption analyses of 18 elements from a single decomposition of alumino-silicate. Chem. Geol. 7:273-284.

Campbell, P. G. C., and A. Tessier. 1984. Paleo-limnological approaches to the study of acid deposition: metal partitioning in lacustrine sediments. Environmental Protection Agency Final Report on the Workshop, Progress and Problems in the Paleolimnological Analysis of the Impact of Acidic Precipitation and Related Pollutants on Lake Ecosystems, S. A. Norton, ed. 40 pp.

Carignan, R., and J. O. Nriagu. 1985. Trace metal deposition and mobility in the sediments of two lakes near Sudbury, Ontario. Geochim. Cosmochim. Acta 49:1753-1764.

Carignan, R., and A. Tessier. 1985. Zinc deposition in acid lakes: the role of diffusion. Science 228:1524-1526.

Charles, D. F. 1982. Studies of Adirondack Mountain (N.Y.) lakes: limnological characteristics and sediment diatom-water chemistry relationships. Doctoral dissertation, Indiana University, Bloomington, Ind.

Charles, D. F. 1984. Recent pH history of Big Moose Lake (Adirondack Mountains, New York, USA) inferred from sediment diatom assemblages. Int. Ver. Theor. Angew. Limnol. Verh. 22:559-566.

Charles, D. F. 1985a. Relationships between surface sediment diatom assemblages and lakewater characteristics in Adirondack lakes. Ecology 66:994-1011.

Charles, D. F. 1985b. A new diatom species, Fragilaria acidobiontica, from acidic lakes in northeastern North America. In Diatoms and Lake Acidity, J. P. Smol, R. W. Battarbee, R. B. Davis, and J. Meriläinen, eds. The Hague, The Netherlands: W. Junk.

Cholnoky, B. J. 1968. Die Okologie der Diatomeen in Binnengewasser. J. Federal Republic of Germany: Lehre, Cramer.

Cleve, P. T. 1891. The diatoms of Finland. Acta Soc. Fauna Flora Fenn. 8:1-68.

Clymo, R. S. 1963. Ion exchange in Sphagnum and its relation to bog ecology. Ann. Bot. 27:309-324.

Clymo, R. S. 1967. Experiments on breakdown of Sphagnum in two bogs: J. Ecol. 53:747-757.

Clymo, R. S. 1978. A model of peat bog growth: Pp. 187-223 of Ecological Studies, O. W. Heal and D. F. Perkins, eds. Vol. 27. Berlin: Springer-Verlag.

Colingsworth, R. F., M. H. Hohn, and G. B. Collins. 1967. Post-glacial physiochemical conditions of Vestaburg Bog, Montcalm County, Michigan, based on diatom analysis. Pap. Mich. Acad. Sci. Arts Lett. 52:19-30.

Colquhoun, J., W. Krester, and M. Pfeiffer. 1984. Acidity status update of lakes and streams in New York State. New York State Department of Environmental Conservation, Albany, N.Y. 139 pp.

Conger, P. S. 1939. The contribution of diatoms to the sediments of Crystal Lake, Vilas County, Wisconsin. Am. J. Sci. 237:324-340.

Crabtree, K. 1969. Post-glacial diatom zonation of limnic deposits in north Wales. Mitt. Int. Ver. Theor. Angew. Limnol. 17:165-171.

Crecelius, E. A. 1975. The geochemical cycle of arsenic in Lake Washington and its relation to other elements. Limnol. Oceanogr. 20:441-451.

Crocker, R. L., and J. Major. 1955. Soil development in relation to vegetation and surface age at Glacier Bay, Alaska. J. Ecol. 43:427-448.

Damman, A. W. H. 1978. Distribution and movement of elements in ombrotrophic peat bogs. Oikos 30:480-495.

Davis, R. B., and D. S. Anderson. 1984. Assessment of trends based on paleolimnological technique. Pp. 4-99 -- 4-105 of The Acidic Deposition Phenomenon and Its Effects: Critical Assessment Review Papers, Vol. 2, A. P. Altshuller and R. A. Linthurst, eds. Washington, D.C.: U.S. Environmental Protection Agency.

Davis, R. B., and D. S. Anderson. 1985. Methods of pH calibration of sedimentary diatom remains for reconstructing history of pH in lakes. Hydrobiologia 120:69-87.

Davis, R. B., and F. Berge. 1980. Atmospheric deposition in Norway during the last 300 years as recorded in SNSF lake sediments II. Diatom stratigraphy and inferred pH. Pp. 270-271 of Ecological Impact of Acid Precipitation. Proceedings of an international conference, Sandefjord, Norway. D. Drabløs and A. Tollan, eds. Sur Nedbørs Virkning Pa Skog Og Fisk (SNSF) Project, Oslo.

Davis, R. B., and S. A. Norton. 1978. Paleolimnologic studies of human impact on lakes in the United States, with emphasis on recent research in New England. Pol. Arch. Hydrobiol. 25:99-115.

Davis, R. B., T. E. Bradstreet, R. Stuckenrath, Jr., and H. W. Borns. 1975. Vegetation and associated environments during the past 14,000 years near Moulton Davis, R. B., S. A. Norton, C. T. Hess, and D. F. Brakke.

Davis, R. B., S. A. Norton, C. T. Hess, and D. F. Brakke. 1983. Paleolimnological reconstruction of the effects of atmospheric deposition of acids and heavy metals on the chemistry and biology of lakes in New England and Norway. Hydrobiologia 103:113-123.

Davis, R. B., C. T. Hess, S. A. Norton, D. W. Hanson, K. D. Hoagland, and D. S. Anderson. 1984. ^{137}Cs and ^{210}Pb dating of sediments from soft water lakes in New England (U.S.A.) and Scandinavia, a failure of ^{137}Cs dating. Chem. Geol. 44:151-185.

Davis, R. B., D. S. Anderson, and F. Berge. 1985. Loss of organic matter, a fundamental process in lake acidification: paleolimnological evidence. Nature 316:436-438.

Davis, R. B., D. S. Anderson, D. F. Charles, and J. N. Galloway. Two-hundred year pH history of Woods, Sagamore, and Panther Lakes in the Adirondack Mountains, New York State, USA. (In preparation.)

Dearing, J. A. 1983. Changing patterns of sediment accumulation in a small lake in Scania, southern Sweden: Hydrobiologia 103:59-64.

Del Prete, A. 1972. Postglacial diatom changes in Lake George, New York. Ph.D. dissertation, Rensalear Polytechnic Institute, Troy, N.Y., 110 pp.

Del Prete, A., and J. N. Galloway. 1983. Temporal trends in the pH of Woods, Sagamore and Panther lakes as determined by an analysis of diatom populations. In The Integrated Lake-Watershed Acidification Study: Proceedings of the ILWAS Annual Review Conference, Palo Alto, Calif.: Electric Power Research Institute.

Del Prete, A., and C. Schofield. 1981. The utility of diatom analysis of lake sediments for evaluating acid precipitation effects on dilute lakes. Arch. Hydrobiol. 91:332-340.

Delorme, L. D., S. R. Esterby, and H. Duthie. 1984. Prehistoric pH trends in Kejimkujik Lake, Nova Scotia. Int. Rev. Gesam. Hydrobiol. 69:41-55.

Delorme, L. D., H. C. Duthie, S. R. Esterby, S. M. Smith, and N. S. Harper. In preparation. Prehistoric inferred pH changes in Batchawana Lake, Ontario from sedimentary diatom assemblages.

Dickman, M., and J. Fortescue. 1984. Rates of lake acidification inferred from sediment diatoms for 8 lakes located north of Lake Superior, Canada. Int. Ver. Theor. Angew. Limnol. Verh. 22:1345-1356.

Dickman, M., S. Dixit, J. Fortescue, R. Barlow, and J. Terasmae. 1983. Diatoms as indicators of the rate of lake acidification. Water Air Soil Pollut. 21:227-241.

Dickson, W. 1980. Properties of acidified waters. Pp. 75-83 of The Ecological Impact of Acid Precipitation. Proceedings of an international conference, Sandefjord, Norway. D. Drabløs and A. Tollan, eds. Sur Nedbørs Virkning Pa Skog Og Fisk (SNSF) project, Oslo.

Digerfeldt, G. 1972. The post-glacial development of Lake Trummen. Regional vegetation history, water level changes, and paleolimnology. Folia Limnol. Scand. 16:1-104.

Digerfeldt, G. 1975. The post-glacial development of Ranviken Bay in Lake Immein. III. Paleolimnology. Geol. Foeren. Stockholm Foerh. 97:13-28.

Digerfeldt, G. 1977. The Flandrian development of Lake Flarken. Regional vegetation history and paleolimnology. Department of Quaternary Geology, University of Lund, Sweden, Report 13. 101 pp.

Dillon, P. J., and R. D. Evans. 1982. Whole-lake lead burdens in sediments of lakes in southern Ontario, Canada. Hydrobiologia 91:121-130.

Driscoll, C. T. 1980. Chemical characterization of some dilute acidified lakes and streams in the Adirondack region of New York State. Ph.D. dissertation, Cornell University, Ithaca, N.Y.

Driscoll, C. T., and R. M. Newton. 1985. Chemical characteristics of "acid sensitive" lakes in the Adirondack region of New York. Environ. Sci. Technol. 19:1018-1024.

Driscoll, C. T., J. P. Baker, J. J. Bisogni, and C. L. Schofield. 1984. Aluminum speciation and equilibria in dilute acidic surface waters of the Adirondack region of New York State. Pp. 55-75 of Geological Aspects of Acid Deposition, O. Bricker, ed. Boston, Mass.: Butterworth Publishers.

Edginton, D. N., and J. A. Robbins. 1976. Records of lead deposition in Lake Michigan sediments since 1800: Environ. Sci. Technol. 10:266-274.

Esterby, S. R., and A. H. El-Shaarawi. 1981a. Likelihood inference about the point of change in a regression regime. J. Hydrol. 53:17-30.

Esterby, S. R., and A. H. El-Shaarawi. 1981b. Inference about the point of change in a regression model. J. Appl. Stat. 30:277-285.

Evans, G. H. 1970. Pollen and diatom analysis of late-quaternary deposits in the Blelham Basin, North Lancashire. New Phytol. 69:821-874.

Evans, G. H., and R. Walker. 1979. The late-quaternary history of the diatom flora of Llyn Clyd and Llyn Glas, two small oligotrophic high mountain tarns in Snowdonia (Wales). New Phytol. 78:221-236.

Evans, H. E., P. J. Smith, and P. J. Dillon. 1983. Anthropogenic zinc and cadmium burdens in sediments of selected southern Ontario lakes. Can. J. Fish. Aquat. Sci. 40:570-579.

Evans, R. D., and P. J. Dillon. 1982. Historical changes in anthropogenic lead fallout in southern Ontario. Hydrobiologia 91:131-137.

Flower, R. J., and R. W. Battarbee. 1983. Diatom evidence for recent acidification of two Scottish lochs. Nature 305:130-133.

Flower, R. J., and R. W. Battarbee. 1985. The morphology and biostratigraphy of Tabellaria quadriseptata (Bacillariophycene) in acid waters and lake sediments in Galloway, Southwest Scotland. Brit. Phycol. J. 20:69-78.

Foged, N. 1969. Diatoms in a post-glacial core from the bottom of the Lake Grane Langsø, Denmark. Dan. Geol. Foren. 19:237-256.

Foged, N. 1972. The diatoms in four postglacial deposits in Greenland. Medd. Groen. 194:1-67.

Ford, M. S. (J.) 1984. The influence of lithology on ecosystem development in New England: a comparative paleoecological study. Doctoral dissertation, University of Minnesota, Minneapolis, Minn.

Ford, M. S. (J.) 1985. The recent history of a naturally acidic lake (Cone P., N.H.). In Diatoms and Lake Acidity, J. P. Smol, R. W. Battarbee, R. Davis, and J. Meriläinen, eds. The Hague, The Netherlands: W. Junk. (In press.)

Fortescue, J. A. C. 1984. Interdisciplinary research for an environmental component (acid rain) in regional geochemical surveys (Wawa area). Algoma District, Ontario Geological Survey Map 80 713, Geochemical Series, compiled 1983.

Fortescue, J. A. C., I. Thompson, M. Dickman, and J. Terasmae. 1981. Multidisciplinary followup of regional pH patterns in lakes north of Lake Superior, District of Algoma, Ontario Geological Survey OFR 5342. Part I, 134 pp., 6 tables, 65 figures.

Fortescue, J. A. C., I. Thompson, M. Dickman, and J. Terasmae. 1984. Multidisciplinary followup of regional pH patterns in lakes north of Lake Superior. District of Algoma, Ontario, Geological Survey OFR 5342. Part II, Information and data listings for lakes studied. 366 pp.

Frey, D. G. 1969. The rationale of paleolimnology. Mitt. Int. Ver. Theor. Angew. Limnol. 17:7-18.

Fritz, S. C., and R. E. Carlson. 1982. Stratigraphic diatom and chemical evidence for acid strip-mine lake recovery. Water Air Soil Pollut. 17:151-163.

Galloway, J. N., and G. E. Likens. 1979. Atmospheric enhancement of metal deposition in Adirondack Lake sediments: Limnol. Oceanogr. 24:427-433.

Galloway, J. N., J. D. Thornton, S. A. Norton, H. L. Volchok, and R. A. N. McLean. 1982. Trace metals in atmospheric deposition: a review and assessment: Atmos. Environ. 16:1677-1700.

Galloway, J. N., C. L. Schofield, N. E. Peters, G. R. Hendrey, and E. R. Altwicker. 1983. Effect of atmospheric sulfur on the composition of three Adirondack lakes. Can. J. Fish. Aquat. Sci. 40:799-806.

Gasse, F., and F. Tekaia. 1983. Transfer functions for estimating paleoecological conditions (pH) from East African diatoms. Hydrobiologia 103:85-90.

Gasse, F., J. F. Talling, and P. Kilham. 1983. Diatom assemblages in East Africa: classification, distribution and ecology. Rev. Hydrobiol. Tropic. 16:3-34.

Gensemer, R. W., and S. S. Kilham. 1984. Growth rates of five freshwater algae in well-buffered acidic media. Can. J. Fish. Aquat. Sci. 41:1240-1243.

Gibson, J. H., J. N. Galloway, C. Schofield, W. McFee, R. Johnson, S. McCorley, N. Dise, and D. Herzog. 1984. Rocky Mountain Acidification Study. U.S. Fish and Wildlife Service, Division of Biological Services, Eastern Energy and Land Use Team. FWS/OBS-80/40.17. 137 pp.

Gorham, E. 1955. On the acidity and salinity of rain: Geochim. Cosmochim. Acta 7:231-239.

Gorham, E. 1958. Atmospheric pollution by hydrochloric acid, Quart. J. R. Meteorol. Soc. 84:274-276.

Gorham, E., P. M. Vitousek, and W. A. Reiners. 1979. The regulation of chemical budgets over the course of terrestrial ecosystem succession. Ann. Rev. Ecol. Syst. 10:53-84.

Gorham, E., S. J. Eisenreich, J. Ford, and M. V. Santelmann. 1985. The chemistry of bog waters. Pp. 339-363 of Chemical Processes in Lakes, W. Stumm, ed. New York: Wiley-Interscience.

Griffin, J. J., and E. D. Goldberg. 1981. Sphericity as a characteristic of solids from fossil fuel burning in a Lake Michigan sediment. Geochim. Cosmochim. Acta 45:763-769.

Groet, S. S. 1976. Regional and local variations in heavy metal concentrations of bryophytes in the northeastern United States. Oikos 27:445-456.

Hakanson, L. 1977. The influence of wind, fetch, and water depth on the distribution of sediments in Lake Vanern, Sweden. Can. J. Earth Sci. 14:397-412.

Hanson, D. W., S. A. Norton, and J. S. Williams. 1982. Modern and paleolimnological evidence for accelerated leaching and metal accumulation in soils in New England caused by atmospheric deposition. Water Air Soil Pollut. 18:227-239.

Hase, A., and R. A. Hites. 1976. On the origin of polycyclic aromatic hydrocarbons in recent sediments: biosynthesis by anaerobic bacteria. Geochim. Cosmochim. Acta 40:1141-1143.

Haworth, E. Y. 1969. The diatoms of a sediment core from Blea Tarn, Langdale. J. Ecol. 57:429-439.

Haworth, E. Y. 1972. Diatom succession in a core from Pickerel Lake, northeastern South Dakota. Geol. Soc. Am. Bull. 83:157-172.

Heit, M., Y. Tan, C. Klusek, and J. C. Burke. 1981. Anthropogenic trace elements and polycyclic aromatic hydrocarbon levels in sediment cores from two lakes in the Adirondack acid lake region. Water Air Soil Pollut. 15:441-464.

Hemond, H. F. 1980. Biogeochemistry of Thoreau's Bog, Concord, Massachusetts. Ecol. Monogr. 50:507-526.

Holdren, G. R., Jr., T. M. Brunelle, G. Matisoff, and M. Wahlen. 1984. Timing the increase in atmospheric sulphur deposition in the Adirondack Mountains. Nature 311:245-248.

Hongve, D. 1978. Buffering of acid lakes by sediments: Verh. Int. Ver. Limnol. 20:743-748.

Hongve, D., and A. Erlandsen. 1979. Shortening of surface sediment cores during sampling. Hydrobiologia 65:283-287.

Howells, G. D. 1983. Acid waters--the effect of low pH and acid associated factors on fisheries. Appl. Biol. 9:143-255.

Hultberg, H., and I. Andersson. 1982. Liming of acidified lakes and streams--aspects on induced physical-chemical and biological changes. Water Air Soil Pollut. 18:311-331.

Hustedt, F. 1927-1966. Die Kieselalgen. Rabenhorsts Kryptogamen-flora von Deutschland, Österreich und der Schweiz. Band 7. Leipzig, Germany: Geest und Portig.

Hustedt, F. 1939. Systematische und ökologische Untersuchungen uber die Diatomeen-Flora von Java, Bali, und Sumatra nach dem Material der Deutschen Limnologischen Sunda-Expedition III. Die ökologischen Factorin und ihr Einfluss auf die Diatomeenflora. Arch. Hydrobiol. Suppl. 16:274-394.

Huttunen, P., and J. Meriläinen. 1983. Interpretation of lake quality from contemporary diatom assemblages. Hydrobiologia 103:91-97.

Iskandar, I. K., and D. R. Keeney. 1974. Concentration of heavy metals in sediment cores from selected Wisconsin lakes. Environ. Sci. Technol. 8:165-170.

Jacobson, G. L., Jr., and H. J. B. Birks. 1980. Soil development on recent end moraines of the Klutlan Glacier, Yukon Territory, Canada. Quat. Res. 14:87-100.

Jeffries, D. S., R. G. Semkin, R. Neureuther, and M. D. Jones. 1983. Data Report: Major ion composition of lakes in the Turkey Lakes Watershed (January 1980-May 1982). Turkey Lakes Watershed, Unpublished Report No. TLW-83-11. 9 pp.

Johannessen, M. 1980. Aluminum--a buffer in acidic waters. Pp. 222-223 of Ecological Impacts of Acid Precipitation, Proceedings of the International Conference, Sandefjord, Norway. D. Drabløs and A. Tolonan, eds. Sur Nedbørs Virkning Pa Skog Og Fisk (SNSF) (Acid Precipitation--Effects on Forest and Fish) project, Oslo.

Johnson, D. W., and D. W. Cole. 1977. Sulfate mobility in an outwash soil in western Washington. Water Air Soil Pollut. 7:489-495.

Johnson, N. M., C. T. Driscoll, J. S. Eaton, G. E. Likens, and W. H. McDowell. 1981. "Acid rain," dissolved aluminum and chemical weathering at the Hubbard Brook Experimental Forest, New Hampshire. Geochim. Cosmochim. Acta 45:1421-1437.

Johnston, S. E. 1980. A comparison of dating methods in laminated lake sediments in Maine. M.S. thesis, University of Maine, Orono, Me. 79 pp.

Johnston, S. E., S. A. Norton, C. T. Hess, R. B. Davis, and R. S. Anderson. 1982. Chronology of atmospheric deposition of acids and metals in New England based on the record in lake sediments. Pp. 117-127 of Energy and Environmental Chemistry, L. H. Keith, ed. Ann Arbor, Mich.: Ann Arbor Science.

Jørgensen, E. G. 1948. Diatom communities in some Danish lakes and ponds. K. Dan. Vidensk. Selsk. Biol. Skr. 5:1-140.

Junge, G. E., and R. J. Werby. 1958. The concentration of chloride, sodium, potassium, calcium and sulfate in rainwater over the United States. J. Meteorol. 15:417-425.

Kahl, J. S., and S. A. Norton. 1983. Metal input and mobilization in two acid-stressed lake watersheds in Maine: Completion Report, Project A-053-ME. Land and Water Resources Center, University of Maine, Orono, Me. 70 pp.

Kahl, J. S., S. A. Norton, and J. S. Williams. 1984. Chronology, magnitude and paleolimnological record of changing metal fluxes related to atmospheric deposition of acids and metals in New England. Pp. 23-35 of Geologic Aspects of Acid Deposition, O. P. Bricker, ed. Ann Arbor, Mich.: Ann Arbor Science.

Kolbe, R. W. 1932. Grundlinien einer allgemeinen Ökologie der Diatomeen. Ergeb. Biol. 8:221-348.

Larsen, R. J. 1984. Worldwide deposition of ^{90}Sr through 1982. EML-430. Environmental Measurements Laboratory, Department of Energy. 168 pp.

Lazerte, B., and P. Dillon. 1984. Relative importance of anthropogenic versus natural sources of acidity in lakes and streams of central Ontario. Can. J. Fish. Aquat. Sci. 41:1664-1677.

Lazrus, A. L., E. Lorange, and I. P. Lodge. 1970. Lead and other ions in U.S. precipitation. Environ. Sci. Technol. 4:55-58.

Lillie, R. A., and J. W. Mason. 1983. Limnological characteristics of Wisconsin lakes. Tech. Bull. No. 138. Wisconsin Department of Natural Resources. 116 pp.

Livett, E. A., J. A. Lee, and J. H. Tallis. 1979. Lead, zinc, and copper analyses of British blanket peats. J. Ecol. 67:865-891.

Livingstone, D. A., K. Bryan, Jr., and R. G. Leaky. 1958. Effects of an arctic environment on the origin and development of freshwater lakes. Limnol. Oceanogr. 3:192-214.

Longmore, M. E., B. M. O'Leary, and C. W. Rose. 1981. Cesium-137 profiles in a near meromictic lake on Great Sandy Island (Fraser Island), Queensland, Australia. Paper presented at the Third International Symposium on Paleolimnology, Joensur, Finland.

Lowe, R. L. 1974. Environmental requirements and pollution tolerance of freshwater diatoms. EPA-670/4-74-005. U.S. Environmental Protection Agency.

Lundquist, G. 1924. Utvecklingshistoriska insjöstudier i Syd-sverige. Sver. Geol. Unders. Arsb. Ser. C 330:1-129.

Madsen, P. P. 1981. Peat bog records of atmospheric mercury deposition. Nature 293:127-130.

Martin, C. W., D. S. Noel, and C. A. Federer. 1984. Effects of forest clearcutting in New England on stream chemistry. J. Environ. Qual. 13:204-210.

Mayer, L. M., F. P. Liotta, and S. A. Norton. 1982. Hypolimnetic redox and phosphorus cycling in hypereutrophic Lake Sebasticook, Maine. Water Res. 16:1189-1196.

Meriläinen, J. 1967. The diatom flora and the hydrogen-ion concentration of the water. Ann. Bot. Fenn. 4:51-58.

Mitchell, M. J., D. H. Landers, and D. F. Brodowski. 1981. Sulfur constituents of sediments and their relationship to lake acidification. Water Air Soil Pollut. 16:351-359.

Mitchell, M. J., M. B. David, and A. J. Uutala. 1985. Sulfur distribution in lake sediment profiles as an index of historical depositional patterns. Hydrobiologia 121:121-127.

National Atmospheric Deposition Program. 1984. National Acid Deposition Program Data Report, Fourth Quarter, 1982. Vol. 4, 227 pp.

Neiboer, E., H. M. Ahmed, K. J. Puckett, and D. H. S. Richardson. 1972. Heavy metal content of lichens in relation to distance from a nickel smelter in Sudbury, Ontario. Lichenologist 5:292-304.

New York State Department of Environmental Conservation. 1978. Lake and pond survey data sheets. Albany, N.Y.

Nilsson, S. I., H. G. Miller, and J. D. Miller. 1982. Forest growth as a possible cause of soil and water acidification: an examination of the concepts. Oikos 39:40-49.

Norton, S. A. 1983. Lake sediment studies: An historical perspective of the chronology and impact of atmospheric deposition. Report of the NATO/CCMS Pilot Study on Air Pollution Control Strategies and Impact Modeling.

Norton, S. A. 1983. Sediment chemistry of four lakes in Rocky Mountain National Park, Colorado. Draft final report. Water Resources Field Support Laboratory. U.S. Department of the Interior, National Park Service, Colorado State University, Fort Collins, Colo.

Norton, S. A. 1984. Paleolimnological assessment of acidic deposition impacts on select acidic lakes: EPA/NCSU Acid Precipitation Program. Final report to U.S. Environmental Protection Agency; contract no. APP-0119-1982, 127 pp.

Norton, S. A. 1985. The sedimentary record of atmospheric pollution in Jerseyfield Lake, Adirondack Mountains, New York. In Acid Deposition--Environmental, Economic, and Policy Issues, D. D. Adams and W. Page, eds. New York: Plenum.

Norton, S. A. Geochemistry of selected Maine peat bogs. Maine Geol. Surv. Bull. 34. 38 pp. (In press.)

Norton, S. A., and C. T. Hess. 1980. Atmospheric deposition in Norway during the last 300 years as recorded in S.N.S.F. lake sediment. 1. Sediment dating and chemical stratigraphy. Pp. 268-269 of Proceedings of International Conference on the Ecological Impact of Acid Precipitation. Sur Nedbørs Virkning Pa Skog Og Fisk (SNSF), Sandefjord, Norway.

Norton, S. A., and R. F. Wright. 1984. Buffering by sediments during liming and reacidification of two acidified lakes in southern Norway. Paper presented at the Third International Symposium on Interaction between Sediment and Water, Geneva, Switzerland, Aug. 28-31.

Norton, S. A., R. F. Dubiel, D. R. Sasseville, and R. B. Davis. 1978. Paleolimnologic evidence for increased zinc loading in lakes of New England, U.S.A. Verh. Int. Ver. Limnol. 20:538-545.

Norton, S. A., C. T. Hess, and R. B. Davis. 1980. Rates of Accumulation of Heavy Metals in Pre- and Post-European Sediments in New England Lakes. Ann Arbor, Mich.: Ann Arbor Science.

Norton, S. A., R. B. Davis, and D. F. Brakke. 1981. Responses of northern New England lakes to atmospheric inputs of acids and heavy metals. Completion Report Project A-048-ME. U.S. Department of the Interior. Land and Water Resources Center, University of Maine at Orono, Orono, Me. 90 pp.

Norton, S. A., D. W. Hanson, and J. S. Williams. 1982. Modern and paleolimnological evidence for accelerated leaching and metal accumulation in soils in New England, caused by atmospheric deposition. Water Air Soil Pollut. 18:227-239.

Norton, S. A., R. B. Davis, and D. S. Anderson. 1985. The distribution and extent of acid and metal precipitation in Northern New England. Final Report. U.S. Fish and Wildlife Service Grant No. 14-16-0009-75-040.

Norton, V. C., and S. J. Herrmann. 1980. Paleolimnology of fresh-water diatoms (Bacillariophyceae) from a sediment core of a Colorado semidrainage mountain lake. Trans. Am. Microscop. Soc. 99:416-425.

Nriagu, J. O. 1983. Arsenic enrichment in lakes near the smelters at Sudbury, Ontario. Geochim. Cosmochim. Acta 47:1523-1526.

Nriagu, J. O., and H. H. Harvey. 1978. Isotopic variations as an index of sulfur pollution in lakes around Sudbury, Ontario. Nature 273:223-224.

Nriagu, J. O., and H. K. Wong. 1983. Selenium pollution of lakes near the smelters at Sudbury, Ontario. Nature 301:55-57.

Nriagu, J. O., A. L. W. Kemp, H. K. T. Wong, and N. Harper. 1979. Sedimentary record of heavy metal pollution in Lake Erie. Geochim. Cosmochim. Acta 43:247-258.

Nriagu, J. O., H. K. T. Wong, and R. D. Coker. 1982. Deposition and chemistry of pollutant metals in lakes around the smelters at Sudbury, Ontario. Environ. Sci. Technol. 16:551-560.

Nygaard, G. 1956. Ancient and recent flora of diatoms and Chrysophyceae in Lake Gribsø. Pp. 32-94 of Studies on Humic, Acid Lake Gribsø, K. Berg and I. C. Peterson, eds. Fol. Limnol. Scand. 8:1-273.

Oden, S. 1968. The acidification of air and precipitation and its consequences in the natural environment. Ecology Committee Bull. No. 1, Swedish National Research Council, Stockholm.

Oehlert, G. W. 1984. Error bars for paleolimnologically inferred pH histories: a first pass. Technical Report Number 267, Series 2. Department of Statistics, Princeton University, Princeton, N.J. 5 pp.

Oldfield, F., R. Thompson, and K. E. Barber. 1978. Changing atmospheric fallout of magnetic particles recorded in recent ombrotrophic peat sections. Science 199:679-680.

Oldfield, F., P. G. Appleby, R. S. Cambray, J. D. Eakins, K. E. Barger, R. E. Battarbee, G. R. Pearson, and J. M. Williams. 1979. ^{210}Pb, ^{137}Cs, and ^{239}Pu profiles in ombrotrophic peat. Oikos 33:40-45.

Oldfield, F., K. Tolonen, and R. Thompson. 1981. History of particulate atmospheric pollution from magnetic measurements in dated Finnish peat profiles. Ambio 10:185-188.

Oliver, B. G., and J. R. M. Kelso. 1983. A role for sediments in retarding the acidification of headwater lakes. Water Air Soil Pollut. 20:379-389.

Oliver, B. G., E. M. Thurman, and R. L. Malcolm. 1983. The contribution of humic substances to the acidity of colored natural waters. Geochim. Cosmochim. Acta 47:2031-2035.

Olson, E. T. 1983. The distribution and mobility of cesium-137 in Sphagnum bogs of northern North America. M.S. thesis, Department of Physics, University of Maine, Orono, Me. 113 pp.

Ottar, B. 1977. International agreement needed to reduce long-range transport of air pollutants in Europe. Ambio 6:262-269.

Ouellet, M., and H. G. Jones. 1983. Paleolimnological evidence for the long-range atmospheric transport of acidic pollutants and heavy metals into the Province of Quebec, eastern Canada. Can. J. Earth Sci. 20:23-36.

Pakarinen, P. 1981a. Metal content of ombrotrophic Sphagnum mosses in NW Europe. Ann. Bot. Fenn. 18:281-292.

Pakarinen, P. 1981b. Nutrient and trace metal content and retention in reindeer lichen carpets of Finnish ombrotrophic bogs. Ann. Bot. Fenn. 18:265-274.

Pakarinen, P., and K. Tolonen. 1976. Regional survey of heavy metals in peat mosses (Sphagnum). Ambio 5:38-40.

Pakarinen, P., and K. Tolonen. 1977. Distribution of lead in Sphagnum fuscum profiles in Finland. Oikos 28:69-73.

Pakarinen, P., K. Tolonen, and J. Soveri. 1980. Distribution of trace metals and sulfur in the surface peat of Finnish raised bogs. Paper presented at the Sixth International Peat Congress. Duluth, Minnesota, Aug. 17-23.

Patrick, R. 1943. The diatoms of Linsley Pond, Connecticut. Proc. Acad. Nat. Sci. Philadelphia 95:53-110.

Patrick, R. 1954. The diatom flora of Bethany Bog. J. Protozool. 1:34-37.

Patrick, R. 1977. Ecology of freshwater diatoms-diatom communities. Pp. 284-332 of The Biology of Diatoms, D. Werner, ed. Botanical Monographs, Volume 13, Oxford, England: Blackwell Scientific Publications.

Patrick, R., and C. W. Reimer. 1966. The Diatoms of the United States. Volume 1. Monogr. Acad. Nat. Sci. Philadelphia. 688 pp.

Patrick, R., and C. W. Reimer. 1975. The Diatoms of the United States, Volume 2, Part 1. Monogr. Acad. Nat. Sci. Philadelphia. 213 pp.

Patrick, R., V. P. Binetti, and S. G. Halterman. 1981. Acid lakes from natural and anthropogenic causes. Science 211:446-448.

Patterson, W. A. 1977. The effect of fire on sedimentation in a small lake in Minnesota. Abstract, p. 211 of Program of the Twentieth Congress of the Society of International Limnologists, Copenhagen.

Pennington, W. 1981. Records of a lake's life in time: the sediments. Hydrobiologia 79:197-219.

Pennington, W. 1984. Long-term natural acidification of upland sites. Pp. 28-46 in Cumbria: Evidence from Post-Glacial Lake Sediments. 52nd Annual Report. Freshwater Biological Association, Ambleside, Cumbria, England.

Pennington, W., E. Y. Haworth, A. P. Bonny, and J. P. Lishman. 1972. Lake sediments in northern Scotland. Phil. Trans. R. Soc. London Ser. B 264:191-294.

Pennington, W. (Mrs. T. G. Tutin), R. S. Cambray, and E. M. Fisher. 1973. Observations on lake sediments using fallout Cs-137 as a tracer. Nature 242:324-326.

Reed, S. 1982. The late-glacial and postglacial diatoms of Heart Lake and Upper Wallface Pond in the Adirondack Mountains, New York. M.S. thesis, San Francisco State University, San Francisco, Calif.

Rehakova, Z. 1983. Diatom succession in the post-glacial sediments of the Komorany Lake, northwest Bohemia, Czechoslovakia. Hydrobiologia 103:241-245.

Renberg, I. 1976. Paleolimnological investigations in Lake Prästsjön. Early Norrland 9:113-159.

Renberg, I. 1978. Paleolimnology and varve counts of the annually laminated sediment of Lake Rudetjärn, northern Sweden. Early Norrland 11:63-92.

Renberg, I., and T. Hellberg. 1982. The pH history of lakes in southwestern Sweden, as calculated from the subfossil diatom flora of the sediments. Ambio 11:30-33.

Renberg, I., and M. Wik. 1984. Dating recent lake sediments by soot particle counting. Verh. Int. Ver. Limnol. 22:712-718.

Reuther, R., R. F. Wright, and U. Forstner. 1981. Distribution and chemical forms of heavy metals in sediment cores from two Norwegian lakes affected by acid deposition. Pp. 318-321 of Proceedings of the International Conference on Heavy Metals in the Environment. Edinburgh, United Kingdom: CEP Consultants Ltd.

Rochester, H. 1978. Late-glacial and postglacial diatom assemblages of Berry Pond, Massachusetts, in relation to watershed ecosystem development. Ph.D. dissertation, Indiana University, Bloomington, Ind.

Rosenqvist, I. Th. 1978. Alternative sources for acidification of river water in Norway. Sci. Total Environ. 10:39-49.

Round, F. E. 1957. The late-glacial and post-glacial diatom succession in the Kentmere Valley. New Phytol. 56:98-126.

Round, F. E. 1961. The diatoms of a core from Esthwaite Water. New Phytol. 60:43-59.

Salomaa, R., and P. Alhonen. 1983. Biostratigraphy of Lake Spitaalijärvi: an ultraoligotrophic small lake in Lauhanvuori, western Finland. Hydrobiologia 103:295-301.

Schindler, D. W., R. H. Hesslein, and R. Wagermann. 1980. Effects of acidification on mobilization of heavy metals and radionuclides from the sediments of a freshwater lake. Can. J. Fish. Aquat. Sci. 37:373-377.

Schofield, C. L. 1965. Water quality in relation to survival of brook trout, Salvelinus fontinalis (Mitchell). Trans. Am. Fish. Soc. 94:227-235.

Schofield, C. L. 1976. Acidification of Adirondack lakes by atmospheric precipitation: extent and magnitude of the problem. Final report, Project F-28-R. New York State Department of Environmental Conservation, Albany, N.Y.

Schofield, C. L. 1984. Surface water chemistry in the ILWAS basins. Pp. 6-1--6-32 in The Integrated Lake-Watershed Acidification Study, Vol. 4, R. A. Goldstein and S. A. Gherini, eds. Palo Alto, Calif.: Electric Power Research Institute.

Schofield, C. L., and J. N. Galoway. 1977. The utility of paleolimnological analyses of lake sediments for evaluating acid precipitation effects on dilute lakes. Research Project Technical Completion Report. Proj. No. A-071-NY. Office of Water Res. Technol. U.S. Department of the Interior.

Sherman, J. W. 1976. Post-pleistocene diatom assemblages in New England lake sediments. Ph.D. dissertation, University of Delaware, Newark, Del.

Shiramata, H., R. W. Elias, C. C. Patterson, and M. Koide. 1980. Chronological variations in concentrations and isotopic compositions of anthropogenic atmospheric lead in sediments of a remote subalpine pond. Geochim. Cosmochim. Acta 44:149-162.

Skei, J., and P. E. Paus. 1979. Surface metal enrichment and partitioning of metals in a dated sediment core from a Norwegian fjord. Geochim. Cosmochim. Acta 433:239-246.

Smith, R. A. 1872. Air and Rain: The Beginnings of a Chemical Climatology. London: Longmans, Green.

Smol, J. P. 1980. Fossil synuracean (Chrysophyceae) scales in lake sediments: a new group of paleoindicators. Can. J. Bot. 58:458-465.

Smol, J. P. 1983. Paleophycology of a high arctic lake near Cape Herschel, Ellesmere Island. Can. J. Bot. 61:2195-2204.

Smol, J. P. 1985a. Chrysophycean microfossils as indicators of lakewater pH. In Diatoms and Lake Acidity, J. P. Smol, R. W. Battarbee, R. B. Davis, and J. Meriläinen, eds. The Hague, The Netherlands: W. Junk.

Smol, J. P. 1985b. Chrysophycean microfossils in paleolimnological studies. Palaeogeogr. Palaeoclimatol. Palaeoecol. (In press.)

Smol, J. P., and M. D. Dickman. 1981. The recent histories of three Canadian Shield lakes: a paleolimnological experiment. Arch. Hydrobiol. 93:83-108.

Smol, J. P., D. F. Charles, and D. R. Whitehead. 1984a. Mallomonadacean (Chrysophyceae) assemblages and their relationships with limnological characteristics in 38 Adirondack (New York) lakes. Can. J. Bot. 62:911-923.

Smol, J. P., D. F. Charles, and D. R. Whitehead. 1984b. Mallomonadacean microfossils provide evidence of recent lake acidification. Nature 307:628-630.

Smol, J. P., R. W. Battarbee, R. B. Davis, and J. Meriläinen, eds. 1985. Diatoms and Lake Acidity. The Hague, The Netherlands: W. Junk. (In press.)

Somers, K. M., and H. H. Harvey. 1984. Alteration of fish communities in lakes stressed by acid deposition and heavy metals near Wawa, Ontario. Can. J. Fish. Aquat. Sci. 41:20-29.

Sreenivasa, M. R., and H. C. Duthie. 1973. The postglacial diatom history of Sunfish Lake, southwestern Ontario. Can. J. Bot. 51:1599-1609.

Steinnes, E. 1977. Atmospheric deposition of trace elements in Norway studied by means of moss analysis. Kjeller, Norway: Institute for Atomenergi. 13 pp. and Appendix.

Stockner, J. G., and W. W. Benson. 1967. The succession of diatom assemblages in the recent sediments of Lake Washington. Limnol. Oceanogr. 12:513-533.

Sweets, P. R. 1983. Differential deposition of diatom frustules in Jellison Hill Pond, Maine. M.S. thesis, University of Maine, Orono, Me. 156 pp.

Tan, Y. L., and M. Heit. 1981. Biogenic and abiogenic polynuclear aromatic hydrocarbons in sediments from two remote Adirondack Lakes. Geochim. Cosmochim. Acta 45:2267-2279.

Taylor, D. R., M. A. Tompkins, S. E. Kirton, T. Mauney, and D. F. S. Natusch. 1982. Analysis of fly ash produced from combustion of refuse-derived fuel and coal mixtures. Environ. Sci. Technol. 16:148-154.

Tolonen, K. 1967. Über die Entwicklung der Moore in finnischen Nordbarelien. Ann. Bot. Fenn. 4:219-416.

Tolonen, K. 1972. On the palaeo-ecology of the Hamptjärn Basin II. Bio- and chemostratigraphy. Early Norrland 1:53-77.

Tolonen, K. 1980. Pollen, algal remains, and macrosubfossils from Lake Galltrask, S. Finland. Ann. Bot. Fenn. 17:394-405.

Tolonen, K., and T. Jaakkola. 1983. History of lake acidification and air pollution studied in sediments in South Finland. Ann. Bot. Fenn. 20:57-78.

Tolonen, K., and J. Meriläinen. 1983. Sedimentary chemistry of a small polluted lake, Galltrask, S. Finland. Hydrobiologia 103:309-318.

Tolonen, K., R. B. Davis, and L. S. Widoff. 1984. Peat accumulation rates in selected Maine peatlands. Maine Geological Survey, Department of Conservation. 109 pp.

Tolonen, K., M. Liukkonen, R. Harjula, and A. Patila. 1985. Acidification of small lakes in Finland studied by means of sedimentary diatom and chrysophyceae remains. In Diatoms and Lake Acidity, J. P. Smol, R. W. Battarbee, R. B. Davis, and J. Meriläinen, eds. The Hague, The Netherlands: W. Junk.

Tolonen, K., R. B. Davis, and L. Widoff. Peat accumlation rates in selected Maine peatlands. Maine Geol. Surv. Bull. (In press.)

Tynni, R. 1972. The development of Lovojärvi on the basis of its diatoms. Aqua Fenn. 74-82.

van Dam, H., and H. Kooyman-van Blokland. 1978. Man-made changes in some Dutch moorland pools, as reflected by historical and recent data about diatoms and macrophytes. Int. Rev. Gesam. Hydrobiol. 63:587-607.

van Dam, H., G. Suurmond, and C. J. F. ter Braak. 1981. Impact of acidification on diatoms and chemistry of Dutch moorland pools. Hydrobiologia 83:425-459.

Vitt, D. H., and S. Bayley. 1984. The vegetation and water chemistry of four oligotrophic basin mires in northwestern Ontario. Can. J. Bot. 62:1485-1500.

Wakeham, S. G., C. Schaffner, and W. Giger. 1980. Polycyclic aromatic hydrocarbons in recent lake sediments. I. Compounds having anthropogenic origins. Geochim. Cosmochim. Acta 44:403-413.

Walker, R. 1978. Diatom and pollen studies of a sediment profile from Melynllyn, a mountain tarn in Snowdownia, North Wales. New Phytol. 81:791-804.

Watanabe, T., and I. Yasuda. 1982. Diatom assemblages in lacustrine sediments of Lake Shibu-ike, L. Misume-ike, L. Naga-ike, L. Kido-ike in Shiga Highland and a new biotic index based on the diatom assemblage for the acidity of lake water. Jpn. J. Limnol. 43:237-245.

Whitehead, D. R., D. F. Charles, S. E. Reed, S. T. Jackson, and M. C. Sheehan. In press. Late-glacial and Holocene acidity changes in Adirondack (N.Y.) lakes. In Diatoms and Lake Acidity, J. P. Smol, R. W. Battarbee, R. B. Davis, and J. Meriläinen, eds. Dordrecht, The Netherlands: W. Junk.

Wright, H. E., Jr. 1981. The role of fire in land/water interactions. In Fire Regimes and Ecosystem Properties, Pp. 421-444 of Proceedings of a conference held December 11-15, 1978, in Honolulu, Hawaii. General Technical Report WO-26, U.S. Department of Agriculture.

Wright, R. F. 1983. Predicting acidification of North American lakes. Report 4/1983, Norwegian Institute for Water Research, Oslo. 165 pp.

Wright, R. F. 1984. Changes in the chemistry of Lake Hovvatn, Norway following liming and reacidification. Report 6/1984, Norwegian Institute for Water Research, Oslo. 68 pp.

Yan, N. 1983. Effects of changes in pH on transparency and thermal regimes of Lohi lake, near Sudbury, Ontario. Can. J. Fish. Aquat. Sci. 40:621-626.

APPENDIXES

APPENDIX

A

Methods for Sampling and Analysis of Red Spruce Data

Red spruce were surveyed during the summer of 1982 at 11 sites in New Hampshire, Vermont, New York, West Virginia, Virginia, and North Carolina. The northern sites were emphasized. Mount Washington, Mount Mansfield, and Whiteface Mountain were selected because they had operating weather stations at their summits. The sites in the southern Appalachians were selected as typical stands through consultation with local foresters. Hunter and Plateau mountains in the Catskill Mountains (New York) were selected on the basis of the stand descriptions of McIntosh and Hurley (1964).

Transects were established with an altimeter at arbitrary elevations on east- and west-facing slopes. On Mount Washington, Mount Mansfield, and Whiteface Mountain, transects 100 m long were established parallel to elevational contours at 50-m intervals through the spruce-fir, transition, and hardwood forests. The elevations of the transects generally ranged from 1100 to 650 m. The elevation of a given transect was adjusted if topographic conditions so dictated. At the other sites an arbitrary elevation judged to be representative of the stand was selected and a transect 600 m long was established. Sampling along the transects was designed to meet three major objectives: (1) to rate the extent of deterioration of spruce in the canopy, (2) to determine the percentage of dead spruce across a variety of size classes, and (3) to obtain increment cores to determine canopy age and for tree ring analysis.

To determine the extent of deterioration of spruce in the canopy, sampling points were established at 25-m intervals by tape if practical, or by pacing. The nearest spruce greater than 10-cm diameter breast height (dbh) in the canopy in each quadrant around the sampling point was rated using a four-point scale as follows: 1, little or no loss of foliage in the upper crown. 2, loss of foliage from the top of the crown and/or some loss of foliage on the outer tips of live branches; total loss of foliage from the upper portion of the crown less than 50 percent. 3, loss of foliage and dieback of the upper crown greater than 50 percent. 4, dead. A zero was recorded if there were no red spruce in a quadrant within an arbitrary distance of the sampling point (usually 12.5 m). At lower elevations in hardwood dominated forests, the sampling intervals were extended in some locations so as to include a reasonable number of spruce.

Results are recorded in Table A.1. At each sampling point, the nearest live red spruce greater than 10-cm dbh in the canopy was cored to estimate canopy ages. Two cores per tree were extracted at 1.2 m above ground surface, parallel to the topographic contours. At each cored tree an estimate of stand basal area was made using a wedge prism (basal area factor = metric 2.5). This estimate has a slight positive bias, as the sampling point was always located within 1 m of a cored tree. The species, dbh, and live/dead status of each stem in the prism were recorded and used to estimate the percentage of live and dead red spruce in each dbh class. To get a sample of dominant trees for tree ring analyses, stands between 800 and 950 m were sampled for red spruce and balsam fir. (See Figures 6.8 to 6.10.) A subset of cores was collected from trees that were in class 1 or 2, had no sign of past injury, and were taller than the neighboring trees. This subset included some cored trees from the intermediate elevation transects as well as spruce from nearby if all spruce at the sampling point were class 3 or 4. Occasionally class 3 trees were sampled if they were the only ones available. Fifteen trees (two cores per tree) that cross-dated well were used for the analyses. This sample is biased in favor of more vigorous trees.

Tree ring index chronologies shown in Figures 6.8 to 6.10 were produced by D. Duvick and T. J. Blasing at Oak Ridge National Laboratories for the Forest Response to Anthropogenic Stress (FORAST) program (McLaughlin et al. 1983). We gratefully acknowledge their contributions. Measured cores were cross-dated with a skeleton-plot technique (Stokes and Smiley 1968) to assure accurate determination of the date of each ring measured.

The time series of annual growth measurements for any given core will typically contain two types of variation:

TABLE A.1 Determination of the Extent of Deterioration in Red Spruce at Selected Sites in the Eastern United States

Site Name	No. Red Spruce Observed on Transect	% Red Spruce Rated 1	% Red Spruce Rated 2	% Red Spruce Rated 3	% Red Spruce Rated 4	Elevation of Transect (m)	Basal Area (m^2/ha) (Live Plus Dead > 1.5 m Tall)
Mt. Washington, East	017	65	12	18	06	1120	50
	019	42	00	32	26	1060	38
	019	53	05	16	26	1000	42
	020	50	15	20	15	0950	34
	020	25	15	35	25	0900	39
	019	53	26	11	11	0850	32
	018	28	22	22	28	0800	34
	016	50	19	12	19	0750	33
	017	41	12	00	47	0700	32
	020	55	30	05	10	0650	34
Mt. Washington, West	020	25	25	30	20	1150	36
	020	15	35	30	20	1100	40
	020	35	10	35	20	1050	39
	020	15	25	40	20	1000	37
	020	55	15	15	15	0950	32
	020	45	15	20	20	0900	37
	020	40	15	05	40	0850	42
	020	60	20	00	20	0800	34
	020	60	15	10	15	0750	32
	020	60	20	00	20	0700	36
Mt. Mansfield, East	015	07	07	47	40	1120	26
	019	37	05	37	21	1060	30
	020	20	15	25	40	1000	31
	017	12	18	29	41	0950	25
	008	50	00	12	38	0900	29
	016	38	25	12	25	0850	24
	018	28	28	00	44	0800	29
	013	46	15	15	23	0750	26
	013	54	15	08	23	0700	28
	012	75	08	00	17	0650	24
Mt. Mansfield, West	020	20	10	35	35	1050	18
	018	11	22	45	22	1000	26

TABLE A.1 (continued)

Site Name	No. Red Spruce Observed on Transect	% Red Spruce Rated 1	% Red Spruce Rated 2	% Red Spruce Rated 3	% Red Spruce Rated 4	Elevation of Transect (m)	Basal Area (m^2/ha) (Live Plus Dead > 1.5 m Tall)
Mt. Mansfield, West	020	35	05	25	35	0950	22
	019	26	21	05	47	0900	30
	020	25	20	15	40	0850	32
	020	35	25	15	25	0800	24
Whiteface Mtn., East	020	10	10	45	35	1110	28
	019	16	42	26	16	1060	31
	017	24	47	18	12	1000	42
	018	56	28	00	17	0950	30
	014	50	29	14	07	0900	28
	014	00	79	00	21	0850	41
	019	00	63	21	16	0800	40
	016	38	56	00	06	0750	33
	020	50	35	05	10	0700	30
	020	45	40	10	05	0650	27
Whiteface Mtn., West	019	26	21	21	32	1220	34
	020	00	20	40	40	1150	34
	014	14	14	29	43	1070	34
	020	25	20	25	30	1000	36
	018	28	44	00	28	0950	42
	017	47	24	06	24	0910	38
	020	70	10	10	10	0850	52
	019	68	26	00	05	0780	40
	018	56	17	00	28	0720	40
	020	75	15	00	10	0670	34
Hunter Mtn.	092	48	21	11	20	1052	30
Plateau Mtn.	100	63	20	07	10	1128	35
Shenandoah Natl. Pk.	088	72	08	00	20	1005	40
G. Washington Natl. For.	100	89	07	00	04	1128	32
Monongahela Natl. For.	100	69	10	02	19	1311	48
Whitetop Mtn.	100	84	10	02	04	1524	37
Roan Mtn.	093	83	04	01	12	1768	35
Mt. Mitchell	082	91	04	01	04	1615	36

(1) high-frequency variance owing to the influence of factors over time intervals of a year or a few years and (2) low-frequency variance caused by longer-term fluctuations involving factors, such as tree age or stand development, or long-term climatic changes that have periods of influence typically extending over several years or decades. To describe the low-frequency variations in growth over the tree's life cycle, cubic spline functions (Blasing et al. 1983, Cook and Peters 1981) were fitted to the ring-width series of each tree over discrete time intervals (i.e., before and after a sudden major release from competition after a known major disturbance such as logging). A spline statistically delineates a weighted moving average of growth over the specified time interval. Dividing each ring-width value by the corresponding value of the spline function produces a series of ring-width indices (Fritts et al. 1971, Fritts 1976, Blasing et al. 1983) that have a mean of 1.0 and a variance that is approximately constant through time. The filtering characteristics of the spline were chosen to remove low-frequency variance characteristic of the slow changes in competitive status and tree age but to preserve intermediate- and high-frequency variance (Nash et al. 1975, Blasing et al. 1983).

REFERENCES

Blasing, T. J., D. N. Duvick, and E. R. Cook. 1983. Filtering the effects of competition from ring width series. Tree Ring Bull. 43:19-30.

Cook, E. R., and K. Peters. 1981. The smoothing spline: A new approach to standardizing forest interior tree-ring width series for dendroclimatic studies. Tree Ring Bull. 41:45-54.

Fritts, H. C. 1976. Tree Rings and Climate. London: Academic Press.

Fritts, H. C., T. J. Blasing, P. B. Hayden, and J. E. Kutzbach. 1971. Multivariate techniques for specifying tree-growth and climate relationships and for reconstructing anomalies in paleoclimate. J. Appl. Meteorol. 10:845-864.

McIntosh, R. P., and R. T. Hurley. 1964. The spruce-fir forests of the Catskill Mountains. Ecology 45:314-326.

McLaughlin, S. B., T. J. Blasing, L. H. Mann, and D. N. Duvick. 1983. Effects of acid rain and gaseous pollutants on forest productivity. J. Air Pollut. Control Assoc. 33:1042-1048.

Nash, T. III, H. C. Fritts, and M. A. Stokes. 1975. A technique for examining non-climatic variation in widths of annual tree rings with special reference to air pollution. Tree Ring Bull. 35:14-20.

Stokes, M. A., and T. L. Smiley. 1968. An Introduction to Tree Ring Dating. Chicago: University of Chicago Press.

APPENDIX

B

Input Sulfate Fluxes to Lakes from Wet-Only Deposition and Output Sulfate Fluxes from Lakes: Data Sources and Methods

Input sulfate fluxes to lakes from wet-only deposition (grams per square meter per year) were calculated as the product of sulfate concentration in wet deposition (grams per cubic meter) and the annual amount of wet deposition (meters per year). Sulfate concentration data from the Canadian Network for Sampling Precipitation (CANSAP) and the Air and Precipitation Network (APN) were provided through the courtesy of the Atmospheric Environment Service, Canada. Sulfate concentration data from the National Acid Deposition Program (NADP) monitoring network in the United States were provided through the courtesy of the Illinois State Water Survey. In total, we examined data from 3 APN stations, 16 CANSAP stations, and 17 NADP stations for the period 1980-1982. The annual average amount of wet deposition at these stations over the same time period was determined with standard meteorological rain gauges. The flux data are summarized in Table B.1. These data were plotted and contoured, and values of sulfate flux from wet-only deposition at each lake were determined from interpolation of the contour plot.

Output sulfate fluxes from lakes (grams per square meter per year) were determined as the product of the concentration of sulfate in lake water (grams per cubic meter) and the net annual average amount of wet deposition (precipitation minus evaporation; meters per year). The following summarizes the sources of data for sulfate concentrations in lakes. (Sample sizes are in parentheses.)

(1) New England: Haines et al. 1983. States surveyed were Massachusetts (34), Connecticut (23), Rhode

TABLE B.1 Average Fluxes of Sulfate in Wet-Only Deposition for the Period 1980 to 1982 in Northeastern North America

Location	Station	Flux (g m^{-2} yr^{-1})
Canada		
APN:		
Baie D'Espoir	47°59′,55°48′	1.48
Kejimkujik	44°26′,65°12′	3.37
Montmorency	47°19′,71°09′	2.52
CANSAP:		
Acadia FES	46°00′,66°22′	2.61
Charlo	48°00′,66°20′	2.85
Chibougamau	49°49′,74°25′	1.75
Fort Chimo	58°06′,68°25′	0.62
Gander	48°57′,54°34′	1.90
Goose	53°19′,60°25′	0.94
Kejimkujik	44°26′,65°02′	4.00
Maniwaki	46°23′,75°58′	2.97
Nitchequon	53°12′,70°54′	3.95
Quebec City	46°48′,71°24′	5.19
Sable Island	43°56′,60°01′	7.99
Saint John	45°19′,65°53′	3.49
Sept Isles	50°13′,66°15′	2.06
Shelburne	43°43′,65°15′	2.87
St. Hubert	45°31′,73°25′	4.32
Stephensville	48°32′,58°33′	2.50
Truro	45°22′,63°16′	2.96
United States Of America		
NADP:		
MA01	41°58′23″,70°01′12″	2.13 (3.01)[a]
MA08	42°21′49″,72°23′27″	3.22 (1.89)
MA13	42°23′02″,71°12′53″	2.16 (3.01)
ME00	42°52′08″,68°00′55″	1.61 (1.55)
ME02	44°06′27″,70°43′44″	1.80 (1.62)
ME09	45°29′23″,69°39′52″	1.20 (1.29)
ME99	44°24′30″,68°14′42″	2.42 (2.40)
NH02	46°56′35″,71°42′12″	2.28 (2.37)
NY08	42°44′02″,76°39′35″	2.70 (2.83)
NY10	42°17′58″,79°23′47″	4.45 (3.28)
NY20	43°58′19″,74°13′25″	3.52 (3.28)
NY51	41°21′00″,74°02′22″	2.35 (2.05)
NY52	43°31′34″,75°56′59″	4.34 (3.13)
NY65	42°06′22″,77°32′08″	2.32 (1.65)
VT01	42°52′34″,73°09′48″	2.28 (2.37)

[a] Flux calculated using the volume of sample and collector diameter given in parentheses.

Island (8), Vermont and New Hampshire (43), and Maine (33); there were many duplicate analyses.

(2) Adirondack Mountains, New York: Charles 1983. Averages of lake profile data were used. Sample size, 51.

(3) Vermont: Vermont Department of Water Resources and Environmental Engineering 1983; averages over 3 years of monthly sampling. Sample size, 49.

(4) Eastern Canada: Clair et al. 1982; including Labrador (105), Newfoundland (94), and Quebec. These were typically for one sample only and included some rivers.

(5) Quebec: Bobee et al. 1983; analyzed mostly headwater lakes in Quebec; the analysis consists of one sample. Sample size including lakes of Quebec sampled in survey (4) was 186.

A value for the sulfate output flux at each lake was estimated in the following way:

1. Average precipitation amounts for 1980-1982 for Canada (contoured, supplied by Atmospheric Environment Service) and discrete U.S. Weather Bureau data for the United States were combined into one contour map from which the average annual amount of precipitation was estimated for each lake.

2. Average annual evaporation data were contoured and estimates of evaporation were made for each lake. Canadian data consisted of actual measurements of evaporation on lakes at 15 different locations in eastern Canada. Evaporation estimates for the northeastern United States were calculated for nonwinter averages of Weather Bureau evaporation pan data at 14 locations. The evaporation pan data were multiplied by a factor of 0.77 to account for the difference between pan evaporation and lake surface evaporation (Kohler et al. 1955).

3. The net annual wet deposition was determined as the wet deposition amount minus evaporation for each lake.

4. The sulfate output flux at each lake was determined as the product of the sulfate concentration of the lake water and the net annual wet deposition.

REFERENCES

Bobee, B., M. Lachance, J. Haermmerli, A. Tessier, J. Charette, and J. Kramer. 1983. Evaluation de la sensibilite a l'acidification des lacs du sud du Quebec et incidences sur le reseau d'acquisition de donnees. Scientific Report No. 157. Ste-Foy, Quebec: University of Quebec.

Charles, D. 1983. Studies of Adirondack Mountain (New York) lakes: limnological characteristics and sediment diatom-water chemistry relationships. Ph.D. dissertation, Indiana University. Ann Arbor, Michigan: University Microfilms.

Clair, T. A., D. R. Engstrom, W. Whitman. 1982. Data report--water quality of surface waters of Newfoundland and Labrador. Environment Canada, Burlington, Ontario. Report No. IWD-AR-WQB-82-32, 55 pp.

Haines, T. A., J. J. Akielagzek, and P. J. Rago. 1983. A regional survey of chemistry of headwater lakes and streams in New England: vulnerability to acidification. Air Pollution and Acid Rain Series. U.S. Fish and Wildlife Service Report.

Kohler, M. A., T. J. Nordenson, and W. E. Fox. 1955. Evaporation from pans and lakes. Research paper 38. U.S. Weather Bureau, Department of Commerce.

Vermont Department of Water Resources and Environmental Engineering. 1983. Vermont acid precipitation program--long-term lake monitoring. Montpelier, Vermont.

APPENDIX C

Characteristics of Bench-Mark Streams

Table C.1 and Figure C.1 give the locations of the U.S. Geological Survey Bench-Mark streams analyzed in Chapter 7. The numbering system used for these basins follows that in Table C.2, which contains data on the chemical composition of these streams, including estimates of time trends in concentrations of common ions using the seasonal Kendall test.

In Table C.2, slopes of trends significant at the 0.1 level are denoted by an asterisk. Slopes of trends significant at the 0.01 level are denoted by a double asterisk. ID denotes insufficient data for estimating trends. N is the number of samples measured. Y/D is the ratio of basin sulfate yield to sulfate deposition. Sulfate output from the Bench-Mark watersheds was estimated as the product of discharge-weighted mean sulfate concentrations and mean basin runoff for the 2-year period from 1980 and 1981. Sulfate deposition at each basin was estimated by linear interpolation of wet-only sulfate deposition data from the National Acid Deposition Program (NADP) for 1980 and 1981.

Table C.3 (which begins on page 464) contains descriptions of 16 basins of the U.S. Geological Survey Bench-Mark streams with alkalinities lower than 500 μeq/L.

TABLE C.1 Key to Location of USGS Bench-Mark Streams (see Figure C.1)

Map Number	Name of Stream	Location
1	Wild River	Gilead, Maine
2	Esopus Creek	Shandaken, New York
3	McDonalds Branch	Lebanon State Forest, New Jersey
4	Young Womans Creek	near Renovo, Pennsylvania
5	Holiday Creek	near Andersonville, Virginia
6	Scape Ore Swamp	near Bishopville, South Carolina
7	Upper Three Runs	near New Ellenton, South Carolina
8	Falling Creek	near Juliette, Georgia
9	Sopchoppy River	near Sopchoppy, Florida
10	Sipsey Fork	near Grayson, Alabama
11	Cypress Creek	near Janice, Mississippi
12	Cataloochee Creek	near Cataloochee, North Carolina
13	Buffalo River	near Flat Woods, Tennessee
14	Washington Creek	Windigo, Michigan
15	Popple River	near Fence, Wisconsin
16	Kawishiwi River	near Ely, Minnesota
17	North Fork Whitewater River	near Elba, Minnesota
18	Castle Creek above Deerfield Reservoir	near Hill City, South Dakota
19	Encampment River above Hog Park Creek	near Encampment, Wyoming
20	Dismal River	near Thedford, Nebraska
21	North Sylamore Creek	near Fifty Six, Arkansas
22	Half Moon Creek	near Malta, Colorado
23	Blue Beaver Creek	near Cache, Oklahoma
24	Kiamichi River	near Big Cedar, Oklahoma
25	Big Creek	Pollock, Louisiana
26	Rio Mora	near Terrero, New Mexico
27	Wet Bottom Creek	near Childs, Arizona
28	Merced River at Happy Isles Bridge	near Yosemite, California
29	Hayden Creek below North Fork	near Hayden Lake, Idaho
30	Andrews Creek	near Mazama, Washington
31	Big Jacks Creek	near Bruneau, Idaho
32	Minam River	Minam, Oregon

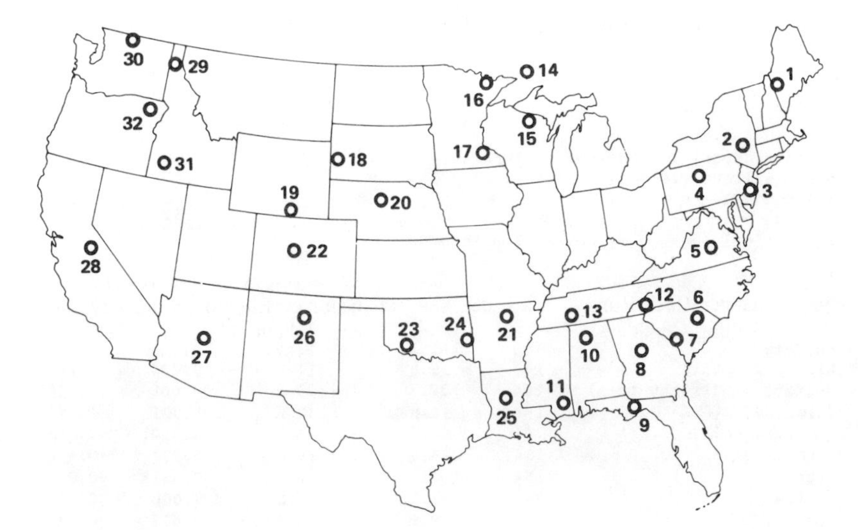

FIGURE C.1 Locations of USGS Bench-Mark streams.

TABLE C.2 Chemical Characteristics and Time Trends of Chemical Components in Bench-Mark Streams

1. WILD RIVER AT GILEAD, ME

Drainage area (km^2) 179.94
Mean discharge (m^3/sec) 4.56
Period of record 1964 - 1983

Sulfate deposition as sulfur (g/m^2.yr) 0.86
Basin sulfate yield as sulfur (g/m^2.yr) 1.31
Ratio (Y/D) 1.51

Statistical Summary of Common Ion Records

Ion	N	Mean Concentration (µeq/l)	Standard Deviation	Trend p	Trend Slope (µeq/l.yr)
Sulfate	70	102.7	19.5	0.002	-1.5**
Nitrate, total	51	3.9	2.5	0.219	0.1
Sulfate + nitrate, total	70	105.6	19.6	0.008	-1.2**
Chloride	67	22.7	12.8	0.123	0.4
Ammonia, total	22	3.2	2.9	ID	ID
Calcium	70	87.6	18.8	0.219	-0.6
Magnesium	70	41.4	15.1	0.539	-0.2
Sodium	69	55.7	14.1	0.607	-0.1
Potassium	70	10.0	3.6	0.609	0.0
Major cations (Ca+Mg+Na+K)	69	194.0	33.0	0.218	-0.8
Alkalinity	70	67.7	38.4	0.024	1.0*
Hydrogen ion	25	442.0	266.4	1.000	-4.4

2. ESOPUS CREEK AT SHANDAKEN, NY

Drainage area (km^2) 154.05
Mean discharge (m^3/sec) 3.79
Period of record 1964 - 1983

Sulfate deposition as sulfur (g/m^2.yr) 1.16
Basin sulfate yield as sulfur (g/m^2.yr) 1.95
Ratio (Y/D) 1.68

Statistical Summary of Common Ion Records

Ion	N	Mean Concentration (µeq/l)	Standard Deviation	Trend p	Trend Slope (µeq/l.yr)
Sulfate	153	162.1	25.9	0.000	-1.8**
Nitrate, total	143	15.8	11.8	0.000	1.1**
Sulfate + nitrate, total	154	177.7	29.5	0.008	-0.8**
Chloride	154	94.8	40.6	0.001	1.7**
Ammonia, total	51	4.3	4.3	0.000	-1.1**
Calcium	154	274.0	59.6	0.334	-0.5
Magnesium	154	100.7	18.9	0.931	-0.0
Sodium	154	93.1	33.2	0.000	2.0**
Potassium	154	8.6	3.6	0.012	0.1*
Major cations (Ca+Mg+Na+K)	154	476.5	99.4	0.073	1.6*
Alkalinity	154	198.7	80.0	0.084	1.2*
Hydrogen ion	89	329.6	480.4	0.619	9.1

3. McDONALDS BRANCH IN LEBANON STATE FOREST, NJ

Drainage area (km^2) 5.98
Mean discharge (m^3/sec) 0.06
Period of record 1963 - 1983

Sulfate deposition as sulfur (g/m^2.yr) 1.08
Basin sulfate yield as sulfur (g/m^2.yr) 0.75
Ratio (Y/D) 0.69

Statistical Summary of Common Ion Records

Ion	N	Mean Concentration (µeq/l)	Standard Deviation	Trend p	Trend Slope (µeq/l.yr)
Sulfate	40	154.6	85.5	0.150	-4.3
Nitrate, total	36	4.5	3.6	0.787	-0.0
Sulfate + nitrate, total	40	158.7	86.0	0.150	-4.3
Chloride	40	99.7	26.1	0.369	0.6
Ammonia, total	16	1.9	1.8	ID	ID
Calcium	40	50.8	27.3	0.072	-1.7*
Magnesium	38	44.8	25.2	0.857	-0.6
Sodium	40	79.7	19.0	0.125	-0.9
Potassium	36	9.1	8.3	0.769	0.2
Major cations (Ca+Mg+Na+K)	40	181.7	65.4	0.281	-2.2
Alkalinity	40	-63.9	37.1	0.472	1.2
Hydrogen ion	24	64227.0	29048.8	0.296	4199.1

4. YOUNG WOMANS CREEK NEAR RENOVO, PA

Drainage area (km^2) 119.61
Mean discharge (m^3/sec) 2.04
Period of record 1967 - 1983

Sulfate deposition as sulfur (g/m^2.yr) 1.16
Basin sulfate yield as sulfur (g/m^2.yr) 1.44
Ratio (Y/D) 1.24

Statistical Summary of Common Ion Records

Ion	N	Mean Concentration (μeq/l)	Standard Deviation	Trend p	Trend Slope (μeq/l.yr)
Sulfate	140	164.0	29.6	0.936	0.1
Nitrate, total	138	16.8	7.9	0.000	0.7**
Sulfate + nitrate, total	140	180.6	30.3	0.041	1.1*
Chloride	139	38.2	13.8	0.000	-0.9**
Ammonia, total	25	1.9	2.5	1.000	-0.0
Calcium	140	203.4	30.1	0.297	-0.4
Magnesium	140	85.8	15.8	0.345	0.2
Sodium	140	42.6	15.9	0.059	-0.3*
Potassium	139	20.6	5.7	0.226	-0.1
Major cations (Ca+Mg+Na+K)	140	352.7	51.0	0.100	-1.1*
Alkalinity	140	131.9	55.4	0.014	1.7*
Hydrogen ion	92	195.8	143.7	0.255	-4.0

5. HOLIDAY CREEK NEAR ANDERSONVILLE, VA

Drainage area (km^2) 22.08
Mean discharge (m^3/sec) 0.23
Period of record 1967 - 1983

Sulfate deposition as sulfur (g/m^2.yr) 1.00
Basin sulfate yield as sulfur (g/m^2.yr) 0.34
Ratio (Y/D) 0.34

Statistical Summary of Common Ion Records

Ion	N	Mean Concentration (μeq/l)	Standard Deviation	Trend p	Trend Slope (μeq/l.yr)
Sulfate	94	60.6	28.6	0.471	-0.3
Nitrate, total	90	6.2	4.4	0.095	-0.3*
Sulfate + nitrate, total	94	66.5	29.4	0.280	-0.6
Chloride	93	58.2	21.6	0.068	-0.6*
Ammonia, total	18	3.2	3.9	ID	ID
Calcium	93	133.5	40.0	0.013	-1.7*
Magnesium	93	88.6	24.6	0.303	0.4
Sodium	94	121.5	30.3	0.005	-1.2**
Potassium	94	20.2	11.7	0.077	-0.3*
Major cations (Ca+Mg+Na+K)	93	362.8	68.5	0.011	-4.1*
Alkalinity	93	230.4	67.0	0.921	0.3
Hydrogen ion	54	502.6	1559.6	0.005	-78.8**

6. SCAPE ORE SWAMP NEAR BISHOPVILLE, SC

Drainage area (km^2) 248.54
Mean discharge (m^3/sec) 2.77
Period of record 1970 - 1983

Sulfate deposition as sulfur ($g/m^2.yr$) 0.89
Basin sulfate yield as sulfur ($g/m^2.yr$) 0.36
Ratio (Y/D) 0.40

Statistical Summary of Common Ion Records

Ion	N	Mean Concentration (μeq/l)	Standard Deviation	Trend p	Trend Slope (μeq/l.yr)
Sulfate	101	63.0	29.6	0.000	3.3**
Nitrate, total	100	12.7	11.3	0.700	0.1
Sulfate + nitrate, total	101	75.3	27.7	0.000	2.8**
Chloride	102	119.6	19.7	0.949	0.0
Ammonia, total	18	2.1	1.8	ID	ID
Calcium	102	43.8	21.2	0.124	0.5
Magnesium	102	44.1	15.2	0.012	0.8*
Sodium	102	124.0	24.3	0.000	1.7**
Potassium	101	11.9	4.5	0.051	-0.2*
Major cations (Ca+Mg+Na+K)	102	224.0	35.1	0.005	2.9**
Alkalinity	102	21.9	31.9	0.200	0.7
Hydrogen ion	80	4699.1	5692.7	0.003	550.2**

7. UPPER THREE RUNS NEAR NEW ELLENTON, SC

Drainage area (km^2) 225.24
Mean discharge (m^3/sec) 3.11
Period of record 1968 - 1983

Sulfate deposition as sulfur ($g/m^2.yr$) 0.88
Basin sulfate yield as sulfur ($g/m^2.yr$) 0.14
Ratio (Y/D) 0.16

Statistical Summary of Common Ion Records

Ion	N	Mean Concentration (μeq/l)	Standard Deviation	Trend p	Trend Slope (μeq/l.yr)
Sulfate	68	21.7	13.1	0.058	0.9*
Nitrate, total	67	12.6	5.6	0.006	0.4**
Sulfate + nitrate, total	68	32.8	15.9	0.012	1.5*
Chloride	70	63.6	13.3	0.000	-1.3**
Ammonia, total	7	2.7	1.8	ID	ID
Calcium	69	27.1	12.9	0.048	0.6*
Magnesium	70	24.6	7.5	0.518	-0.1
Sodium	69	57.8	10.4	0.006	0.8**
Potassium	70	6.2	2.8	0.111	0.1
Major cations (Ca+Mg+Na+K)	70	118.1	18.4	0.006	1.7**
Alkalinity	70	23.2	20.4	0.149	0.6
Hydrogen ion	35	1539.2	1033.4	0.008	131.8**

8. FALLING CREEK NEAR JULIETTE, GA

Drainage area (km^2) 186.93
Mean discharge (m^3/sec) 1.50
Period of record 1968 - 1983

Sulfate deposition as sulfur (g/m^2.yr) 0.81
Basin sulfate yield as sulfur (g/m^2.yr) 0.44
Ratio (Y/D) 0.55

Statistical Summary of Common Ion Records

Ion	N	Mean Concentration (μeq/l)	Standard Deviation	Trend p	Trend Slope (μeq/l.yr)
Sulfate	94	101.1	72.4	0.000	3.7**
Nitrate, total	95	4.5	5.5	0.000	0.2**
Sulfate + nitrate, total	93	103.1	72.7	0.000	3.8**
Chloride	102	94.4	27.9	0.051	0.7*
Ammonia, total	43	3.3	4.0	0.386	0.1
Calcium	102	470.9	123.5	0.002	-5.3**
Magnesium	102	391.9	98.5	0.166	-1.8
Sodium	102	327.6	101.1	0.013	3.2*
Potassium	102	33.9	10.9	0.131	-0.2
Major cations (Ca+Mg+Na+K)	102	1224.3	294.7	0.529	-2.4
Alkalinity	102	975.2	314.5	0.187	-3.8
Hydrogen ion	80	202.3	472.6	0.007	-27.6**

9. SOPCHOPPY RIVER NEAR SOPCHOPPY, FL

Drainage area (km^2) 264.08
Mean discharge (m^3/sec) 4.44
Period of record 1965 - 1983

Sulfate deposition as sulfur (g/m^2.yr) 0.91
Basin sulfate yield as sulfur (g/m^2.yr) 0.90
Ratio (Y/D) 0.99

Statistical Summary of Common Ion Records

Ion	N	Mean Concentration (μeq/l)	Standard Deviation	Trend p	Trend Slope (μeq/l.yr)
Sulfate	74	94.4	56.2	0.003	3.0**
Nitrate, total	54	4.8	4.9	0.730	-0.0
Sulfate + nitrate, total	71	95.4	57.9	0.006	3.0**
Chloride	74	127.5	22.6	0.195	0.2
Ammonia, total	37	3.0	1.7	0.894	0.0
Calcium	74	946.8	743.3	0.796	1.9
Magnesium	74	174.3	120.0	1.000	-0.1
Sodium	74	96.9	20.9	0.038	0.7*
Potassium	72	6.3	3.6	0.893	-0.0
Major cations (Ca+Mg+Na+K)	74	1224.2	876.4	0.863	6.5
Alkalinity	74	960.1	901.5	0.931	-1.9
Hydrogen ion	39	14070.2	31289.3	0.810	452.9

10. SIPSEY FORK NEAR GRAYSON, AL

Drainage area (km^2) 233.27
Mean discharge (m^3/sec) 4.41
Period of record 1967 - 1983

Sulfate deposition as sulfur (g/m^2.yr) 0.95
Basin sulfate yield as sulfur (g/m^2.yr) 0.81
Ratio (Y/D) 0.85

Statistical Summary of Common Ion Records

Ion	N	Mean Concentration (µeq/l)	Standard Deviation	Trend p	Trend Slope (µeq/l.yr)
Sulfate	67	85.7	17.9	0.011	1.3*
Nitrate, total	64	5.2	5.4	0.000	0.2**
Sulfate + nitrate, total	67	90.7	19.0	0.008	1.7**
Chloride	67	41.3	13.1	0.117	-0.5
Ammonia, total	14	7.3	11.4	ID	ID
Calcium	67	665.4	214.3	0.482	-4.3
Magnesium	67	95.7	21.3	0.627	0.1
Sodium	67	54.0	17.0	0.256	-0.5
Potassium	67	22.9	9.5	0.256	0.2
Major cations (Ca+Mg+Na+K)	67	837.9	227.8	0.627	-3.8
Alkalinity	67	670.2	221.3	0.176	-7.1
Hydrogen ion	53	126.6	309.0	0.726	3.1

11. CYPRESS CREEK NEAR JANICE, MS

Drainage area (km^2) 135.15
Mean discharge (m^3/sec) 2.72
Period of record 1967 - 1983

Sulfate deposition as sulfur (g/m^2.yr) 0.73
Basin sulfate yield as sulfur (g/m^2.yr) 0.49
Ratio (Y/D) 0.67

Statistical Summary of Common Ion Records

Ion	N	Mean Concentration (µeq/l)	Standard Deviation	Trend p	Trend Slope (µeq/l.yr)
Sulfate	48	44.4	28.7	0.499	0.9
Nitrate, total	46	5.3	4.5	0.389	-0.2
Sulfate + nitrate, total	48	47.0	29.8	0.759	0.3
Chloride	48	93.9	22.4	0.243	0.7
Ammonia, total	8	2.1	1.7	ID	ID
Calcium	48	52.9	22.4	0.157	-1.1
Magnesium	47	32.2	13.2	0.759	-0.1
Sodium	48	93.5	25.7	0.499	-0.7
Potassium	47	11.6	4.8	1.000	0.0
Major cations (Ca+Mg+Na+K)	48	189.7	36.7	0.243	-1.8
Alkalinity	48	41.0	32.2	0.296	1.6
Hydrogen ion	27	2223.7	3986.6	0.483	376.5

12. CATALOOCHEE CREEK NEAR CATALOOCHEE, NC

Drainage area (km^2) 127.38
Mean discharge (m^3/sec) 3.17
Period of record 1967 - 1983

Sulfate deposition as sulfur (g/m^2.yr) 1.09
Basin sulfate yield as sulfur (g/m^2.yr) 0.35
Ratio (Y/D) 0.32

Statistical Summary of Common Ion Records

Ion	N	Mean Concentration (μeq/l)	Standard Deviation	Trend p	Trend Slope (μeq/l.yr)
Sulfate	61	25.6	13.7	0.011	1.0*
Nitrate, total	54	9.6	5.6	0.000	0.4**
Sulfate + nitrate, total	61	33.9	16.0	0.001	1.2**
Chloride	62	20.3	9.6	0.607	-0.4
Ammonia, total	25	1.3	1.2	1.000	-0.0
Calcium	62	48.1	13.5	0.072	0.5*
Magnesium	62	26.7	9.3	0.368	0.2
Sodium	62	49.0	8.5	0.607	0.2
Potassium	62	13.9	5.9	0.001	0.4**
Major cations (Ca+Mg+Na+K)	62	137.6	23.3	0.029	1.3*
Alkalinity	62	80.2	20.6	0.607	-0.2
Hydrogen ion	32	169.1	201.3	1.000	24.1

13. BUFFALO RIVER NEAR FLAT WOODS, TN

Drainage area (km^2) 1157.28
Mean discharge (m^3/sec) 21.34
Period of record 1963 - 1983

Sulfate deposition as sulfur (g/m^2.yr) 1.16
Basin sulfate yield as sulfur (g/m^2.yr) 0.85
Ratio (Y/D) 0.74

Statistical Summary of Common Ion Records

Ion	N	Mean Concentration (μeq/l)	Standard Deviation	Trend p	Trend Slope (μeq/l.yr)
Sulfate	101	84.2	27.6	0.639	0.3
Nitrate, total	92	16.2	11.0	0.000	1.3**
Sulfate + nitrate, total	102	100.6	38.4	0.021	1.4*
Chloride	103	50.4	17.6	0.043	0.9*
Ammonia, total	20	2.5	1.7	ID	ID
Calcium	103	642.2	115.6	0.269	-1.3
Magnesium	103	149.0	24.9	0.014	-0.9*
Sodium	103	52.8	11.8	0.065	0.5*
Potassium	102	19.6	5.3	0.007	0.3**
Major cations (Ca+Mg+Na+K)	103	865.2	133.2	0.623	-1.1
Alkalinity	103	700.5	163.8	0.075	-2.6*
Hydrogen ion	61	83.3	75.2	0.447	-3.9

14. WASHINGTON CREEK AT WINDIGO, MI

Drainage area (km^2) 34.17
Mean discharge (m^3/sec) 0.45
Period of record 1967 - 1983

Sulfate deposition as sulfur (g/m^2.yr) 0.48
Basin sulfate yield as sulfur (g/m^2.yr) 0.84
Ratio (Y/D) 1.74

Statistical Summary of Common Ion Records

Ion	N	Mean Concentration (μeq/l)	Standard Deviation	Trend p	Trend Slope (μeq/l.yr)
Sulfate	82	128.2	54.7	0.533	0.8
Nitrate, total	81	7.5	4.5	0.002	-0.4**
Sulfate + nitrate, total	82	135.6	55.3	0.610	0.3
Chloride	82	112.4	77.1	0.283	0.5
Ammonia, total	13	1.6	0.9	ID	ID
Calcium	83	930.5	288.9	0.010	-5.7*
Magnesium	83	455.9	147.3	0.008	-3.7**
Sodium	83	146.0	64.5	0.413	0.4
Potassium	83	14.2	5.3	0.000	-0.6**
Major cations (Ca+Mg+Na+K)	83	1546.7	489.5	0.000	-9.5**
Alkalinity	83	1233.4	439.8	0.004	-7.8**
Hydrogen ion	54	53.5	52.2	0.001	-6.8**

15. POPPLE RIVER NEAR FENCE, WI

Drainage area (km^2) 359.87
Mean discharge (m^3/sec) 3.42
Period of record 1965 - 1983

Sulfate deposition as sulfur (g/m^2.yr) 0.59
Basin sulfate yield as sulfur (g/m^2.yr) 0.79
Ratio (Y/D) 1.34

Statistical Summary of Common Ion Records

Ion	N	Mean Concentration (μeq/l)	Standard Deviation	Trend p	Trend Slope (μeq/l.yr)
Sulfate	152	164.9	65.5	0.096	-1.1*
Nitrate, total	149	12.3	11.0	0.000	-0.7**
Sulfate + nitrate, total	152	176.9	69.2	0.016	-2.0*
Chloride	148	52.5	25.7	0.000	1.2**
Ammonia, total	24	4.0	2.7	0.015	0.0*
Calcium	153	956.3	296.9	0.075	-2.7*
Magnesium	153	801.7	263.0	0.546	-0.6
Sodium	152	65.6	17.2	0.746	-0.1
Potassium	152	20.1	6.9	0.007	0.2**
Major cations (Ca+Mg+Na+K)	153	1850.3	557.0	0.311	-2.6
Alkalinity	153	1572.9	586.5	0.445	-3.3
Hydrogen ion	96	89.3	194.8	0.000	-13.5**

16. KAWISHIWI RIVER NEAR ELY, MN

Drainage area (km^2) 655.02
Mean discharge (m^3/sec) 6.00
Period of record 1966 - 1983

Sulfate deposition as sulfur (g/m^2.yr) 0.43
Basin sulfate yield as sulfur (g/m^2.yr) 0.43
Ratio (Y/D) 1.01

Statistical Summary of Common Ion Records

Ion	N	Mean Concentration (μeq/l)	Standard Deviation	Trend p	Trend Slope (μeq/l.yr)
Sulfate	62	103.6	23.7	0.586	-0.5
Nitrate, total	56	6.2	8.1	0.192	0.2
Sulfate + nitrate, total	62	109.2	24.1	0.384	-0.4
Chloride	58	19.7	12.1	0.226	0.5
Ammonia, total	8	3.9	5.9	ID	ID
Calcium	62	170.1	26.6	0.103	-0.9
Magnesium	62	119.0	16.4	0.514	-0.0
Sodium	62	47.6	11.6	0.663	-0.0
Potassium	62	10.1	4.1	0.022	-0.3*
Major cations (Ca+Mg+Na+K)	62	346.9	36.6	0.231	-0.8
Alkalinity	62	186.5	38.4	0.276	-1.7
Hydrogen ion	27	98.7	82.9	0.767	3.8

17. NORTH FORK WHITEWATER RIVER NEAR ELBA, MN

Drainage area (km^2) 261.49
Mean discharge (m^3/sec) 1.16
Period of record 1967 - 1983

Sulfate deposition as sulfur (g/m^2.yr) 0.61
Basin sulfate yield as sulfur (g/m^2.yr) 0.70
Ratio (Y/D) 1.15

Statistical Summary of Common Ion Records

Ion	N	Mean Concentration (μeq/l)	Standard Deviation	Trend p	Trend Slope (μeq/l.yr)
Sulfate	147	335.1	69.7	0.058	1.8*
Nitrate, total	146	172.4	90.9	0.000	5.3**
Sulfate + nitrate, total	147	506.3	136.0	0.000	8.5**
Chloride	147	216.7	97.1	0.000	11.3**
Ammonia, total	25	7.9	11.8	0.681	-6.3
Calcium	147	3369.1	608.8	0.000	24.1**
Magnesium	147	1944.0	357.0	0.000	18.3**
Sodium	147	216.3	59.2	0.000	8.2**
Potassium	144	58.5	37.3	0.000	-1.5**
Major cations (Ca+Mg+Na+K)	147	5600.6	924.2	0.000	48.2**
Alkalinity	147	4918.0	860.7	0.001	28.8**
Hydrogen ion	97	9.2	9.9	1.000	0.0

18. CASTLE CREEK ABOVE DEERFIELD RESERVOIR NEAR HILL CITY, SD

Drainage area (km^2) 214.89
Mean discharge (m^3/sec) 0.28
Period of record 1965 - 1983

Sulfate deposition as sulfur ($g/m^2.yr$) 0.22
Basin sulfate yield as sulfur ($g/m^2.yr$) 0.10
Ratio (Y/D) 0.47

Statistical Summary of Common Ion Records

Ion	N	Mean Concentration (µeq/l)	Standard Deviation	Trend p	Trend Slope (µeq/l.yr)
Sulfate	161	163.4	64.4	0.000	3.0**
Nitrate, total	156	11.6	13.6	0.211	0.1
Sulfate + nitrate, total	161	174.1	68.7	0.000	3.4**
Chloride	165	37.8	23.1	0.193	-0.2
Ammonia, total	21	5.4	2.9	ID	ID
Calcium	168	2688.1	324.1	0.000	11.4**
Magnesium	167	2454.6	158.9	0.001	7.6**
Sodium	167	79.3	37.8	0.987	0.0
Potassium	166	35.3	12.1	0.045	-0.2*
Major cations (Ca+Mg+Na+K)	168	5265.2	385.0	0.000	25.6**
Alkalinity	168	4956.1	411.7	0.584	2.0
Hydrogen ion	108	7.3	9.4	0.482	-0.0

19. ENCAMPMENT RIVER ABOVE HOG PARK CREEK, NEAR ENCAMPMENT, WY

Drainage area (km^2) 188.22
Mean discharge (m^3/sec) 3.06
Period of record 1966 - 1983

Sulfate deposition as sulfur ($g/m^2.yr$) 0.41
Basin sulfate yield as sulfur ($g/m^2.yr$) 0.62
Ratio (Y/D) 1.49

Statistical Summary of Common Ion Records

Ion	N	Mean Concentration (µeq/l)	Standard Deviation	Trend p	Trend Slope (µeq/l.yr)
Sulfate	105	85.5	47.7	0.000	4.9**
Nitrate, total	92	4.4	3.2	0.009	0.2**
Sulfate + nitrate, total	103	89.0	48.4	0.000	5.1**
Chloride	107	19.9	16.5	0.903	-0.0
Ammonia, total	18	5.9	3.2	ID	ID
Calcium	112	366.2	82.5	0.056	-1.3*
Magnesium	113	123.1	39.8	0.474	0.2
Sodium	112	98.3	35.1	0.772	0.1
Potassium	111	21.8	6.1	0.058	-0.2*
Major cations (Ca+Mg+Na+K)	112	609.9	132.1	0.828	-0.1
Alkalinity	112	497.3	121.4	0.628	-0.5
Hydrogen ion	62	103.0	162.4	0.033	-15.4*

20. DISMAL RIVER NEAR THEDFORD, NE

Drainage area (km^2) 2485.44
Mean discharge (m^3/sec) 5.41
Period of record 1967 - 1983

Sulfate deposition as sulfur (g/m^2.yr) 0.33
Basin sulfate yield as sulfur (g/m^2.yr) 0.16
Ratio (Y/D) 0.51

Statistical Summary of Common Ion Records

Ion	N	Mean Concentration (μeq/l)	Standard Deviation	Trend p	Trend Slope (μeq/l.yr)
Sulfate	105	149.8	36.1	0.007	2.0**
Nitrate, total	106	27.7	12.8	0.001	0.5**
Sulfate + nitrate, total	105	176.3	39.8	0.000	3.2**
Chloride	108	32.3	13.1	0.008	-0.7**
Ammonia, total	18	5.9	3.1	ID	ID
Calcium	109	1138.6	65.6	0.599	0.0
Magnesium	109	290.2	28.8	0.659	-0.3
Sodium	108	306.5	24.4	0.239	0.3
Potassium	108	130.6	11.0	0.019	-0.4*
Major cations (Ca+Mg+Na+K)	109	1868.1	86.4	0.614	1.0
Alkalinity	109	1641.0	87.0	0.001	-5.3**
Hydrogen ion	58	29.3	15.3	0.967	0.1

21. NORTH SYLAMORE CREEK NEAR FIFTY SIX, AR

Drainage area (km^2) 150.42
Mean discharge (m^3/sec) 1.19
Period of record 1966 - 1983

Sulfate deposition as sulfur (g/m^2.yr) 0.89
Basin sulfate yield as sulfur (g/m^2.yr) 0.47
Ratio (Y/D) 0.53

Statistical Summary of Common Ion Records

Ion	N	Mean Concentration (μeq/l)	Standard Deviation	Trend p	Trend Slope (μeq/l.yr)
Sulfate	161	119.1	38.3	0.002	1.9**
Nitrate, total	122	5.2	6.4	0.000	0.3**
Sulfate + nitrate, total	161	123.0	39.5	0.000	2.2**
Chloride	161	53.5	23.5	0.000	1.4**
Ammonia, total	28	4.3	3.8	0.310	0.0
Calcium	164	2242.7	273.9	0.001	-7.3**
Magnesium	163	504.7	108.6	0.167	-1.1
Sodium	162	63.6	18.8	0.185	0.3
Potassium	163	21.4	4.8	0.000	-0.2**
Major cations (Ca+Mg+Na+K)	164	2840.9	364.0	0.013	-7.1*
Alkalinity	164	2663.2	385.5	0.000	-13.5**
Hydrogen ion	106	10.4	6.1	0.001	-0.6**

22. HALFMOON CREEK NEAR MALTA, CO

Drainage area (km^2) 61.10
Mean discharge (m^3/sec) 0.74
Period of record 1965 - 1983

Sulfate deposition as sulfur ($g/m^2.yr$) 0.44
Basin sulfate yield as sulfur ($g/m^2.yr$) 0.55
Ratio (Y/D) 1.26

Statistical Summary of Common Ion Records

Ion	N	Mean Concentration (μeq/l)	Standard Deviation	Trend p	Trend Slope (μeq/l.yr)
Sulfate	153	116.6	41.6	0.265	0.7
Nitrate, total	154	9.7	10.5	0.008	0.2**
Sulfate + nitrate, total	153	126.0	44.6	0.396	0.6
Chloride	150	21.2	13.9	0.000	-1.7**
Ammonia, total	27	5.8	3.4	0.614	0.1
Calcium	158	473.5	93.2	0.018	-1.5*
Magnesium	158	295.6	76.3	0.117	-1.0
Sodium	158	64.4	20.2	0.188	-0.2
Potassium	156	16.3	4.8	0.325	0.1
Major cations (Ca+Mg+Na+K)	158	850.6	176.5	0.003	-3.1**
Alkalinity	158	690.2	158.8	0.002	-3.7**
Hydrogen ion	91	73.0	71.4	0.042	4.6*

23. BLUE BEAVER CREEK NEAR CACHE, OK

Drainage area (km^2) 63.69
Mean discharge (m^3/sec) 0.25
Period of record 1967 - 1983

Sulfate deposition as sulfur ($g/m^2.yr$) 0.38
Basin sulfate yield as sulfur ($g/m^2.yr$) 0.63
Ratio (Y/D) 1.64

Statistical Summary of Common Ion Records

Ion	N	Mean Concentration (μeq/l)	Standard Deviation	Trend p	Trend Slope (μeq/l.yr)
Sulfate	68	335.9	101.3	0.750	-0.6
Nitrate, total	62	8.1	7.8	1.000	-0.1
Sulfate + nitrate, total	68	343.3	101.0	0.750	-0.4
Chloride	68	223.8	96.2	1.000	0.0
Ammonia, total	9	5.9	2.5	ID	ID
Calcium	68	725.3	218.4	0.068	7.7*
Magnesium	68	361.1	98.4	0.111	-3.6
Sodium	68	556.0	198.7	0.495	0.7
Potassium	67	39.9	12.4	0.924	-0.0
Major cations (Ca+Mg+Na+K)	68	1683.6	470.3	0.219	7.1
Alkalinity	68	1100.1	396.3	0.076	10.6*
Hydrogen ion	28	103.0	106.3	0.826	8.8

24. KIAMICHI RIVER NEAR BIG CEDAR, OK

Drainage area (km^2) 103.82
Mean discharge (m^3/sec) 2.09
Period of record 1967 - 1983

Sulfate deposition as sulfur ($g/m^2.yr$) 0.80
Basin sulfate yield as sulfur ($g/m^2.yr$) 0.73
Ratio (Y/D) 0.92

Statistical Summary of Common Ion Records

Ion	N	Mean Concentration (μeq/l)	Standard Deviation	Trend p	Trend Slope (μeq/l.yr)
Sulfate	93	69.6	31.2	0.155	1.3
Nitrate, total	83	4.5	4.6	0.794	0.0
Sulfate + nitrate, total	93	73.5	31.3	0.247	0.7
Chloride	93	63.5	19.1	1.000	-0.0
Ammonia, total	8	5.0	0.9	ID	ID
Calcium	94	67.2	27.5	0.579	-0.4
Magnesium	93	65.0	30.9	0.057	0.9*
Sodium	93	92.1	22.1	0.922	-0.0
Potassium	94	20.2	6.6	0.922	0.0
Major cations (Ca+Mg+Na+K)	94	248.9	70.9	0.379	0.8
Alkalinity	94	105.5	56.8	0.282	-1.3
Hydrogen ion	47	435.2	807.0	0.258	47.6

25. BIG CREEK AT POLLOCK, LA

Drainage area (km^2) 132.04
Mean discharge (m^3/sec) 1.47
Period of record 1966 - 1983

Sulfate deposition as sulfur ($g/m^2.yr$) 0.68
Basin sulfate yield as sulfur ($g/m^2.yr$) 0.29
Ratio (Y/D) 0.42

Statistical Summary of Common Ion Records

Ion	N	Mean Concentration (μeq/l)	Standard Deviation	Trend p	Trend Slope (μeq/l.yr)
Sulfate	121	47.4	30.1	0.000	3.5**
Nitrate, total	120	6.6	4.8	0.605	-0.0
Sulfate + nitrate, total	121	51.0	32.0	0.000	3.2**
Chloride	124	121.0	23.5	0.082	0.7*
Ammonia, total	29	5.1	3.3	0.094	0.0*
Calcium	124	107.5	28.2	0.814	0.2
Magnesium	123	58.2	23.3	0.172	0.3
Sodium	124	183.8	35.8	0.300	0.7
Potassium	124	33.1	8.6	0.041	0.3*
Major cations (Ca+Mg+Na+K)	124	384.1	58.8	0.038	2.3*
Alkalinity	124	182.8	51.4	0.008	-2.2**
Hydrogen ion	95	1019.8	1682.2	0.000	234.4**

26. RIO MORA NEAR TERRERO, NM

Drainage area (km^2) 137.73
Mean discharge (m^3/sec) 0.76
Period of record 1967 - 1983

Sulfate deposition as sulfur ($g/m^2.yr$) 0.43
Basin sulfate yield as sulfur ($g/m^2.yr$) 0.43
Ratio (Y/D) 1.01

Statistical Summary of Common Ion Records

Ion	N	Mean Concentration (μeq/l)	Standard Deviation	Trend p	Trend Slope (μeq/l.yr)
Sulfate	111	177.1	54.6	0.083	1.6*
Nitrate, total	87	4.6	3.4	0.000	0.2**
Sulfate + nitrate, total	111	180.5	54.9	0.039	1.9*
Chloride	108	22.6	17.4	0.064	0.4*
Ammonia, total	25	4.7	2.3	0.180	0.1
Calcium	113	813.2	170.7	0.103	-2.4
Magnesium	113	145.1	38.1	0.142	0.7
Sodium	112	60.3	20.5	0.240	0.3
Potassium	112	13.3	5.0	0.000	0.3**
Major cations (Ca+Mg+Na+K)	113	1036.9	212.4	0.963	-0.2
Alkalinity	113	806.5	174.6	0.024	-4.1*
Hydrogen ion	62	22.3	28.2	0.819	-0.6

27. WET BOTTOM CREEK NEAR CHILDS, AZ

Drainage area (km^2) 94.24
Mean discharge (m^3/sec) 0.37
Period of record 1968 - 1983

Sulfate deposition as sulfur ($g/m^2.yr$) 0.50
Basin sulfate yield as sulfur ($g/m^2.yr$) 0.37
Ratio (Y/D) 0.74

Statistical Summary of Common Ion Records

Ion	N	Mean Concentration (μeq/l)	Standard Deviation	Trend p	Trend Slope (μeq/l.yr)
Sulfate	98	188.2	79.5	0.396	-0.8
Nitrate, total	77	5.2	6.3	0.033	0.2*
Sulfate + nitrate, total	97	192.3	80.6	0.539	-1.3
Chloride	100	262.5	144.6	0.004	4.9**
Ammonia, total	26	5.7	3.0	0.307	-791.2
Calcium	100	1367.1	632.4	0.687	-4.1
Magnesium	100	663.9	318.9	0.016	8.6*
Sodium	100	873.3	444.6	0.074	11.5*
Potassium	100	31.9	10.1	0.195	0.2
Major cations (Ca+Mg+Na+K)	100	2936.2	1364.3	0.328	18.9
Alkalinity	100	2429.7	1255.9	0.227	19.5
Hydrogen ion	74	21.6	21.7	0.008	-2.4**

28. MERCED RIVER AT HAPPY ISLES BRIDGE NEAR YOSEMITE, CA

Drainage area (km^2) 468.61
Mean discharge (m^3/sec) 14.74
Period of record 1968 - 1983

Sulfate deposition as sulfur ($g/m^2.yr$) 0.56
Basin sulfate yield as sulfur ($g/m^2.yr$) 0.48
Ratio (Y/D) 0.86

Statistical Summary of Common Ion Records

Ion	N	Mean Concentration (μeq/l)	Standard Deviation	Trend p	Trend Slope (μeq/l.yr)
Sulfate	65	31.6	19.8	0.679	-0.2
Nitrate, total	62	4.6	5.4	0.571	0.1
Sulfate + nitrate, total	65	34.5	21.2	0.756	-0.3
Chloride	71	92.9	60.1	0.211	0.8
Ammonia, total	42	3.8	3.2	0.531	0.1
Calcium	72	118.4	44.3	0.030	1.1*
Magnesium	68	19.6	10.8	0.257	0.7
Sodium	71	89.0	34.6	0.445	0.4
Potassium	72	10.2	3.6	0.038	0.2*
Major cations (Ca+Mg+Na+K)	72	239.4	87.4	0.014	2.7*
Alkalinity	72	105.0	47.9	0.850	0.1
Hydrogen ion	57	363.4	408.9	0.010	55.0**

29. HAYDEN CREEK BELOW NORTH FORK, NEAR HAYDEN LAKE, ID

Drainage area (km^2) 56.96
Mean discharge (m^3/sec) 0.71
Period of record 1967 - 1983

Sulfate deposition as sulfur ($g/m^2.yr$) 0.45
Basin sulfate yield as sulfur ($g/m^2.yr$) 0.40
Ratio (Y/D) 0.89

Statistical Summary of Common Ion Records

Ion	N	Mean Concentration (μeq/l)	Standard Deviation	Trend p	Trend Slope (μeq/l.yr)
Sulfate	83	63.3	27.8	0.402	0.9
Nitrate, total	79	5.7	5.6	0.021	0.3*
Sulfate + nitrate, total	83	68.2	29.6	0.246	1.5
Chloride	73	16.7	8.7	0.751	0.1
Ammonia, total	17	4.4	3.0	ID	ID
Calcium	89	420.4	87.8	0.645	0.4
Magnesium	89	220.2	42.5	0.624	-0.2
Sodium	88	62.1	11.4	0.232	0.2
Potassium	88	15.3	3.8	0.063	0.1*
Major cations (Ca+Mg+Na+K)	89	720.8	132.9	0.205	1.6
Alkalinity	89	625.2	131.8	0.908	-0.3
Hydrogen ion	53	91.5	114.8	0.376	-5.6

30. ANDREWS CREEK NEAR MAZAMA, WA

Drainage area (km^2)	57.22
Mean discharge (m^3/sec)	0.91
Period of record	1972 - 1983

Sulfate deposition as sulfur ($g/m^2.yr$)	0.32
Basin sulfate yield as sulfur ($g/m^2.yr$)	0.32
Ratio (Y/D)	1.00

Statistical Summary of Common Ion Records

Ion	N	Mean Concentration (μeq/l)	Standard Deviation	Trend p	Trend Slope (μeq/l.yr)
Sulfate	77	50.6	29.5	0.052	-3.1*
Nitrate, total	79	5.0	5.9	0.014	0.2*
Sulfate + nitrate, total	77	54.7	30.4	0.079	-2.8*
Chloride	78	15.3	13.7	0.010	-1.2**
Ammonia, total	24	5.1	3.7	0.516	-0.7
Calcium	81	320.9	75.2	0.611	0.9
Magnesium	80	76.8	19.8	0.163	0.5
Sodium	80	95.1	22.2	0.827	0.1
Potassium	80	13.8	4.1	0.337	0.0
Major cations (Ca+Mg+Na+K)	81	510.7	112.6	0.582	1.2
Alkalinity	81	456.4	124.8	0.472	-1.1
Hydrogen ion	82	44.0	50.8	0.591	1.0

31. BIG JACKS CREEK NEAR BRUNEAU, ID

Drainage area (km^2)	655.02
Mean discharge (m^3/sec)	0.25
Period of record	1967 - 1983

Sulfate deposition as sulfur ($g/m^2.yr$)	0.09
Basin sulfate yield as sulfur ($g/m^2.yr$)	0.02
Ratio (Y/D)	0.28

Statistical Summary of Common Ion Records

Ion	N	Mean Concentration (μeq/l)	Standard Deviation	Trend p	Trend Slope (μeq/l.yr)
Sulfate	67	144.2	38.5	0.057	1.7*
Nitrate, total	58	14.2	17.0	0.503	0.2
Sulfate + nitrate, total	67	156.6	40.9	0.012	2.4*
Chloride	67	124.4	32.7	0.000	3.2**
Ammonia, total	11	5.7	2.9	ID	ID
Calcium	67	550.8	95.3	0.035	3.1*
Magnesium	67	227.8	44.8	0.328	0.8
Sodium	67	457.8	89.2	0.381	0.7
Potassium	67	101.5	18.4	0.035	-0.7*
Major cations (Ca+Mg+Na+K)	67	1337.9	219.3	0.198	5.5
Alkalinity	67	1078.9	235.1	0.027	8.4*
Hydrogen ion	37	22.8	51.0	0.024	-4.9*

32. MINAM RIVER AT MINAM, OR

Drainage area (km^2) 621.36
Mean discharge (m^3/sec) 12.23
Period of record 1967 - 1983

Sulfate deposition as sulfur ($g/m^2.yr$) 0.33
Basin sulfate yield as sulfur ($g/m^2.yr$) 0.40
Ratio (Y/D) 1.22

Statistical Summary of Common Ion Records

Ion	N	Mean Concentration (μeq/l)	Standard Deviation	Trend p	Trend Slope (μeq/l.yr)
Sulfate	99	42.8	34.4	0.000	3.5**
Nitrate, total	85	4.2	5.3	0.266	0.0
Sulfate + nitrate, total	93	43.3	37.0	0.000	3.7**
Chloride	88	17.7	11.0	0.487	-0.1
Ammonia, total	21	3.7	1.6	ID	ID
Calcium	103	287.8	64.7	0.330	0.8
Magnesium	103	118.4	35.4	0.361	-0.4
Sodium	103	96.5	20.7	0.976	0.0
Potassium	103	27.8	7.4	0.601	-0.0
Major cations (Ca+Mg+Na+K)	103	530.9	117.8	0.361	1.1
Alkalinity	103	451.5	110.9	0.005	-4.0**
Hydrogen ion	61	62.3	49.0	0.076	6.8*

TABLE C.3 Basin Characteristics of Bench-Mark Streams

The following entries are descriptions of 16 basins of the U.S. Geological Survey Bench-Mark streams for which alkalinities are lower than 500 μeq/L. Numbering of the basins is keyed to Table C.1 and Figure C.1.

1. Wild River, Gilead, Maine (References 1, 4, 5, 10, 11)

Geology Predominantly Littleton formation paragneiss with quartz andesine biotite, muscovite. Minor units: lime - silicate rocks, Concord quartz monzonite, mica schist - micaceous quartzite till at lower elevations. Some pyrite present.

Soils Hermon, Lyman, Peru, Berkshire; (Lithic Haplorthods, Aquic Fragiorthods, Typic Haplorthods).

Vegetation Hardwood forest on lower slopes with pockets of conifers.

Land use/disturbance National forest, selective logging.

2. Esopus Creek, Shandaken, New York (References 4, 11, 22)

Geology Sandstone bedrock, till covered (outwash and alluvium in valleys).

Soils Arnot-Oquaga-Lackawanna, rock outcrop, Lithic Dystochrepts Typic Fragiochrepts, Typic Dystrochrepts.

Vegetation Forested/hardwoods and conifers.

Land use/disturbance Communities, scattered homes, sewage-treatment plant, reservoir on tributary.

3. McDonalds Branch, Lebanon State Forest, New Jersey (References 4, 11, 12, 17)

Geology Deep quartz sands and gravels of Miocene/ Pliocene age (Cohansey formation).

Soils Quartzipsamments, Aquods, Histosols.

Vegetation Pine-oak forest ~100 years old, pitch pine, lowland vegetation, cedar swamp, hardwood swamp (aggrading forest).

Land use/disturbance Fire 1963 (minor), a few unpaved roads.

4. Young Womans Creek, Renovo, Pennsylvania (References 4, 11, 18, 21)

Geology Sandstone/shale, some glacial outwash.

Soils Mostly Dekalb series pH, 4-5; percent base saturation 4-20, average ~12 (Ochrepts)

Vegetation Hardwood forest.

Land use/disturbance Secondary roads, fire roads, logging, few people.

5. Holiday Creek, Andersonville, Virginia (References 4, 6, 11, 23)

Geology Wissahickon formation (oligoclase - biotite schist); minor area of hornblende gneiss.

Soils Cecil Series (Typic Hapludults), Mantio Series (Lithic Dystrochrepts), Nason Series (Typic Hapludult), Tatum Series (Typic Hapludult)

Vegetation Mostly hardwood forest, some recently planted pine.

Land use/disturbance Selective cutting, roads.

6. Scape Ore Swamp, Bishopville, South Carolina (References 4, 11, 19)

Geology Unconsolidated sediments.

Soils Ultisols dominant with some sandy Entisols (probably Psamments). Major soil series: Norfolk, Lakeland, Ruston, Gilead. Organic soils line the stream.

Vegetation 50 percent pine-oak forest, 50 percent agriculture.

Land use/disturbance Agricultural land use.

7. Upper Three Runs, New Ellenton, South Carolina (References 4, 11, 15)

Geology Costal plain sediments, Barnwell sand (Eocene).

Soils Ultisols and Entisols.

Vegetation 50 percent pine-oak forest, 50 percent agriculture.

Land use/disturbance Agricultural land use.

11. Cypress Creek, Janice, Mississippi (References 3, 4, 11, 14)

Geology Coastal plain sands, gravels, silts. Pascagoula and Hattiesburg formations (Miocene).

Soils Aquults, Udults, minor Aquents, Saprists.

Vegetation Pine forests.

Land use/disturbance Minimal agriculture, some timber harvesting; military uses - bombing, tanks, etc.; prescribed burning, herbicide application.

12. Cataloochee Creek, Cataloochee, North Carolina (References 4, 7, 11)

Geology Sandstones, quartzite, phyllite (Longarm quartzite (may contain minor pyrite), Roaring Fork sandstone, Thunderhead sandstone).

Soils No data.

Vegetation Forest: mostly second growth oaks, hickory, tulip, poplar.

Land use/disturbance Originally cleared for agriculture or timber harvesting, some mining; otherwise little information available.

16. Kawishiwi River, Ely, Minnesota (References 4, 11)

Geology Gabbro, granite, greenstone.

Soils No data.

Vegetation Forested.

Land use/disturbance Few roads, selective cutting.

19. Encampment River, Encampment, Wyoming (References 4, 11, 13)

Geology Granite, metasedimentary rocks in the basin.

Soils Orthents, Ochrepts, rock outcrop, minor Borolls.

Vegetation Coniferous forest: lodgepole pine, spruce, fir; alpine at high elevation.

Land use/disturbance Gauge moved, clear cutting, grazing (shift from sheep to cattle).

24. Kiamichi River, Big Cedar, Oklahoma (References 2, 4, 11, 20)

Geology Shales, sandstones of Moyers and Tenmile Creek Formations (minor Wildhorse Mountain Formation).

Soils Hapludults, Paleudults, Hapludalfs, Paleudalfs.

Vegetation Hardwood forest, pine.

Land use/disturbance Small agricultural holdings (corn, pasture); roads, a few dwellings, logging.

25. Big Creek, Pollack, Louisiana (References 4, 8, 11, 16)

Geology Pleistocene sands and gravels.

Soils Udults, some Udalfs.

Vegetation Pine forest.

Land use/disturbance Small community, few roads, logging.

28. Merced River, Yosemite, California (References 4, 9, 11)

Geology Half-dome granodiorite.

Soils No data.

Vegetation Shrubs, some forest (fir, pine, sequoia), alpine.

Land use/disturbance Heavy backpacking.

30. Andrews Creek, Mazama, Washington (References 4, 11)

Geology Mostly granite rocks.

Soils No data.

Vegetation Fir, cedar, hemlock forest.

Land use/disturbance North Cascades primitive area; two hiking trails.

32. Minam River, Minam, Oregon (References 4, 11)

Geology Predominantly basalt.

Soils No data.

Vegetation Ponderosa pine to 1700 m; lodgepole, white fir, larch above 1700 m.

Land use/disturbance Some water use for irrigation. Most is wilderness area.

REFERENCES

1. Billings, M. P., and K. Fowler-Billings. 1975. Geology of the Gorham Quadrangle, New Hampshire-Maine. New Hampshire Department of Resources and Economic Development. Bull. no. 6, 120p.
2. Briggs, G. 1973. Geology of the eastern part of the Lynn Mountain syncline, LeFlore County, Oklahoma. Oklahoma Geol. Survey Circular 75, 34p.
3. Brown, G. F. 1944. Geology and ground water resources of the Camp Shelby area. Mississippi State Geol. Survey Bull. 58, 72p.
4. Cobb, E. D., and J. E. Biesecker. 1971. The national hydrologic Bench-Mark network. U.S. Geol. Survey Circ. 460-D, 38p.
5. Fisher, I. S. 1962. Petrological and structure of the Bethel Area, Maine. Geol. Soc. Amer. Bull. 73:1395- 1420.
6. Jonas, A. I. 1932. Kyanite in Virginia: geology of the Kyanite Belt of Virginia. Virginia Geol. Survey Bull. 38, 52p.
7. King, P. B. 1964. Geology of the central Great Smoky Mountains, Tennessee. U.S. Geol. Survey Report 349-C, 146p.
8. Maher, J. C. 1941. Groundwater resources of Grant and LaSalle parishes, Louisiana. Dept. of Minerals, Louisiana Geological Survey; Geol. Bull. No. 20. 91p.
9. Peck, D. L. 1980. Geologic map of the Merced Peak Quadrangle. U.S. Geological Survey. Geologic Map GQ-1531.
10. Rourke, R. V., J. A. Feruerda, and K. J. LaFlamme. 1977. The soils of Maine. Revision of Miscellaneous Pub. 676. Orono, Maine: University of Maine, LSA Experimental Station.
11. Smith, R. A., and R. B. Alexander. 1983. Evidence for acid-precipitation-induced trends in stream chemistry at hydrologic Bench-Mark stations. U.S. Geol. Survey Circ. 910, 12p.
12. Turner, R. S. 1983. Bio-geochemistry of trace elements in the McDonalds Branch Watershed, New Jersey Pine Barrens. Ph.D. dissertation, University of Pennsylvania. Ann Arbor, Michigan: University Microfilms.
13. U.S. Department of Agriculture, Forest Service. Medicine Bow National Forest, Encampment, Wyoming. Unpublished data.

14. U.S. Department of Agriculture, Forest Service. Wiggins, Mississippi. Unpublished data.
15. U.S. Department of Agriculture, Soil Conservation Service. Aiken, South Carolina. Unpublished data.
16. U.S. Department of Agriculture, Soil Conservation Service, Colfax, Louisiana. Unpublished data.
17. U.S. Department of Agriculture, Soil Conservation Service. 1971. Soil survey of Burlington County, New Jersey, 120p.
18. U.S. Department of Agriculture, Soil Conservation Service. 1966. Soil survey of Clinton County, Pennsylvania. 141p.
19. U.S. Department of Agriculture, Soil Conservation Service. 1959. Soil survey of Lee County, South Carolina. 85p.
20. U.S. Department of Agriculture, Soil Conservation Service. 1983. Soil survey of LeFlore County, Oklahoma, 209p.
21. U.S. Department of Agriculture, Soil Conservation Service. 1958. Soil survey of Potter County, Pennsylvania, 101p.
22. U.S. Department of Agriculture, Soil Conservation Service. 1979. Soil survey of Ulster County, New York. 273p.
23. U.S. Department of Agriculture, Soil Conservation Service. 1966. Soils in relation to woodland use, Buckingham-Appomatox State Forest, 54p. and maps.

APPENDIX

D

Historical Correction Factors for Alkalinity and Acid Status of Surface Waters

Assumptions Regarding Protolytes

The assumptions made in the following derivations are (1) that the protolytes consist only of carbonate species (e.g., CO_2^*, HCO_3^-, CO_3^{2-}) and acid and base (H^+, OH^-) and (2) that the system is closed; that is, that total carbonate (C_t) is constant ($C_t = [CO_2^*] + [HCO_3^-] + [CO_3^{2-}]$, where $CO_2^* = CO_2(aq) + H_2CO_3$).

Assumption (1) is assessed later in this discussion. Assumption (2) has to be applied in the derivations (using C_t) in order to be able to relate alkalinity, pH, and CO_2 acidity. Actually the condition that must be met for assumption (2) to be valid is that pH and acidity be measured for the same C_t. This condition is not required for alkalinity because changes in C_t would not alter alkalinity.

Fundamental Definitions

Alkalinity ([Alk]) is defined for the ion balance condition for the solubility of CO_2; thus for all the protolyte species, HCO_3^-, CO_3^{2-}, H^+, and OH^-, the ion balance condition is

$$[H^+] = [OH^-] + [HCO_3^-] + 2[CO_3^{2-}], \qquad (1a)$$

and for this condition [Alk] = 0. [Alk], the "excess base," is then defined as

*Authored by James R. Kramer.

$$[Alk] = [OH^-] - [H^+] + [HCO_3^-] + 2[CO_3^{2-}]. \tag{1b}$$

CO_2 acidity ([Acy]) is defined for the ion balance condition of the "neutral" salt, $MHCO_3$. For this stoichiometry, the ion species are M^+, HCO_3^-, CO_3^{2-}, H^+, OH^-; in addition the stoichiometry of $MHCO_3$ results in $[M^+] = C_t = [CO_2^*] + [HCO_3^-] + [CO_3^{2-}]$. Substituting the stoichiometric condition into the ion balance expression ($[M^+] + [H^+] = [HCO_3^-] + 2[CO_3^{2-}] + [OH^-]$) results in

$$[CO_2^*] + [H^+] = [CO_3^{2-}] + [OH^-]. \tag{2a}$$

CO_2 acidity, ([Acy]), = 0 for the condition of Eq. (2a). Thus [Acy] is defined as the "excess acid" relative to the $MHCO_3$ stoichiometry:

$$[Acy] = [CO_2^*] + [H^+] - [OH^-] - [CO_3^{2-}]. \tag{2b}$$

Examination of Eqs. (1b) and (2b) shows that [Alk] and [Acy] can be positive or negative and are precisely zero at the endpoints as defined for the specific solution conditions (e.g., Eqs. (1a) and (2a)).

An alternative way of viewing the stoichiometric condition for alkalinity is to imagine a titration of strong acids, defined by their anions, ΣA, to strong bases, defined by their cations, ΣC, in the presence of CO_2. The ion balance expressions, rearranged, are

$$\Sigma C - \Sigma A = [HCO_3^-] + 2[CO_3^{2-}] + [OH^-] - [H^+] \tag{3a}$$

or

$$\Sigma C - \Sigma A = [Alk]. \tag{3b}$$

Thus the alkalinity endpoint is the condition in which strong acids equal strong bases with variable amounts of CO_2 in solution. If there were no dissolved CO_2 in solution, then Eq. (3a) would reduce to

$$\Sigma A - \Sigma C = [OH^-] - [H^+], \tag{3c}$$

where the endpoint would be $[H^+] = [OH^-] = 10^{-7}$ eq/L, the more common endpoint for titration of a strong acid with a strong base.

By examining Eq. (3a) and comparing it with Eq. (3c) one can qualitatively deduce that the amount of dissolved

CO_2 present will change the alkalinity endpoint. The quantitative expressions for the pHs of endpoints for alkalinity and CO_2 acidity are obtained by solution for H^+ ion concentration in Eqs. (1a) and (2a).

The following expressions or approximations are used in the following derivations:

$$[CO_2^*] = C_t[H^+]^2/D, \quad (4a)$$
$$[HCO_3^-] = C_t[H^+]K_1/D, \quad (4b)$$
$$[CO_3^{2-}] = C_tK_1K_2/D, \quad (4c)$$

where

$$D = [H^+]^2 + [H^+]K_1 + K_1K_2, \quad (4d)$$

where

$$K_1 = [H^+][HCO_3^-]/[CO_2^*],$$

$$K_2 = [H^+][CO_3^{2-}]/[HCO_3^-],$$

and

$$K_w = [H^+][OH^-].$$

For pHs less than 7 (and usually less than 8.3), Eq. (1a) reduces to

$$[H^+] = [HCO_3^-] \quad (5a)$$

because $[OH^-]$ and $[CO_3^{2-}]$ are negligible relative to the other terms. Substitution of the approximation of Eq. (4b) (i.e., $[CO_3^{2-}]$ is negligible so $[HCO_3^-] = C_tK_1/([H^+] + K_1)$) into Eq. (5a), rearrangement, and solution by the quadratic equation give

$$[H^+] = \{(K_1^2 + 4C_tK_1)^{1/2} - K_1\}/2. \quad (5b)$$

For $-\log K_1 = 6.34$, K_1^2 is maximally only 1 percent of $4C_tK_1$ for $C_t > 10^{-5}$ eq/L. Thus Eq. (5b) becomes

$$[H^+] = (C_tK_1)^{1/2} - K_1/2. \quad (5c)$$

From Eq. (5c), it is obvious that the alkalinity endpoint is not fixed but varies with total carbonate (C_t). For example, the alkalinity endpoint pH is 5.18 for C_t = 100 μeq/L (typical of a dilute unbuffered water), whereas the endpoint pH = 4.52 for C_t is 2000 μeq/L (typical of a solution in equilibrium with $CaCO_3$).

The endpoint pH for CO_2 acidity is determined by solving for H^+ ion concentration in Eq. (2a). For pH > 7, and $[OH^-] > [H^+]$, Eq. (2a) becomes

$$[CO_2^*] = [CO_3^{2-}] + [OH^-]. \quad (6a)$$

Substituting the valid approximations for the pH range 7 to 9 of $[CO_2^*] = C_t[H^+]/([H^+ + K_1)$ and $[CO_3^{2-}] = C_tK_2/([H^+] + K_2)$, into Eq. (6a), rearranging, and eliminating negligible terms give

$$[H^+] = [K_1(K_2 + K_w/C_t)]^{1/2} \quad (6b)$$

Thus the endpoint pH for CO_2 acidity varies with total carbonate; for example, the endpoint pH is 8.09 for C_t = 100 μM (molal), typical of poorly buffered waters, and is 8.31 for C_t = 2000 μM, typical of water in equilibrium with $CaCO_3$.

There are many colorimetric and electrometric recipes for titration to a fixed pH. If the endpoint pH of the specific technique (defined as pH_x) is known, the value obtained for the incorrect endpoint can be adjusted to the value for the definitions of alkalinity and CO_2 acidity by the following equations:

$$[Alk] = [Alk]_x + [HCO_3^-]_x + 2[CO_3^{2-}]_x + [OH^-]_x - [H^+]_x \quad (7a)$$

and

$$[Acy] = [Acy]_x + [CO_2^*]_x + [H^+]_x - [OH^-]_x - [CO_3^{2-}], \quad (7b)$$

where the subscript x refers to concentrations of the various constituents at the titration endpoint of the pH = x.

Thus, the correct expression is merely the measured value plus the difference from zero of the alkalinity or of the CO_2 acidity concentration terms at the titration endpoint. For example, this can readily be ascertained by writing sums for alkalinity. The value of $[Alk]_x$ is equal to the change in its constituents to get to the correct endpoint (e) from initial conditions (i) plus that required to get to the titration endpoint (x); or (assuming the case $[Alk] = [HCO_3] - [H^+]$ for simplicity of illustration only):

$$[Alk]_x = \text{[initial conditions]} - \text{[correct endpoint condition]} + \text{[correct endpoint condition]} - \text{[titration endpoint condition]}$$

or

$$[Alk]_x = ([HCO_3^-]_i - [H^+]_i) - ([HCO_3^-]_e - [H^+]_e) + ([HCO_3^-]_e - [H^+]_e) - ([HCO_3^-]_x - [H^+]_x). \quad (7c)$$

Remembering that [Alk] = 0 at the correct endpoint, then for this example $[HCO_3^-]_e = [H^+]_e$, and Eq. (7c) becomes

$$[Alk]_x = [Alk] - [HCO_3^-]_x + [H^+]_x,$$

or, on rearranging

$$[Alk] = [Alk]_x + [HCO_3^-]_x - [H^+]_x. \quad (7d)$$

The corrections for the endpoint using Eqs. (7a) and (7b) are carried out by reformulation in terms of C_t. Thus Eqs. (7a) and (7b) become:

$$[Alk] = [Alk]_x + C_t([H^+]_x K_1 - 2K_1K_2)/D_x + K_w/[H^+]_x - [H_x^+] \quad (7e)$$

and

$$[Acy] = [Acy]_x + (C_t[H^+]_x^2 - K_1K_2)/D_x + [H^+]_x - K_w/[H^+]_x, \quad (7f)$$

where $D_x = D$ for $[H^+] = [H_x^+]$.

If $[MO] = [Alk]_x$ for methyl orange titration ($pH_x = 4.04$ or 4.19), and if $[AC] = [Acy]_x$ for phenolphthalien titration ($pH_x = 8.25$), then Eqs. (7e) and (7f) become

$$[Alk] = [MO] + 4.987 \times 10^{-3} C_t - 9.120 \times 10^{-5}, \; pH_x = 4.04 \quad (7g)$$

$$[Alk] = [MO] + 7.030 \times 10^{-3} C_t - 6.457 \times 10^{-5}, \; pH_x = 4.19 \quad (7h)$$

and

$$[Acy] = [Ac] + 3.905 \times 10^{-3} C_t - 1.773 \times 10^{-6} \quad (7i)$$

where [Alk], [Acy], [MO], and [AC] are in equivalents per liter and C_t is in molal units; other values used to determine the coefficients are $-\log K_1 = 6.34$, $-\log K_2 = 10.33$, and $-\log K_w = 14$.

For a given set of data, [Alk] and [Acy] are determined in an iterative fashion. First the approximate values of [Alk] and [Acy] are obtained for $C_t = 0$; then convergence to a constant value of C_t is obtained by substituting $C_t = [Alk] + [Acy]$ back into Eqs. (7g)-(7i).

If Gran functions or modified Gran functions are used to determine [Alk] and [Acy], the correct endpoint is automatically obtained because of the nature of the function. For any region of an acid/base titration curve, [Alk] and [Acy] are valid because they are statements of ion balance for the entire system. Thus for a titration of acid with volume (v) and concentration (C_a) relative to the correct endpoint volume (v_a) for sample volume (V_s), the expression for titration is obtained from ion balance considerations as

$$(V_s + v)([HCO_3^-] + 2[CO_3^{2-}] + [OH^-] - [H^+]) = (v - v_a)C_a. \quad (8a)$$

(A similar expression can be written for a base titration.) The Gran function employs the more acidic titration region beyond the endpoint where $[H^+] >> ([HCO_3^-] + 2[CO_3^{2-}] + [OH^-])$. Thus Eq. (8a) becomes

$$(V_s + v)[H^+] = (v_a - v)C_a. \quad (8b)$$

Equation (8b) may be rearranged into

$$F_1 = a_0 + a_1 v,$$

where $F_1 = (V_s + v)[H^+]$, and a_0 and a_1 are linear-regression coefficients for a titration data subset of v-pH values that do not violate the condition of excess $[H^+]$. The equivalence point volume is then obtained from $v_a = -a_0/a_1$, because $v_a = v$ when $F_1 = 0$; and $[Alk] = v_a C_a/V_s$. If correctly used, the Gran function gives the true alkalinity. However, close attention must be paid to certain details because Eq. (8b) is valid only at some distance from the equivalance point; thus, small errors in the function can produce large errors in v_a on large extrapolation. First, the titration should be carried out in a constant ionic strength medium owing to the large changes in acid, which would alter the activity coefficient of H^+ ion and would also probably change the pH electrode response. Second, the pH response must be linear with respect to the $\log[H^+]$ for the subset of data used (i.e., $pH = a + b \log[H^+]$). Quite often lack of attention to these details and/or violation of the condition for use of Eq. (8b) leads to an underestimation of alkalinity.

The modified Gran function uses data points on both sides of the equivalence point by considering Eq. (8a), which has no titration conditions. To carry out the analysis, Eq. (8b) is modified so that the carbonate terms are expressed as a function of C_t. Once again C_t is obtained iteratively from C_t = [Alk] + [Acy] (Kramer 1982).

Colorimetric pH Correction

The colorimetric pH correction can be carried out quantitatively only if the acid/base properties of the solution (e.g., alkalinity), the kind of colorimetric dye needed to obtain its concentration and dissociation constant, and the respective indicator and sample volumes are known.

Statistical correlations of electrometric pH and colorimetric pH have limited value because they do not incorporate the above factors. For example, a water sample of high alkalinity would not be altered by the pH indicator solution; hence the colorimetric pH would read as the correct pH; whereas a low-alkalinity solution might be greatly modified by the colorimetric indicator. Furthermore, the pH effect of one dye would vary compared to the effect of another indicator. Since indicators are (were) used for specified pH intervals, different colorimetric pH readings must take into account the use of different indicators.

In this report colorimetric pH, MO alkalinity, and phenolphthalein (AC) free CO_2 were determined for historical lake data from New York, New Hampshire, and Wisconsin. Alkalinity and acidity (and C_t) were obtained by using Eqs. (7g) and (7h).* pH was determined

*The free CO_2 titration for historical Wisconsin data used Na_2CO_3 instead of NaOH, requiring a modification of Eqs. (7e)-(7h).

three different ways by using the following pairs of data: [Alk]/[Acy], [Alk]/pH_c, and [Acy]/pH_c, where pH_c refers to the colorimetric pH measurement. Derivation of the three equations considers the ion balance condition before and after addition of indicator dye. Before the addition of dye, we may invoke Eq. (3a):

$$\Sigma C - \Sigma A = [Alk] = B. \qquad (9a)$$

On addition of indicator of volume v_I, concentration C_I, and acid dissociation constant K_I to a sample of volume V_s, a new ion balance condition occurs because of the interaction of the dye protolyte with the solution and dilution owing to mixing of dye and solution. Indicator dyes historically were added as the "neutral" salt, MId, so that the stoichiometric condition gives $[M^+] = [Id^-] + [HId]$. The ion balance that results is

$$[M^+] + [B]' = [Id^-] + [Alk_c], \qquad (9b)$$

where [B]' is the diluted difference of strong cation, $[Alk_c]$ is the alkalinity altered by dilution and reaction with the indicator, and the subscript c refers to the solution condition after addition of colorimetric pH indicator. On substitution for $[M^+]$ and [B]' and considering dilution, Eq. (9a) becomes

$$[HId] + [Alk]' = [Alk_c]. \qquad (9c)$$

Inspection of Eq. (9c) reveals the conditions under which the colorimetric indicator will change the system. If alkalinity is much larger than the dye concentration, there will be no alteration of the solution by the addition of dye. Additionally, if $[H^+]$ or $[OH^-]$ is large compared with the dye concentration, then the H^+ ion will be marginally affected. Dye indicator solution concentrations range from about 10^{-4} to 10^{-5} M in sample solution. Thus alkalinity concentrations of about absolute (10^{-3} to 10^{-4} eq/L) or less would be altered by the indicator solution.

For pH less than about 8, [Alk] is approximated as $[HCO_3^-] - [H^+]$, and $C_t \sim [CO_2^*] + [HCO_3^-]$. Thus, the relationship between C_t and [Alk] becomes

$$[Alk] = C_t K_1/([H^+] + K_1) - [H^+], \qquad (9d)$$

and rearrangement gives a quadratic equation for $[H^+]$:

$$[H^+]^2 + [H^+](K_1 + [Alk]) + K_1([Alk] - C_t) = 0. \quad (9e)$$

Equation (9c), the ion balance statement, can be modified for dilution, concentrations, etc., to give

$$C_I v_I[H_C^+]/([H^+] + K_I) + V_s[Alk] = C_t V_s \{K_1/([H_C^+] + K_1)\} - [H_C^+](V_s + v_I), \quad (9f)$$

where $[H_C^+]$ equals the colorimetrically determined H^+ ion concentration. Equation (9f) is solved for C_t and substituted into Eq. (9d), which is then solved for the actual sample H^+ ion concentration by using the quadratic formula

$$\begin{aligned}[H^+] = 0.5\,(-([Alk] + K_1) + \{([Alk] + K_1)^2 \\ + 4([H_C^+] + K_1)/V_s\ [[H_C^+](V_s + v_I) \\ + V_s([Alk] + C_I v_I[H_C^+]/([H_C^+] + K_1) \\ - [Alk]V_s K_1/([H^+] + K_1)]\})1/2\end{aligned} \quad (10)$$

Thus Eq. (10) uses $[Alk]/pH_c$ data to determine sample H^+ ion concentration.

Equation (9e) can be modified by substituting $[Alk] = C_t - [Acy]$ so that pH may be determined from [Acy]/pHc data. Again Eq. (9f) is solved for C_t, which is substituted into modified Eq. (9e) to give

$$[H^+] = [-B + (B^2 + 4K_1[Acy])^{1/2}]/2, \quad (11)$$

where

$$B = K_1[Acy]/[H_C^+] - (V_s + v_I)([H_C^+] + K_1)/V_s - C_I v_I([H_C^+] + K_1)/V_s([H_C^+] + K_c) + K_1.$$

The third means of determining actual solution pH is from [Alk]/[Acy] data. Substitution of $C_t = [Alk] + [Acy]$ into Eq. (9e), simplification, and solution of the quadratic equation give

$$[H^+] = \{-(K_1 + [Alk]) + [([Alk] + K_1)^2 + 4K_1[Acy]]^{1/2}\}/2. \quad (12)$$

In addition to three independent estimates of pH using Eqs. (10 to 12), one may obtain an independent estimate of alkalinity from [Acy] and the actual solution pH. C_t is obtained from the definition of acidity (Eq. (2b)), substitution of Eqs. (4a) and (4c), and rearrangement to give

$$C_t = D([Acy] - [H^+] + K_w/[H^+])/([H^+]^2 - K_1K_2) \quad (13)$$

Then the second estimate of alkalinity is obtained by the difference of C_t (Eq. (13)) and the value for [Acy].

Other Protolyte Effects

An important assumption for the above calculations and reconstructions is that only carbonate protolytes are present in significant amounts. There are no data available to permit assessment of the importance of other protolytes. Therefore some limiting conditions are considered. Imagine another (organic) protolyte of total concentration C_0 and dissociation constant K_0 such that $4 < -\log K_0 < 9$. These limits are imposed because if $-\log K_0$ lies outside the pH titration bounds of 4 to 9 units it will behave as a strong-acid anion or a strong base. In this regard Oliver et al. (1983) have proposed an empirical relationship between $-\log K_0$ and solution pH for organic acid functional groups:

$$-\log K_0 = 0.958 + 0.90 \text{ pH} - 0.039(\text{pH})^2.$$

If this relationship is used to estimate the other protolyte acid dissociation constants, the contributions of organic protolytes to alkalinity should be negligible in samples with pH less than about 4.6. There could be, however, protolytes other than organics that would be reactive in the titration ranges. For example, most aluminosilicates and metal oxides would be reactive. In this case the total amount of other reactive protolytes must be compared with alkalinity or acidity if we assume that the acid dissociation constant is intermediate in the titration range. The critical region then for other protolytes to have maximum effect and thus negate the carbonate assumption is when the reacting carbonates are minimal and at intermediate pH where H^+ ion would have a negligible buffering effect. This condition would be near the equivalence point for alkalinity, which would be near a pH of 5. At this condition either organic protolytes or aluminum hydrolysis might be another important factor. Oliver et al. (1983) estimated the ratio of total organic protolyte (C_0) to dissolved organic carbon (DOC) to be about 10 (units of μM C_0/mg DOC). If all the organic protolytes were reactive at pH of 5, then about 1 mg/L DOC (a "clear" lake) would be equal to the carbonate buffer intensity at its minimum condition. For the same condition, one would need about 550 μg/L total aluminum because one half of the aluminum would be in the form of the aquo ion and the other half in the form of $AlOH^{2+}$ (pK = 5 at pH of 5), and only one half equivalent per mole contributes to alkalinity.

REFERENCES

Kramer, J. R. 1982. Alkalinity and acidity. Pp. 128 ff. in Water Analysis, Vol. 1: Inorganic Species, R. A. Minear and L. H. Keith, eds. New York: Academic Press.

Oliver, B. G., E. M. Thurman, and R. L. Malcolm. 1983. The contributions of humic substances to the acidity of colored natural waters. Geochim. Cosmochim. Acta 47: 2031-2035.

APPENDIX

E
Physical and Chemical Characteristics of Some Lakes in North America for Which Sediment-Diatom Data Exist

FIGURE E.1 Locations of 27 lakes in Regions A and B for which data on diatoms in sediments are available. The number next to each lake corresponds to the order in which it appears in Tables E.1 to E.5.

TABLE E.1 Lake Location and Morphometric Data

Lake	Latitude/ Longitude	Surface Area (ha)	Watershed Area (ha[a])	Maximum Depth (m)	Elevation (m)
Adirondack Park, N.Y.					
1 Honnedaga Lake	43°30′15″N 74°49′00″W	331	1050	58	667
2 Seventh Lake	43°44′17″N 74°45′50″W	254	–	27	544
3 Woodhull Lake	43°34′56″N 74°59′25″W	451	1100	33	573
4 Panther Lake	43°41′00″N 74°55′16″W	18	121	7.0	557
5 Sagamore Lake	43°45′58″N 74°37′43″W	66	4893	23.0	580
6 Woods Lake	43°51′56″N 74°57′20″W	26	212	12.0	601
7 Big Moose Lake	43°49′02″N 74°51′23″W	515	8760	22	556
8 Deep Lake	43°36′58″N 74°39′52″W	11	72	24	789
9 Lake Arnold	44°07′45″N 73°56′25″W	0.4	14	2.5	1150
10 Upper Wallface Pond	44°08′47″N 74°03′15″W	5.5	58	9.0	948
11 Little Echo Pond	44°18′20″N 74°21′25″W	0.8	–	4.5	482
N. New England					
12 Branch Pond, Vt.	43°04′52″N 73°01′06″W	21	218	10.0	802
13 E. Chairback. Pond, Me.	45°27′00″N 69°16′35″W	16.2	75	17.7	462
14 Klondike Pond, Me.	45°55′39″N 68°56′06″W	2	150	2.8	1044
15 Ledge Pond, Me.	44°54′47″N 70°32′22″W	2.4	23	7.3	893
16 Mountain Pond (Rangeley) Me.	44°53′38″N 70°38′50″W	16	160	11.0	726
17 Lake Solitude, N.H.	43°18′25″N 72°04′05″W	2	14	7.0	750
18 Speck Pond, Me.	44°33′49″N 70°58′33″W	4	39	11	1123
19 Tumbledown Pond, Me.	44°45′47″N 70°32′47″W	3.6	49	6.7	813
20 Unnamed Pond, Me.	45°11′21″N 68°11′08″W	6	23	6.7	141
21 Cone Pond, N.H.	43°54′ N 71°36′ W	4.8	54	8.4	468

TABLE E.1 (continued)

Lake	Latitude/ Longitude	Surface Area (ha)	Watershed Area (ha[a])	Maximum Depth (m)	Elevation (m)
Ontario, Canada					
Algonquin Prov. Park					
22 Delano Lake	45°30′ N 78°36′ W	30	140	20	454
23 Found Lake	45°33′ N 78°39′ W	13	1.4	33	442
24 Jake Lake	45°33′ N 78°33′ W	9	35	13	421
N.E. of Lake Superior					
25 B	48°04′ N 85°03′ W	25	300	3.8[b]	–
26 CS	48°09′ N 84°07′ W	9	30	11[b]	427
27 Batchawana Lake (N. Basin)	47°03′55″N 84°23′30″W	5.9	18.1	11	497
Rocky Mt. Natl. Pk., Colo.					
Emerald Lake	40°17′32″N 105°40′05″W	1.9	205	17.0	3110
Lake Haiyaha	40°17′31″N 105°39′43″W	3.9	404	10.0	3110
Lake Husted	40°30′33″N 105°63′58″W	4.1	107	8.3	3380
Lake Louise	40°30′29″N 105°37′10″W	2.3	111	6.0	3365

[a] Not including lake-surface area.
[b] Depth at water-chemistry sampling point.

TABLE E.2 Recent Surface-Water Chemistry Data I

Lake	pH[a]	Alkalinity (μeq/L)[b]	Conductivity (uS/cm at 25°C)	Color (Pt-Co units)	Total P (μg/L)	Chlorophyll a (μg/L)	Reference
Adirondack Park, N.Y.							
1 Honnedaga Lake	4.7-4.8	10[c]	–	very low	–	–	Del Prete and Schofield 1981
	3.9-6.0	0-60[c]	22-39	–	–	–	Schofield 1965
	4.8	−15	28	–	–	–	Colquhoun et al. 1984
2 Seventh Lake	6.4-7.0	–	–	–	–	–	Del Prete and Schofield 1981
	7.4	141	44	–	–	–	Colquhoun et al. 1984
3 Woodhull Lake	5.1-5.3	–	–	–	–	–	Del Prete and Schofield 1981
	5.7	13	–	–	–	–	New York State Department of Environmental Conservation, 1978
	5.6	3	20	–	–	–	Colquhoun et al. 1984
4 Panther Lake	6.2	147	–	clear	–	–	Galloway et al. 1983
	6.6;4.7-7.0[d]	–	–	–	–	–	Davis et al., in preparation
5 Sagamore Lake	5.6	31	–	brown	–	–	Galloway et al. 1983
	5.6;4.8-6.3[d]	–	–	–	–	–	Davis et al., in prepration
6 Woods Lake	4.7	−10	–	clear	–	–	Galloway et al. 1983
	4.8;4.5-5.0[d]	–	–	–	–	–	Davis et al., in preparation
7 Big Moose Lake	4.6-5.0	−10	33	15 (Charles, unpublished data)	–	–	Driscoll 1980
8 Deep Lake	4.7	−18	29	7.5	3.5	1.7	Charles 1982
9 Lake Arnold	4.8	−13	21	20	4.2	1.0	Charles 1982
10 Upper Wallface Pond	4.9-5.0	−2	20	17	3.4	1.2	Charles 1982
11 Little Echo Pond	4.2	−39	31	>50 (Charles, unpublished, data)	33.0	1.1	Driscoll and Newton 1985; C. Driscoll, Syracuse University, unpublished data

TABLE E.2 (continued)

Lake	pH[a]	Alkalinity (μeq/L)[b]	Conductivity (uS/cm at 25°C)	Color (Pt-Co units)	Total P (μg/L)	Chlorophyll a (μg/L)	Reference
N. New England							
12 Branch Pond, Vt.	4.8	−18	20	25	11.3	3.3	Norton et al. 1981; Davis and Norton, University of Maine, personal communication
13 E. Chairback Pond, Me.	5.2	−11	15	2	2.6	1.0	Norton et al. 1981; Davis and Norton, University of Maine, personal communication
14 Klondike Pond, Me.	6.2	28	13	22	5.1	3.6	Norton et al. 1981; Davis and Norton, University of Maine, personal communication
15 Ledge Pond, Me.	5.6	−2	21	34	8.8	3.3	Norton et al. 1981; Davis and Norton, University of Maine, personal communication
16 Mountain Pond (Rangeley), Me.	6.1	16	20	13	5.8	2.6	Norton et al. 1981; Davis and Norton, University of Maine, personal communication
17 Lake Solitude, N.H.	4.8	−20	23	8	8.0	7.7	Norton et al. 1981; Davis and Norton, University of Maine, personal communication
18 Speck Pond, Me.	5.2	−11	20	15	9.0	5.2	Norton et al. 1981; Davis and Norton, University of Maine, personal communication
19 Tumbledown Pond, Me.	5.1	−8	16	4	6.2	0.9	Norton et al. 1981; Davis and Norton, University of Maine, personal communication
20 Unnamed Pond, Me.	4.8	−16	16	1	2.8	0.8	Norton et al. 1981; Davis and Norton, University of Maine, personal communication.

21 Cone Pond, N.H.	4.6	−27	25	5	5.8	0.65	Norton et al. 1981; Davis and Norton, University of Maine, personal communication
	4.5-4.8	−22	21	<5	–	–	Ford 1985.
Ontario, Canada							
Algonquin Prov. Park							
22 Delano Lake[f]	6.0	84	32	–	–	–	Algonquin Fisheries Assessment Unit; Ontario Ministry of Natural Resources, Acidification Study Group-Dorset
23 Found Lake[f]	6.8	104	78	14	5	1.1	Ontario Ministry of Natural Resources, Acidification Study Group-Dorset
24 Jake Lake[f]	6.2	72	145	20	–	–	Ontario Ministry of Natural Resources, Acidification Study Group-Dorset
N.E. of Lake Superior							
25 B	5.2	–	26	–	–	–	Fortescue et al. 1984
26 CS	6.6	120	46	–	–	–	Fortescue et al. 1984
27 Batchawana Lake	6.1	31	18	–	–	–	Jeffries et al. 1983
Rocky Mt. Natl. Pk., Colo.[g]							
Emerald Lake	6.2	28	10	–	0.2[h]	–	Baron 1983
Lake Haiyaha	6.4	43	11	–	0.2[h]	–	Baron 1983
Lake Husted	6.9	58	12	–	0.3[h]	–	Baron 1983
Lake Louise	6.8	129	23	–	0.6[h]	–	Baron 1983

[a] pH meter reading unless otherwise indicated.
[b] Gran plot titration unless otherwise indicated.
[c] Colorimetric indicator.
[d] Mean; range.
[e] Data for 27 July 1982.
[f] Data for 1982.
[g] Data for May-December, 1981.
[h] Concentration of soluble reactive phosphorus.

TABLE E.3 Recent Surface-Water Chemistry Data II

Lake	Ion (μeq/L)							Total Al	
	Ca	Mg	Na	K	SO_4	NO_3	Cl	(μg/L)	Reference
Adirondack Park, N.Y.									
1 Honnedaga Lake	80	50	[a]	[a]	130	9	48	–	Berg 1966
2 Seventh Lake	240	–	–	–	–	–	–	–	Colquhoun et al. 1984
3 Woodhull Lake	–	–	–	–	–	–	–	–	–
4 Panther Lake[b]	207	52	41	12	123	23.4	12.6	30	Schofield, in press
5 Sagamore Lake[b]	147	55	35	12	163	24.1	16.6	100	Schofield, in press
6 Woods Lake[b]	73.1	18.6	19.3	6.2	126	19.4	9.1	100	Schofield, in press
7 Big Moose Lake[c]	91	27	23	11	145	–	–	253	Driscoll 1980
8 Deep Lake[d]	66	26	13	7	145	12	5	667	Charles 1982
9 Lake Arnold[d]	84	18	15	2	120	10	9	462	Charles 1982
10 Upper Wallface Pond[d]	65	17	14	1	110	0	3	192	Charles 1982
11 Little Echo Pond[e]	18	8	2	5	91	0	7	–	Driscoll and Newton 1985; C. Driscoll, Syracuse University, unpublished data
N. New England									
12 Branch Pond, Vt.	45	25	26	10	95	0	7	182	Norton et al. 1981; Norton, University of Maine, personal communication
13 E. Chairback, Pond, Me.	36	27	28	4	76	0	19	103	Norton et al. 1981; Norton, University of Maine, personal communication

14	Klondike Pond, Me.	42	3	27	5	42	8	11	61	Norton et al. 1981; Norton, University of Maine, personal communication
15	Ledge Pond, Me.	104	25	24	6	98	0	13	286,301	Norton et al. 1981; Norton, University of Maine, personal communication
16	Mountain Pond (Rangeley), Me.	78	34	36	6	97	–	16	81,97	Norton et al. 1981; Norton, University of Maine, personal communication
17	Lake Solitude, N.H.	54	19	31	4	101	0	14	138,116	Norton et al. 1981; Norton, University of Maine, personal communication
18	Speck Pond, Me.	48	17	17	6	–	–	–	344	Norton et al. 1981; Norton, University of Maine, personal communication
19	Tumbledown Pond, Me.	51	26	22	3	90	0	10	68,156	Norton et al. 1981; Norton, University of Maine, personal communication
20	Unnamed Pond, Me.	20	10	35	6	61	0	26	114,36	Norton et al. 1981; Norton and Haines, University of Maine, personal communication
21	Cone Pond N.H.	41	16	29	2	144	0	10	313,359	Norton et al. 1981; Norton, University of Maine, personal communication
		37	17	37	24	114	UD[f]	15	510	Ford 1984.
Ontario, Canada										
Algonquin Prov. Park										
22	Delano Lake[g]	140	58	26	8	180	3	8	10	Ontario Ministry of Natural Resources-Acidification Study Unit, Dorset

TABLE E.3 (continued)

Lake	Ion (μeq/L)							Total Al	
	Ca	Mg	Na	K	SO_4	NO_3	Cl	(μg/L)	Reference
23 Found Lake[g]	205	105	291	15	160	–	–	4	Ontario Ministry of Natural Resources-Acidification Study Unit, Dorset
24 Jake Lake[g]	255	82	813	13	190	–	–	43	Ontario Ministry of Natural Resources-Acidification Study Unit, Dorset
N.E. of Lake Superior									
25 B	75	–	–	–	–	190	–	–	Fortescue et al. 1984
26 CS	–	–	–	–	–	–	–	–	
27 Batchawana Lake[h]	114	34	12	6	105	1	9	–	Jeffries et al. 1983
Rocky Mt. Natl. Pk., Colo.[i]									
Emerald Lake	48	12	14	3	30	30	4	–	Baron 1983; Baron, National Park Service, Colorado State University, personal communication

Lake Haiyaha	59	15	18	3	35	21	5	–	Baron 1983; Baron, National Park Service, Colorado State University, personal communication
Lake Husted	88	18	30	2	17	1	8	–	Baron 1983; Baron, National Park Service, Colorado State University, personal communication
Lake Louise	141	22	29	5	35	24	4	–	Baron 1983; Baron, National Park Service, Colorado State University, personal communication

[a] Sum of Na and K is 53 μeq/L.
[b] Average annual concentrations in outlets, January 1 to December 31, 1980; Aluminum concentrations are typical summer values.
[c] Data for July 1978.
[d] Data for August 1979, surface sample.
[e] Data for July 27 1982; dissolved organic carbon was 13.4 ppm; monomeric aluminum was 30 μg/L and organic aluminum was 20 μg/L.
[f] UD = Undectectable.
[g] Data for 1982; Found Lake and Jake Lake have roads within their watersheds.
[h] Data for June 11, 1980, except for Ca, which is for June 11, 1982.
[i] Data for 1981.

TABLE E.4 Dating and Diatom Data[a]

Lake	Year Core Taken	Dating Information	Profile of Dominant Taxa	Profiles of Percents of pH Categories	pH Inference Methods	Equation Calibration Data Set and Region	Error Estimate for pH Inferences	Reference[b]
Adirondack Park, N.Y.								
1 Honnedaga Lake	1976	none	Y	N	α	Del Prete and Schofield 1981, Scandanvia and Adirondacks	N	1
2 Seventh Lake	1976	none	N	N	α	Del Prete and Schofield 1981, Scandanvia and Adirondacks	N	1
3 Woodhull Lake	1976	Cs-137	N	N	α	Del Prete and Schofield 1981, Scandanvia and Adirondacks	N	1
4 Panther Lake	1980	Pb-210 pollen/charcoal	Y	Y	α, B, MR, TC, PCA	Charles 1985 Adirondacks	SE/95% CI	2
	1977	none	Y	Y	α	Same as Honnedaga Lake	N	3
5 Sagamore Lake	1978	Pb-210 pollen/charcoal	Y	Y	α, B, MR, TC, PCA	Charles 1985 Adirondacks	SE/95% CI	2
	1977	none	Y	Y	α	Same as Honnedaga Lake	N	3
6 Woods Lake	1980	Pb-210 pollen/charcoal	Y	Y	α, B, MR, TC, PCA	Charles 1985 Adirondacks	SE/95% CI	2

		1977	none	Y	Y	α	Same as Honnedaga Lake	N	3
7	Big Moose Lake	1982	Pb-210 pollen/charcoal	Y	Y	α, B, MR, TC	Charles 1985 Adirondacks	SE/95% CI	4
8	Deep Lake	1983	Pb-210 pollen/charcoal	Y	Y	α, B, MR, TC	Charles 1985 Adirondacks	SE/95% CI	5
9	Lake Arnold	1982	Pb-210 pollen	Y	Y	α, B, MR, TC	Charles 1985 Adirondacks	SE/95% CI	5
10	Upper Wallface Pond	1983	Pb-210	Y	Y	α, B, MR, TC	Charles 1985 Adirondacks	SE/95% CI	5
11	Little Echo Pond	1982	Pb-210 pollen	Y	Y	α, B, MR, TC	Charles 1985 Adirondacks	SE/95% CI	5
N. New England									
12	Branch Pond, Vt.	1980	Pb-210 pollen/charcoal	Y	Y	MR, B, α PCA, TAX	Davis and Anderson 1985, N. England	SE	6
13	E. Chairback Pond, Me.	1979	Pb-210 pollen/charcoal	Y	Y	MR, B, α PCA, TAX	Davis and Anderson 1985, N. England	SE	6, 7
14	Klondike Pond, Me.	1978	Pb-210 pollen/charcoal	Y	Y	MR, B, α PCA, TAX	Davis and Anderson 1985, N. England	SE	7
15	Ledge Pond, Me.	1979	Pb-210 pollen/charcoal	Y	Y	MR, B, α PCA, TAX	Davis and Anderson 1985, N. England	SE	6, 7
16	Mountain Pond, (Rangeley), Me.	1979	Pb-210 pollen/charcoal	Y	Y	MR, B, α PCA, TAX	Davis and Anderson 1985, N. England	SE	6
17	Lake Solitude, N.H.	1979	Pb-210 pollen/charcoal	Y	Y	MR, B, α PCA, TAX	Davis and Anderson 1985, N. England	SE	6, 7
18	Speck Pond, Me.	1978	Pb-210	Y	Y	MR, B, α	Davis and Anderson	SE	7

TABLE E.4 (continued)

Lake	Year Core Taken	Dating Information	Profile of Dominant Taxa	Profiles of Percents of pH Categories	pH Inference Methods	Equation Calibration Data Set and Region	Error Estimate for pH Inferences	Reference[b]
		pollen/charcoal			PCA, TAX	1985, N. England		
19 Tumbledown Pond, Me.	1978	Pb-210 pollen/charcoal	Y	Y	MR, B, α PCA, TAX	Davis and Anderson 1985, N. England	SE	7
20 Unnamed Pond, Me.	1979	Pb-210 pollen/charcoal	Y	Y	MR, B, α PCA, TAX	Davis and Anderson 1985, N. England	SE	8
21 Cone Pond, N.H.	1978	Pb-210	Y	N	ω, PCA, MR	Davis and Anderson 1985, N. England; Charles 1985 Adirondacks	SE/95% CI	9
Ontario, Canada								
Algonquin Prov. Park								
22 Delano Lake	1978	pollen	Y	N	–	–	–	10
23 Found Lake	1978	pollen	Y	N	–	–	–	10
24 Jake Lake	1978	pollen	Y	N	–	–	–	10
N.E. of Lake Superior								
25 B		Cs-137 pollen	N	N	α	Fortescue et al. 1984, Algoma region, N. of L. Superior	95% CI	11
26 CS		Cs-137 pollen	N	N	α	Fortescue et al. 1984, Algoma region, N. of L. Superior	95% CI	11

27	Batchawana Lake	1981	pollen	Y	Y	α	Norton et al. 1981 (Figure 9-1)	N	12
	Rocky Mt. Natl. Pk, Colo.								
	Emerald Lake	1982	Pb-210	Y	Y	B	Renberg and Hellberg 1982, Scandinavia; Beeson, National Park Service, Colorado State University, unpublished; Rocky Mts.	N	13
	Lake Hiaiyaha	1982	Pb-210	Y	Y	B	Renberg and Hellberg 1982, Scandinavia; Beeson, National Park Service, Colorado State University, unpublished; Rocky Mts.	N	13
	Lake Husted	1982	Pb-210	Y	Y	B	Renberg and Hellberg 1982, Scandinavia; Beeson, National Park Service, Colorado State University, unpublished; Rocky Mts.	N	13

TABLE E.4 (continued)

Lake	Year Core Taken	Dating Information	Profile of Dominant Taxa	Profiles of Percents of pH Categories	pH Inference Methods	Equation Calibration Data Set and Region	Error Estimate for pH Inferences	Reference[b]
Lake Louise	1982	Pb-210	Y	Y	B	Renberg and Hellberg 1982, Scandinavia; Beeson, National Park Service, Colorado State University, unpublished; Rocky Mts.	N	13

[a] Key to symbols:

Y, yes
N, no
α, index alpha
B, index B
MR, multiple regression of pH categories
SE, standard error
95% CI, 95 percent confidence interval
TC, multiple regression of percent of diatoms in taxa clusters
PCA, multiple regression of principal component of taxa data sets
TAX, multiple regression of common taxa
ω, index omega

[b] Key to references: 1 Del Prete and Schofield 1981; 2 Davis et al in prepration; 3 Del Prete and Galloway 1983; 4 Charles 1984; 5 D. F. Charles, Indiana University, unpublished data; 6 Norton et al. 1985; 7 Davis et al. 1983, and R. Davis, University of Maine, personal communication; 8 R. Davis and D. Anderson, University of Maine personal communication; 9 Ford 1985; 10 Smol and Dickman 1981; 11 Fortescue et al. 1984 Dickman et al. 1983, Dickman and Fortescue 1984; 12 Delorme et al. in preparation; 13 Besson 1984, and D.R. Besson, National Park Service, Fort Collins, Colorado, personal communication.

TABLE E.5 Summary of Diatom-Inferred pH Changes

Lake	Current Lake-Water pH	Surface-Sediment-Inferred pH	Inferred Background pH (pre-1800)	Approximate Magnitude of Diatom Inferred pH Change from Background to Recent	Approximate Onset of Inferred pH Change(s)	Period of Most Rapid Inferred pH Change	Most Probable Cause(s) of Inferred pH Change	Reference and Comments
Adirondack Park, N.Y.								
1 Honnedaga Lake	4.7-4.8	5.2	6.1	1.1	—	Post-1940	Acid deposition; watershed disturbance	pH's based on analysis of surface, 3.0-4.0 and 18.0-19.0 cm levels using Adirondack calibrated equations (Charles 1985; Charles, Indiana University unpublished data; Del Prete and Schofield 1981).
2 Seventh Lake	6.4-7.0	6.5	6.6	a	a	a	—	General decline in pH from 20 to 5 cm (0.4 pH unit), then an increase of similiar magnitude over the top 5 cm (Del Prete and Schofield 1981).
3 Woodhull Lake	5.1-5.3	6.0	6.1	a	a	a	—	No obvious trend in pH (Del Prete and Schofield 1981).
4 Panther Lake[b]	6.2	6.1	6.4	a	a	a	—	Decrease from about 5-10 cm toward surface (Del Prete and Galloway, 1983.)

TABLE E.5 (continued)

Lake		Current Lake-Water pH	Surface-Sediment-Inferred pH	Inferred Background pH (pre-1800)	Approximate Magnitude of Diatom Inferred pH Change from Background to Recent	Approximate Onset of Inferred pH Change(s)	Period of Most Rapid Inferred pH Change	Most Probable Cause(s) of Inferred pH Change	Reference and Comments
5	Sagamore Lake[b]	5.6	6.3	6.1	a	a	a	—	No overall significant trend in pH (Del Prete and Galloway, 1983).
6	Woods Lake[b]	4.7	4.8	5.2	0.3-0.5	—	—	—	pH increases around 20 cm, then decreases to the surface (Del Prete and Galloway, 1983).
7	Big Moose Lake	4.6-5.0	4.8	5.8	1.0	1950s	1955-1970	Acid deposition	Charles 1984
8	Deep Lake	4.7	4.3	5.0	0.7	1940s-1950s	1950-1970	Acid deposition, logging	Increase in percent acidobiontic from 30% pre-1800 to 1945 to 70-80% in 1970s (Charles, Indiana University, unpublished data).
9	Lake Arnold	4.8	4.5	4.8	a	a	a	—	Major increase in *Fragilaria acidobiontica* toward surface of core, suggesting increase in lake-water aluminum concentration (Charles, Indiana University, unpublished data).

No.	Lake								
10	Upper Wallface Pond	4.9-5.0	4.7	5.1	0.4	1930s-1940s	1930-1970	Acid deposition	Major increase in *Fragilaria acidobiontica* and other acidobiontics and steady decline in pH from 1930-1970 (Charles, Indiana University, unpublished data).
11	Little Echo Pond	4.3	4.1-4.7	4.7	a	a	a	—	Major increase (to >90%) in *Asterionella ralfsii* v. *americana* at top of core (1.5 cm, 1956), suggesting a change within the lake, not necessarily related to acidification (Charles, Indiana University, unpublished data).
	N. New England								
12	Branch Pond, Vt.	4.7	4.5	4.8	0.3	1930	1930s-1960s	Watershed disturbance, acid deposition	Decrease of about 0.3 pH unit in 1930s, increase 0.2 unit in 1940s, decrease 0.2 in 1960s (Norton et al. 1985; Davis and Anderson, University of Maine, personal communication).
13	E. Chairback Pond, Me.	5.2	5.0-5.1	4.8-4.9	1835, 1930, 1940	1835-1940	1820-1910	Watershed disturbance?	pH increases from 4.8-4.9 at 1800 to 5.3 at 1875-1900, possibly from increased nutrient and/or base cation flux due to lumbering, then decreases to ~5.0 in the 1930s and 1920s. Forest fire 1825; timber harvests 1914-1979. No

TABLE E.5 (continued)

Lake	Current Lake-Water pH	Surface-Sediment-Inferred pH	Inferred Background pH (pre-1800)	Approximate Magnitude of Diatom Inferred pH Change from Background to Recent	Approximate Onset of Inferred pH Change(s)	Period of Most Rapid Inferred pH Change	Most Probable Cause(s) of Inferred pH Change	Reference and Comments
								significant change from 1920s to 1979 (Norton et al. 1985; Davis and Anderson, University of Maine, personal communication).
14 Klondike Pond, Me.[b]	6.2	5.5	5.7	a	a	a	—	Possibly slight decrease starting ~1960. Top two or three core intervals with pH ~0.2 less than other post-1800 intervals. Maximum change in pH in core <0.3 pH units (Davis et al. 1983; Davis and Anderson, personal communication).
15 Ledge Pond, Me.	5.6	4.4-4.7	5.1-5.3	0.6-0.7	1885, 1970	1965-present	Acid deposition	From pre-1700 to ~1880, pH ~5.2; pH then decreases to 4.9-5.0, with increased rate of decrease from ~1965 (Norton et al. 1985; Davis and Anderson, University of Maine, personal communication).

16	Mountain Pond (Rangeley), Me.	6.1	5.0	5.0-5.1	a	1870-1900	1870-1900	Watershed disturbance	pH increase of 0.3-0.6 units correlated with lumbering 1870-1900; lake now returned to about same inferred pH (~5.0) as before lumbering (Norton et al. 1985; Davis and Anderson, University of Maine, personal communication).
17	Lake Solitude, N.H.	4.8	5.0	4.9-5.2	a	a	a	—	No pH change detected (Norton et al. 1985; Davis and Anderson, University of Maine, personal communication).
18	Speck Pond, Me.	5.2	4.9	5.1	a	a	a	—	Possible decrease of ~0.2 pH unit 1915-1935, then no change to present (Davis et al. 1983; Davis and Anderson, University of Maine, personal communication).
19	Tumbledown Pond, Me.	5.2	5.1	5.0	a	a	a	—	Davis et al. 1983; Battarbee 1984; Davis and Anderson, University of Maine, personal communication
20	Unnamed Pond, Me.[b]	4.8	5.1	5.1	a	a	a	—	Possible higher pH by ~0.2 unit from mid-1800's to ~1930 due to disturbance (Davis and Anderson, University of Maine, personal communication).
21	Cone Pond, N.H.	4.5-4.8	4.6-4.7	4.5-4.8	a	a	a	—	No obvious recent trend in pH (Ford, 1985).

TABLE E.5 (continued)

Lake	Current Lake-Water pH	Surface-Sediment-Inferred pH	Inferred Background pH (pre-1800)	Approximate Magnitude of Diatom Inferred pH Change from Background to Recent	Approximate Onset of Inferred pH Change(s)	Period of Most Rapid Inferred pH Change	Most Probable Cause(s) of Inferred pH Change	Reference and Comments
Ontario, Canada Algonquin Prov. Park								
22 Delano Lake	6.0	—	—	—	—	—	—	Inferred pH not calculated for any of the 3 lakes. Interpretation of percent profiles of dominant diatom and chrysophyte taxa indicates no major pH changes have occurred. Changes in flora have occurred but are apparently most closely related to road construction near the lakes (Smol and Dickman, 1981).
23 Pound Lake	6.8	—	—	—	—	—	—	
24 Jake Lake	6.2	—	—	—	—	—	—	
N.E. of L. Superior								
25 B	5.2	4.8	6.4-6.6	1.6	Post-1954	1960s-1970s	Acid deposition, much from smelters at Wawa, Ontario	Fortescue et al. 1984 and Dickman and Fortescue, 1984. Onset of pH change corresponds with increased local mining and smelting activity from 1949-1958 (Somers and Harvey 1984).

26 CS	5.2	6.4	7.2	1.0	Mid-1950s	1960s	Acid deposition, much from smelters at Wawa, Ontario	Relative importance of local vs. distant sources of acid deposition is unknown.
27 Batchawana Lake (Core 79)	6.1	6.2	6.0	—	—	—	—	No obvious trend in diatom-inferred pH; maximum range in pH fluctuation is 5.8-6.4 (Delorme et al. in preparation).
Rocky Mt. Natl. Pk., Colo.								
Emerald Lake	6.2	6.9	7.2[d]	No statistically significant change	—	—	—	Inferred pH ranged from 6.8 to 7.2; no major trends.
Lake Haiyaha	6.4	6.8	6.8[d]	No statistically significant change	—	—	—	Inferred pH varied from 6.8 to 6.9 within core.
Lake Husted	6.9	6.8	6.6	No statistically significant change	—	—	—	Percent alkaliphils increase from 20% to 50% from 16 cm (1820) to surface of core.
Lake Louise	6.8	6.9	6.9	No statistically significant change	—	—	—	Inferred pH varied from 6.7 to 7.2.

[a]pH changes are within or close to standard error of the inference techniques. Significant pH trends or differences may exist, but they cannot be determined without further statistical analysis.

[b]Panther, Sagamore, and Woods lake sediment cores have also been studied by Davis et al. (in preparation). Unpublished results are not presented here but may be seen with permission (contact R. B. Davis, University of Maine at Orono).

[c]Column entries for these lakes based on Davis et al. (1983) supplemented by more recent unpublished data, including updated diatom taxonomy with distinction of *Tabellaria quadriseptata*.

[d]Mid-1800s.

REFERENCES

Baron, J. 1983. Comparative water chemistry of four lakes in Rocky Mountain National Park. Water Resour. Bull. 19:897-902.

Battarbee, R. W. 1984. Diatom analysis and the acidification of lakes. Phil. Trans. R. Soc. London. 305:451-477.

Beeson, D. R. 1984. Historical fluctuations in lake pH as calculated by diatom remains in sediments from Rocky Mountain National Park. Bull. Ecol. Soc. Am. 65:135.

Berg, C. O. 1966. Middle Atlantic States. In Limnology in North America, D. G. Frey, ed. The University of Wisconsin Press, Madison, pp. 191-237.

Charles, D. F. 1982. Studies of Adirondack Mountain (N.Y.) lakes: limnological characteristics and sediment diatom-water chemistry relationships. Doctoral dissertation, Indiana University, Bloomington, Ind.

Charles, D. F. 1984. Recent pH history of Big Moose Lake (Adirondack Mountains, New York, USA) inferred from sediment diatom assemblages. Int. Ver. Theor. Angew. Limnol. Verh. 22:559-566.

Charles, D. F. 1985. Relationships between surface sediment diatom assemblages and lakewater characteristics in Adirondack lakes. Ecology 66:994-1011.

Colquhoun, J., W. Krester, and M. Pfeiffer. 1984. Acidity status update of lakes and streams in New York State. New York State Department of Environmental Conservation, Albany, N.Y. 139 pp.

Davis, R. B., and D. S. Anderson. 1985. Methods of pH calibration of sedimentary diatom remains for reconstructing history of pH in lakes. Hydrobiologia 120:69-87.

Davis, R. B., S. A. Norton, C. T. Hess, and D. F. Brakke. 1983. Paleolimnological reconstruction of the effects of atmospheric deposition of acids and heavy metals on the chemistry and biology of lakes in New England and Norway. Hydrobiologia 103:113-123.

Davis, R. B., D. S. Anderson, D. F. Charles, and J. N. Galloway. Two-hundred year pH history of Woods, Sagamore, and Panther Lakes in the Adirondack Mountains, New York State, USA (in preparation).

Delorme, L. D., H. C. Duthie, S. R. Esterby, S. M. Smith, and N. S. Harper. Prehistoric inferred pH changes in Batchawana Lake, Ontario from sedimentary diatom assemblages (in preparation).

Del Prete, A., and J. N. Galloway. 1983. Temporal trends in the pH of Woods, Sagamore and Panther lakes as determined by an analysis of diatom populations. In The Integrated Lake-Watershed Acidification Study: Proceedings of the ILWAS Annual Review Conference, Electric Power Research Institute, Palo Alto, Calif.

Del Prete, A., and C. Schofield. 1981. The utility of diatom analysis of lake sediments for evaluating acid precipitation effects on dilute lakes. Arch. Hydrobiol. 91:332-340.

Dickman, M., and J. Fortescue. 1984. Rates of lake acidification inferred from sediment diatoms for 8 lakes located north of Lake Superior, Canada. Int. Ver. Theor. Angew. Limnol. Verh. 22:1345-1356.

Dickman, M., S. Dixit, J. Fortescue, R. Barlow, and J. Terasmae. 1983. Diatoms as indicators of the rate of lake acidification. Water Air Soil Pollut. 21:227-241.

Driscoll, C. T. 1980. Chemical characterization of some dilute acidified lakes and streams in the Adirondack region of New York state. Ph.D. dissertation, Cornell University, Ithaca, N.Y.

Driscoll, C. T., and R. M. Newton. 1985. Chemical characteristics of Adirondack lakes. Environ. Sci. Technol. 19:1018-1023.

Ford, M. S. (J.) 1984. The influence of lithology on ecosystem development in New England: a comparative paleoecological study. Doctoral dissertation, University of Minnesota, Minneapolis.

Ford, M. S. (J.) 1985. The recent history of a naturally acidic lake (Cone P., N.H.). In Diatoms and Lake Acidity, J. P. Smol, R. W. Battarbee, R. Davis, and J. Merilainen (eds.). W. Junk, The Hague, The Netherlands (in press).

Fortescue, J. A. C., I. Thompson, M. Dickman, and J. Terasmae. 1984. Multidisciplinary followup of regional pH patterns in lakes north of Lake Superior, District of Algoma, Ontario Geological Survey OFR 5342 Part II, Information and data listings for lakes studied, 366 pp.

Galloway, J. N., C. L. Schofield, N. E. Peters, G. R. Hendrey, and E. R. Altwicker. 1983. Effect of atmospheric sulfur on the composition of three Adirondack lakes. Can. J. Fish. Aquat. Sci. 40:799-806.

Jeffries, D. S., R. G. Semkin, R. Neureuther, and M. D. Jones. 1983. Data Report: Major ion composition of lakes in the Turkey Lakes Watershed (January 1980 - May 1982). Turkey Lakes Watershed, unpublished Report No. TLW-83-11, 9 pp.

New York State Department of Environmental Conservation. 1978. Lake and pond survey data sheets. Albany, N.Y.

Norton, S. A., R. B. Davis, and D. F. Brakke. 1981. Responses of northern New England lakes to atmospheric inputs of acids and heavy metals. Completion Report Project A-048-ME. U.S. Department of the Interior. Land and Water Resources Center, University of Maine at Orono. 90 pp.

Norton, S. A., R. B. Davis, and D. S. Anderson. 1985. The distribution and extent of acid and metal precipitation in Northern New England. Final Report. U.S. Fish and Wildlife Service Grant No. 14-16-0009-75-040.

Renberg, I., and T. Hellberg. 1982. The pH history of lakes in southwestern Sweden, as calculated from the subfossil diatom flora of the sediments. Ambio 11:30-33.

Schofield, C. L. 1965. Water quality in relation to survival of brook trout, Salvelinus fontinalis (Mitchell). Trans. Am. Fish. Soc. 94:227-235.

Schofield, C. L. Surface water chemistry in the ILWAS basins. Water Air Soil Pollut. (in press).

Smol, J. P., and M. D. Dickman. 1981. The recent histories of three Canadian Shield lakes: a paleolimnological experiment. Arch. Hydrobiol. 93:83-108.

Somers, K. M., and H. H. Harvey. 1984. Alteration of fish communities in lakes stressed by acid deposition and heavy metals near Wawa, Ontario. Can. J. Fish. Aquat. Sci. 41:20-29.